The IMA Volumes in Mathematics and its Applications

Volume 149

Series Editors
Douglas N. Arnold Arnd Scheel

For other titles published in this series, go to
www.springer.com/series/811

Mihai Putinar Seth Sullivant
Editors

Emerging Applications of Algebraic Geometry

Editors
Mihai Putinar
Department of Mathematics
University of California
Santa Barbara, CA 93106
USA
http://www.math.ucsb.edu/~mputinar/

Seth Sullivant
Department of Mathematics
North Carolina State University
Raleigh, North Carolina 27695
USA
http://www4.ncsu.edu/~smsulli2/

ISSN: 0940-6573
ISBN: 978-1-4419-1881-9 e-ISBN: 978-0-387-09686-5
DOI: 10.1007/978-0-387-09686-5

Mathematics Subject Classification (2000): 13P10, 14P10, 44A60, 52B12, 52C45, 62-09, 62H17, 90C22, 90C26, 92-08, 92D20, 93B25, 93B28

Camera-ready copy provided by the IMA.

9 8 7 6 5 4 3 2 1

springer.com

FOREWORD

This IMA Volume in Mathematics and its Applications

EMERGING APPLICATIONS OF ALGEBRAIC GEOMETRY

contains papers presented at two highly successful workshops, Optimization and Control, held January 16–20, 2007, and Applications in Biology, Dynamics, and Statistics, held March 5–9, 2007. Both events were integral parts of the 2006-2007 IMA Thematic Year on "Applications of Algebraic Geometry." We are grateful to all the participants for making these workshops a very productive and stimulating events. The organizing committee for the first workshop are Dimitris Bertsimas (Sloan School of Management, Massachusetts Institute of Technology), J. William Helton (Mathematics, University of California - San Diego), Jean Bernard Lasserre (LAAS-CNRS, France), and Mihai Putinar (Mathematics, University of California - Santa Barbara), and for the second workshop, Serkan Hosten (Mathematics, San Francisco State University), Lior Pachter (Mathematics, University of California - Berkeley), and Seth Sullivant (Mathematics, North Carolina State University). The IMA thanks them for their excellent work.

We owe special thanks to Mihai Putinar and Seth Sullivant for their superb role as editors of these proceedings. We take this opportunity to thank the National Science Foundation for its support of the IMA.

Series Editors

Douglas N. Arnold, Director of the IMA

Arnd Scheel, Deputy Director of the IMA

PREFACE

Algebraic geometry, that noble and refined branch of pure mathematics, resurfaces into novel applications. The present volume contains only a portion of these emerging new facets of modern algebraic geometry, reflected by the interdisciplinary activities hosted during the academic year 2006–2007 by the IMA- the Institute for Mathematics and its Applications, at the University of Minnesota, Minneapolis.

What has algebraic geometry to do with non-convex optimization, the control theory of complicated engineering systems, phylogenetic tree reconstruction, or the statistical analysis of contingency tables? The answer is: quite a lot. And all this is due to innovative ideas and connections discovered in the last decade between pure and applied mathematics. The reader of the present volume will find detailed and informative answers to all of the above questions, presented in a self-contained form by experts working on the boundary of algebraic geometry and one specific area of its applications.

Two major conferences organized at IMA: *Optimization and Control* (January 2007) and *Applications in Biology, Dynamics, and Statistics* (March 2007) gathered mathematicians of all denominations, engineers, biologists, computer scientists and statisticians, linked by the common use of a few basic methods of algebraic geometry. Among the most important of these methods: Positivstellensätze over the real field, elimination theory, and invariant theory, all with a computational/numerical component predominant in the background.

In order to explain why optimization and control theory appear naturally connected to real algebraic geometry one has to go back a century ago and recall that both the positivity of regular functions (polynomials for instance) and of linear transformations in a inner product space are the byproduct of a universal structure: they are representable by simple operations involving sums of squares in the respective algebras. Heuristically speaking, when proving inequalities one most often completes squares. It was Hilbert who unveiled the central role of sums of squares decompositions (see his 17-th problem) and their logical ramifications. On the algebraic side, his ideas were crystalized and developed by Emil Artin into the theory of real closed fields. Second, Hilbert's footprint was cast in the spectral theorem for self-adjoint linear transformations, which is essentially equivalent to an infinite sum of squares decomposition of a specific bilinear form. Third, Hilbert's program of revising the logical foundations of mathematics found a solid validation in the discovery due to Tarski, that any proposition of the first order involving only real variables is decidable (in particular implying that one can verify that every system of polynomial inequalities in $\mathbb{R}^d$ has a solution). Progress in real algebraic geometry rendered Tarski's theorem even more precise with the discovery of the so-called Positivstellensätze, that is theorems which quantify in algebraic certificates the existence of solutions of polynomial inequalities (e.g.: $b^2 - c \geq 0$ implies that there exists a real x satisfying $x^2 - 2bx + c = 0$).

It comes then as no surprise that the extremal problem

$$p^* = \min\{p(u);\ u \in S\},$$

where p is a polynomial and S is a subset of $\mathbb{R}^d$ defined by finitely many polynomial inequalities can be converted into algebraic identities which depend only on the coefficients of p and the minimal value p^*. At this point optimization theorists (and by extension control theory engineers) became aware that one can solve such extremal problems by global algebraic means, rather than by, for instance, the time consuming inner point technique.

An unexpected turn into this history was marked by an application of duality theory of locally convex spaces leading to the simplification of the known Positivstellensätze, by interpreting them as solvability criteria for some associated moment problems. In their own, moment problems have a glorious past and an unexpected vitality. Once the relation between real algebraic geometry and moment problems was clarified, this has greatly enhanced the applications of real algebraic geometry to optimization. These days we witness a proliferation of relaxation methods in polynomial minimization problems based on the introduction of new, virtual moment variables. The articles contained in this volume amply illustrate this new direction of research.

The emergence of algebraic geometry as a tool in statistics began with the work of Diaconis and Sturmfels whose research led to the construction of Markov chains for performing Fisher's exact test. They showed that the local moves in these Markov chains are the generators of an associated toric ideal. It was quickly realized that the zero set of this toric ideal in the probability simplex is precisely the set of distributions in the corresponding log-linear model.

The observation that the zero set of a toric ideal in the probability simplex is a statistical model led to the realization that many other statistical models have the structure of a semi algebraic set which can be exploited for various purposes. The foremost examples are hierarchical and graphical models, which include as special cases many of the statistical models used in computational biology.

Of course, this description of the emergence of algebraic geometry as a tool in statistics and its applications is brief and highly incomplete. In particular, it leaves out the algebraic developments in the design of experiments, the use of polynomial dynamical systems over finite fields to investigate gene regulatory networks, polyhedral tools for parametric inference, optimization problems in statistical disclosure limitation, and a slew of other connections between statistics and algebraic geometry, some of which are highlighted in this volume.

A short description of the articles that appear in the volume follows. For a general perspective, more details and a better description we recommend the reader to consult the introductions to any one of the contributions.

1. The article **Polynomial optimization on odd-dimensional spheres** by John d'Angelo and Mihai Putinar contains a basic Positivstellensatz with supports on an odd dimensional sphere, written in terms of hermitian sums of squares and obtained with methods of functional analysis, complex analysis and complex geometry.
2. The survey **Engineering systems and free semi-algebraic geometry** by Mauricio de Oliveira, John W. Helton, Scott McCullough and Mihai Putinar illustrates the ubiquity of positivity computations in a finitely generated free *-algebra in the study of dimensionless matrix inequalities arising in engineering systems.
3. The article **Algebraic statistics and contingency table problems: estimation and disclosure limitation** by Adrian Dobra, Stephen Fienberg, Alessandro Rinaldo, Aleksandra Slavkovic, and yi Zhou describes the use of algebraic techniques for contingency table analysis, and provides many open problems, especially in the area of statistical disclosure limitation.
4. Nicholas Eriksson's article **Using invariants for phylogenetic tree reconstruction** describes some of the mathematical challenges in using the polynomials that vanish on probability distributions of phylogenetic models as tools for inferring phylogenetic trees.
5. Abdul Jarrah and Reinhard Laubenbacher's article **On the algebraic geometry of polynomial dynamical systems** describes their work with collaborators on using dynamical systems over finite fields to model biological processes, as well as the mathematical advanced made on studying the dynamical behavior of special classes of these dynamical systems.
6. Jean Bernard Lasserre, Monique Laurent and Philip Rostalski in their article **A unified approach to computing real and complex zeros of zero dimensional ideals** introduce a powerful and totally new idea of finding the support of ideals of finite codimension in the polynomial ring. They propose a relaxation based on finitely many moments as auxiliary variables.
7. Monique Laurent's ample survey **Sums of squares, moment matrices and optimization of polynomials** is an invaluable source, as the only one of this kind, for the rapidly growing relaxation methods aimed at solving in optimal time minimization problems on polynomial functions restricted to semi-algebraic sets. All presented in a very clear, systematic way, with a spin towards the computational complexity, algorithmic aspects and numerical methods.

8. Claus Scheiderer's survey **Positivity and sums of squares: a guide to recent results** contains a lucid and comprehensive overview of areas of real algebraic geometry contingent to the concept of positivity. This unique synthesis brings the subject up to our days, is informative for all readers and it is accessible to non-experts.
9. Konrad Schmüdgen's **Non-commutative real algebraic geometry: some basic concepts and first ideas** is an essay about the foundations of real algebraic geometry over a *-algebra, based on the author's recent results in enveloping algebras of Lie algebras.
10. Bernd Sturmfels' article **Open problems in algebraic statistics** provides a summary of his lecture in the workshop *Applications in Biology, Dynamics, and Statistics* and covers a number of mathematical open problems whose solutions would greatly benefit this emerging area.

The completion of this volume could have not been achieved without the enthusiasm and professionalism of all contributors. The invaluable technical support of the experienced staff at IMA led to a high quality production of the book, in record time. We warmly thank them all.

Mihai Putinar
Department of Mathematics
University of California at Santa Barbara
http://math.ucsb.edu/~mputinar/

Seth Sullivant
Department of Mathematics
North Carolina State University
http://www4.ncsu.edu/~smsulli2/

CONTENTS

POLYNOMIAL OPTIMIZATION ON ODD-DIMENSIONAL SPHERES

JOHN P. D'ANGELO* AND MIHAI PUTINAR†

Abstract. The sphere S^{2d-1} naturally embeds into the complex affine space $\mathbb{C}^d$. We show how the complex variables in $\mathbb{C}^d$ simplify the known Striktpositivstellensätze, when the supports are resticted to semi-algebraic subsets of odd dimensional spheres. We also illustrate the subtleties involved in trying to control the number of squares in a Hermitian sum of squares.

Key words. Positive polynomial, Hermitian square, unit sphere, plurisubharmonic function, Cauchy-Riemann manifold.

AMS(MOS) subject classifications. Primary 14P10, Secondary 32A70.

1. Introduction. A deep observation going back to the work of Tarski in the 1930-ies implies the decidability (in a very strong, algorithmic sense) of the statement "there exists a real solution $x \in \mathbb{R}^n$ of a system of real polynomial inequalities". What is called today the algebraic certificate that such a system has a solution is condensed into the basic Positiv- and Null-stellensätze, discovered by Stengle in the 1970-ies (see for instance for details [23]).

All these topics are relevant to and depend on the structure of the convex cone $\mathbb{R}_+(K)$ of positive polynomials on a given basic semi-algebraic set $K \subset \mathbb{R}^n$ and on algebraic refinements of it, such as the pre-order associated to a system of defining equations of K and the corresponding quadratic module, both regarded as convex cones of the polynomial algebra. Thus, it comes as no surprise that duality techniques of functional analysis (i.e. the structure of non-negative functionals on these convex cones) reflect at the level of purely algebraic statements the mighty powers coming from a different territory. More precisely the spectral theorem for commuting tuples of self-adjoint operators simplifies some of the known Positivstellensätze, see for instance [30, 26].

The present note is an illustration of the latter transfer between Hilbert space techniques and real algebraic geometry. We specialize to basic semi-algebraic sets $K \subset S^{2d-1} \subset \mathbb{R}^{2d}$of an odd-dimensional sphere, and prove that the expected Positivstellensatz (i.e. the structure theorem of a positive polynomial on such a set) can further be simplified by means of the induced complex variables $\mathbb{R}^{2d} = \mathbb{C}^d$. To this aim we greatly benefit from the

*Department of Mathematics, University of Illinois at Urbana-Champaign, Urbana, IL 61801 (jpda@math.uiuc.edu). The first author was partially supported by the National Science Foundation grant DMS-0500765.

†Mathematics Department, University of California, Santa Barbara, CA 93106 (mputinar@math.ucsb.edu). The second author was supported in part by the Institute for Mathematics and its Applications (Minneapolis) with funds provided by the National Science Foundation and by the NSF grant DMS-0701094.

works of several functional analysts interested into some tuples of Hilbert space operators known as spherical isometries, and separately from the works of geometers and complex analysts dealing with Cauchy-Riemann manifolds and proper holomorphic maps between Euclidean balls. The following pages reveal some fascinating new frontiers for future research, such as the rigidity phenomenon discussed in the last section.

2. Preliminaries. Let $\mathbb{C}^d$ denote complex Euclidean space with Euclidean norm given by $|z|^2 = \sum_{j=1}^d |z_j|^2$. The unit, odd dimensional sphere

$$S^{2d-1} = \{z \in \mathbb{C}^d;\ |z| = 1\}$$

is a particularly important example of a Cauchy-Riemann (usually abbreviated CR) manifold. This note will show how one can study problems of polynomial optimization over semi-algebraic subsets of S^{2d-1} by using the induced Cauchy-Riemann structure. Our results can be regarded as multivariate analogues of classical phenomena about positive trigonometric polynomials, known for a long time in dimension one ($d = 1$). They are also related to results concerning proper holomorphic mappings between balls in different dimensional complex Euclidean spaces and the geometry of holomorphic vector bundles. See [9] for an exposition of these connections.

A polynomial map $p : \mathbb{C}^d \times \mathbb{C}^d \to \mathbb{C}$ is called *Hermitian symmetric* if

$$p(z, \overline{w}) = \overline{p(w, \overline{z})}$$

for all z and w. By polarization one can recover a Hermitian symmetric polynomial from its real values $p(z, \overline{z})$. We therefore work on the diagonal (where $w = z$) and let $\mathcal{H} \subset \mathbb{C}[z, \overline{z}]$ denote the space of Hermitian symmetric polynomials on $\mathbb{C}^d$. Note that $\mathcal{H}$ is a real algebra, naturally isomorphic to the polynomial algebra $\mathbb{R}[x, y]$, where $z = x + iy \in \mathbb{R}^d + i\mathbb{R}^d$. Henceforth we will freely identify a Hermitian symmetric polynomial $P(z, \overline{z})$ with its real form $p(x, y) = P(x + iy, x - iy)$.

We denote by $\Sigma^2\mathcal{H}$ the convex cone consisting of sums of squares of Hermitian polynomials. We denote by $\Sigma_h^2\mathcal{H}$ the convex cone consisting of polynomials which are squared norms of (holomorphic) polynomial mappings. Thus $R \in \Sigma_h^2\mathcal{H}$ if there exist polynomials $p_j \in \mathbb{C}[z]$ such that

$$R(z, \overline{z}) = \sum_{j=1}^{m} |p_j(z)|^2.$$

See [11] and [1] for various characterizations of $\Sigma_h^2\mathcal{H}$. We have the containment

$$\Sigma_h^2\mathcal{H} \subset \Sigma^2\mathcal{H},$$

simply because

$$|p|^2 = \left(\frac{p+\overline{p}}{2}\right)^2 + \left(\frac{p-\overline{p}}{2i}\right)^2 = u^2 + v^2,$$

where u and v are the real and imaginary parts of p. The containment is strict as illustrated by the following two examples.

EXAMPLE a). In one variable we define a polynomial R by

$$R(z, \overline{z}) = (z + \overline{z})^2 = 4x^2.$$

It is evidently a square but not in $\Sigma_h^2\mathcal{H}$. Note that the zero set of an element in $\Sigma_h^2\mathcal{H}$ must be a complex variety and thus cannot be the imaginary axis.

EXAMPLE b). In two variables we define $R(z, \overline{z}) = (|z_1|^2 - |z_2|^2)^2$. Again R lies in $\Sigma^2\mathcal{H}$ but not in $\Sigma_h^2\mathcal{H}$. Here one can observe that elements of $\Sigma_h^2\mathcal{H}$ must be plurisubharmonic but that R is not. In 4.3 we will show additionally that R cannot be written as a squared norm on the unit sphere.

In this paper we are primarily concerned with optimization on the sphere. We therefore first let $I = I(S^{2d-1})$ be the ideal of $\mathcal{H}$ consisting of all polynomials vanishing on S^{2d-1}. We then define

$$\mathcal{H}(S^{2d-1}) = \mathcal{H}/I,$$

and regard it as a space of polynomial functions defined on the sphere. As a matter of fact, each real-valued polynomial p has a representative in $\mathcal{H}(S^{2d-1})$, when p is regarded as a function on the sphere.

In analogy with the above notations we denote by $\Sigma^2\mathcal{H}(S^{2d-1})$ the convex cone consisting of sums of squares of Hermitian polynomials on the sphere. We denote by $\Sigma_h^2\mathcal{H}(S^{2d-1})$ the convex hull of Hermitian squares:

$$\Sigma_h^2\mathcal{H}(S^{2d-1}) = \text{co}\{|p(z)|^2;\ p \in \mathbb{C}[z]\} \mod I.$$

A polynomial that is positive on the sphere must agree with the squared norm of a holomorphic polynomial mapping there. In the final section we naturally ask what is the minimum number of terms in the representation as a squared norm. This difficult question is particularly natural for at least three reasons: there is considerable literature on this problem in the real case, the Hermitian case is closely connected with difficult work on the classification of proper holomorphic mappings between balls, and finally because number-theoretic analogues such as Waring's problem have appealed to mathematicians for centuries. We consider only the second issue here. In Section 5 we provide a striking non-trivial example illustrating some of the subtleties.

Let us return to the general situation and recall a classical one-dimensional result which is guiding our investigation. We include its elementary proof for convenience.

LEMMA 2.1 (Riesz-Fejér). *A non-negative trigonometric polynomial is the squared modulus of a trigonometric polynomial.*

Proof. Let $p(e^{i\theta}) = \sum_{-d}^{d} c_j e^{ij\theta}$ and assume that $p(e^{i\theta}) \geq 0, \quad \theta \in [0, 2\pi]$. Since p is real-valued $c_{-j} = \overline{c_j}$ for all j. We set $z = |z|e^{i\theta}$ and extend p to the rational function defined by $p(z) = \sum_{-d}^{d} c_j z^j$. It follows that $p(z) = \overline{p(1/\overline{z})}$; furthermore its zeros and poles are symmetrical (in the sense of Schwarz) with respect to the unit circle.

Write $z^d p(z) = q(z)$. Then q is a polynomial of degree $2d$ whose modulus $|q|$ satisfies $|q| = |p| = p$ on the unit circle. In view of the mentioned symmetry one finds

$$q(z) = cz^{\nu} \prod_j (z - \lambda_j)^2 \prod_k (z - \mu_k)(z - 1/\overline{\mu_k}),$$

where c is a constant, $|\lambda_j| = 1$ and $0 < |\mu_k| < 1$.

Evaluating on the circle and using $|\zeta^2| = |\zeta|^2$ we obtain

$$p(e^{i\theta}) = |p(e^{i\theta})| = |q(e^{i\theta}| = |c| \prod_j |e^{i\theta} - \lambda_j|^2 \prod_k \frac{|e^{i\theta} - \mu_k|^2}{|\mu_k|^2},$$

and hence p is the squared modulus of a trigonometric polynomial. □

This fundamental lemma has deeply influenced twentieth century functional analysis. For instance the Riesz-Fejér Lemma is equivalent to the spectral theorem for unitary operators; see [29].

When invoking duality, the above is not less interesting. It was in this form that Riesz-Fejér Lemma was first generalized to an arbitrary dimension.

LEMMA 2.2. *Let $L \in \mathcal{H}(S^{2d-1})'$ be a linear functional which is non-negative on $\Sigma_h^2 \mathcal{H}(S^{2d-1})$. Then L is represented by a positive Borel measure supported on the sphere.*

The proof has implicitly appeared in the works of Ito [17], Yoshino [33], Lubin [22] and Athavale [3], all dealing with subnormality criteria for commuting tuples of bounded linear operators. Without aiming at completeness, here is the main idea.

Proof (Sketch). Let L be a non-negative functional on $\Sigma_h^2 \mathcal{H}(S^{2d-1})$. Fix a polynomial $p \in \mathbb{C}[z]$ and consider the functional

$$f(r_1^2, \ldots, r_d^2) \mapsto L(f|p(z)|^2), \quad f \in \mathbb{R}[r_1^2, \ldots, r_d^2],$$

where $r_j^2 = |z_j|^2$. Since

$$1 - |z_j|^2 = \sum_{k \neq j} |z_k|^2,$$

$$L\Big(\prod_j[(1-r_j^2)^{n_j}r_j^{2m_j}]|p|^2\Big) \geq 0, \qquad n_j, m_j \geq 0.$$

By a classical result of Haviland, see for instance [2], there exists a positive Borel measure $\mu_{|p|^2}$ on the simplex Δ defined by

$$\Delta = \{(r_1^2, ..., r_d^2);\ r_1^2 + ... + r_d^2 = 1\},$$

with the property

$$L(f|p(z)|^2) = \int_\Delta f d\mu_{|p|^2}.$$

The total mass of $\mu_{|p|^2}$ is $L(|p|^2)$.

By polarization, one can define complex valued measures by

$$L(fp\overline{q}) = \int_\Delta f d\mu_{p\overline{q}},\ \ f \in \mathbb{R}[r_1^2, ..., r_d^2],\ \ p, q \in \mathbb{C}[z],$$

so that the sesqui-linear kernel $(p, q) \mapsto \mu_{p\overline{q}}$ is positive semi-definite.

In short, the functional L can be extended to the linear space of functions (on the sphere) of the form

$$F(r, z) = \sum_{|\alpha| \leq n} c_\alpha(r) z^\alpha,$$

where $c_\alpha(r)$ are bounded, Borel measurable functions on the simplex Δ. The extended functional $\tilde{L}$ still satisfies

$$\tilde{L}(|F(r, z)|^2) \geq 0.$$

Next we pass to polar coordinates $z_j = r_j\omega_j$, $|\omega_j| = 1$ and remark that multiplication by ω_j satisfies the isometric condition

$$\tilde{L}(|\omega_j F(r, z)|^2) = \tilde{L}(|F(r, z)|^2).$$

Thus, we can further extend the functional $\tilde{L}$ to all polynomials in r and $\omega, \overline{\omega}$, so that

$$\tilde{L}(|\omega_j^{-1} F(r, z)|^2) = \tilde{L}(|F(r, z)|^2)$$

and

$$\tilde{L}(|p(r, \omega, \overline{\omega})|^2) \geq 0.$$

We refer to [33] or [31] for the details how this extension is constructed. By rewriting the latter positivity condition we have in particular

$$\tilde{L}(|h(z, \overline{z})|^2) \geq 0,\ \ h \in \mathbb{C}[z, \overline{z}],$$

whence, by the Stone-Weierstrass Theorem and the Riesz Representation Theorem, the functional $\tilde{L}$ is represented by a positive Borel measure, supported on the sphere.

The representing measure is unique by the Stone-Weierstrass Theorem. □

3. A Striktpositivstellensatz. We now turn to the basic question considered in this paper. We are given a finite set of real polynomials in $2d$ variables $p, q_1, ..., q_r$, or equivalently, Hermitian symmetric polynomials in d complex variables. We suppose that $p(z, \overline{z})$ is strictly positive on the subset of S^{2d-1} where each q_j is nonnegative. Can we write p as a weighted sum of squared norms with q_i as weights, as the real affine Striktpositivstellensatz (see for instance [23]) suggests? The answer is yes, and we can offer at least two different reasons why it is so.

THEOREM 3.1. *Let $p, q_1, ..., q_r \in \mathbb{R}[x, y]$, where $x + iy = z \in \mathbb{C}^d$. If*

$$(|z| = 1,\ q_i(z, \overline{z}) \geq 0, 1 \leq i \leq r) \Rightarrow (p(z, \overline{z}) > 0),$$

then

$$p \in \Sigma_h^2 + q_1\Sigma_h^2 + ... + q_r\Sigma_h^2 + I(S^{2d-1}).$$

First we discuss the history of such Hermitian squares decompositions, in the case where there are no constraints. A Hermitian symmetric polynomial p is called bihomogeneous of degree (m, m) if

$$p(\lambda z, \overline{\lambda z}) = |\lambda|^{2m} p(z, \overline{z})$$

for all complex numbers λ and all $z \in \mathbb{C}^d$. The values of a bihomogeneous polynomial are determined by its values on the sphere. When p is bihomogeneous and strictly positive on the sphere, Quillen [28] proved that there is an integer k and a homogeneous polynomial vector-valued mapping $h(z)$ such that

$$|z|^{2k} p(z, \overline{z}) = |h(z)|^2.$$

This result was discovered independently by the first author and Catlin [6] in conjunction with the first author's work on proper mappings between balls in different dimensions. The proof in [6] uses the Bergman projection and some facts about compact operators, and it generalizes to provide an isometric imbedding theorem for certain holomorphic vector bundles [7].

It is worth noting that the integer k and the number of components of h can be arbitrarily large, even for polynomials p of total degree four in two variables. The result naturally fits into the phenomena encoded into the old or recent Positivestellensätze, see for instance [23]. For the specific case of Hermitian polynomials on spheres see [8] for considerable discussion and generalizations.

Using a process of bihomogenization, Catlin and the first author (see [6, 8] and [9]) proved that if p is arbitrary (not necessarily bihomogeneous) and strictly positive on the sphere, then p agrees with a squared norm on the sphere; in other words, $p \in \Sigma_h^2 + I(S^{2d-1})$. Thus Theorem 1 holds when there are no constraints. Our proof of Theorem 1 first considers the case of no constraints, but we approach this case in a completely different manner.

Strict positivity is required for these results. The polynomial $(|z_1|^2 - |z_2|^2)^2$ is bihomogeneous and nonnegative everywhere, but there is no element in Σ_h^2 agreeing with it on the sphere. See Example 4.3 below.

Proof of Theorem 1. Suppose first that no q_i's are present and assume by contradiction that $p \notin \Sigma_h^2$, all regarded as elements of $\mathcal{H}(S^{2d-1})$. Since the constant function 1 belongs to the algebraic interior of the convex cone $\mathcal{H}(S^{2d-1})$, the separation lemma due to Eidelheit-Kakutani [12, 18] provides a linear functional $L \in \mathcal{H}(S^{2d-1})'$, satisfying both $L(1) > 0$ and

$$L(p) \leq 0 \leq L|_{\Sigma_h^2}.$$

According to Lemma 2, there exists a positive Borel measure μ, supported on the unit sphere, which represents L. Therefore

$$0 \geq L(p) = \int p d\mu > 0,$$

a contradiction.

The proof of the general case is similar, with the difference that we have to prove that the support of the measure μ is contained in the nonnegativity set defined by the functions q_i. To this aim, fix an index i, and remark that

$$\int q_i |p|^2 d\mu \geq 0$$

for all $p \in \mathbb{C}[z]$. Now, by the first case, every positive polynomial $P(z, \overline{z})$ is in the convex hull of the Hermitian squares, whence

$$\int q_i P(z, \overline{z}) d\mu \geq 0$$

whenever $P(z, \overline{z}) > 0$ on the sphere, that is whenever $P(z, \overline{z}) \geq 0$ on the sphere. In view of Stone-Weierstrass Theorem, every continuous functions f on the sphere can be uniformly approximated by real polynomials. In particular, we infer

$$\int q_i f^2 d\mu \geq 0, \quad f \in C(S^{2d-1}).$$

But this inequality holds only if the support of μ is contained in the nonnegativity set $q_i(z, \overline{z}) \geq 0$. □

4. Examples.

4.1. Optimization on the closed disk. The following simple example shows that Hermitian sums of squares do not suffice as positivity certificates on more general semi-algebraic sets. Specifically, let

$$P(z,\overline{z}) = 1 - \frac{4}{3}|z|^2 + a|z|^4,$$

with $\frac{1}{3} < a$. Note that

$$P(z,\overline{z}) = \left(1 - \frac{2}{3}|z|^2\right)^2 + \left(a - \frac{4}{9}\right)|z|^4,$$

and hence $P \in \Sigma^2\mathcal{H}$ when $a \geq \frac{4}{9}$. Hence we assume $\frac{1}{3} < a < \frac{4}{9}$. The polynomial $1 - \frac{4}{3}t + at^2$ is decreasing for $0 < t < 1$ when $a < \frac{2}{3}$; therefore $|z| \leq 1$ implies $P(z,\overline{z}) \geq 1 + a - \frac{4}{3} > 0$.

On the other hand,

$$P \notin \Sigma_h^2 + (1 - |z|^2)\Sigma_h^2.$$

To see that P is not in this set, we apply the hereditary calculus. See [1] for details. We replace z with a contractive operator T and replace $\overline{z}$ with T^*. We follow the usual convention of putting all T^*'s to the left of the powers of T. If P were in this set, we would obtain

$$\|T\| \leq 1 \;\Rightarrow\; p(T,\overline{T}) \geq 0.$$

In particular let T be the 2×2 Jordan block with 1 above the diagonal. We obtain a contradiction by computing that $P(T,T^*)$ is the diagonal matrix with eigenvalues 1 and $-\frac{1}{3}$.

On the other hand, the larger convex cone $\Sigma^2 + (1 - |z|^2)\Sigma_h^2$ is appropriate in this case, see [25, 27].

4.2. Squared norms. Recall that $\Sigma_h^2\mathcal{H}$ denotes the convex cone consisting of polynomials which are squared norms of (holomorphic) polynomial mappings. In all dimensions the zero set of an element in $\Sigma_h^2\mathcal{H}$ must be a complex variety.

Suppose $R(z,\overline{z}) \geq 0$ for all z. Even in one dimension we cannot conclude that $R \in \Sigma_h^2\mathcal{H}$. We noted earlier, where $x = \mathrm{Re}(z)$, the example

$$R(z,\overline{z}) = (z + \overline{z})^2 = 4x^2.$$

The zero set of R is the imaginary axis, which has no complex structure. In one dimension of course, the zero set of an element in $\Sigma_h^2\mathcal{H}$ must be either all of $\mathbb{C}$ or a finite set.

Things are more complicated and interesting in higher dimensions. Consider the following example from [8]. Define a Hermitian bihomogeneous polynomial in three variables by

$$p(z,\overline{z}) = (|z_1 z_2|^2 - |z_3|^4)^2 + |z_1|^8.$$

This polynomial p is nonnegative for all z, and its zero set is the complex plane given by $z_1 = z_3 = 0$ with z_2 arbitrary. Yet p is not a sum of squared moduli; even more striking is that p cannot be written as the quotient $\frac{|a(z)|^2}{|b(z)|^2}$ where a and b are sums of squared moduli. See [10] for additional information on this example and several tests for deciding whether a nonnegative polynomial R can be written as a quotient of squared norms. See [32] for a necessary and sufficient condition involving the zeroes of R.

We give an additional example in one dimension. Define p by

$$p(z,\overline{z}) = 1 + bz^2 + \overline{b}\overline{z}^2 + c|z|^2 + |z|^4.$$

The condition for being a quotient of squared norms is that one of the following three statements holds:

$$c > 2|b|^2 - 2,$$

$$b = 0, \quad c > -2,$$

$$|b| = 1, \quad c = 0.$$

The condition for being nonnegative is simpler: $c \geq 2|b| - 2$.

4.3. Proof of Example b). We claimed earlier that the polynomial $(|z_1|^2 - |z_2|^2)^2$ is bihomogeneous and nonnegative everywhere, but there is no element in Σ_h^2 agreeing with it on the sphere.

Proof. Put $R(z,\overline{z}) = (|z_1|^2 - |z_2|^2)^2$, and let $V(R)$ denote its zero set. We note that $V(R) \cap S^{2n-1}$ is the torus T defined by $|z_1|^2 = |z_2|^2 = \frac{1}{2}$. Suppose for some polynomial mapping $z \to P(z)$ we have $R = |P|^2$ on the unit sphere. Note first that the zero set of $|P|^2$ is a complex variety. We have $|P(z)|^2 = 0$ for $z \in T$. We claim that P is identically zero. For each fixed z_2 with $|z_2| = 1$, the vector-valued polynomial mapping $z_1 \to P(z_1, z_2)$ vanishes on the circle $|z_1|^2 = \frac{1}{2}$ and hence vanishes identically. Since z_2 was an arbitrary point with $|z_2|^2 = \frac{1}{2}$ we conclude that the mapping $(z_1, z_2) \to P(z_1, z_2)$ vanishes whenever $z_1 \in \mathbf{C}$ and z_2 lies on a circle. By symmetry it also vanishes with the roles of the variables switched. It follows that the zero set of P (which is a complex variety) is at least three real dimensions, and hence P vanishes identically. Since R does not vanish identically on the sphere we obtain a contradiction. □

4.4. Example. There exist non-negative polynomials R such that R is not in $\Sigma_h^2\mathcal{H}$, yet there is a positive integer N for which $R^N \in \Sigma_h^2\mathcal{H}$. The bihomogeneous polynomial R_λ given by

$$R_\lambda(z,\overline{z}) = (|z_1|^2 + |z_2|^2)^4 - \lambda|z_1 z_2|^4$$

satisfies this property whenever $\lambda < 8$. See [11] and [32]. For $\lambda < 16$, $R_\lambda > 0$ on the sphere. By Theorem 1 it agrees with a squared norm *on the sphere.*

5. Ranks of Hermitian forms on spheres and spectral values. Let R be a Hermitian symmetric polynomial. In this section we consider how many terms are needed to write R as an Hermitian sum of squares on the unit sphere. As we mentioned in the introduction, the real variables analogue of this problem is already quite appealing. It corresponds to the case when certain Hermitian forms are diagonal, and hence things are much easier. We therefore begin with a simple real variables example, observe an interesting phenomenon, and then turn to its Hermitian version.

Let p be a homogeneous polynomial of several real variables. Suppose $p(x) > 0$ for all x in the nonnegative orthant, the set where each $x_j \geq 0$. By a classical theorem of Polya, there is an integer d such that the polynomial f defined by $(\sum_i x_i)^d p(x) = f(x)$ has only positive coefficients. See [8] for considerable discussion of Polya's classical theorem and its Hermitian analogues. Here we will discuss a simple example where we are concerned with the number of terms involved.

Consider the one parameter family of real polynomials on $\mathbf{R}^2$ defined by $p_\lambda(x,y) = x^2 - \lambda xy + y^2$. Each p_λ is homogeneous of degree 2, and hence is determined by its values on the line given by $x + y = 1$. We ask when we can find a polynomial f with nonnegative coefficients and which agrees with p_λ on this line. We also want to know the minimum number N_λ of terms f must have. For $\lambda > 2$, the polynomial has negative values, and hence cannot be a sum of terms with positive coefficients. The same conclusion holds at the border case when $\lambda = 2$. When $\lambda < 2$, Polya's theorem guarantees that such an f exists. The number N_λ tends to infinity as λ tends to two. The value $\lambda = 1$ plays a surprising special role. On the line $x + y = 1$ we can write

$$x^2 - xy + y^2 = \frac{x^3 + y^3}{x + y} = x^3 + y^3.$$

Thus $N_\lambda = 2$ when $\lambda = 1$. On the other hand, $N_\lambda > 2$ for $0 < \lambda < 1$. Thus the minimum number of squares needed is not monotone in λ. This striking phenomenon also holds in the Hermitian case; we could create the Hermitian analogue simply by writing $x = |z_1|^2$ and $y = |z_2|^2$.

Let now R denote a Hermitian symmetric polynomial. Suppose $R \geq 0$ as a function. We write R in the form

$$R(z, \overline{z}) = \sum c_{\alpha\beta} z^\alpha \overline{z}^\beta$$

and we know that R is itself a squared norm if and only if the matrix of coefficients $(c_{\alpha\beta})$ is non-negative definite. In this case there is an integer N and holomorphic polynomials f_j such that

$$R(z, \overline{z}) = \sum_{j=1}^{N} |f_j(z)|^2 = |f(z)|^2.$$

By elementary linear algebra, the minimum possible N equals the rank of the matrix of coefficients. Thus the global problem is easy.

Things are considerably different on the sphere. For example, with the right choice of nonnegative constants c_j and integer K, the expression

$$1 - \sum_{j=1}^{K} c_j |z|^{2j}$$

will be strictly positive on the sphere while the underlying matrix of coefficients will have arbitrarily many negative eigenvalues. Suppose however that we write R as a squared norm $|f|^2$ on the sphere. What can we say about N, the rank of the coefficient matrix of $|f|^2$? This problem is difficult. The following example and the fairly detailed sketches of the proofs provide an accurate illustration of the subtleties involved.

5.1. Example. Let $n = 2$. Given N, is there a polynomial or rational function g from $\mathbf{C}^2$ to $\mathbf{C}^N$ such that $|g(z)|^2 = 1 - |\zeta z_1 z_2|^2$ on the sphere?

0) If $|\zeta|^2 \geq 4$, then for all N, the answer is no.

1) If $N = 1$, then the answer is yes only when $\zeta = 0$.

2) If $N = 2$, the answer is yes precisely when one of the following holds: $\zeta = 0, |\zeta|^2 = 1, |\zeta|^2 = 2, |\zeta|^2 = 3$.

3) For each ζ with $|\zeta|^2 < 4$, there is a smallest N_ζ for which the answer is yes. The limit as $|\zeta|$ tends to 2 of N_ζ is infinity.

Proof. We merely indicate the main ideas in the proofs and refer for complete details to some published articles. In general we are seeking a holomorphic polynomial mapping g such that

$$|g_1(z)|^2 + ... + |g_N(z)|^2 + |\zeta|^2 |z_1 z_2|^2 = 1$$

on the unit sphere. The components of g and the additional term $\zeta z_1 z_2$ define a holomorphic mapping from the n ball to the $N+1$ ball which maps the sphere to the sphere. Either such a map is constant or proper. We now consider the results for small N.

0) The maximum of $|\zeta z_1 z_2|^2$ on the sphere is 1 when $|z_1|^2 = |z_2|^2 = \frac{1}{2}$. Hence $|\zeta|^2 \leq 4$ must hold if the question has a positive answer. We claim that $|\zeta|^2 = 4$ cannot hold either. Suppose $|\zeta|^2 = 4$ and g exists. Then we would have

$$|g(z)|^2 + 4|z_1|^2|z_2|^2 = 1 = (|z_1|^2 + |z_2|^2)^2$$

on the sphere, and hence

$$|g(z)|^2 = (|z_1|^2 - |z_2|^2)^2$$

on the sphere. By Example 3.b), no such g exists.

1) The only proper mappings from the 2-ball to itself are automorphisms, hence linear fractional transformations. Hence the term $\zeta z_1 z_2$ can arise only if $\zeta = 0$. When $\zeta = 0$ we may of course choose $g(z)$ to be (z_1, z_2).

2) This result follows from Faran's classification [14] of the proper holomorphic rational mappings from B_2 to B_3. First we mention that maps g and h are spherically equivalent if there are automorphisms u, v of the domain and target balls such that $h = vgu$. If g existed, then there would be a proper polynomial mapping h from B_2 to B_3 with the monomial $\zeta z_1 z_2$ as a component. It follows from Faran's classification that h would have to be spherically equivalent to one of the four mappings:

$$\begin{aligned} h(z_1, z_2) &= (z_1, z_2, 0) \\ h(z_1, z_2) &= (z_1, z_1 z_2, z_2^2) \\ h(z_1, z_2) &= (z_1^2, \sqrt{2} z_1 z_2, z_2^2) \\ h(z_1, z_2) &= (z_1^3, \sqrt{3} z_1 z_2, z_2^3). \end{aligned}$$

These four mappings provide the four possible values for $|\zeta|$. It turns out, however, that one can say more. In this case one can prove, via an analysis of the possible denominators, that h must be *unitarily* equivalent to one of these four maps. Since h must have the component $\zeta z_1 z_2$, one can compute that the only possible values of $|\zeta|$ are the four that occur in these formulas.

3) This conclusion appears in [8].

We discuss the situation further. There are certain *spectral* values ζ^* for which the value N_{ζ^*} is smaller than that of N_ζ for some ζ with $|\zeta| < |\zeta^*|$. If $\zeta = 1$ we can solve the problem with $N = 2$; take $g(z) = (z_1^2, z_2)$ for example. Yet, if $|\zeta| < 1$, then we cannot solve the problem when $N = 2$ unless $\zeta = 0$. The proof relied on Faran's determination of all proper (rational) mappings from B_2 to B_3. There are 4 spherical equivalence classes; each class contains a monomial map, but there are no families of maps. If we allow one larger target dimension, then we can get a one-parameter family of maps:

$$f(z) = (z_1, z_2^2, \cos(t) z_1 z_2, \sin(t) z_1 z_2^2, \sin(t) z_1^2 z_2).$$

From this formula we see that we can recover all values of $|\zeta|$ up to unity, but not beyond. We omit the details. The phenomenon that certain discrete values become possible before smaller values do continues as we increase N. If $N = 4$, for example, the answer is yes for $0 \leq |\zeta|^2 \leq 2$ and the following additional values for $|\zeta|^2$:

$$\frac{7}{2}, \frac{10}{3}, \frac{8}{3}, \frac{5}{2}.$$

We satisfy ourselves with explicit maps where the constants $\sqrt{\frac{7}{2}}$ and $\sqrt{\frac{10}{3}}$ arise as coefficients of $z_1 z_2$:

$$f(z)=\left(z_1^7, z_2^7, \sqrt{\frac{7}{2}}z_1z_2, \sqrt{\frac{7}{2}}z_1^5z_2, \sqrt{\frac{7}{2}}z_1z_2^5\right)$$

$$f(z)=(z_1^5, z_2^5, \sqrt{\frac{10}{3}}z_1z_2, \sqrt{\frac{5}{3}}z_1^4z_2\sqrt{\frac{5}{3}}z_1z_2^4).$$

Both these maps are proper from the two ball to the five ball.

The results in Example 4.1 illustrate clearly the difference between finding a representation as a squared norm of rank N and finding any representation as a squared norm. We close with an explanation of why we called this section *spectral values*. Given the polynomial $R(z,\overline{z})$, we solve the problem $R = |f|^2$ on the sphere as in [6] or [8]. We add a variable t to homogenize R; call the result R_h. We may choose C large enough such that the function

$$R_h(z,t,\overline{z},\overline{t}) + C(|z|^2 - |t|^2)^m$$

is strictly positive on the sphere in $\mathbf{C}^{n+1}$. The underlying matrix of coefficients need not be non-negative definite. We then invoke [6] or [28] to find an integer d such that, after multiplication by $(|z|^2 + |t|^2)^d$, the underlying form is positive definite. We then dehomogenize and evaluate on the sphere. Thus the isolated values of $|\zeta|$ for which we can solve the problem in Example 4.1 are in fact vanishing eigenvalues of a Hermitian form; for nearby values of ζ the eigenvalues may become negative. If we multiply by higher powers of $|z|^2 + |t|^2$, then we can make these eigenvalues positive (an open condition), but other eigenvalues will generally vanish. Given ζ with $|\zeta| < 4$, there always exists an N_ζ, but N_ζ depends on ζ, rather than only on the dimension and degree of R. See [8] for lengthy discussion.

We close by mentioning that the proof of the Positivstellensatz cannot provide effective information on N_λ based upon only dimension and degree. One must take into account precise information about the size of R, and even then things are delicate. Example 4.1 relies on Faran's deep work. No general classification of proper polynomial mappings between balls exists that gives precise information on the relationship between the degree and the target dimension. See [9] for more information.

REFERENCES

[1] J. Agler and J.E. McCarthy, Pick interpolation and Hilbert function spaces. Graduate Studies in Mathematics. **44**. Providence, RI: American Mathematical Society (2002).

[2] N.I. Akhiezer, The Classical Moment Problem. Oliver and Boyd, Edinburgh and London (1965).

[3] A. Athavale, Holomorphic kernels and commuting operators. Trans. Amer. Math. Soc. **304**:101–110 (1987).

[4] J.A. Ball and T.T. Trent, Unitary colligations, reproducing kernel Hilbert spaces, and Nevanlinna-Pick interpolation in several variables. J. Funct. Anal. **157**(1):1–61 (1998).

[5] A. BARVINOK, Integration and optimization of multivariate polynomials by restriction onto a random subspace. Foundations of Computational Mathematics **7**:229–244 (2007).

[6] D.W. CATLIN AND J.P. D'ANGELO, A stabilization theorem for Hermitian forms and applications to holomorphic mappings, Math. Res. Lett. **3**(2):149–166 (1995).

[7] ———, An isometric imbedding theorem for holomorphic bundles. Math. Res. Lett. **6**(1):43–60 (1999).

[8] J.P. D'ANGELO, Inequalities from Complex Analysis, Carus Mathematical Monographs, No. **28**, Mathematical Assn. of America (2002).

[9] ———, Proper holomorphic mappings, positivity conditions, and isometric imbedding J. Korean Math Soc. **40**(3):341–371 (2003).

[10] ———, Complex variables analogues of Hilbert's seventeenth problem, International Journal of Mathematics **16**:609–627 (2005).

[11] J.P. D'ANGELO AND D. VAROLIN, Positivity conditions for Hermitian symmetric functions, Asian J. Math. **8**:215–232 (2004).

[12] M. EIDELHEIT, Zur Theorie der konvexen Mengen in linearen normierten Räumen. Studia Math. **6**:104–111 (1936).

[13] J. ESCHMEIER AND M. PUTINAR, Spherical contractions and interpolation problems on the unit ball. J. Reine Angew. Math. **542**:219–236 (2002).

[14] J. FARAN, Maps from the two-ball to the three-ball, Inventiones Math. **68**:441–475 (1982).

[15] C. FOIAS AND A.E. FRAZHO, The commutant lifting approach to interpolation problems. Operator Theory: Advances and Applications, **44**. Basel. Birkhäuser (1990).

[16] J.W. HELTON, S. MCCULLOUGH, AND M. PUTINAR, Non-negative hereditary polynomials in a free *-algebra. Math. Zeitschrift **250**:515–522 (2005).

[17] T. ITO, On the commutative family of subnormal operators, J. Fac. Sci. Hokkaido Univ. **14**:1–15 (1958).

[18] S. KAKUTANI, Ein Beweis des Satzes von M. Eidelheit über konvexe Mengen. Proc. Imp. Acad. Tokyo **13**, 93–94 (1937).

[19] M.G. KREIN, M.A. NAIMARK, The method of symmetric and Hermitian forms in the theory of separation of the roots of algebraic equations. (Translated from the Russian by O. Boshko and J. L. Howland). Linear Multilinear Algebra **10**:265–308 (1981).

[20] J.E. MCCARTHY AND M. PUTINAR, Positivity aspects of the Fantappiè transform. J. d'Analyse Math. **97**:57–83 (2005).

[21] J.B. LASSERRE, Global optimization with polynomials and the problem of moments. SIAM J. Optim. **11**(3):796–817 (2001).

[22] A. LUBIN, Weighted shifts and commuting normal extension, J. Austral. Math. Soc. **27**:17–26 (1979).

[23] A. PRESTEL AND C.N. DELZELL, Positive polynomials. From Hilbert's 17th problem to real algebra. Springer Monographs in Mathematics. Berlin: Springer (2001).

[24] A. PRESTEL, Representation of real commutative rings. Expo. Math. **23**:89–98 (2005).

[25] M. PUTINAR, Sur la complexification du problème des moments. C. R. Acad. Sci., Paris, Série I **314**(10):743–745 (1992).

[26] ———, Positive polynomials on compact semi-algebraic sets. Indiana Univ. Math. J. **42**(3):969–984 (1993).

[27] ———, On Hermitian polynomial optimization. Arch. Math. **87**:41–5 (2006).

[28] D.G. QUILLEN, On the representation of Hermitian forms as sums of squares. Invent. Math. **5**:237–242 (1968).

[29] F. RIESZ AND B.SZ.-NAGY, Functional analysis. Transl. from the 2nd French ed. by Leo F. Boron. Reprint of the 1955 orig. publ. by Ungar Publ. Co., Dover Books on Advanced Mathematics. New York: Dover Publications, Inc. (1990).

[30] K. Schmüdgen, The K-moment problem for compact semi-algebraic sets, Math. Ann. **289**:203–206 (1991).

[31] B.Sz.-Nagy and C. Foias, Analyse harmonique des opérateurs de l'espace de Hilbert. Budapest: Akademiai Kiado; Paris: Masson et Cie, 1967.

[32] D. Varolin, Geometry of Hermitian algebraic functions: quotients of squared norms. American J. Math., to appear.

[33] T. Yoshino, On the commuting extensions of nearly normal operators, Tohoku Math. J. **25**:163–172 (1973).

ENGINEERING SYSTEMS AND FREE SEMI-ALGEBRAIC GEOMETRY*

MAURICIO C. DE OLIVEIRA†, J. WILLIAM HELTON‡,
SCOTT A. MCCULLOUGH§, AND MIHAI PUTINAR¶

(To Scott Joplin and his eternal RAGs)

Abstract. This article sketches a few of the developments in the recently emerging area of real algebraic geometry (in short RAG) in a free* algebra, in particular on "noncommutative inequalities". Also we sketch the engineering problems which both motivated them and are expected to provide directions for future developments. The free* algebra is forced on us when we want to manipulate expressions where the unknowns enter naturally as matrices. Conditions requiring positive definite matrices force one to noncommutative inequalities. The theory developed to treat such situations has two main parts, one parallels classical semialgebraic geometry with sums of squares representations (Positivstellensätze) and the other has a new flavor focusing on how noncommutative convexity (similarly, a variety with positive curvature) is very constrained, so few actually exist.

1. Introduction. This article sketches a few of the developments in the recently emerging area of real algebraic geometry in a free* algebra, and the engineering problems which both motivated them and are expected to provide directions for future developments. Most linear control problems with mean square or worst case performance requirements lead directly to matrix inequalities (MIs). Unfortunately, many of these MIs are badly behaved and unsuited to numerics. Thus engineers have spent considerable energy and cleverness doing non-commutative algebra to convert, on an ad hoc basis, various given MIs into equivalent better behaved MIs.

A classical core of engineering problems are expressible as linear matrix inequalities (LMIs). Indeed, LMIs are the gold standard of MIs, since they are evidently convex and they are the subject of many excellent numerical packages. However, for a satisfying theory and successful numerics a convex MI suffices and so it is natural to ask:

How much more restrictive are LMIs than convex MIs?

It turns out that the answer depends upon whether the MI is, as is the case for systems engineering problems, fully characterized by perfor-

*The authors received partial support respectively from the Ford Motor Co., NSF award DMS-0700758 and the Ford Motor Co., NSF award DMS-0457504 and NSF award DMS-0701094.

†Mechanics and Aerospace Engineering Department, UC at San Diego, La Jolla, CA 92093 (mauricio@ucsd.edu).

‡Department of Mathematics, University of California at San Diego, La Jolla, CA 92093 (helton@math.ucsd.edu).

§Department of Mathematics, University of Florida, Gainesville, FL 32611-8105 (sam@math.ufl.edu).

¶Department of Mathematics, University of California, Santa Barbara, CA 93106 (mputinar@math.ucsb.edu).

mance criteria based on L^2 and signal flow diagrams (as are most textbook classics of control). Such problems have the property we refer to as "*dimension-free*".

Indeed, there are two fundamentally different classes of linear systems problems: dimension free and dimension dependent. A dimension free MI is a MI where the unknowns are g-tuples of matrices which appear in the formulas in a manner which respects matrix multiplication. Dimension dependent MIs have unknowns which are tuples of numbers.

The results presented here suggest the surprising conclusion that for dimension free MIs convexity offers no greater generality than LMIs. Indeed, we conjecture:

Dimension free convex problems are equivalent to an LMI.

The key ingredient in passing from convex MIs to LMIs and proving their equivalence lies in the recently blossoming and vigorously developing direction of semi-algebraic in a free $*$ algebra; i.e., semi-algebraic geometry with variables which, like matrices, do not commute. Indeed at this stage there are two main branches of this subject. One includes non-commutative Positivstellensätze which characterize things like one polynomial p being positive where another polynomial q is positive. The other classifies situations with prescribed curvature.

As of today there are numerous versions of the Positivstellensätze for a free $*$- algebra, with typically cleaner statements than in the commutative case. For instance, in the non-commutative setting, positive polynomials are sums of squares. Through the connection between convexity and positivity of a Hessian, non-commutative semi-algebraic dictates a rigid structure for polynomials, and even rational functions, in non-commuting variables. For instance, a noncommutative polynomial p has second derivative p'' which is again a polynomial. Further, if p is matrix convex (as defined below), then p'' is matrix positive (also defined below) and is thus a sum of squares. It is a bizarre twist that p'' can be sum of squares only if p has degree at most two (see §3. The authors suspect that this is a harbinger of a very rigid structure in a free $*$-algebra for "irreducible varieties" whose curvature is either nearly positive or nearly negative; but this is a tale for another day.

A substantial opportunity for noncommutative algebra and symbolic computation lies in numerical computation for problems whose variables are naturally matrices. The goal is to exploit this special structure to accelerate and to increase the allowable size of computation. This is the subject of Section 9.

This survey is not intended to be comprehensive. Rather its purpose is to provide some snippets of results in non-commutative semi-algebraic geometry and their related computer algebra and numerical algorithms, and of motivating engineering problems with the idea of entertaining and even piquing the readers interest in the subject. In particular, this article draws heavily from [HP07] and [HPMV]. Sometimes we shall abbreviate the word noncommutative to NC.

As examples of other important directions and themes, some of which are addressed in other articles in this volume, there is a non-commutative algebraic geometry based on the Weyl algebra and corresponding computer algebra implementations, for example, Gröbner basis generators for the Weyl algebra are in the standard computer algebra packages such as Plural/Singular. A very different and elegant area is that of rings with a polynomial identity, in short PI rings, e.g. $N \times N$ matrices for fixed N. While most PI research concerns identities, there is one line of work on polynomial inequalities, indeed sums of squares, by Procesi-Schacher [PS76]. A Nullstellensätz for PI rings is discussed in [Ami57].

As indicated LMIs play a large role in this paper, so now we describe them precisely.

1.1. LMIs and noncommutative LMIS. Since they play a central role in engineering and the study of convexity in the free $*$ setting, we digress, in the next subjection to define the notion of an LMI.

Given $d \times d$ symmetric (real entry) matrices $\Lambda_0, \Lambda_1, \ldots, \Lambda_g$, the function $L : \mathbb{R}^g \to S_d(\mathbb{R})$ given by

$$L(x) = \sum_{j=0} \Lambda_j x_j$$

is a classical linear pencil; and the inequality $L(x) \succeq 0$ is the classical (commutative) linear matrix inequality. Here $(x_1, \ldots, x_g) \in \mathbb{R}^g$.

In the non-commutative (dimension free) setting it is natural to substitute $X \in \mathbb{S}_n(\mathbb{R}^g)$ for the x above, obtaining the non-commutative version of a linear pencil. Namely, for each n a function $L : \mathbb{S}_n(\mathbb{R}^g) \to \mathbb{SR}^{n \times n}$

$$L_n(X) = L(X) = \sum \Lambda_j \otimes X_j.$$

The inequality $L(X) \succeq 0$ is what we will generally mean by LMI. And, as with polynomials, when we discuss LMIs and linear pencils it will be understood in the non-commutative sense.

EXAMPLE 1.1. For $x := (x_1, x_2)$ being either commuting or noncommuting variables L written as

$$L(x) := \begin{pmatrix} 1 & 0 \\ 0 & 1 \end{pmatrix} + \begin{pmatrix} 2 & 3 \\ 3 & 0 \end{pmatrix} x_1 + \begin{pmatrix} 3 & 5 \\ 2 & 0 \end{pmatrix} x_2$$

denotes a linear pencil or NC linear pencil. For $X := (X_1, X_2)$ with $X_j \in \mathbb{R}^{n \times n}$

$$L(X) = \begin{pmatrix} 1 & 0 \\ 0 & 1 \end{pmatrix} + \begin{pmatrix} 2 & 3 \\ 3 & 0 \end{pmatrix} \otimes X_1 + \begin{pmatrix} 3 & 5 \\ 2 & 0 \end{pmatrix} \otimes X_2$$

$$= \begin{pmatrix} I_n + 2X_1 + 3X_2 & 3X_1 + 5X_2 \\ 3X_1 + 2X_2 & I_n \end{pmatrix}.$$

For $X := (X_1, X_2)$ with $X_j \in \mathbb{R}$, the set of solutions to $L(X) \succeq 0$ is

$$\mathcal{C} := \{(X_1, X_2) : \quad 1 + 2X_1 + 3X_2 - (3X_1 + 5X_2)(3X_1 + 2X_2) \succeq 0.\}, \quad (1.1)$$

This last equivalence follows from taking an appropriate Schur complement which we now recall. The *Schur complement* of a matrix (with pivot γ^{-1}) is defined by

$$SchurComp \begin{pmatrix} \alpha & \beta \\ \beta^* & \gamma \end{pmatrix} := \alpha - \beta\gamma^{-1}\beta^*.$$

A key fact is: if γ is invertible, then the matrix is positive semi-definite if and only if $\gamma > 0$ and its Schur complement is positive semi-definite.

EXAMPLE 1.2. Apply this to the LMI in our example to obtain (1.1) for $X := (X_1, X_2)$ with $X_j \in \mathbb{R}^{n \times n}$ and X_j symmetric.

The Schur complement of $L(x)$ using the other pivot is the "rational expression"

$$I_n - (3X_1 + 2X_2)(I_n + 2X_1 + 3X_2)^{-1}(3X_1 + 5X_2).$$

1.2. Outline. The remainder of the survey is organized as follows. We expand upon the connection between systems engineering problems and dimension free MIs in Section 2. Convexity in the non-commutative (namely equal free $*$) setting is formalized in Section 3. This section also contains a brief glimpse into the *NCAlgebra* package. *NCAlgebra*, and the related *NCGB* (stands for non-commutative Gröbner basis) [HdOSM05] do symbolic computation in a free $*$- algebra and greatly aided the discovery of the results discussed in this survey. *NCAlgebra* and *NCGB* are free, but run under Mathematica which is not. Section 4 describes the engineering necessity for having a theory of matrix-valued non-commutative polynomials whose coefficients are themselves polynomials in non-commuting variables; much of the analysis of Section 3 carries over naturally in this setting. The shockingly rigid structure of convex rational functions is described in Section 5, with a sketch of proofs behind this "curvature oriented" non-commutative semi-algebraic geometry in §6. Section 7 discusses numerics designed to take advantage of matrix variables. Sections 8 gives the solution to the H^∞ control problem stated in §4.

Sections 9 describes noncommutative semi-algebraic geometry aimed at positivity and Positivstellensäten, this is an analogue of classical semi-algebraic geometry which is elegant though it does not have direct engineering applications.

1.3. Acknowledgments. The authors are grateful to Igor Klep for his comments generally and specifically for considerable help with Section 9 and allowing us to include his forthcoming Theorem 9.7.

2. Dimension free engineering: the map between systems and algebra. This section illustrates how linear systems problems lead to semi-algebraic geometry over a free or nearly free $*$- algebra and the role of convexity in this setting. The discussion will also inform the necessary further directions in the developing theory of non-commutative semi-algebraic needed to fully treat engineering problems.

In the engineering literature, the action takes place over the real field. Thus in much of this article, and in particular in this section, we restrict to real scalars. However,we do break from the engineering convention in that we will use A^* to denote the transpose of a (real entries) matrix and at the same time the usual involution on matrices with complex entries. Context will evidently determine the meaning.

The inner product of vectors in a real Hilbert space will be denoted $u \cdot v$.

2.1. Linear systems. A *linear system* $\mathfrak{F}$ is given by the linear differential equations

$$\frac{dx}{dt} = Ax + Bu,$$
$$y = Cx,$$

with the vector

- $x(t)$ at each time t being in the vector space $\mathcal{X}$ called the *state space*,
- $u(t)$ at each time t being in the vector space $\mathcal{U}$ called the *input space*,
- $y(t)$ at each time t being in the vector space $\mathcal{Y}$ called the *output space*,

and A, B, C being linear maps on the corresponding vector spaces.

2.2. Connecting linear systems. Systems can be connected in incredibly complicated configurations. We describe a simple connection and this goes along way toward illustrating the general idea. Given two linear systems $\mathfrak{F}$, $\mathfrak{G}$, we describe the formulas for connecting them as follows.

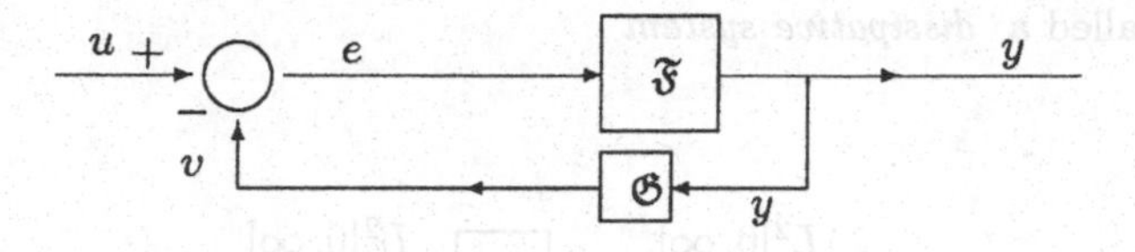

Systems $\mathfrak{F}$ and $\mathfrak{G}$ are respectively given by the linear differential equations

$$\frac{dx}{dt} = Ax + Be, \qquad \frac{d\xi}{dt} = a\,\xi + b\,w,$$
$$y = Cx, \qquad v = c\,\xi.$$

The connection diagram is equivalent to the algebraic statements

$$w = y \qquad \text{and} \qquad e = u - v.$$

The *closed loop system* is a new system whose differential equations are

$$\begin{aligned}\frac{dx}{dt} &= Ax - Bc\xi + Bu,\\ \frac{d\xi}{dt} &= a\,\xi + b\,y = a\,\xi + b\,Cx,\\ y &= Cx.\end{aligned}$$

In matrix form this is

$$\begin{aligned}\frac{d}{dt}\begin{pmatrix} x \\ \xi \end{pmatrix} &= \begin{pmatrix} A & -Bc \\ bC & a \end{pmatrix}\begin{pmatrix} x \\ \xi \end{pmatrix} + \begin{pmatrix} B \\ 0 \end{pmatrix} u,\\ y &= \begin{pmatrix} C & 0 \end{pmatrix}\begin{pmatrix} x \\ \xi \end{pmatrix},\end{aligned} \tag{2.1}$$

where the state space of the closed loop systems is the direct sum '$\mathcal{X} \oplus \mathcal{Y}$' of the state spaces $\mathcal{X}$ of $\mathfrak{F}$ and $\mathcal{Y}$ of $\mathfrak{G}$. The moral of the story is:

System connections produce a new system whose coefficients are matrices with entries which are polynomials or at worst "rational expressions" in the coefficients of the component systems.

Complicated signal flow diagrams give complicated matrices of polynomials or rationals. Note in what was said the dimensions of vector spaces and matrices never entered explicitly; the algebraic form of (2.1) is completely determined by the flow diagram. Thus, such linear systems lead to *dimension free* problems.

2.3. Energy dissipation. We have a system $\mathfrak{F}$ and want a condition which checks whether

$$\int_0^\infty |u|^2\,dt \ge \int_0^\infty |\mathfrak{F}u|^2\,dt, \qquad x(0) = 0,$$

holds for all input functions u, where $\mathfrak{F}u = y$ in the above notation . If this holds $\mathfrak{F}$ is called a *dissipative system*

$L^2[0,\infty]$ → $\mathfrak{F}$ → $L^2[0,\infty]$

The energy dissipative condition is formulated in the language of analysis, but it converts to algebra (or at least an algebraic inequality) because of the following construction, which assumes the existence of a "potential energy" like function V on the state space. A function V which satisfies $V \geq 0$, $V(0) = 0$, and

$$V(x(t_1)) + \int_{t_1}^{t_2} |u(t)|^2 dt \quad \geq \quad V(x(t_2)) + \int_{t_1}^{t_2} |y(t)|^2 dt$$

for all input functions u and initial states x_1 is called a *storage function.* The displayed inequality is interpreted physically as

potential energy now + energy in $\geq$ *potential energy then + energy out.*

Assuming enough smoothness of V, we can manipulate this integral condition to obtain first a differential inequality and then an algebraic inequality, as follows:

$$0 \geq \frac{V(x(t_2)) - V(x(t_1))}{t_2 - t_1} \quad + \quad \frac{1}{t_2 - t_1} \int_{t_1}^{t_2} |y(t)|^2 - |u(t)|^2 dt,$$

$$0 \geq \nabla V(x(t_1)) \cdot \frac{dx}{dt}(t_1) \quad + \quad |y(t_1)|^2 - |u(t_1)|^2.$$

Substituting $\frac{dx}{dt}(t_1) = Ax(t_1) + Bu(t_1)$ and $y = Cx$ gives

$$0 \geq \nabla V(x(t_1)) \cdot (Ax(t_1) + Bu(t_1)) \quad + \quad |Cx(t_1)|^2 - |u(t_1)|^2.$$

The system is dissipative if this inequality holds for all $u(t_1)$, $x(t_1)$ which can occur when it runs (starting at $x(0) = 0$). Denote $x(t_1)$ by x and $u(t_1)$ by u. With these notations, the inequality becomes

$$0 \geq \nabla V(x) \cdot (Ax + Bu) \quad + \quad |Cx|^2 - |u|^2, \tag{2.2}$$

All vectors $u(t_1)$ in $\mathcal{U}$ can certainly occur as an input and if all $x(t_1)$ can occur we call the system *reachable* . In the case of linear systems, V can be chosen quadratic of the form $V(x) = \langle Ex, x\rangle$ with $E \succeq 0$ and $\nabla V(x) = 2Ex$.

THEOREM 2.1. *The linear system A, B, C is dissipative if inequality (2.2) holds for all $u \in \mathcal{U}, x \in \mathcal{X}$. Conversely, if A, B, C is reachable, then dissipativity implies inequality (2.2) holds for all $u \in \mathcal{U}, x \in \mathcal{X}$.*

2.3.1. Riccati inequalities. In the linear case, we may substitute $\nabla V(x) = 2Ex$ in (2.2) to obtain

$$0 \geq 2Ex \cdot (Ax + Bu) + |Cx|^2 - |u|^2,$$

for all u, x. Thus,

$$0 \geq \max_u \left([EA + A^*E + C^*C]x \cdot x + 2B^*Ex \cdot u - |u|^2 \right). \tag{2.3}$$

Since the maximizer in u is $u = B^*Ex$,

$$0 \geq 2Ex \cdot Ax + 2|B^*Ex|^2 + |Cx|^2 - |B^*Ex|^2.$$

This last inequality is conveniently expressed as

$$0 \geq [EA + A^*E + EBB^*E + C^*C]x \cdot x.$$

Thus the classical *Riccati matrix inequality*

$$0 \succeq EA + A^*E + EBB^*E + C^*C \qquad \textit{with} \quad E \succeq 0 \tag{2.4}$$

insures dissipativity of the system; and, it turns out, is also implied by dissipativity when the system is reachable.

2.3.2. Schur complements and linear matrix inequalities. Using Schur complements, the Ricatti inequality of equation (2.4) is equivalent to the inequality

$$L(E) := \begin{pmatrix} EA + A^*E + C^*C & EB \\ B^*E & -I \end{pmatrix} \preceq 0. \tag{2.5}$$

Here A, B, C describe the system and E is an unknown matrix. If the system is reachable, then A, B, C is dissipative if and only if $L(E) \preceq 0$ and $E \succeq 0$.

The key feature in this reformulation of the Ricatti inequality is that $L(E)$ is linear in E, so inequality $L(E) \preceq 0$ is a Linear Matrix Inequality (LMI) in E. This is more general than was introduced in §1.1 since the coefficients A, B, C are themselves matrices rather than scalars.

2.4. The basic questions. What we have seen so far are the basic components of how one produces matrix inequalities (MI's) from engineering problems. It is in fact a very mechanical procedure. The trouble is that one typically produces messy MI's having no apparent good properties (see §4.1 for an example). We would like for them to be convex or to transform to a convex matrix inequality, justifying the claim that major issues in linear systems theory are:

1. *Are convex matrix inequalities more general than LMIs?*
2. *Which problems convert to a convex matrix inequality? How does one do the conversion?*
3. *Find numerics which will solve large convex problems. How do you use special structure, such as most unknowns are matrices and the formulas are all built of noncommutative rational functions?*

The mathematics here aims toward helping an engineer who writes a toolbox which other engineers will use for designing systems, like control systems. What goes in such toolboxes is algebraic formulas with matrices A, B, C unspecified and reliable numerics for solving them when a user does specify A, B, C as matrices. A user who designs a controller for a helicopter puts in the mathematical systems model for his helicopter and puts in matrices, for example, A is a particular $R^{8\times 8}$ matrix etc. Another user who designs a satellite controller might have a 50 dimensional state space and of course would pick completely different A, B, C. Essentially any matrices of any compatible dimensions can occur and our claim that our algebraic formulas are convex in the ranges we specify must be true.

The toolbox designer faces two completely different tasks. One is manipulation of algebraic inequalities; the other is numerical solutions. Often the first is far more daunting since the numerics is handled by some standard package although for numerics problem size is a demon. Thus there is a great need for algebraic theory.

Much of this paper bears on the first question when the unknowns are matrices, which though not fully solved has already motivated the construction of a considerable amount of noncommutative semi-algebraic geometry outlined in this survey. Section 7 bears on Question 3.

3. Convexity in a free algebra.

Convexity of functions, domains and their close relative, positive curvature of varieties, are very natural notions in a $*$-free algebra. Shockling, convex polynomials and rational functions have a structure so rigid as to be nearly trivial. Indeed, a polynomial whose zero variety has positive curvature on a Zaraski open set is itself convex. In this section we survey what is known about convex polynomials in a free $*$-algebra. In later sections we treat convex rational functions, varieties with some positive curvature, and polynomials whose coefficients are themselves formally letters as in equation (2.5).

Let $\mathbb{R}\langle x\rangle$ denotes the free $*$-algebra in indeterminates $x = (x_1, ..., x_g)$, over the real field. Elements of $\mathbb{R}\langle x\rangle$ are non-commutative polynomials. There is a natural involution on $\mathbb{R}\langle x\rangle$ which reverses the order of multiplication $(fp)^* = p^* f^*$. In particular $x_j^* = x_j$ and for this reason the variables are *symmetric*. It is also possible to allow for non-symmetric variables by introducing the g additional variables x_j^*, but in the literature we are summarizing typically x_j can be taken either free or symmetric with no change in the conclusion. Thus for expositional purposes we will stick with symmetric variables in this survey, except for Section 9.

Let $\mathbb{S}_n(\mathbb{R}^g)$ denote g-tuples $X = (X_1, \ldots, X_g)$ of symmetric $n \times n$ matrices. Non-commutative polynomials are naturally evaluated at an $X \in \mathbb{S}_n(\mathbb{R}^g)$ by substitution. The involution on $\mathbb{R}\langle x\rangle$ is compatible with transpose of matrices in that $p(X)^* = p^*(X)$. A polynomial p is *symmetric* if $p = p^*$. Thus, if p is symmetric, then $p(X)^* = p(X)$.

A symmetric polynomial p is *matrix convex*, or simply *convex* for short, if for each positive integer n, each pair of tuples $X \in \mathbb{S}_n(\mathbb{R}^g)$ and $Y \in \mathbb{S}_n(\mathbb{R}^g)$, and each $0 \leq t \leq 1$,

$$p(tX + (1-t)Y) \preceq tp(X) + (1-t)p(Y). \tag{3.1}$$

Even in one-variable, convexity in the noncommutative setting differs from convexity in the commuting case because here Y need not commute with X. For example, to see that the polynomial $p = x^4$ is not matrix convex, let

$$X = \begin{pmatrix} 4 & 2 \\ 2 & 2 \end{pmatrix} \text{ and } Y = \begin{pmatrix} 2 & 0 \\ 0 & 0 \end{pmatrix}$$

and compute

$$\frac{1}{2}X^4 + \frac{1}{2}Y^4 - \left(\frac{1}{2}X + \frac{1}{2}Y\right)^4 = \begin{pmatrix} 164 & 120 \\ 120 & 84 \end{pmatrix}$$

which is not positive semi-definite. On the other hand, to verify that x^2 is a matrix convex polynomial, observe that

$$\begin{aligned} &tX^2 + (1-t)Y^2 - (tX + (1-t)Y)^2 \\ &\qquad = t(1-t)(X^2 - XY - YX + Y^2) = t(1-t)(X-Y)^2 \succeq 0. \end{aligned}$$

It is possible to automate checking for convexity, rather than depending upon lucky choices of X and Y as was done above. The theory described in [CHSY03], sketched later in §6, leads to and validates a symbolic algorithm for determining regions of convexity of noncommutative rational functions (noncommutative rationals are formally introduced in §5) which is currently implemented in NCAlgebra.

We introduce now a sample NCAlgebra command, leaving a more detailed discussion for later (see §6.2). Noncommutative multiplication will be denoted **. The command is

NCConvexityRegion[Function F, Variable x].

Let us illustrate it on the example $p(x) = x^4$ with $x = x^*$.

```
In[1]:=  SetNonCommutative[x]};
In[2]:=   NCConvexityRegion[ x**x**x**x, x ]}
Out[2]:= { {2, 0, 0},  {0, 2 }, {-2, 0} }
```

which we interpret as saying that $p(x) = x^4$ is convex on the set of matrices X for which the the 3×3 block matrix

$$\begin{pmatrix} 2 & 0 & 0 \\ 0 & 0 & 2 \\ 0 & -2 & 0 \end{pmatrix} \tag{3.2}$$

is positive semi-definite. Thus, we conclude that p is *nowhere* convex.

This is a simple special case of the following theorem.

THEOREM 3.1 ([HM03]). *Every convex symmetric polynomial in the free algebra* $\mathbb{R}\langle x\rangle$ *has degree two or less.*

A symmetric polynomial q is *matrix positive* or *positive* for short if for each n and $X \in \mathbb{S}_n(\mathbb{R}^g)$, $q(X) \succeq 0$. As we shall see convexity of p is equivalent to its "second directional derivative" being a positive polynomial. More generally, a symmetric polynomial whose k^{th} derivative is nonnegative has degree at most k (see Theorem 3.6 below).

It turns out that even if p is convex only on an open non-commutative domain, then in fact it is convex everywhere. To state the result we need to introduce some notation and terminology.

Let $\mathcal{P}$ denote a subset of $\mathbb{R}\langle x\rangle$ consisting of symmetric polynomials. Define the matrix nonnegativity domain $\mathcal{D}(\mathcal{P})$ of $\mathcal{P}$ to be the sequence of sets $(\mathcal{D}(\mathcal{P})_n)_{n=1}^{\infty}$ where

$$\mathcal{D}(\mathcal{P}) = \{X \in \mathbb{S}_n(\mathbb{R}^g) : p(X) \succeq 0, \quad q \in \mathcal{P}\}.$$

THEOREM 3.2 ([HM03]). *Suppose* $\mathcal{P}$ *is a set of symmetric polynomials whose matrix nonnegativity domain* $\mathcal{D}(\mathcal{P})$ *contains open sets in all large enough dimensions; i.e., there is an* n_0 *so that for each* $n \geq n_0$ *the set* $\mathcal{D}(\mathcal{P})_n$ *contains an open set. If* $p \in \mathbb{R}\langle x\rangle$ *is symmetric and convex on* $\mathcal{D}(\mathcal{P})$*, then* p *has degree at most two.*

The proofs will be sketched shortly.

3.1. Some history of convex polynomials. The earliest related results we know of are due to Karl Löwner who studied a class of real analytic functions in one real variable called matrix monotone functions, which we shall not define here. Löwner gave integral representations and these have developed beautifully over the years. The impact on our story comes a few years later when Löwner's student Klaus [Kra36] introduced matrix convex functions f in one variable. Such a function f on $[0, \infty] \subset \mathbb{R}$ can be represented as $f(t) = tg(t)$ with g matrix monotone, so the representations for g produce representations for f. It seems to be a folk theorem that the one variable version of Theorem 3.1 was known as a consequence of more general results on these matrix convex functions. Modern references are [OST07], [Uch05]. Frank Hansen has extensive deep work on matrix convex and monotone functions whose definition in several variables is different than the one we use here, see[HT07]; and for a more recent reference see [Han97].

3.2. The proof of theorem 3.1 and its ingredients. Just as in the commutative case, convexity of a symmetric $p \in \mathbb{R}\langle x\rangle$ is equivalent to positivity of its Hessian. Unlike the commutative case, positive non-commutative polynomials are sums of squares. Combinatorial considerations say that a Hessian which is also a sum of squares must come from a polynomial of degree two. In the remainder of this section we flesh out this argument, introducing the needed techniques and results.

The proof of Theorem 3.2 requires different, though certainly related, machinery which is discussed in §6.

3.2.1. Noncommutative derivatives. For a polynomial $p \in \mathbb{R}\langle x\rangle$ define the kth*-directional derivative* : by

$$p^{(k)}(x)[h] = \frac{d^k}{dt^k} p(x+th)\Big|_{t=0}.$$

Note that $p^{(k)}(x)[h]$ is homogeneous of degree k in h.

More formally, we regard the directional derivative $p'(x)[h] \in \mathbb{R}\langle x, h\rangle$ as a polynomial in $2g$ free symmetric (i.e. invariant under *) variables $(x_1, \ldots, x_g, h_1, \ldots, h_g)$; In the case of a word $w = x_{j_1}x_{j_2}\cdots x_{j_n}$ the derivative is:

$$w'[h] = h_{j_1}x_{j_2}\cdots x_{j_n} + x_{j_1}h_{j_2}x_{j_3}\cdots x_{j_n} + \ \ldots\ + x_{j_1}\cdots x_{j_{n-1}}h_{j_n}$$

and for a polynomial $p = p'(x)[h] = \sum p_w w$ the derivative is

$$p'(x)[h] = \sum p_w w'[h].$$

If p is symmetric, then so is p'.

For $X, H \in \mathbb{S}_n(\mathbb{R}^g)$ observe that

$$p'(X)[H] = \lim_{t\to 0} \frac{p(X+tH) - p(X)}{t}.$$

Alternately, with $q(t) = p(X+tH)$,

$$p'(X)[H] = q'(0).$$

Likewise for a polynomial $p \in \mathbb{R}\langle x\rangle$, the *Hessian* $p''(x)[h]$ of $p(x)$ can be thought of as the formal second directional derivative of p in the "direction" h. Equivalently, the Hessian of $p(x)$ can also be defined as the part of the polynomial

$$r(x)[h] := p(x+h) - p(x)$$

in the free algebra in the symmetric variables that is homogeneous of degree two in h.

If $p'' \neq 0$, that is, if degree $p \geq 2$, then the degree of $p''(x)[h]$ as a polynomial in the $2g$ variables $x_1, \ldots, x_g, h_1 \ldots, h_g$ is equal to the degree of $p(x)$ as a polynomial in $x_1, \ldots, x_g$. Likewise for k^{th} derivatives.

EXAMPLE 3.3. The first (non-commutative) derivative of $p(x) = x_2x_1x_2$ is

$$p'(x)[h] = \frac{d}{dt}[(x_2+th_2)(x_1+th_1)(x_2+th_2)]\Big|_{t=0} = h_2x_1x_2 + x_2h_1x_2 + x_2x_1h_2.$$

EXAMPLE 3.4. The one variable $p(x) = x^4$ has first derivative

$$p'(x)[h] = hxxx + xhxx + xxhx + xxxh.$$

Note each term is linear in h and h replaces each occurrence of x once and only once.

The Hessian, or second derivative, of p is

$$p''(x)[h] = hhxx + hhxx + hxhx + hxxh + hxhx + xhhx + \\ + xhhx + xhxh + hxxh + xhxh + xxhh + xxhh,$$

which simplifies to

$$p''(x)[h] = 2hhxx + 2hxhx + 2hxxh + 2xhhx + 2xhxh + 2xxhh.$$

Note each term is degree two in h and h replaces each pair of x's exactly once. Likewise $p^{(3)}(x)[h] = 6(hhhx+hhxh+hxhh+xhhh)$ and $p^{(4)}(x)[h] = 24hhhh$ and $p^{(5)}(x)[h] = 0$.

EXAMPLE 3.5. The Hessian of $p = x_1^2 x_2$ is $p''(x)[h] = h_1^2 x_2 + h_1 x_1 h_2 + x_1 h_1 h_2$.

Theorem 3.1 is the $k = 2$ case of the following result.

THEOREM 3.6 ([HP07]). *Every symmetric polynomial $p \in \mathbb{R}\langle x \rangle$ whose k^{th} derivative is a matrix positive polynomial has degree k or less.*

Proof. See [HP07] for the full proof or [HM03] for case of $k = 2$. The very intuitive proof based upon a little non-commutative semi-algebraic geometry is sketched in the next subsubsection. □

3.2.2. A little Noncommutative semi-algebraic geometry. A central theme of semi-algebraic geometry are positivstellensätz which, in the simplest forms, represent polynomials which are positive, or positive on a domain. It turns out positivstellensätzae in the free $*$ setting generally have cleaner statements than in the commutative case. Proofs of Theorems 3.1 and 3.2 require a non-commutative positivstellensätz.

Recall, a symmetric polynomial p is **matrix positive polynomial** or simply **positive** provided $p(X_1, \cdots, X_g)$ is positive semidefinite for every $X \in \mathbb{S}_n(\mathbb{R}^g)$ (and every n). An example of a matrix positive polynomial is a **Sum of Squares** of polynomials, meaning an expression of the form

$$p(x) = \sum_{j=1}^{c} h_j(x)^* h_j(x).$$

Substituting $X \in \mathbb{S}_n(\mathbb{R}^g)$ gives $p(X) = \sum_{j=1}^{c} h_j(X)^* h_j(X) \succeq 0$. Thus p is positive. Remarkably these are the only positive non-commutative polynomials.

THEOREM 3.7. *Every matrix positive polynomial is a sum of squares.*

As noted above, this non-commutative behavior is much cleaner than that of conventional "commutative" semi-algebraic geometry. See [Par00, Las01] for a beautiful treatment of applications of commutative semialgebraic geometry. This theorem is just a sample of the structure of noncommutative semialgebraic geometry, the topic of §9.

Suppoe $p \in \mathbb{R}\langle x \rangle$ is (symmetric and) convex and $Z, H \in \mathbb{S}_n(\mathbb{R}^g)$ and $t \in \mathbb{R}$ are given. In the definition of convex, choosing $X = Z + tH$ and $Y = Z - tH$, it follows that

$$0 \preceq p(Z+tH) + p(Z-tH) - 2p(Z),$$

and therefore

$$0 \preceq \lim_{t \to 0} \frac{p(X+tH) + p(X-tH) - 2p(X)}{t^2} \to p''(X)[H].$$

Thus the Hessian of p is matrix positive and since, in the noncommutative setting, positive polynomials are sums of squares we obtain the following theorem.

PROPOSITION 3.8. *If p is matrix convex, then its Hessian $p''(x)[h]$ is a sum of squares.*

3.2.3. Proof of Theorem 3.2 by example. Example 3.4 serves to illustrate the proof of Theorem 3.2 in the case $k = 2$.

EXAMPLE 3.9. The one-variable polynomial $p = x^4$ is not matrix convex.

Here is a sketch of the proof based upon Proposition 3.8.

If $p(x) = x^4$ is matrix convex, then $p''(x)[h]$ is matrix positive and therefore, by Proposition 3.8, there exists a ℓ and polynomials $f_1(x,h), \ldots, f_\ell(x,h)$ such that

$$\begin{aligned} p''(x)[h] &= hhxx + hxhx + hxxh + xhhx + xhxh + xxhh \\ &= f_1(x,h)^* f_1(x,h) + \cdots + f_\ell(x,h)^* f_\ell(x,h). \end{aligned}$$

One can show that each $f_j(x,h)$ is linear in h. On the other hand, some term $f_i^* f_i$ contains $hhxx$ and thus f_i contains hx^2. Let m denote the largest ℓ such that some f_j contains the term hx^ℓ. Then $m \geq 1$ and for such j, the product $f_j^* f_j$ contains the term $hx^{2m}h$ which can't be cancelled out, a contradiction. □

The proof of the more general, order k derivative, is similar, see [HP07].

4. A bit of engineering reality: coefficients in an algebra. A level of generality which most linear systems problems require is polynomials p or noncommutative rational functions (to be discussed later) in two classes of variables, say a and x, rather than x alone. As the example in Subsection 4.1 below illustrates, the x play the role of unknowns and the a the role of systems parameters and we are interested in matrix convexity in

x over ranges of the variable(s) a. Describing this setup fully takes a while, as one can see in [CHSY03] where it is worked out. An engineer might look at [CHS06], especially the first part which describe a computational noncommutative algebra attack on convexity, it seems to be the most intuitive read on the subject at hand.

In this section some sample results from [HHLM] and a motivating engineering example are presented.

In [HHLM] one shows that second derivatives of a symmetric polynomial $p(a, x)$ in x determine convexity in x and that convexity in the x variable on some "open set" of a, x implies that p has degree 2 or less in x.

THEOREM 4.1. *If $P(a, x)$ is a symmetric $d \times d$ matrix with polynomial entries $p_{ij}(a, x)$, then convexity in x for all X and all A satisfying some strict algebraic inequality of the form $g(A) \succ 0$, implies each p_{ij} has degree 2 or less.*

Proof. See [HP07] survey combined with [HHLM]. □

Assume a $d \times d$ matrix of polynomials $P(a, x)$ has degree 2 in x. There are tests (not perfect) to see where in the a variable $P(X, A)$ is negative semi-definite for all X. Equivalently, to see where P is convex in x, see [HP07].

The following is a further example of results from [HHLM].

THEOREM 4.2. *A symmetric $p(a, x)$ is convex in x and concave in a if and only if*

$$p(a, x) = L(a, x) + R(x)^* R(x) - S(a)^* S(a),$$

where $L(a, x)$ has degree at most one in each of x and a and R and S are vectors which are linear in x and a respectively.

Note that $R(x)^* R(x)$ is a homogeneous of degree two sum of squares.

The subsection below gives a flavor of how two types of variables a, x as well as matrices with noncommutative polynomial entries arise naturally in engineering applications. It continues the discussions of Section 2.

4.1. A sample engineering messy algebra problem. Here is a basic engineering problem, the *standard problem of H^∞ control*:

Make a given system dissipative by designing a feedback law.

To be more specific, we are given a signal flow diagram:

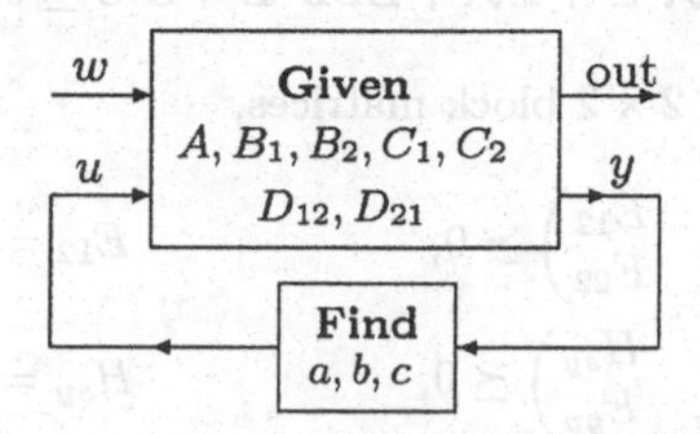

where the given system is

$$\frac{ds}{dt} = As + B_1 w + B_2 u,$$
$$\text{out} = C_1 s + D_{12} u + D_{11} w,$$
$$y = C_2 s + D_{21} w,$$

$$D_{21} = I, \qquad D_{12} D_{12}' = I, \qquad D_{12}' D_{12} = I, \qquad D_{11} = 0.$$

The assumptions on D are to simplify calculations. In practice one needs something messier.

We want to find an unknown system

$$\frac{d\xi}{dt} = a\xi + by, \qquad u = c\xi,$$

called the *controller*, which makes the system dissipative over every finite horizon. Namely:

$$\int_0^T |w(t)|^2 dt \geq \int_0^T |out(t)|^2 dt, \quad s(0) = 0.$$

So a, b, c are the critical unknowns.

4.1.1. Conversion to algebra. The dynamics of the "closed loop" system has the form

$$\frac{d}{dt}\begin{pmatrix} s \\ \xi \end{pmatrix} = \mathcal{A}\begin{pmatrix} s \\ \xi \end{pmatrix} + \mathcal{B}w,$$
$$\text{out} = \mathcal{C}\begin{pmatrix} s \\ \xi \end{pmatrix} + \mathcal{D}w,$$

where $\mathcal{A}, \mathcal{B}, \mathcal{C}, \mathcal{D}$ are "2×2 block matrices" whose entries are polynomials in the $A's, B's, \cdots, a, b, c$ etc. The storage function inequality which corresponds to energy dissipation (see Subsection 2.3) has the form

$$H := \mathcal{A}^* E + E\mathcal{A} + E\mathcal{B}\mathcal{B}^* E + \mathcal{C}^* \mathcal{C} \preceq 0. \tag{4.1}$$

Expressing E and H as 2×2 block matrices,

$$E = \begin{pmatrix} E_{11} & E_{12} \\ E_{21} & E_{22} \end{pmatrix} \succeq 0, \qquad E_{12} = E_{21}^*,$$
$$H = \begin{pmatrix} H_{ss} & H_{sy} \\ H_{ys} & E_{yy} \end{pmatrix} \preceq 0, \qquad H_{sy} = H_{ys}^*,$$

the challenge of the algebraic inequality of equation (4.1) is to find $H \preceq 0$ where the entries of H are the polynomials:

$$H_{ss} = E_{11}\,A + A^*\,E_{11} + C_1^*\,C_1 + E_{12}^*\,b\,C_2 + C_2^*\,b^*\,E_{12}^* + E_{11}\,B_1\,b^*\,E_{12}^* + \\ + E_{11}\,B_1\,B_1^*\,E_{11} + E_{12}\,b\,b^*\,E_{12}^* + E_{12}\,b\,B_1^*\,E_{11},$$

$$H_{sz} = E_{21}\,A + \frac{1}{2}a^*\,(E_{21} + E_{12}^*) + c^*\,C_1 + E_{22}\,b\,C_2 + c^*\,B_2^*\,E_{11}^* + \\ + \frac{1}{2}E_{21}\,B_1\,b^*\,(E_{21} + E_{12}^*) + E_{21}\,B_1\,B_1^*\,E_{11}^* + \\ + \frac{1}{2}E_{22}\,b\,b^*\,(E_{21} + E_{12}^*) + E_{22}\,b\,B_1^*\,E_{11}^*,$$

$$H_{zs} = A^*\,E_{21}^* + C_1^*\,c + \frac{1}{2}(E_{12} + E_{21}^*)\,a + E_{11}\,B_2\,c + \\ + C_2^*\,b^*\,E_{22}^* + E_{11}\,B_1\,b^*\,E_{22}^* + E_{11}\,B_1\,B_1^*\,E_{21}^* + \\ + \frac{1}{2}(E_{12} + E_{21}^*)\,b\,b^*\,E_{22}^* + \frac{1}{2}(E_{12} + E_{21}^*)\,b\,B_1^*\,E_{21}^*,$$

$$H_{zz} = E_{22}\,a + a^*\,E_{22}^* + c^*\,c + E_{21}\,B_2\,c + c^*\,B_2^*\,E_{21}^* + E_{21}\,B_1\,b^*\,E_{22}^* + \\ + E_{21}\,B_1\,B_1^*\,E_{21}^* + E_{22}\,b\,b^*\,E_{22}^* + E_{22}\,b\,B_1^*\,E_{21}^*.$$

Here A, B_1, B_2, C_1, C_2 are known and the unknowns are a, b, c and for E_{11}, E_{12}, E_{21} and E_{22}. If one can find E, then it turns out that there are explicit formulas for a, b, c in terms of E.

We very much wish that these inequalities (4.1) are convex in the unknowns (so that numerical solutions will be reliable). But our key inequality above is not convex in the unknowns.

The key question: *Is there is a set of noncommutative convex inequalities whose set of solutions is equivalent to those of* (4.1)*?*

This is a question in algebra not in numerics and we leave it as a nearly impossible challenge to the reader. Later in §8 we give the answer. There are many ways to derive the solution for this as well as a broad class of related problems, but all appeal to very special structure. An issue driving our development of $*-$ free semi-algebraic geometry is how to give a general theory which solves this and many other examples as a special case. It is clear that such a theory must includes the possibility to change of variables, identifying which non-convex problems can be converted to convex problems, and automating the conversion when possible.

4.2. What is needed for engineering. Many linear systems problems which are "dimension free" readily convert to non-commutative inequalities on $d \times d$ matrices of polynomials of the form $P(a, x) \preceq 0$ as the example in Section 4.1 illustrates. Often the inequality $P(a, x) \preceq 0$ can be simplified by various means such as solving for some variables and substituting to get other hopefully simpler inequality $R(a, x) \preceq 0$. In fact what one gets by standard manipulations in all circumstances (to our knowledge) is matrices $R(a,x)$ with noncommutative rational expressions as entries. Thus there is the need to generalize Theorem 4.1 from polynomials to rational expressions. The notion of a noncommutative rational function is given in the following section where Theorem 5.3 gives solid support for our conjecture that convex noncommutative rational functions $R(a,x)$ have a surprisingly simple structure in x.

This very strong conclusion is bad news for engineers because it says convexity for dimension free problems is much rarer than for dimension dependent problems. We emphasize that the result does not preclude transformation, by change of variable say, to achieve convexity and understanding such transformation is a challenging mostly open area.

5. Convexity for Noncommutativerationals. This section describes the extension of the convex noncommutative polynomial theorem, Theorem 3.1, to symmetric noncommutative rational functions, $r = r(x)$, of the x variable alone which are convex near the origin, see [HMV06]. The results provide further evidence for the rigidity of convexity in the noncommutative (dimension free) setting.

5.1. Noncommutativerational functions. We shall discuss the notion of a noncommutative rational function in terms of rational expressions. We refer to [HMV06, Section 2 and Section 16] for details. In what follows, the casual reader can ignore the technical condition, "analytic at 0", which we include for the sake of precision.

A **noncommutative rational expression analytic at** 0 is defined recursively. Non-commutative polynomials are noncommutative rational expressions as are all sums and products of noncommutative rational expressions. If r is a noncommutative rational expression and $r(0) \neq 0$, then the inverse of r is a rational expression analytic at 0.

The notion of the **formal domain of a rational expression** r, denoted $\mathcal{F}_{r,\text{for}}$, and the evaluation $r(X)$ of the rational expression at a tuple $X \in \mathbb{S}_n(\mathbb{R}^g) \cap \mathcal{F}_{r,\text{for}}$ are also defined recursively[1]. Example (5.1) below is illustrative.

An example of a noncommutative rational expression is the Riccati expression for discrete-time systems:

$$r = a^*xa - x + c^*c + (a^*xb + c^*d)(I - d^*d - b^*xb)^{-1}(b^*xa + d^*c)$$

Here some variables are symmetric some are not. A difficulty is two different expressions, such as

$$r_1 = x_1(1 - x_2x_1)^{-1} \quad \text{and} \quad r_2 = (1 - x_1x_2)^{-1}x_1,$$

[1]The formal domain of a polynomial p is all of $\mathbb{S}_n(\mathbb{R}^g)$ and $p(X)$ is defined just as before. The formal domain of sums and products of rational expressions is the intersection of their respective formal domains. If r is an invertible rational expression analytic at 0 and $r(X)$ is invertible, then X is in the formal domain of r^{-1}.

that can be converted into each other with algebraic manipulation represent the same rational function. Thus it is necessary to specify an equivalence relation on rational expressions to arrive at what are typically called **noncommutative rational functions**. (This is standard and simple for commutative (ordinary) rational functions.) There are many alternate ways to describe the noncommutative rational functions and they go back 50 years or so in the algebra literature. The simplest one for our purposes is **evaluation equivalence** — two rational expressions r_1 and r_2 are evaluation equivalent if $r_1(X) = r_2(X)$ for all $X \in \mathcal{F}_{r_1,\text{for}} \cap \mathcal{F}_{r_2,\text{for}}$. For engineering purposes one need not be too concerned, since what happens is that two expressions r_1 and r_2 are equivalent whenever the usual manipulations you are accustomed to with matrix expressions convert r_1 to r_2.

For $\mathfrak{r}$ a rational function, that is, an "equivalence class of rational expressions r", define its **domain** by

$$\mathcal{F}_{\mathfrak{r},\text{for}} := \cup_{\{r \text{ represents } \mathfrak{r}\}} \mathcal{F}_{r,\text{for}}.$$

Let $\mathcal{F}^0_{\mathfrak{r},\text{for}}$ denote the arcwise connected component of $\mathcal{F}_{\mathfrak{r},\text{for}}$ containing 0 (and similarly for $\mathcal{F}^0_{r,\text{for}}$). We call $\mathcal{F}^0_{\mathfrak{r},\text{for}}$ the **principal component** of $\mathcal{F}_{\mathfrak{r},\text{for}}$. Henceforth we do not distinguish between the rational functions $\mathfrak{r}$ and rational expressions r, since this causes no confusion. We give several examples.

EXAMPLE 5.1.

$$r(x_1, x_2) = (1 + x_1 - (3 + x_2)^{-1})^{-1},$$

where we take $x_1 = x_1^*, x_2 = x_2^*$ is a symmetric noncommutative rational expression. The domain $\mathcal{F}_{r,\text{for}}$ is

$$\cup_{n>0}\{(X_1, X_2) \in \mathbb{SR}^{n\times n}(\mathbb{R}^2)\colon 1{+}X_1{-}(3+X_2)^{-1} \text{ and } 3{+}X_2 \text{ are invertible}\}.$$

Its principal component $\mathcal{F}^0_{r,\text{for}}$ is

$$\cup_{n>0}\{(X_1, X_2) \in \mathbb{SR}^{n\times n}(\mathbb{R}^2)\colon \quad 1{+}X_1{-}(3+X_2)^{-1} \succ 0 \text{ and } 3{+}X_2 \succ 0\}.$$

EXAMPLE 5.2. We return to convexity checker command and illustrate it on

$$F((a, b, r), (x, y)) := -(y + a^*xb)(r + b^*xb)^{-1}(y + b^*xa) + a^*xa. \quad (5.1)$$

where $x = x^*, y = y^*$. Here we are viewing F as a function of two classes of variables (see Section 4). An application of the command **NCConvexityRegion**[F, $\{x, y\}$] outputs the list

$$\{-2\,(r + b^*xb)^{-1},\ 0,\ 0,\ 0\}.$$

This output has the meaning that whenever A, B, R are fixed matrices, the function F is "x, y-matrix concave" on the domain of matrices X, and Y

$$\mathcal{G}_{A,B,R} := \{(X,Y) : (R + B^* X B)^{-1} \succ 0\}.$$

The command NCConvexityRegion also has an important feature which, for this problem, assures us no domain bigger than

$$\bar{\mathcal{G}}_{A,B,R} := \{(X,Y) : R + B^* X B \succeq 0\}$$

is a "domain of concavity" for F. The algorithm is discussed briefly in §6. For details and proof of the last assertion, see [CHSY03].

5.2. Convexity vs LMIs. Now we restrict from functions $r(a,x)$ in two types of variables to $r(x)$ of only one type. The following theorem characterizes symmetric noncommutative rational functions (in x) which are convex near the origin in terms of an LMI. The more general $r(a,x)$ has not been worked out.

THEOREM 5.3 ([HMV06]). *Suppose $r = r(x)$ is a noncommutative symmetric rational function which is convex (in x) near the origin. Then*

(1) r has a representation

$$r(x) = r_0 + r_1(x) + \ell(x)\ell(x)^* + \Lambda(x)(I - L(x))^{-1}\Lambda(x)^*, \tag{5.2}$$

where

$$L(x), \quad \ell(x), \quad \Lambda(x), \quad r_0 + r_1(x)$$

are linear pencils in $x_1, \cdots, x_g$ satisfying

$$L(0) = 0, \quad \ell(0) = 0, \quad \Lambda(0) = 0, \quad r_1(0) = 0.$$

In addition L and r_1 are symmetric, for example, $L(x)$ has the form $L(x) = A_1 x_1 + \cdots + A_g x_g$ for symmetric matrices A_j.
Thus for γ any real number $r - \gamma$ is a Schur complement of the noncommutative linear pencil

$$\mathcal{L}_\gamma(x) := \begin{pmatrix} -1 & 0 & \ell(x)^* \\ 0 & -(I - L(x)) & \Lambda(x)^* \\ \ell(x) & \Lambda(x) & r_0 - \gamma + r_1(x) \end{pmatrix}.$$

(2) The principal component of the domain of r is a convex set, indeed it is the positivity set of the pencil $I - L(x)$. Indeed this holds for any r of the form (5.2), subject to a minimality type condition on $\mathcal{L}_\gamma$.

This correspondence between properties of the pencil and properties of r yields

COROLLARY 5.4. *For any $\gamma \in \mathbb{R}$, the principal component, $\mathcal{G}^0_\gamma$, of the set of solutions X to the NCMI*

$$r(X) \prec \gamma I$$

equals the set of solutions to a NCLMI based on a certain linear pencil $\mathcal{L}_\gamma(x)$.

That is, **numerically solving matrix inequalities based on r is equivalent to numerically solving a NCLMI associated to r.**

5.3. Proof of Corollary 5.4. By item (2) of Theorem 4.2 the upper 2×2 block of $\mathcal{L}_\gamma(X)$ is negative definite if and only if $I - L(X) \succ 0$ if and only if X is in the component of 0 of the domain of r. Given that the upper 2×2 block of $\mathcal{L}_\gamma(X)$ is negative definite, by the LDL^* (Cholesky) factorization , $0 \succ \mathcal{L}_\gamma(X)$ is negative definite if and only if $\gamma I \succ r(X)$. □

6. Ideas behind some proofs and the convexity checker algorithm. The proofs of Theorem 3.2, the results from Section 4 on polynomials in two classes of variables, and many of the results on rational functions exposited in the previous section begin, just as in the case of everywhere convex polynomials, with the observation that matrix convexity of a noncommutative rational function on a *noncommutative convex domain* is equivalent to its noncommutative second directional derivative being matrix positive. This link between convexity and positivity remains in the noncommutative setting. While we will not define carefully the notion of a noncommutative convex domain, a special example is the $\epsilon > 0$ neighborhood of 0 which is the sequence of sets $(N_{\epsilon,n})_n$ where

$$N_{\epsilon,n} = \{X \in \mathbb{S}_n(\mathbb{R}^g) : \sum X_j^2 \preceq \epsilon^2 I_n\}.$$

The phrase, *for X near zero*, is shorthand for *in some noncommutative ϵ neighborhood of* 0.

Dealing with polynomials in variables (a, x) which are convex (on a domain) in the variable x only, requires noncommutative partial derivatives. The informal definition of the kth *partial derivative* of a noncommutative rational function $r(x)$ with respect to x in the direction h is defined by

$$\frac{\partial^k}{\partial^k x} r(a,x)[h] = \frac{d^k}{dt^k} r(a, x + th)\Big|_{t=0} \tag{6.1}$$

When there are no a variables, we write, as one would expect, $r'(x)[h]$ and $r''(x)[h]$ instead of $\frac{\partial}{\partial x} r(x)$ and $\frac{\partial^k}{\partial^k x} r(x)$.

6.1. The middle matrix. There is a canonical representation of rational functions $q(b)[h]$ which are homogeneous of degree 2 in h as a matrix product,

$$q(b)[h] = V(b)[h]^* Z(b) V(b)[h]. \tag{6.2}$$

In the case that $r(a,x)$ is a polynomial of degree d and $q(a,x)[h] := \frac{\partial^k}{\partial^k x} r(a,x)[h]$, then (a,x) constitutes b and $V(a,x)[h]$ is a (column) vector whose entries are monomials of the form $h_j m(a,x)$ where $m(a,x)$ is a monomial in the variables (a,x) of degree at most $d-1$ (each such monomial appearing exactly once), where $Z(a,x)$ is a matrix whose entries are polynomials in (a,x). The matrix Z is unique, up to the order determined by the ordering of the $h_j m(a,x)$ in $V(a,x)[h]$. The matrix $Z(a,x)$ is the called **middle matrix** and $V(a,x)[h]$ is the **border** or **tautological vector**. The following basic noncommutative principle, which we state very informally, is key to many of the proofs many of the results presented in this survey.

PRINCIPLE 6.1. A variety of very weak hypotheses on positivity of $q(a,x)[h]$ imply positivity of the middle matrix ([CHSY03] or [HMV06]).

Indeed, for a polynomial $p(a,x)$, the condition $\frac{\partial^k}{\partial^k x} p(A,X)[H] \succeq 0$ for X near 0 and all A and H is far more than needed to imply $Z(A,X) \succeq 0$ for X near 0.

A key ingredient of the principle is the CHSY-Lemma.

LEMMA 6.2 (CHSY). *Let ℓ be given and let $\nu = g\sum_0^\ell g^j$. There is a κ so that if $(X,v) \in \mathbb{S}_n(\mathbb{R}^g) \times \mathbb{R}^n$ and the set $\{m(X)v : m \in \mathbb{R}\langle x\rangle$ is a monomial of degree $\ell\}$ is linearly independent, then the codimension of $\{V(X)[H] : H \in \mathbb{S}_n(\mathbb{R}^g)\}$ in $\mathbb{R}^{n\nu}$ is at most κ (independent of n).*

Consider the perpetually reoccurring example, $p(x) = x^4$. The decomposition of equation $p''(x)[h]$ is given by

$$p''(x)[h] = 2\begin{pmatrix} h & xh & x^2h \end{pmatrix} \begin{pmatrix} x^2 & x & 1 \\ x & 1 & 0 \\ 1 & 0 & 0 \end{pmatrix} \begin{pmatrix} h \\ hx \\ hx^2 \end{pmatrix}. \tag{6.3}$$

It is evident that the middle matrix for the polynomial $p(x) = x^4$ is not positive semi-definite (for any X) its Hessian is not positive semi-definite near 0 and hence, in view of Principle 6.1 p is not convex. This illustrates the idea behind the proof of Theorem 3.2 and the idea also applies to Theorem 4.1 in (a,x).

6.2. Automated convexity checking. The example above of $p(x) = x^4$ foreshadows the layout of our convexity checking algorithm.

Convexity Checker Algorithm for an noncommutative rational r:

1. Compute symbolically the Hessian $q(a,x)[h] := \frac{\partial^2 r}{\partial x^2}(a,x)[h]$.
2. Represent $q(a,x)[h]$ as $q(a,x)[h] = V(a,x)[h]^* Z(a,x) V(a,x)[h]$.
3. Apply the noncommutative LDL decomposition to the matrix $Z(a,x) = LDL^*$. The diagonal matrix $D(a,x)$ has the form $D = diag\ \{\rho_1(a,x), \ldots, \rho_c(a,x)\}$.
4. If $D(A,X) \succeq 0$ (that is, each $\rho_j(A,X) \succeq 0$), then the Hessian $q(A,X)[H]$ is positive semidefinite for all H. Thus a set $\mathcal{D}$ where r is matrix convex is given by

$$\mathcal{D} = \{(A, X) : \quad \rho_j(A, X) \succeq 0, \; j = 1, \ldots, c\}. \tag{6.4}$$

(In the example (6.3) $D(X)$ equals (3.2) which is not diagonal. In particular, convexity fails.)

The surprising and deep fact is that (under very weak hypotheses) the closure of $\mathcal{D}$ is the largest possible domain of convexity. See [CHSY03] for the proof. Extensions of the result and streamlined proofs can be found in [HMV06]. See also [KVV07]

It is hard to imagine a precise "convexity region algorithm" not based on noncommutative calculations, the problem being that matrices of practical size often have thousands of entries and so would lead to calculations with huge numbers of polynomials in thousands of variables.

6.3. Proof of Theorem 5.3. The proof consists of several stages. It is interesting that the technique for the first stage, which yields an initial representation for r as a Schur Complement of a linear pencil, is classical. In fact, the following representation of any symmetric noncommutative rational function r is the symmetric version of the one due originally to Kleene, Schützenberger, and Fliess (who were motivated by automata and formal languages, and bilinear systems; see [BR84] for a good survey), and further studied recently by Beck [Bec01], see also [BDG96, LZD96], and by Ball–Groenewald–Malakorn [BGM06a, BGM06b, BGM05].

THEOREM 6.3. *If r is a noncommutative symmetric rational function which is analytic at the origin, then r has a symmetric minimal* **Descriptor, or Recognizable Series, Realization**. *Namely,*

$$r(x) = r_0 + C(J - \sum_{j=1}^{g} \mathcal{A}_j x_j)^{-1} C^*, \tag{6.5}$$

where and $\mathcal{A}_j \in \mathbb{R}^{n \times n}$ are symmetric and J a signature matrix; i.e., J is symmetric and $J^2 = I$.

Here minimality means that $CJ\mathcal{A}_{i_1} \cdots J\mathcal{A}_{i_k} v = 0$ for all words $i_1 \cdots i_k$ in the indices $1, \cdots, g$ implies $v = 0$.

Of course in general the above symmetric realization is not monic, i.e., $J \neq I$. The second stage of the proof uses the convexity of r near the origin, more precisely, the positivity of the noncommutative Hessian, to force J to be within rank one of I. For notational ease, let $L_{\mathcal{A}}(x) = \sum \mathcal{A}_j x_j$. The Hessian of $r(x)$ is then,

$$r''(x)[h] = 2\Gamma(x)^* L_{\mathcal{A}}[h](J - L_{\mathcal{A}}(x))^{-1} L_{\mathcal{A}}[h]\Gamma(x)$$

where $\Gamma(x) = (J - L_{\mathcal{A}}(x))^{-1} C^*$. The heuristic argument is that there is an $X \in \mathbb{S}_n(\mathbb{R}^g)$ (with n as large as necessary) close to 0 and a vector v so that $\Gamma(X)v$ has components $z_1, \ldots, z_d \in \mathbb{R}^n$ which are independent. A minimality hypothesis on the descriptor realization allows for an argument similar to that of the CHSY-Lemma to prevail with the conclusion that

$\{L_A[H]\Gamma(X)v : H \in \mathbb{S}_n(\mathbb{R}^g)\}$ has small codimension. Indeed, with n large enough, the restriction on this codimension implies that J can have at most one negative eigenvalue; i.e., is nearly positive definite. From here, algebraic manipulations give Theorem 5.3 item (1).

The third stage of the proof — establishing item (2) — was quite gruelling in [HMV06], but it is subsumed now under the following fairly general singularities theorem for various species of minimal noncommutative realizations.

THEOREM 6.4 ([KVV07]). *Suppose*

$$r(x) = d(x) + C(x)(I - \sum_{j=1}^{g} A_j x_j)^{-1} B(x), \tag{6.6}$$

where $d(x)$ is a noncommutative polynomial, $A_j \in \mathbb{R}^{n \times n}$, and $B(x) = \sum B_{j_1 \dots j_r} x_{j_1} \cdots x_{j_r}$ and $C(x) = \sum C_{j_1 \dots j_l} x_{j_1} \cdots x_{j_l}$ are $n \times 1$ and $1 \times n$ matrix valued noncommutative polynomials, homogeneous of degrees r and l, respectively. Assume the "minimality type" conditions:

$$\operatorname*{span}_{k \geq 0;\, 1 \leq i_1, \dots, i_k, j_1, \dots, j_r \leq g} \operatorname{ran} A_{i_1} \cdots A_{i_k} B_{j_1 \dots j_r} = \mathbb{R}^n,$$

$$\bigcap_{k \geq 0;\, 1 \leq i_1, \dots, i_k\, j_1, \dots, j_l \leq g} \ker C_{j_1 \dots j_l} A_{i_1} \cdots A_{i_k} = 0.$$

Then

$$\mathcal{F}^0_{r,for} = \{(X_1, \dots, X_g) : \quad \det(I - A_1 \otimes X_1 - \cdots A_g \otimes X_g) \neq 0\}.$$

The proof is based on the formalism of noncommutative backward shifts and thus the theorem applies more generally to matrix-valued noncommutative rational functions.

7. Numerics and symbolics for matrix unknowns. In this section we discuss some ideas for combining symbolic and numerical computations to solve equations and optimization problems involving matrix unknowns. We focus on the big picture rather than on the details.

7.1. Unconstrained zero finding. The problem is, given a NC rational function $f(a, x)$ and A, find X such that $f(A, X) = 0$. A conceptual algorithm proceeds as follows.

ALGORITHM 7.1 (Newton-Rapson with line search). *Let X_0 and $\epsilon > 0$ be given and set $k = 0$.*

1. Compute $\frac{\partial}{\partial x} f(a, x)[h]$ symbolically.

2. Find H_k satisfying the linear equation

$$f(A, X_k) + \frac{\partial}{\partial x} f(A, X_k)[H_k] = 0, \tag{7.1}$$

3. Find $\alpha_k \in R$ *such that* $X_{k+1} = X_k + \alpha_k H_{x_k}$ *satisfies* $\|f(A, X_{k+1})\| < \|f(A, X_k)\|$.

4. Stop if $\|f(A, X_{k+1})\| \leq \epsilon$. *Otherwise increment k and go to (1).*

The step (3) is called a **line search** and is often performed on a nonnegative real valued function ϕ which has the key property that $\phi(A, X) = 0$, if $f(A, X) = 0$. In (3) we took $\phi(A, X) = \|f(A, X_{k+1})\|$. Typically a line search selects α_k to approximately achieve

$$\min_{\alpha \in R} \phi(A, X_k + \alpha H_k).$$

Under certain conditions the above algorithm will converge to X^* satisfying $f(A, X^*) = 0$. For example, once very near a minimizer of ϕ, this is rapidly convergent even with α set to 1. The analysis of these convergence conditions is standard (see [GM82], for instance) and will not be pursued any further in here. For very large problems (many unknown variables) there are two main show stoppers and they both relate to the **linear subproblem**, (7.1). The first (widely known) one is the numerical solution of (7.1). The second is the actual construction of the linear subproblem; this can consume large amounts of time and memory. *We feel this second issue is an excellent opportunity for the subject of computer algebra* and which is the emphasis of this section. When f is a NC rational function it is possible to take advantage of noncommutative algebra and organize the problem as a problem in g noncommuting variables rather than a problem in $g\,n(n+1)/2$ commuting variables. Indeed, the structure of problems like (7.1) is revealed in the next theorem. See [CHS06, dOH06b] for more details.

THEOREM 7.2. *Let* $f(a, x)$ *be a NC rational function of* (a, x). *Equation* (7.1) *is a* Generalized Sylvester Equation, *that is, it can be represented in the form*

$$f(a, x_k) + \sum_i^{Syl} r_i(a, x_k)\, h_k\, s_i(a, x_k) + \sum_j^{SylT} t_j(a, x_k)\, h_k^*\, u_j(a, x_k) = 0 \quad (7.2)$$

where the coefficient r_i, s_i, t_j, *and* u_j *are rational functions of* a *and* x *and the sums are both finite.*

This representation is not unique as is well illustrated by the following examples. Also these examples illustrate noncommutative symbolic computation which we believe is essential to exploiting the special structure in constructing the linear subproblem. The relevant symbolic calculations are carried out using *NCAlgebra*. Here `a**b` stands for noncommutative multiplication, `tp[ ]` is an involution and we think of `tp[x]` as the "transpose" of `x`. The command `NCExpand` expands expressions while `NCCollect[x**h+b**h, h]` collects on `h` to produce `(x+b)**h`, and `DirectionalD[f,x,h]` takes the noncommutative directional derivative of f wrt. x in direction h.

EXAMPLE 7.3.

1. For the quadratic function $f((a,b,c),x) = a\,x + x\,a^* + x\,b\,x + c$ the left hand side of (7.1) is

$$f((a,b,c),x) + a\,h + h\,a^* + x\,b\,h + h\,b\,x \tag{7.3}$$

and is computed in NCAlgebra as:

```
In[6]:= f[x_] = a ** x + x ** tp[a] + x ** b ** x + c;
In[7]:= Sylvester1 = f[x] + DirectionalD[f[x], x, h];
Out[7]= f[x] + a ** h + h ** tp[a] + h ** b ** x +
    + x ** b ** h
```

The coefficients of (7.2) are

$$r_1 = a, \qquad s_1 = I \qquad r_2 = x\,b, \qquad s_2 = I,$$
$$r_3 = s_1^*, \qquad s_3 = r_1^*, \qquad r_4 = s_2^*, \qquad s_4 = r_2^*.$$

2. A different representation of the type (7.2) for (7.3) is obtain by collecting on h. In NCAlgebra:

```
In[10]:= Sylvester2 = NCCollect[Sylvester1, h];
Out[11]= f[x]+ h ** (b ** x + tp[a])+(a + x ** b) ** h
```

Now we have different coefficients

$$r_1 = a + x\,b, \qquad s_1 = I \qquad r_2 = s_1^*, \qquad s_2 = r_1^*.$$

3. The rational function $f((a,c),x) = a\,x\,a^* - x + c + x(I-x)^{-1}x$ produces a representation (7.2) which has coefficients

$$r_1 = s_1^* = a, \qquad r_2 = -I, \qquad s_2 = I - (I-x)^{-1}x,$$
$$r_3 = x(I-x)^{-1}, \qquad s_3 = I + (I-x)^{-1}x.$$

as produced by *NCAlgebra* `NCExpand/NCCollect`:

```
In[13]:= f[x_] = a ** x ** tp[a] - x + c +
    + x ** inv[1 - x] ** x;
In[14]:= Sylvester3 = f[x] +
    + NCCollect[NCExpand[DirectionalD[f[x],x,h]],h];
Out[14]= f[x] - h ** (1 - inv[1 - x] ** x) +
    + a ** h ** tp[a] +
    + x ** inv[1-x] ** h ** (1 + inv[1-x] ** x)
```

4. The same rational function also produces a representation (7.2) with the following coefficients.

$$r_1 = s_1^* = a, \quad r_2 = -s_2 = -\sqrt{2}I, \quad r_3 = s_3^* = I + x(I-x)^{-1}.$$

after some manipulation.

We define *syl* and *sylT* to be *the Sylvester indices* of equation (7.2). That is the number of terms in each of the summations indicated in equation (7.2). For instance, in example (1) $syl = 4$ and $sylT = 0$. Note that the representation (7.2) is not unique (e.g. examples (1-2) and (3-4) above) so each representation may have its own pair of Sylvester indices (e.g. $syl = 4$ in example (1) and $syl = 2$ in example (2)). It turns out that finding a representation with small, or smallest, Sylvester indices is important for numerics. Note also that the coefficients of the Sylvester equation in example (4) are symmetric while the ones in example (3) are not, which may be of importance.

7.2. Optimization with matrix inequality constraints. What often occurs in engineering is an optimization problem subject to matrix inequality constraints. One proceeds by writing down first order Karush-Kuhn-Tucker (KKT) type optimality conditions which are a set of equations of the form $F(a, x) = 0$ but with matrix positivity side conditions, which are then solved numerically often using a barrier type method. We shall illustrate our NC symbolic approach with the following broad and important class of problems:

$$\min_X \{\mathrm{Trace}(C^* X) : \quad f(A, X) \preceq 0\} \tag{7.4}$$

in which $f(a, x)$ is a symmetric noncommutative function. A linear cost function is assumed without loss of generality[2]. Some examples are as follows:

1. Riccati inequality: maximize $\mathrm{Trace}(X)$ such that $A^*X + XA - XBX + C \succeq 0$ and $X \succeq 0$.
2. Problem (1) is not in the form (7.4) but can be easily reformulated as

$$\min_X \{-\mathrm{Trace}(X) : \quad \{XBX - A^*X - XA - C, -X\} \preceq 0\}$$

which is in the form (7.4) for $C = -I$ and

$$f(a, b, c, x) = \{x\,b\,x - a^*x - xa - c,\ -x\}.$$

3. Static output feedback stabilization: minimize $\mathrm{Trace}(X_1)$ such that

$$(A + BX_2D)^*X_1 + X_1\,(A + BX_2D) + C \preceq 0 \quad \textit{and} \quad X_1 \succeq 0.$$

This problem is in the form (7.4) for $C = \{I, 0\}$ and

$$f(a, b, c, x) = \{(a + bx_2d)^*x_1 + x_1\,(a + bx_2d) + c,\ -x_1\}.$$

[2] If the cost is not linear one may add a variable so that $\min_X g(X) = \min_{X,\mu}\{\mu : \mu \geq g(X)\}$. The former problem has a linear cost function.

Many other examples of problems of the class (7.4) can be found in [BEGFB94, SIG98]. Symbolic NC algorithms are discussed in [dOH06a] that can manipulate and produce instances of problems in the form (7.4) in systems and control engineering.

In the above examples braces are used to represent a *vector* of NC functions. Inequalities should be taken as applied to each entry of the vector, e.g.

$$\{x_1, x_2\} \preceq 0 \quad \Leftrightarrow \quad \{x_1 \preceq 0, x_2 \preceq 0\}.$$

We do this also with certain specific NC functions, namely inverses and ln det, where we define

$$\{x_1, \cdots, x_g\}^{-1} := \{x_1^{-1}, \cdots, x_2^{-1}\}$$

and

$$\ln\det\{x_1, \cdots, x_2\} := \ln\det x_1 + \cdots + \ln\det x_2.$$

Note transposes * convert columns of symbols to rows of the transposed symbols.

Now we sketch a popular class of methods for computing a solution to optimization problems of the form (7.4) called interior-point methods. (They follow the outline at the beginning of this subsection.) In order to relate one of many variants of such methods to Algorithm 7.1, let us first write problem (7.4) in the equivalent form

$$\min_{X,Y}\{\mathrm{Trace}(C^* X) : \quad f(A, X) + Y = 0, \quad Y \succeq 0\} \tag{7.5}$$

after introduction of the "slack variable" Y. Consider now the incorporation of a "barrier function" associated with Y into the objective function

$$\phi(X, Y) := \mathrm{Trace}(C^* X) - \mu \log\det Y$$

used to produce the auxiliary problem

$$\min_{X,Y}\{\phi(X, Y) : \quad f(A, X) + Y = 0\}. \tag{7.6}$$

In a nutshell, the idea behind the $-\mu \log\det Y$ barrier function is that this blows up as $Y \succ 0$ gets closer and closer to having a zero eigenvalue. Thus when $\mu > 0$ any iterative algorithm initialized with a positive definite value for the slack variable, i.e. $Y \succ 0$, will keep Y positive definite as long as the objective function is minimized (in a numerical implementation this may require taking short steps).

Now introduce a self-adjoint "Lagrange multiplier" z (i.e. $z = z^*$) and define the NC function

$$g(a, x, z) = \nabla_x \,\mathrm{trace}\left(z f(a, x)\right) = \nabla_{h_x} \mathrm{trace}\left(z \frac{\partial}{\partial x} f(a, x)[h_x]\right). \tag{7.7}$$

Note that in order to manipulate the above expression symbolical we need implement a symbolic operator 'trace', which obeys the familiar linear and cyclic properties present in the Trace functional on square matrices. The KKT optimality conditions for problem (7.6) are

$$c + g(a, x, z) = 0, \qquad z - \mu y^{-1} = 0, \qquad f(a, x) + y = 0. \tag{7.8}$$

For fixed A, C, μ these equations are of the form $F(X, Y, Z) = 0$ in unknowns X, Y, Z and can be solved for X, Y, Z by Algorithm 7.1, the trouble being keeping Y negative semidefinite. However, doing Algorithm 7.1 with a on α to minimize ϕ handles this [3]. Note $Y \succeq 0$ automatically implies $Z \succeq 0$.

Now we turn to the parameter μ which was inserted into the problem. A solution to the original problem (7.4) is found by approximately solving problems of the form (7.8) for a sequence of decreasing positive values of μ. Note that as $\mu \to 0$ the equations (7.8) with $Y \succeq 0$ coincide with the KKT conditions of the modified problem (7.5).

7.3. The linear subproblem and symbolic computation. At this stage in applying Algorithm 7.1 the burning issue is to write down a formula for the linearization of the optimality equations (7.8). It is straightforward to see that the linearization of (7.8), for a given $\mu > 0$, $y_k \succ 0$ and $z_k \succ 0$, are the Generalized Sylvester Equations

$$\begin{aligned} c + g(a, x_k, z_k) + \frac{\partial}{\partial x} g(a, x_k, z_k)[h_x] + g(a, x_k, h_z) &= 0, \\ z_k - \mu y_k^{-1} + h_z + \mu y_k^{-1} h_y y_k^{-1} &= 0, \\ f(a, x_k) + y_k + \frac{\partial}{\partial x} f(a, x_k)[h_x] + h_y &= 0, \end{aligned} \tag{7.9}$$

where we have used the fact that g is linear in z.

In particular problems this must be computed concretely. Our goal in this subsection is to convince the reader that one can carry this out and should symbolically while keeping the matrix variables intact; in other words there is a big advantage to keeping the computations dimension free. Of course one can always take the dimension dependent approach, disaggregating the variables once the size of the matrices are specified. In this case each entry of each matrix is viewed as a variable. If the matrices

[3]For fairness in advertising we reveal that what is often done in practice is a line search on a more complicated function. For instance, in [VS99] the following function is used

$$\phi(X, Y) = \text{Trace}(C^* X) - \mu \log \det Y + \beta \, \| f(A, X) + Y \|^2$$

where there are two adjustable parameters μ and β. The additional penalty term present in ϕ is used with a sufficiently large $\beta > 0$ so as to reenforce the equality constraint in (7.5).

are, say 300×300, then the computations will involve about 10^5 variables on which symbolic computation is a joke.

While we have algorithms [CHS06, dOH06b] and have implementations on classes of problems including broad classes of functions f, we shall confine our illustrations to the Riccati inequality optimization

$$f(a,b,c,x) = \{x\,b\,x - a^*x - xa - c, -x\}$$

defined previously as an example.

The first aspect is the NCAlgebra computation of the function $g(a,x,y)$ of equation (7.7), and one easily gets using `DirectionalD` and a command which invokes the cyclic property of *trace.*:

$$\begin{aligned} g(a,b,c,x,z) &= \nabla_{h_x} \operatorname{trace}(z_1\,(h_x\,b\,x + x\,b\,h_x - a^*h_x - h_x a) - z_2 h_x) \\ &= (b\,x - a)z_1 + z_1(x\,b - a^*) - z_2. \end{aligned}$$

Once g has been computed symbolically, it is straightforward to plug f and g into (7.9) and get concrete formulas. Here, in the Riccati inequality example this gives:

$$\begin{aligned} -I + (b\,x_k - a)z_{1_k} + z_{1_k}(x_k\,b - a^*) - z_{2k} \qquad & \\ +b\,h_x z_{1_k} + z_{1_k}h_x\,b + (b\,x_k - a)h_{z_1} + h_{z_1}(x_k\,b - a^*) - h_{z_2} &= 0, \\ z_{1_k} - \mu y_{1_k}^{-1} + h_{z_1} + \mu y_{1_k}^{-1} h_{y_1} y_{1_k}^{-1} &= 0, \\ z_{2_k} - \mu y_{2_k}^{-1} + h_{z_2} + \mu y_{2_k}^{-1} h_{y_2} y_{2_k}^{-1} &= 0, \\ x_k\,b\,x_k - a^*x_k - x_k a - c + y_{1_k} + (x\,b - a^*)h_x + h_x(b\,x - a) + h_{y_1} &= 0, \\ -x_k + y_{2_k} - h_x + h_{y_2} &= 0. \end{aligned}$$

which can all be computed in NCAlgebra using little more than `DirectionalD`. For example the third equation is gotten from

```
In[36]:= f2[y_,z_] := z - mu inv[y];
In[37]:= Sylvester22 =
      = f2[y2,z2] + DirectionalD[f2[y2,z2], y2, hy2] +
        + DirectionalD[f2[y2,z2], z2, hz2];
Out[37] = f2[y2,z2] + hz2 + mu inv[y2] ** hy2 ** inv[y2]
```

These Sylvester equations are often "reduced" by solving for some of the unknowns *apriori*. For instance, solving for h_y and h_z

$$h_y = -f(a,x_k) - y_k - \frac{\partial}{\partial x} f(a,x_k)[h_x],$$

$$\begin{aligned} h_z &= \mu y_k^{-1} - z_k - \mu y_k^{-1} h_y y_k^{-1} \\ &= \mu y_k^{-1}\left[2y_k + f(a,x_k)\right] y_k^{-1} - z_k + \mu y_k^{-1} \frac{\partial}{\partial x} f(a,x_k)[h_x] y_k^{-1} \end{aligned}$$

as a function of h_y we obtain, for some properly defined $q(a, x_k, y_k, z_k)$, the reduced Generalized Sylvester Equation

$$\begin{aligned} q(a, x_k, y_k, z_k) + \frac{\partial}{\partial x} g(a, x_k, z_k)[h_x] + \\ + \mu g \left(a, y_k^{-1} \frac{\partial}{\partial x} f(a, x_k)[h_x] y_k^{-1} \right) = 0, \end{aligned} \tag{7.10}$$

whose only unknown is h_x. Again note that (7.10) can be computed automatically with NCAlgebra mainly because solving symbolically for the h's is straightforward. Jumping a bit ahead we mention that, for this example, a version of Algorithm 7.1 can be constructed in which step (3), the line search, is not needed.

7.4. Numerical linear solvers. The symbolics of the preceding subsection are run at the beginning of an optimization computation and the formulas one gets are stored effectively symbolically (they are extremely short compared to prevailing methods where the matrices have been disaggregated). Next comes the numerical iteration and at each step the coefficients in our Sylvester linear problem

$$S(H) = Q$$

e.g. (7.1), (7.9) and (7.10), are matrices, say $n \times n$. So far we are exploiting the matrix structure of our original problem.

Now comes the challenge which to a large extent is open. *Find efficient numerical Sylvester linear solvers.* One can ignore the special Sylvester structure of these linear equations, that is, one can "vectorizes" the computation, then one has "unstructured" linear equations in n^2 variables, so for $n = 300$ even storing and accessing the matrix scales like 300^4 which is outlandish. (Our problems are certainly not sparse.) On the other hand, if we keep the Sylvester structure, then storing iterates $H_{k+1} := S(H_k)$ scales like n^2 and computing these iterates scales like $2(syl + sylT)n^3$, which is cheap. This disposes one to linear solvers based on such iteration, for example, Conjugate-Gradient type algorithms. What matters is getting good accuracy with not too many iterations.

We have implemented versions of the Conjugate-Gradient algorithm for the classes of problems described here. They are still being tested and tried with various "preconditioners". Some examples run (on one a gigahertz PC with 1 gigabyte of RAM) on problems of remarkable size, eg. 300×300, but some do not. A likely divide is how well conditioned the original problem is, in which case preconditioners can play a big role.

When f is not convex, then various difficulties of classical type emerge that must be combated. For example, one may have to modify the linear subproblem so that it become positive definite, which is traditionally done by adding a multiple of the identity to the coefficient matrix (see for instance [VS99, BHN99]). The Generalized Sylvester Equation (7.10) structure survives most such modifications.

7.4.1. Matrix convexity of f and the linear subproblem. Seriously effecting the numerics is whether or not the purely linear part of (7.10) in h_x is given by a positive semidefinite operator. We check both terms of this linear part. For the first term use the definition of the function g in (7.7) and set $z = y_k^{-1} \frac{\partial}{\partial x} f(a, x_k)[h_x] y_k^{-1}$ to get

$$\text{trace}\,(h_x^* \, g\,(a, x, z)) = \text{trace}\left(y_k^{-1} \frac{\partial}{\partial x} f(a, x_k)[h_x] y_k^{-1} \frac{\partial}{\partial x} f(a, x_k)[h_x] \right). \quad (7.11)$$

If all variables are substituted with matrices with $y_k \to Y_k$ a positive definite matrix, then the right side is clearly positive for all A, X_k, H_k. For the second term note that

$$\begin{aligned} \text{trace}\left(h_x^* \frac{\partial}{\partial x} g(a, x_k, z_k)[h_x] \right) &= \frac{\partial}{\partial x} \text{trace}\,(h_x^* \, g(a, x_k, z_k))\,[h_x] \\ &= \text{trace}\left(z_k \frac{\partial^2}{\partial^2 x} f(a, x_k)[h_x] \right). \end{aligned}$$

Therefore, from the discussion in Section 3, if f is NC convex, then

$$\frac{\partial^2}{\partial^2 x} f(A, X_k)[H_x] \succeq 0$$

whenever there are matrix substitutions $\{a, x_k\} \to \{A, X_k\}$. Thus

$$\text{Trace}\left(Z_k \frac{\partial^2}{\partial^2 x} f(A, X_k)[H_x] \right) \geq 0 \quad \text{for all } H_x, Z_k \succeq 0.$$

The conclusion is that when the NC function $f(a, x)$ is convex the purely linear part of the Generalized Sylvester Equation (7.10) is positive semidefinite.

As an aside note that the notion of positivity, expressed in the above inequalities and earlier in this section and in Section 3, can be formalized at the level of symbolics. We do not describe this here.

8. Answers to the free sample. Here is an answer to the standard problem of H^∞ control which was stated in §4.1. Recall the key question: *Is there is a set of noncommutative convex inequalities with an equivalent set of solutions?*

This is a question in algebra and the answer after a lot of work is yes. The path to success is:

1. *Firstly, one must eliminate unknowns and change variables to get a new set of inequalities $\mathcal{K}$.*
2. *Secondly, one must check that $\mathcal{K}$ is "convex" in the unknowns.*

This outline transcends our example and applies to very many situations. Issue (2) is becoming reasonably understood, for as we saw earlier, a convex polynomial with real coefficients has degree two or less, so these are trivial to identify. Issue (2), changing variables, is still a collection of isolated tricks with which mathematical theory has not caught up. For the particular problem in our example we shall not derive the solution since it is long. However, we do state the classical answer in the next subsection.

8.1. Solution to the problem. The textbook solution is as follows, due to Doyle-Glover- Kargonekar-Francis. It appeared in [DGKF89] which won the 1991 annual prize for the best paper to appear in an IEEE journal. Roughly speaking it was deemed the best paper in electrical engineering in that year.

We denote

$$DGKF_X := (A - B_2C_1)'X + X(A - B_2C_1) + X(\gamma^{-2}B_1B_1' - B_2^{-1}B_2')X,$$
$$DGKF_Y := A^{\times}Y + YA^{\times\prime} + Y(\gamma^{-2}C_1'C_1 - C_2'C_2)Y,$$

where $A^{\times} := A - B_1C_2$.

THEOREM 8.1 ([DGKF89]). *There is a system solving the control problem if there exist solutions*

$$X \succeq 0 \quad \textit{and} \quad Y \succ 0$$

to inequalities the

$$DGKF_Y \preceq 0 \ \textit{and} \ DGKF_X \preceq 0$$

which satisfy the coupling condition

$$X - Y^{-1} \prec 0.$$

This is if and only if provided $Y \succ 0$ is replaced by $Y \succeq 0$ and Y^{-1} is interpreted correctly.

This set of inequalities while not usually convex in X, Y are convex in the new variables $W = X^{-1}$ and $Z = Y^{-1}$, since $DGKF_X$ and $DGKF_Y$ are linear in W and Z and $X - Y^{-1} = W^{-1} - Z$ has second derivative $2W^{-1}HW^{-1}HW^{-1}$ which is non negative in H for each $W^{-1} = X \succ 0$. These inequalities are also equivalent to LMIs which we do not write down.

9. Classical RAG extended to free-* algebras. At this point one might think of the emerging area of free *- semi-algebraic geometry as having two main paths. One is an analog of classical commutative semi-algebraic geometry and focuses on general polynomial inequalities generally and Positivstellensätze - algebraic identities involving sums of squares - in particular. This noncommutative semi-algebraic geometry is the focus of this section where we shall sketch some basic ideas behind the emerging theory of inequalities involving polynomials on a free $*$-algebra. As we shall see, for strict inequalities the theory gives satisfying theorems, while when vanishing occurs (as in the real Negativstellensatz) the results are less definitive.

The second area of free $*$- semi-algebraic geometry has little analog classically and focuses on noncommutative functions with positive second derivatives or more generally varieties (not in this paper) whose curvature meets inequality constraints. Such convexity issues have been the main topic of this paper so far. The proofs can be done with the "middle matrix" representation in §6, which is a type of special Positivstellensatz for quadratics, thereby avoiding the much more generally applicable Positivstellensätze discussed in this section. However, the theory of noncommutative semi-algebraic geometry exposited in this section is expanding rapidly and may someday find applications.

9.1. Sums of squares in a free $*$-algebra. Let $\mathbb{R}\langle x, x^*\rangle$ denote the algebra of polynomials with real coefficients, in the free variables $x_1, \ldots, x_g, x_1^*, \ldots, x_g^*$. These variables do not commute, but they are associated with an involution:

$$(fq)^* = q^* f^*, \quad (x_j)^* = x_j^*.$$

Thus, we are now breaking with the convention of the rest of this survey in that the variables x_j are now **not** symmetric; i.e., $x^* \neq x$. We will call $\mathbb{R}\langle x, x^*\rangle$ the *real free $*-$ algebra* on generators x, x^*. Let Σ^2 denote the cone of sums of squares:

$$\Sigma^2 = \text{co}\{f^* f;\ f \in \mathbb{R}\langle x, x^*\rangle\},$$

where "co" denotes convex hull.

9.2. A basic technique. Call a linear functional $L \in \mathbb{R}\langle x, x^*\rangle'$ *symmetric* provided that $L(f) = L(f^*)$ for all $f \in \mathbb{R}\langle x, x^*\rangle$. A symmetric linear functional $L \in \mathbb{R}\langle x, x^*\rangle'$ satisfying $L|_{\Sigma^2} \geq 0$ produces a positive semi-definite bilinear form

$$\langle f, q\rangle = L(q^* f)$$

on $\mathbb{R}\langle x, x^*\rangle$. A standard use of Cauchy-Schwarz inequality shows that the set of null-vectors

$$N = \{f \in \mathbb{R}\langle x, x^*\rangle;\ \langle f, f\rangle = 0\}$$

is a vector subspace of $\mathbb{R}\langle x, x^*\rangle$. Whence one can endow the quotient $\mathcal{D} = \mathbb{R}\langle x, x^*\rangle/N$ with a positive definite Hermitian form, and pass to the Hilbert space completion H, with $\mathcal{D}$ the dense subspace of H generated by $\mathbb{R}\langle x, x^*\rangle$. The separable Hilbert space H carries the multiplication operators $M_{x_j} : \mathcal{D} \longrightarrow \mathcal{D}$:

$$M_{x_j} f = x_j f, \quad f \in \mathcal{D},\ 1 \leq j \leq n.$$

One verifies from the definition that each M_{x_j} is well defined and

$$\langle M_{x_j} f, q\rangle = \langle x_j f, q\rangle = \langle f, x_j^* q\rangle, \quad f, q \in \mathcal{D}.$$

Thus $M_{x_j}^* = M_{x_j^*}$. The vector 1 is still cyclic, in the sense that the linear span $\vee_{p\in\mathbb{R}\langle x,x^*\rangle} p(M, M^*)1$ is dense in H. The above is known in the operator theory community as the *Gelfand-Naimark-Segal construction.*

LEMMA 9.1. *There exists a bijective correspondence between symmetric positive linear functionals, namely*

$$L \in \mathbb{R}\langle x, x^*\rangle' \quad \textit{and} \quad L|_{\Sigma^2} \geq 0,$$

and g-tuples of unbounded linear operators T with a cyclic vector ξ, established by the formula

$$L(f) = \langle f(T, T^*)\xi, \xi\rangle, \quad f \in \mathbb{R}\langle x, x^*\rangle.$$

We stress that the above operators do not commute, and might be unbounded. The calculus $f(T, T^*)$ is the noncommutative functional calculus: $x_j(T) = T_j$, $x_j^*(T) = T_j^*$.

An important feature of the above correspondence is that it can be restricted by the degree filtration. Specifically, let $\mathbb{R}\langle x, x^*\rangle_k = \{f;\ \deg f \leq k\}$, and similarly, for a quadratic form L as in the lemma, let $\mathcal{D}_k$ denote the finite dimensional subspace of H generated by the polynomials of $\mathbb{R}\langle x, x^*\rangle_k$. Define also

$$\Sigma_k^2 = \Sigma^2 \cap \mathbb{R}\langle x, x^*\rangle_k.$$

Start with a symmetric functional $L \in \mathbb{R}\langle x, x^*\rangle'_{2k}$ satisfying $L|_{\Sigma_{2k}^2} \geq 0$. One can still construct a finite dimensional Hilbert space H, as the completion of $\mathbb{R}\langle x, x^*\rangle_k$ with respect to the inner product $\langle f, q\rangle = L(q^* f)$, $f, q \in \mathbb{R}\langle x, x^*\rangle_k$. The multipliers

$$M_{x_j} : \mathcal{D}_{k-1} \longrightarrow H, \qquad M_{x_j} f = x_j f,$$

are well defined and can be extended by zero (on the orthogonal complement of $\mathcal{D}_{k-1}$) to the whole H. Let

$$N(k) = \dim \mathbb{R}\langle x, x^*\rangle_k = 1 + (2g) + (2g)^2 + \ldots + (2g)^k = \frac{(2g)^{k+1} - 1}{2g - 1}.$$

In short, we have proved the following specialization of the main Lemma.

LEMMA 9.2. *Let symmetric functional $L \in \mathbb{R}\langle x, x^*\rangle'_{2k}$ satisfy $L|_{\Sigma_{2k}^2} \geq 0$. There exists a Hilbert space of dimension less than or equal to $N(k)$ and an g-tuple of linear operators M on H, with a distinguished vector $\xi \in H$, such that*

$$L(p) = \langle p(M, M^*)\xi, \xi\rangle, \quad p \in \mathbb{R}\langle x, x^*\rangle_{2k-2}. \tag{9.1}$$

Note that we do not exclude in the above lemma $L = 0$, in which case one can take $\xi = 0$.

9.3. Positivstellensätze. This subsection gives an indication of various free *-algebra analogs to the classical theorems characterizing polynomial inequalities in a purely algebraic way. We will start with an easily stated and fundamental Nichtnegativstellensatz.

THEOREM 9.3 ([Hel02]). *Let $p \in \mathbb{R}\langle x, x^*\rangle_d$ be a noncommutative polynomial. If $p(M, M^*) \succeq 0$ for all g-tuples of linear operators M acting on a Hilbert space of dimension at most $N(k)$, $2k \geq d+2$, then $p \in \Sigma^2$.*

Proof. Note that a polynomial p satisfying the hypothesis automatically satisfies $p = p^*$. The only necessary technical result we need is the closedness of the cone Σ_k^2 in the Euclidean topology of the finite dimensional space $\mathbb{R}\langle x, x^*\rangle_k$. This is done as in the commutative case, using Carathédodory's convex hull theorem, more exactly, every polynomial of Σ_k^2 is a convex combination of at most $\dim \mathbb{R}\langle x, x^*\rangle_k + 1$ polynomials. On the other hand the positive functionals on Σ_k^2 separate the points of $\mathbb{R}\langle x, x^*\rangle_k$. See for details [HMP04].

Assume that $p \notin \Sigma^2$ and let $k \geq (d+2)/2$, so that $p \in \mathbb{R}\langle x, x^*\rangle_{2k-2}$. Once we know that Σ_{2k}^2 is a closed cone, we can invoke Minkowski separation theorem and find a symmetric functional $L \in \mathbb{R}\langle x, x^*\rangle'_{2k}$ providing the strict separation:

$$L(p) < 0 \leq L(f), \quad f \in \Sigma_{2k}^2.$$

According to Lemma 9.2 there exists a tuple M of operators acting on a Hilbert space H of dimension $N(k)$ and a vector $\xi \in H$, such that

$$0 \leq \langle p(M, M^*)\xi, \xi\rangle = L(p) < 0,$$

a contradiction. □

When compared to the commutative framework, this theorem is stronger in the sense that it does not assume a strict positivity of p on a well chosen "spectrum". Variants with supports (for instance for spherical tuples M : $M_1^* M_1 + \ldots + M_g^* M_g \preceq I$) of the above result are discussed in [HMP04].

To draw a very general conclusion from the above computations: when dealing with positivity in a free-* algebra, the standard point evaluations (or more precisely prime spectrum evaluations) of the commutative case are replaced by matrix evaluations of the free variables. The positivity can be tailored to "evaluations in a supporting set". The results pertaining to the resulting algebraic decompositions are called Positivstellensätze, see again [PD01] for details in the commutative setting. We state below an illustrative and generic result, from [HM04], for sums of squares decompositions in a free $*$-algebra.

THEOREM 9.4. *Let $p = p^* \in \mathbb{R}\langle x, x^*\rangle$ and let $q = \{q_1, ..., q_k\} \subset \mathbb{R}\langle x, x^*\rangle$ be a set of symmetric polynomials, so that*

$$QM(q) = \mathrm{co}\{f^* q_k f;\ f \in \mathbb{R}\langle x, x^*\rangle,\ 0 \le i \le k\},\ q_0 = 1,$$

contains $1 - x_1^* x_1 - ... - x_g^* x_g$. If for all tuples of linear bounded Hilbert space operators $X = (X_1, ..., X_g)$ subject to the conditions

$$q_i(X, X^*) \succeq 0,\ 1 \le i \le k, \tag{9.2}$$

we have

$$p(X, X^*) \succ 0,$$

then $p \in QM(q)$.

Henceforth, call $QM(q)$ the **quadratic module** generated by the set of polynomials q.

Some interpretation is needed in degenerate cases, such as those where no bounded operators satisfy the relations $q_i(X, X^*) \succeq 0$. Suppose for example, if ϕ_i denotes the defining relations for the Weyl algebra and the q_i include $-\phi^*\phi$. In this case, we would say $p(X, X^*) \succ 0$, since there are no X satisfying $q(X, X^*)$, and voila $p \in QM(q)$ as the theorem says.

Proof. Assume that p does not belong to the convex cone $QM(q)$. Since the latter contains the constants in its algebraic interior, by Minkovski's separation principle there exists a symmetric linear functional $L \in \mathbb{R}\langle x, x^*\rangle'$, such that

$$L(p) \le 0 \le L(f),\ \ f \in QM(q).$$

Define the Hilbert space H associated to L, and remark that the left multipliers M_{x_i} on $\mathbb{R}\langle x, x^*\rangle$ give rise to linear bounded operators (denoted by the same symbols) on H. Then

$$q_i(M, M^*) \succeq 0,\ \ 1 \le i \le k,$$

by construction, and

$$0 < \langle p(M, M^*)1, 1\rangle = L(p) \le 0,$$

a contradiction. See for full definitions and more details [HM04] or the survey [HP07]. □

A paradigm practical question with matrix inequalities is:

Given a NC symmetric polynomial $p(a, x)$ and a $n \times n$ matrix tuple A, find $X \ge 0$ if possible which makes $p(A, X) \succeq 0$.

This fails in a region defined by a noncommutative symmetric polynomial $q(a, x)$ for a given matrix tuple A, means that $p(A, X) \not\succeq 0$ for any X satisfying $q(A,X) \succeq 0$. There is a great thrust of research aimed at numerical solution of such problems (see §7), but it is not clear how Posivstellensätze can aid with solving this particular problem. The next theorem informs us that the main problem here is the matrix coefficients A, in that the theorem gives a "certificate of infeasibility" for the problem when the coefficients are real numbers rather than polynomials.

THEOREM 9.5 (The Klep-Schweighofer Nirgendsnegativsemidefinitheitsstellensatz [KS07]).

Let $p = p^* \in \mathbb{R}\langle x, x^*\rangle$ and let $q = \{q_1, ..., q_k\} \subset \mathbb{R}\langle x, x^*\rangle$ be a set of symmetric polynomials, so that $QM(q)$ contains $1 - x_1^* x_1 - ... - x_g^* x_g$. If for all tuples of linear bounded Hilbert space operators $X = (X_1, ..., X_g)$ subject to the conditions

$$q_i(X, X^*) \succeq 0, \ 1 \le i \le k,$$

we have

$$p(X, X^*) \not\succeq 0,$$

then there exists an integer r and $h_1, \ldots, h_r \in \mathbb{R}\langle x, x^*\rangle$ with $\sum_{i=1}^r h_i^* f h_i \in 1 + QM(q)$.

Proof. The thread of the argument again uses a separating linear functional and the GNS construction, but adorned with clever constructions. □

9.4. Quotient algebras. The results from Section 9.3 allow a variety of specializations to quotient algebras. In this subsection we consider a two sided ideal $\mathcal{I}$ of $\mathbb{R}\langle x, x^*\rangle$ which need not be invariant under $*$. Then one can replace the quadratic module QM in the statement of the Positivstellensätze with $QM(q)+\mathcal{I}$, and apply similar arguments as above. For instance, the next simple observation can be deduced.

COROLLARY 9.6. *Assume, in the hypotheses of Theorem 9.4, that the relations* (9.2) *include some relations of the form $r(X, X^*) = 0$, even with r not symmetric, then*

$$p \in QM(q) + \mathcal{I}_r \tag{9.3}$$

where $\mathcal{I}_r$ denotes the two sided ideal generated by r.

Proof. This follows immediately from $p \in QM(q, \ -r^* r)$ which is a consequence of Theorem 9.4 and the fact

$$QM(q, \ -r^* r) \subset QM(q) + \mathcal{I}_r.$$

□

For instance, we can look at the situation where $r(x) := [x_i, x_j]$ as insisting on positivity of $q(X)$ only on commuting tuples of operators, in which case the ideal $\mathcal{I}$ generated by $[x_j^*, x_i^*]$, $[x_i, x_j]$ is added to $QM(q)$.

The classical commuting case is captured by the corollary applied to the "commutator ideal": $\mathcal{I}_{[x_j^*, x_i^*],\ [x_i, x_j],\ [x_i, x_j^*]}$ for $i, j = 1, \cdots, g$ which requires testing only on commuting tuples of operators drawn from a commuting C^* algebra. The classical Spectral Theorem, then converts this to testing only on $\mathbb{R}^g$, cf [HP07].

In a very different vein of proof is the theorem due to Igor Klep (private communication) below. A special case, $z = [x_1 + x_1^*, x_2 + x_2^*]$, was stated without proof in [HP07].

THEOREM 9.7 (I. Klep). *Let $\mathcal{I}$ be the two sided ideal of $\mathbb{R}\langle x, x^* \rangle$ generated by $z - 1$ with $z^* = -z$. Then $\mathcal{I} + \Sigma^2 = \mathbb{R}\langle x, x^* \rangle$.*

Proof. First of all, $-z^2 = z^* z \in \Sigma^2$, so $-1 = z(z-1) - z^2 \in \mathcal{I} + \Sigma^2$. For a symmetric polynomial $s \in \mathbb{R}\langle x, x^* \rangle$,

$$s = \left(\frac{s+1}{2}\right) - \left(\frac{s-1}{2}\right), \tag{9.4}$$

showing that $\mathcal{I} + \Sigma^2$ contains all symmetric polynomials Sym $\mathbb{R}\langle x, x^* \rangle$. Also, for $j \in \mathcal{I}$,

$$j = (j + j^*) - j \in \operatorname{Sym} \mathbb{R}\langle x, x^* \rangle + \mathcal{I} \subset J + \Sigma^2.$$

Let $t \in \mathbb{R}\langle x, x^* \rangle$ be an arbitrary skew symmetric polynomial. Then $(-t)(z-1) = -tz + t \in \mathcal{I}$. Likewise, $(z-1)t = zt - t \in \mathcal{I}$ and thus by the above, $tz + t = (zt - t)^* \in \mathcal{I}^* \subset \mathcal{I} + \Sigma^2$. Adding the first and the last relation yields $t \in \mathcal{I} + \Sigma^2$.

As every polynomial f is a sum of a symmetric and a skew symmetric polynomial ($f = \frac{f+f^*}{2} + \frac{f-f^*}{2}$), this concludes the proof. □

9.5. A Nullstellensatz. With similar techniques (well chosen, separating, $*$-representations of the free algebra) and a rather different "dilation type" of argument, one can prove a series of Nullstellensätze.

We state for information one of them. For an early version see [HMP05].

THEOREM 9.8. *Let $q_1(x), \ldots, q_m(x) \in \mathbb{R}\langle x \rangle$ be polynomials not depending on the x_j^* variables and let $p(x, x^*) \in \mathbb{R}\langle x, x^* \rangle$. Assume that for every g tuple X of linear operators acting on a finite dimensional Hilbert space H, and every vector $v \in H$, we have:*

$$(q_j(X)v = 0,\ 1 \leq j \leq m) \ \Rightarrow\ (p(X, X^*)v = 0).$$

Then p belongs to the left ideal $\mathbb{R}\langle x, x^* \rangle q_1 + \ldots + \mathbb{R}\langle x, x^* \rangle q_m$.

Again, this proposition is stronger than its commutative counterpart. For instance there is no need of taking higher powers of p, or of adding a sum of squares to p. Note that here $\mathbb{R}\langle x \rangle$ has a different meaning than earlier, since, unlike previously, the variables are nonsymmetric.

We refer the reader to [HMP07] for the proof of Theorem 9.8. However, we say a few words about the intuition behind it. We are assuming

$$q_j(X)v = 0, \forall j \quad \Longrightarrow \quad p(X, X^*)v = 0.$$

On a very large vector space, if X is determined on a small number of vectors, then X^* is not heavily constrained; it is almost like being able to take X^* to be a completely independent tuple Y. If it were independent, we would have

$$q_j(X)v = 0, \forall j \quad \Longrightarrow \quad p(X, Y)v = 0.$$

Now, in the free algebra $\mathbb{R}\langle x, y\rangle$, it is much simpler to prove that this implies $p \in \sum_j^m \mathbb{R}\langle x, y\rangle\, q_j$, as required. We isolate this fact in a separate lemma.

LEMMA 9.9. *Fix a finite collection $q_1, \ldots, q_m$ of polynomials in noncommuting variables $\{x_1, \ldots, x_g\}$ and let p be a given polynomial in $\{x_1, \ldots, x_g\}$. Let d denote the maximum of the* $\deg(p)$ *and* $\{\deg(q_j) : 1 \leq j \leq m\}$.

There exists a real Hilbert space $\mathcal{H}$ of dimension $\sum_{j=0}^d g^j$, such that, if

$$p(X)v = 0$$

whenever $X = (X_1, \ldots, X_g)$ is a tuple of operators on $\mathcal{H}$, $v \in \mathcal{H}$, and

$$q_j(X)v = 0 \text{ for all } j,$$

then p is in the left ideal generated by $q_1, \ldots, q_m$.

Proof. (of Lemma). We sketch a proof based on an idea of G. Bergman, see [HM04].

Let $\mathcal{I}$ be the left ideal generated by $q_1, \ldots, q_m$ in $F = \mathbb{R}\langle x_1, \ldots, x_g\rangle$. Define $\mathcal{V}$ to be the vector space $F/\mathcal{I}$ and denote by $[f]$ the equivalence class of $f \in F$ in the quotient $F/\mathcal{I}$. Define X_j on the vector space $F/\mathcal{I}$ by $X_j[f] = [x_j f]$ for $f \in F$, so that $x_j \mapsto X_j$ implements a quotient of the left regular representation of the free algebra F.

If $\mathcal{V} := F/\mathcal{I}$ is finite dimensional, then the linear operators $X = (X_1, \ldots, X_g)$ acting on it can be viewed as a tuple of matrices and we have, for $f \in F$,

$$f(X)[1] = [f].$$

In particular, $q_j(X)[1] = 0$ for all j. If we do not worry about the dimension counts, by assumption, $0 = p(X)[1]$, so $0 = [p]$ and therefore $p \in \mathcal{I}$. Minus the precise statement about the dimension of $\mathcal{H}$ this establishes the result when $F/\mathcal{I}$ is finite dimensional.

Now we treat the general case where we do not assume finite dimensionality of the quotient. Let $\mathcal{V}$ and $\mathcal{W}$ denote the vector spaces

$$\mathcal{V} := \{[f] : f \in F,\ \deg(f) \leq d\},$$

$$\mathcal{W} := \{[f] : f \in F,\ \deg(f) \leq d-1\}.$$

Note that the dimension of $\mathcal{V}$ is at most $\sum_{j=0}^{d} g^j$. We define X_j on $\mathcal{W}$ to be multiplication by x_j. It maps $\mathcal{W}$ into $\mathcal{V}$. Any linear extension of X_j to the whole $\mathcal{V}$ will satisfy: if f has degree at most d, then $f(X)[1] = [f]$. The proof now proceeds just as in the part 1 of the proof above. □

With this observation we can return and finish the proof of Theorem 9.8. Since X^* is dependent on X, an operator extension with properties stated in the lemma below gives just enough structure to make the above free algebra Nullstellensatz apply.

LEMMA 9.10. *Let $x = \{x_1, \ldots, x_m\}$, $y = \{y_1, \ldots, y_m\}$ be free, noncommuting variables. Let H be a finite dimensional Hilbert space, and let X, Y be two m-tuples of linear operators acting on H. Fix a degree $d \geq 1$.*

Then there exists a larger Hilbert space $K \supset H$, an m-tuple of linear transformations $\tilde{X}$ acting on K, such that

$$\tilde{X}_j|_H = X_j, \quad 1 \leq j \leq g,$$

and for every polynomial $p \in \mathbb{R}\langle x, x^* \rangle$ of degree at most d and vector $v \in H$,

$$p(\tilde{X}, \tilde{X}^*)v = 0 \ \Rightarrow p(X, Y)v = 0.$$

For the construction of the larger Hilbert space $K \supset H$ and $\tilde{X}$ on it, see the proof in [HMP07].

Here is a theorem which could be regarded as a very different type of noncommutative Nullstellensatz.

THEOREM 9.11 (Theorem 2.1 [KS08]). *Let $p = p^* \in \mathbb{R}\langle x, x^* \rangle_d$ be a non-commutative polynomial satisfying $\operatorname{tr} p(M, M^*) = 0$ for all g-tuples of linear operators M acting on a Hilbert space of dimension at most d. Then p is a sum of commutators of noncommutative polynomials.*

9.6. A typical noncommutative phenomenon. We end this subsection with an example which goes against any intuition we would carry from the commutative case, see [HM04].

EXAMPLE 9.12. Let $q = (x^*x + xx^*)^2$ and $p = x + x^*$ where x is a single variable. Then, for every matrix X and vector v (belonging to the space where X acts), $q(X)v = 0$ implies $p(X)v = 0$; however, there does not exist a positive integer m and $r, r_j \in \mathbb{R}\langle x, x^* \rangle$, so that

$$p^{2m} + \sum r_j^* r_j = qr + r^* q. \tag{9.5}$$

Moreover, we can modify the example to add the condition $q(X)$ is positive semi-definite implies $p(X)$ is positive semi-definite and still not obtain this representation.

Proof. Since $A := XX^* + X^*X$ is self-adjoint, $A^2v = 0$ if and only if $Av = 0$. It now follows that if $q(X)v = 0$, then $Xv = 0 = X^*v$ and therefore $p(X)v = 0$.

For $\lambda \in \mathbb{R}$, let

$$X = X(\lambda) = \begin{pmatrix} 0 & \lambda & 0 \\ 0 & 0 & 1 \\ 0 & 0 & 0 \end{pmatrix}$$

viewed as an operator on $\mathbb{R}^3$ and let $v = e_1$, where $\{e_1, e_2, e_3\}$ is the standard basis for $\mathbb{R}^3$.

We begin by calculating the first component of even powers of the matrix $p(X)$. Let $Q = p(X)^2$ and verify,

$$Q = \begin{pmatrix} \lambda^2 & 0 & \lambda \\ 0 & 1+\lambda^2 & 0 \\ \lambda & 0 & 1 \end{pmatrix}. \tag{9.6}$$

For each positive integer m there exist a polynomial p_m so that

$$Q^m e_1 = \begin{pmatrix} \lambda^2(1+\lambda p_m(\lambda)) \\ 0 \\ \lambda(1+\lambda p_m(\lambda)) \end{pmatrix} \tag{9.7}$$

which we now establish by an induction argument. In the case $m = 1$, from equation (9.6), it is evident that $p_1 = 0$. Now suppose equation (9.7) holds for m. Then, a computation of $QQ^m e_1$ shows that equation (9.7) holds for $m+1$ with $p_{m+1} = \lambda(p_m + \lambda + \lambda p_m)$. Thus, for any m,

$$\lim_{\lambda\to 0} \frac{1}{\lambda^2} < Q^m e_1, e_1 >= \lim_{\lambda\to 0}(1 + \lambda p_m(\lambda)) = 1. \tag{9.8}$$

Now we look at q and get

$$q(X) = \begin{pmatrix} \lambda^4 & 0 & 0 \\ 0 & (1+\lambda^2)^2 & 0 \\ 0 & 0 & 1 \end{pmatrix}.$$

Thus

$$\lim_{\lambda\to 0} \frac{1}{\lambda^2}(< r(X)^* q(X) e_1, e_1 > + < q(X) r(X) e_1, e_1 >) = 0.$$

If the representation of equation (9.5) holds, then apply $< \cdot\, e_1, e_1 >$ to both sides and take λ to 0. We just saw that the right side is 0, so the left side is 0, which because

$$< \sum r_j(X)^* r_j(X) e_1, e_1 > \geq 0$$

forces

$$\lim_{\lambda \to 0} \frac{1}{\lambda^2} < Q^m e_1, e_1 > \leq 0$$

a contradiction to equation (9.8). Hence the representation of equation (9.5) does not hold.

The last sentence claimed in the example is true when we use the same polynomial q and replace p with p^2. □

9.7. Non-free Semi-algebraic geometry. A series of natural structures:

• Weyl algebra, enveloping algebras of Lie algebras (see Schmüdgen's article in this volume, and separately the work of Cimpric [Ci]),
• the classical area of PI rings, (e.g. $N \times N$ matrices for fixed N, as studied by Procesi-Schacher [PS76], or the Nullstellensatz for PI rings as discussed in [Ami57]),

are calling for a general framework incorporating all known Positiv- and Nullstellensätze in the literature. Such a construct is missing at the time of writing the present survey.

REFERENCES

[Ami57] S.A. Amitsur. A generalization of Hilbert's Nullstellensatz. *Proceedings of the American Mathematical Society*, **8**:649–656, 1957.

[BDG96] C.L. Beck, J.C. Doyle, and K. Glover. Model reduction of multidimensional and uncertain systems. *IEEE Transactions on Automatic Control*, **41**(10):1406–1477, 1996.

[Bec01] C.L. Beck. On formal power series representations for uncertain systems. *IEEE Transactions on Automatic Control*, **46**(2):314–319, 2001.

[BEGFB94] S.P. Boyd, L.El Ghaoui, E. Feron, and V. Balakrishnan. *Linear Matrix Inequalities in System and Control Theory.* SIAM, Philadelphia, PA, 1994.

[BGM05] J.A. Ball, G. Groenewald, and T. Malakorn. Structured noncommutative multidimensional linear systems. *SIAM Journal On Control And Optimization*, **44**(4):1474–1528, 2005.

[BGM06a] J.A. Ball, G. Groenewald, and T. Malakorn. Bounded real lemma for structured noncommutative multidimensional linear systems and robust control. *Multidimensional Systems And Signal Processing*, **17**(2–3):119–150, July 2006.

[BGM06b] J.A. Ball, G. Groenewald, and T. Malakorn. Conservative structured noncommutative multidimensional linear systems. In *The state space method generalizations and applications*, volume 161 of *Operator Theory Advances and Applications*, pages 179–223. Birkhäuser-Verlag, Basel-Boston-Berlin, 2006.

[BHN99] R.H. Byrd, M.E. Hribar, and J. Nocedal. An interior point algorithm for large scale nonlinear programming. *SIAM Journal on Optimization*, **9**(4):877–900, 1999.

[BR84] J. Berstel and C. Reutenauer. *Rational series and their languages.* EATCS Monographs on Theoretical Computer Science. Springer-Verlag, Berlin-New York, 1984.

[CHS06] J.F. Camino, J.W. Helton, and R.E. Skelton. Solving matrix inequalities whose unknowns are matrices. *SIAM Journal on Optimization*, **17**(1):1–36, 2006.

[CHSY03] J.F. Camino, J.W. Helton, R.E. Skelton, and J. Ye. Matrix inequalities: a symbolic procedure to determine convexity automatically. *Integral Equation and Operator Theory*, **46**(4):399–454, 2003.

[Ci] J. Cimpric Maximal quadratic modules on *-rings. *Algebr. Represent. Theory.* to appear.

[DGKF89] J.C. Doyle, Keith Glover, P.P. Khargonekar, and B.A. Francis. State-space solutions to standard H_2 and H_∞ control problems. *IEEE Transactions on Automatic Control*, **34**(8):831–847, 1989.

[dOH06a] M.C. de Oliveira and J.W. Helton. Computer algebra tailored to matrix inequalities in control. *International Journal of Control*, **79**(11):1382–1400, November 2006.

[dOH06b] M.C. de Oliveira and J.W. Helton. A symbolic procedure for computing semidefinite program duals. In *Proceedings of the 45th IEEE Conference on Decision and Control*, pages 5192–5197, San Diego, CA, 2006.

[GM82] P.E. Gill and W. Murray. *Practical Optimization.* Academic Press, 1982.

[Han97] F. Hansen. Operator convex functions of several variables. *Publ. RIMS, Kyoto Univ.*, **33**:443–464, 1997.

[HdOSM05] J.W. Helton, M.C. de Oliveira, M. Stankus, and R.L. Miller. *NCAlgebra and NCGB*, 2005 release edition, 2005. Available at `http://math.ucsd.edu/~ncalg`.

[Hel02] J.W. Helton. 'Positive' noncommutative polynomials are sums of squares. *Annals of Mathematics*, **156**(2):675–694, September 2002.

[HHLM] J.W. Helton, P.C. Lim, and S.A. McCullough. in preparation.

[HM03] J.W. Helton and S. McCullough. Convex noncommutative polynomials have degree two or less. *SIAM Journal on Matrix Analysis and Applications*, **25**(4):1124–1139, 2003.

[HM04] J.W. Helton and S.A. McCullough. A positivstellensatz for noncommutative polynomials. *Transactions of the American Mathematical Society*, **356**(9):3721–3737, 2004.

[HMP04] J.W. Helton, S.A. McCullough, and M. Putinar. A noncommutative positivstellensatz on isometries. *Journal für die reine und angewandte Mathematik*, **568**:71–80, March 2004.

[HMP05] J.W. Helton, S.A. McCullough, and M. Putinar. Non-negative hereditary polynomials in a free *-algebra. *Mathematische Zeitschrift*, **250**(3):515–522, July 2005.

[HMP07] J.W. Helton, S. McCullough, and M. Putinar. Strong majorization in a free *-algebra. *Mathematische Zeitschrift*, **255**(3):579–596, March 2007.

[HMV06] J.W. Helton, S.A. McCullough, and V. Vinnikov. Noncommutative convexity arises from linear matrix inequalities. *Journal Of Functional Analysis*, **240**(1):105–191, November 2006.

[HP07] J.W. Helton and M. Putinar. Positive polynomials in scalar and matrix variables, the spectral theorem and optimization; in vol. *Operator Theory, Structured Matrices and Dilations.* A volume dedicated to the memory of T. Constantinescu (M. Bakonyi et al., eds.), Theta, Bucharest, pp. 229–306, 2007.

[HPMV] J.W. Helton, M. Putinar, S. McCullough, and V. Vinnikov. Convex matrix inequalities versus linear matrix inequalities. Preprint.

[HT07] F. HANSEN AND J. TOMIYAMA. Differential analysis of matrix convex functions. *Linear Algebra And Its Applications*, **420**(1):102–116, January 2007.

[Kra36] F. KRAUS. Uber konvexe matrixfunctionen. *Math. Zeit.*, **41**:18–42, 1936.

[KS07] I. KLEP AND M. SCHWEIGHOFER. A nichtnegtaivstellensatz for polynomials in noncommuting variables. *Israel Journal of Mathematics*, **161**(1):17–27, 2007.

[KS08] I. KLEP AND M. SCHWEIGHOFER. Connes' embedding conjecture and sums of Hermitian squares. *Advances in Mathematics*, **217**:1816–1837, 2008. to appear.

[KVV07] D.S. KALIUZHNYI-VERBOVETSKYI AND V. VINNIKOV. Singularities of noncommutative rational functions and minimal factorizations, 2007. Preprint.

[Las01] J.B. LASSERRE. Global optimization with polynomials and the problem of moments. *SIAM Journal on Optimization*, **11**(3):796–817, 2001.

[LZD96] W.-M LU, K. ZHOU, AND J.C. DOYLE. Stabilization of uncertain linear systems: an lft approach. *IEEE Transactions on Automatic Control*, **41**(1):50–65, 1996.

[OST07] H. OSAKA, S. SILVESTROV, AND J. TOMIYAMA. Monotone operator functions, gaps and power moment problem. *Mathematica Scandinavica*, **100**(1):161–183, 2007.

[Par00] P.A. PARRILO. *Structured Semidefinite Programs and Semialgebraic Geometry Methods in Robustness and Optimization*. PhD thesis, California Institute of Technology, Passadena, CA, May 2000.

[PD01] A. PRESTEL AND C.N. DELZELL. *Positive polynomials. From Hilbert's 17th problem to real algebra*. Springer Monographs in Mathematics. Springer, Berlin, 2001.

[PS76] C. PROCESI AND M. SCHACHER. A non-commutative real Nullstellensatz and Hilbert's 17th problem. *Ann. of Math.*, **104**(2):395–406, 1976.

[SIG98] R.E. SKELTON, T. IWASAKI, AND K. GRIGORIADIS. *A Unified Algebraic Approach to Control Design*. Taylor & Francis, London, UK, 1998.

[Uch05] M. UCHIYAMA. Operator monotone functions and operator inequalities. *Sugaku Expositions*, **18**(1):39–52, 2005.

[VS99] R.J. VANDERBEI AND D.F. SHANNO. An interior-point algorithm for nonconvex nonlinear programming. *Computational Optimization and Applications*, **13**:231–252, 1999.

ALGEBRAIC STATISTICS AND CONTINGENCY TABLE PROBLEMS: LOG-LINEAR MODELS, LIKELIHOOD ESTIMATION, AND DISCLOSURE LIMITATION

ADRIAN DOBRA*, STEPHEN E. FIENBERG†, ALESSANDRO RINALDO‡, ALEKSANDRA SLAVKOVIC§, AND YI ZHOU¶

Abstract. Contingency tables have provided a fertile ground for the growth of algebraic statistics. In this paper we briefly outline some features of this work and point to open research problems. We focus on the problem of maximum likelihood estimation for log-linear models and a related problem of disclosure limitation to protect the confidentiality of individual responses. Risk of disclosure has often been measured either formally or informally in terms of information contained in marginal tables linked to a log-linear model and has focused on the disclosure potential of small cell counts, especially those equal to 1 or 2. One way to assess the risk is to compute bounds for cell entries given a set of released marginals. Both of these methodologies become complicated for large sparse tables. This paper revisits the problem of computing bounds for cell entries and picks up on a theme first suggested in Fienberg [21] that there is an intimate link between the ideas on bounds and the existence of maximum likelihood estimates, and shows how these ideas can be made rigorous through the underlying mathematics of the same geometric/algebraic framework. We illustrate the linkages through a series of examples. We also discuss the more complex problem of releasing marginal and conditional information. We illustrate the statistical features of the methodology on two examples and then conclude with a series of open problems.

Key words. Conditional tables, marginal tables, Markov bases, maximum likelihood estimate, sharp bounds for cell entries, toric ideals.

AMS(MOS) subject classifications. 13P10, 62B05, 62H17, 62P25.

1. Introduction.

Polynomials abound in the specification of statistical models and inferential methods. In particular, many common statistical procedures involve finding the solution to polynomial equations. Thus, in

*Department of Statistics, University of Washington, Box 354322, Seattle, WA 98195-4322 (adobra@u.washington.edu).

†Department of Statistics, Machine Learning Department and Cylab, Carnegie Mellon University, Pittsburgh, PA 15213-3890 (fienberg@stat.cmu.edu). Supported in part by NSF grants EIA9876619 and IIS0131884 to the National Institute of Statistical Sciences, and NSF Grant DMS-0631589 and Army contract DAAD19-02-1-3-0389 to Carnegie Mellon University.

‡Department of Statistics, Carnegie Mellon University, Pittsburgh, PA 15213-3890 (arinaldo@stat.cmu.edu). Supported in part by NSF Grant DMS-0631589 and a grant from the Pennsylvania Department of Health through the Commonwealth Universal Research Enhancement Program to Carnegie Mellon University.

§Department of Statistics, Pennsylvania State University, University Park, PA 16802 (sesa@stat.psu.edu). Supported in part by NSF grants EIA9876619 and IIS0131884 to the National Institute of Statistical Sciences and SES-0532407 to Pennsylvania State University.

¶Machine Learning Department, Carnegie Mellon University, Pittsburgh, PA 15213-3890 (yizhou@stat.cmu.edu). Supported by Army contract DAAD19-02-1-3-0389 to Carnegie Mellon University.

retrospect, we should not be surprised at the sudden emergence of a wide array of papers linking statistical methodology to modern approaches to computational algebraic geometry. But the fact is that these connections are a relatively recent development in the statistical literature and they have led to the use of the terminology "algebraic statistics" to describe this linkage.

Contingency tables are arrays of non-negative integers arising from cross-classifying n objects based on a set of k criteria or categorical variables (see [1, 32]). Each entry of a contingency table, or cell, is a non-negative integer indicating the number of times a given configuration of the classifying criteria has been observed in the sample. Log-linear models form a class of statistical models for the joint probability of cell entries. Our work has focused on three interrelated classes of problems: (1) geometric characterization of log-linear models for cell probabilities for contingency tables, (2) estimation of cell probabilities under log-linear models, and (3) disclosure limitation strategies associated with contingency tables which protect against the identification of individuals associated with counts in the tables.

The disclosure limitation literature for contingency table data is highly varied, e.g., see [18], but over the past decade a substantial amount of it has focused on the risk-utility tradeoff where risk has been measured either formally or informally in terms of information contained in marginal tables and risk has focused on disclosure potential of small cell counts, especially those equal to 1 or 2 (for details, see [25, 26]). Among the ways considered for assessing risk have been the computation of bounds for cell entries, e.g., see [11, 12, 13, 14, 15], and the enumeration of possible table realizations, e.g., see [26].

Recent advances in the field of algebraic statistics have provided novel and broader mathematical tools for log-linear models and, more generally, the analysis of categorical data. We outline below the most relevant aspects of the algebraic statistics formalism, which essentially involves a representation, through polynomials and polyhedral objects, of the interaction between the set of all possible configurations of cell probabilities, known as the *parameter space*, and the set of all observable arrays of non-negative entries summing to n and satisfying certain linear relationships to be described below, known as the *sample space*.

2. Some technical details for bounds and MLEs. We can describe both the determination of cell bounds associated to the release of marginal tables and the problem of nonexistence of the MLE within the same geometric/algebraic framework.

2.1. Technical specifications and geometrical objects. Consider k categorical random variables, $X_1, \ldots, X_k$, where each X_i takes value on the finite set of categories $[d_i] \equiv \{1, \ldots, d_i\}$. Letting $\mathcal{D} = \bigotimes_{i=1}^{k} [d_i]$, $\mathbb{R}^{\mathcal{D}}$ is the vector space of k-dimensional arrays of the format $d_1 \times \ldots \times d_k$, with a total of $d = \prod_i d_i$ entries. The cross-classification of n independent and identically distributed realizations of $(X_1, \ldots, X_k)$ produces a random integer-valued array $\mathbf{n} \in \mathbb{R}^{\mathcal{D}}$, called a k-way *contingency table*, whose coordinate entry $n_{i_1,\ldots,i_k}$ is the number of times the label combination, or *cell*, $(i_1, \ldots, i_k)$ is observed in the sample (see [1, 32] for details). The probability that a given cell appears in the sample is

$$p_{i_1,\ldots,i_k} = Pr\{(X_1, \ldots, X_k) = (i_1, \ldots, i_k)\}, \qquad (i_1, \ldots, i_k) \in \mathcal{D},$$

and we denote the corresponding array in $\mathbb{R}^{\mathcal{D}}$ with $\mathbf{p}$. It will often be convenient to order the cells in some prespecified way (e.g., lexicographically) and to treat $\mathbf{n}$ and $\mathbf{p}$ as vectors in $\mathbb{R}^d$ rather than arrays. For example, for a 3-way contingency table $\mathbf{n}$ with $d_1 = d_2 = d_3 = 2$, or a $2 \times 2 \times 2$ table, we will use interchangeably the array notation $\mathbf{n} = (n_{111}, n_{112}, \ldots, n_{222})$ and the vector notation $\mathbf{n} = (n_1, n_2, \ldots, n_8)$. A hierarchical log-linear model is a probabilistic model specifying the set of dependencies, or maximal *interactions*, among the k variables of interest. One can think of a log-linear model as a simplicial complex Δ on $[k] = \{1, \ldots, k\}$, whose facets indicate the groups of interacting variables. For example, for a $2 \times 2 \times 2$ table, the model $\Delta = \{\{1,2\},\{3\}\}$ specifies an interaction between the first and second variable, while the third is independent of the other two. Similarly, $\Delta = \{\{1,2\},\{1,3\},\{2,3\}\}$, the model of *no-2nd-order interaction*, postulates an interaction between all pairs of variables. In accordance to the notation adopted in the statistical literature, we will also write these models as [12][3] and [12][13][23], respectively.

A sub-class of log-linear models that enjoys remarkable properties and deserves special attention is the class of decomposable models: Δ is said to be decomposable if there exists a decomposition $\Delta = (\Delta_1, S, \Delta_2)$ with $\Delta_1 \cup \Delta_2 = \Delta$ and $\Delta_1 \cap \Delta_2 = 2^S$, and Δ_i is either a simplex or decomposable, for each $i = 1, 2$. Decomposable models are the simplest log-linear models for which the statistical tasks described in this article become straightforward. The smallest decomposable model with a non-trivial separation is $\Delta = \{\{1,2\},\{2,3\}\}$, where $S = \{2\}$.

For any given log-linear model Δ, the vector of cell probabilities $\mathbf{p}$ is a point in the interior of the standard $(d-1)$-simplex such that $\log \mathbf{p}$ belongs to the row span of some $m \times d$ matrix A, called the *design matrix*, which depends only on Δ (and not on the random realization $\mathbf{n}$). Clearly, for every Δ, there are many choices for A, but we may always assume A to be 0–1. For an example, see Table 1.

Once we specify a model Δ through its design matrix A, we consider the vector $\mathbf{t} = \mathrm{A}\mathbf{n}$ of *margins* or *marginal tables*. From an inferential standpoint, the vector $\mathbf{t}$ is all that a statistician needs to know in order to study Δ: in statistics, $\mathbf{t}$ is called a *minimal sufficient statistics*. In fact, although many different tables may give rise to the same margins $\mathbf{t}$, they are indistinguishable in the sense that they all provide the same information on Δ. See, e.g. [1, 30, 32] for details. In general, we can

TABLE 1
A design matrix for the model of no-2nd-order interaction for a $2 \times 2 \times 2$ table. The first line displays the label combinations ordered lexicographically.

111	112	121	122	211	212	221	222
1	1	0	0	0	0	0	0
0	0	1	1	0	0	0	0
0	0	0	0	1	1	0	0
0	0	0	0	0	0	1	1
1	0	0	0	1	0	0	0
0	1	0	0	0	1	0	0
0	0	1	0	0	0	1	0
0	0	0	1	0	0	0	1
1	0	1	0	0	0	0	0
0	1	0	1	0	0	0	0
0	0	0	0	1	0	1	0
0	0	0	0	0	1	0	1

choose the design matrices in such a way that the coordinates of the vector $\mathbf{t}$ are the marginal sums of the array $\mathbf{n}$ with respect to the coordinates specified by the co-facets of Δ. For the example in Table 1, it is easy to see that the first coordinate of $\mathbf{t}$ is $n_{111}+n_{112}$, which we will write in marginal notation as n_{11+}, the "+" symbol referring to the variable over which the summation is taken. For example, for the $I \times J \times K$ table and the model of no-2nd-order interaction, the minimal sufficient statistics are the three sets of two-dimensional marginal sums, $\{n_{ij+}\}$, $\{n_{i+k}\}$, and $\{n_{+jk}\}$, so that $\mathbf{t}$ is an integer valued random vector of dimension $IJ + IK + JK$.

Parameter Space. The parameter space refers to the set of all probability points $\mathbf{p}$ in the standard $(d-1)$-simplex such that $\log \mathbf{p}$ belongs to the row span of A. In algebraic statistics the parameter space is defined by the solution set of certain polynomial maps. Specifically, the parameter space is a smooth hyper-surface of points satisfying binomial equations, in fact a *toric variety* [41]. For a given design matrix A, the toric variety describing the associated log-linear model is the set of all probability vectors $\mathbf{p}$ such that $\mathbf{p}^{\mathbf{z}^+} - \mathbf{p}^{\mathbf{z}^-} = 0$ for all integer valued vectors $\mathbf{z}$ in kernel(A), where $\mathbf{z}^+ = \max(\mathbf{z}, 0)$, $\mathbf{z}^- = -\min(\mathbf{z}, 0)$, the operations being carried element-wise, and $\mathbf{p}^{\mathbf{z}} = \prod_i p_i^{z_i}$. For a 2×2 table and the model of independence, the toric variety is the familiar surface of independence (see, e.g. [1]) depicted in Figure 1. For further details on the algebraic geometry of other aspects of 2×2 tables, see [37, 40, 3]. The advantage of the algebraic statistics representation for the parameter space over the traditional log-linear representation based on logarithms, is that the points on the toric variety are allowed to be on the relative boundary of the simplex. This implies that the toric variety includes not only the points $\mathbf{p}$ such that $\log \mathbf{p}$ belongs to the row span of A, but also points in its sequential closure.

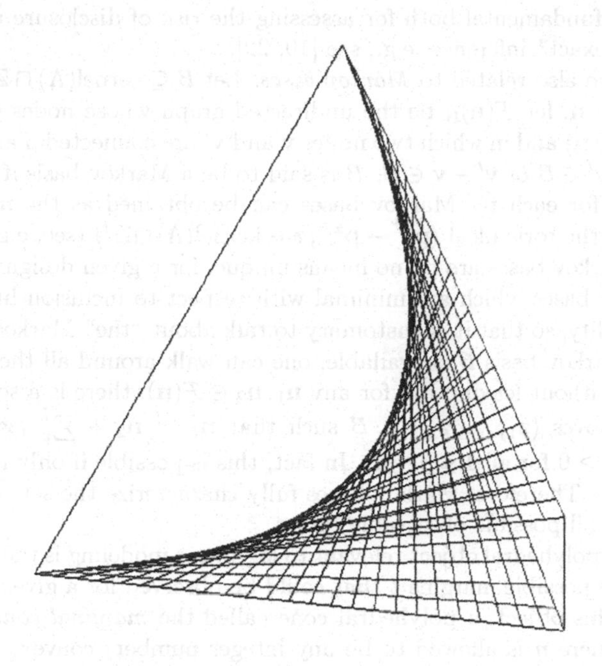

FIG. 1. *Surface of independence for the 2×2 table. The tetrahedron represents the set of all probability distributions $\mathbf{p} = (p_{11}, p_{12}, p_{21}, p_{22})$ for a 2×2 table, while the enclosed surface identifies the probability distributions satisfying the equation $p_{11}p_{22} = p_{12}p_{21}$, i.e. the toric variety for the model of independence.*

Sample Space. The sample space is the set of possible observable contingency tables, namely the set of all non-negative integer-valued arrays in $\mathbb{R}^{\mathcal{D}}$ with entries summing to n. Virtually all data-dependent objects encountered in the study of log-linear models are polyhedra (see, e.g., [44]).

In particular, for a given log-linear model and a set of margins $\mathbf{t}$, consider the polytope

$$\mathrm{P}_{\mathbf{t}} = \{\mathbf{x} \in \mathbb{R}^d : \mathbf{t} = \mathrm{A}\mathbf{x}, \mathbf{x} \geq 0\}$$

of all real-valued non-negative tables having the same margins $\mathbf{t}$, computed using the design matrix A. The set of all integer points inside $\mathrm{P}_{\mathbf{t}}$ is the *fiber* of $\mathbf{t}$. Formally, if $\mathbf{n}$ is any table such that $\mathbf{t} = \mathrm{A}\mathbf{n}$, then the fiber of $\mathbf{t}$ is $\mathcal{F}(\mathbf{n}) \equiv \{\mathbf{v} \in \mathbb{N}^d : \mathbf{n} - \mathbf{v} \in \mathrm{kernel(A)}\}$. Thus the fiber is the portion of the sample space associated with the same set of margins. Since margins are also minimal sufficient statistics, $\mathcal{F}(\mathbf{n})$ consists precisely of all the possible tables that would provide the same information on the unknown underlying vector $\mathbf{p}$ as the observed table. These tables form the support of the conditional distribution given the margins, often called the *exact distribution*, because it does not depend on the model parameters. Properties of the fiber are fundamental both for assessing the risk of disclosure and for conducting "exact" inference, e.g., see [10, 22].

Fibers are also related to *Markov bases.* Let $\mathcal{B} \subseteq \text{kernel}(\mathrm{A}) \cap \mathbb{Z}^d$ and, for each table $\mathbf{n}$, let $\mathcal{F}(\mathbf{n})_{\mathcal{B}}$ be the undirected graph whose nodes are the elements of $\mathcal{F}(\mathbf{n})$ and in which two nodes $\mathbf{v}$ and $\mathbf{v}'$ are connected if and only if either $\mathbf{v} - \mathbf{v}' \in \mathcal{B}$ or $\mathbf{v}' - \mathbf{v} \in \mathcal{B}$. $\mathcal{B}$ is said to be a Markov basis if $\mathcal{F}(\mathbf{n})_{\mathcal{B}}$ is connected for each $\mathbf{n}$. Markov bases can be obtained as the minimal generators of the toric ideal $\langle \mathbf{p}^{\mathbf{z}^+} - \mathbf{p}^{\mathbf{z}^-}, \mathbf{z} \in \text{kernel}(\mathrm{A}) \cap \mathbb{Z}^d \rangle$ (see, e.g. [10]). Although Markov bases are by no means unique, for a given design matrix A all Markov bases which are minimal with respect to inclusion have the same cardinality, so that it is customary to talk about "the" Markov basis. When the Markov basis $\mathcal{B}$ is available, one can walk around all the points in the fiber without leaving $\mathrm{P}_{\mathbf{t}}$: for any $\mathbf{n}_1, \mathbf{n}_2 \in \mathcal{F}(\mathbf{n})$, there is a sequence of Markov moves $(\mathbf{z}_1, \ldots, \mathbf{z}_L) \in \mathcal{B}$ such that $\mathbf{n}_1 = \mathbf{n}_2 + \sum_{i=1}^{L} \mathbf{z}_i$, with $\mathbf{n}_1 + \sum_{i=1}^{j} \mathbf{z}_i \geq 0$ for all $1 \leq j \leq L$. In fact, this is possible if only if $\mathcal{B}$ is a Markov basis. Therefore, Markov bases fully characterize the set of fibers associated to all possible observable tables.

Another polyhedral object relevant to log-linear modeling is the convex hull of all the possible margins $\mathbf{t}$ that could be observed for a given design matrix A. This object, a polyhedral cone called the *marginal cone*, is an unbounded (here n is allowed to be any integer number) convex set consisting of all the linear combinations of the columns of A with nonnegative coefficients, i.e.,

$$\mathrm{C}_{\mathrm{A}} = \{\mathbf{y} \in \mathbb{R}^m : \mathbf{y} = \mathrm{A}\mathbf{x}, \ \mathbf{x} \in \mathbb{R}^d, \mathbf{x} \geq 0\}.$$

As a result, the marginal cone comprises the set of all possible observable sufficient statistics and also their expectations, which are the points $\mathrm{A}(n\mathbf{p})$ with $\mathbf{p}$ ranging over the toric variety associated to A.

To summarize, we use the design matrix A and the marginal tables $\mathbf{t}$ to obtain geometric representations of the parameter and sample space for log-linear models. On one hand, A determines a system of polynomial equations that encode the dependencies among the random variables in the table. The solution set of these equations is the variety representing the parameter space as a compact subset of the simplex. On the other hand, every point $\mathbf{t}$ in the marginal cone C_{A} determines the polytope $\mathrm{P}_{\mathbf{t}}$, which in turn contains the fiber, i.e., the portion of the sample space that is relevant for both statistical inference and disclosure limitation.

2.2. Maximum likelihood estimation. The method of *maximum likelihood* (ML) is a standard approach to parameter estimation which chooses as the estimate of $\mathbf{p}$ the point in the simplex which maximizes the probability of the observed data $\mathbf{n}$ as a function of the parameter $\mathbf{p}$.

The maximum likelihood estimate (MLE) $\widehat{\mathbf{p}}$ is said to exist when $\widehat{\mathbf{p}}$ lies on the interior of the simplex. In this case, $\widehat{\mathbf{p}}$ is the unique point such that $\log \widehat{\mathbf{p}}$ is in the row range of A and satisfies the marginal constraints

$A\widehat{\mathbf{p}} = \frac{1}{n}\mathbf{t}$ (see, e.g., Haberman [30]). In particular, for decomposable log-linear models, the entries of the MLE are rational functions of the sample size n and the entries of the marginal vector $\mathbf{t}$, and we can compute them easily. Furthermore, the MLE of a decomposable model exists if and only if $\mathbf{t} > 0$. In contrast, for non-decomposable models, there is no closed-form expression for the MLE, which we can only evaluate numerically, and positivity of the observed marginals is only a necessary condition.

More generally, Eriksson et al. [20] show that existence of the MLE is equivalent to requiring that the marginal tables $\mathbf{t}$ belong to the interior of the marginal cone C_A. Not only is this condition simple to interpret, but it also reduces the problem of detecting nonexistence of the MLE to a linear optimization program over a convex set. Furthermore, we can use the same geometric formalism to characterize cases in which the MLE does not exist, a circumstance that occurs whenever $\mathbf{t}$ lies on the boundary of C_A. In fact, for any point $\mathbf{t}$ in the marginal cone, the polytope $P_{\mathbf{t}}$ containing the fiber is never empty and it intersects always the toric variety describing the model implied by A at one point $\widehat{\mathbf{p}}^e$ [35]. The first condition implies $A\widehat{\mathbf{p}}^e = \frac{1}{n}\mathbf{t}$ and the second that $\widehat{\mathbf{p}}^e$ is in the closure of the log-linear parameter space. If $\mathbf{t}$ is in the interior of C_A, then these are precisely the defining conditions for the MLE, hence $\widehat{\mathbf{p}}^e = \widehat{\mathbf{p}}$. If $\mathbf{t}$ is instead a point on the boundary of C_A, $\widehat{\mathbf{p}}^e$ will have some zero coordinates and will be the MLE of a restricted log-linear model at the boundary of the parameter space, an *extended MLE*. Notice that an extended MLE does not possess a representation as the logarithm of a point in the simplex belonging to the row range of A. Nonetheless, it is a well defined point on the toric variety. The extended MLE realizes, both statistically and geometrically, the connection between the sample space and the parameter space.

For example, the pattern of zero cells in Table 2(a) leads to the nonexistence of the MLE under the model of no-second-order interaction even though the entries in the margins are strictly positive. We obtained this table, along with others providing novel examples of "pathological" configurations of sampling zeros, using polymake (Gawrilow and Joswig [28]). The example in Table 2(b) is sparser than the one in Table 2(a) but the MLE exists in the former case and not in the latter.

2.3. Bounds for cell counts. Public/government agencies collect high quality, multi-dimensional census and survey data and they generate databases that they do not make fully accessible to the public. These categorical databases are often represented in tabular form as large and sparse contingency tables with small counts. The release of partial information from such databases is of public utility and typically consists of publishing marginal and conditional tables. Users can cumulate the released information and translate it in upper and lower bounds for cell counts. If these bounds are close for a particular cell, then an intruder could learn the

TABLE 2

(a): *Configurations of zero cells that cause nonexistence of the MLE (a facial set) for the model of no-second-order interaction without producing null margins.*
(b): *Example of a table with many sampling zeros but for which the MLE for the model of no-second-order interaction is well defined. Cells with entries + indicate positive entries. Source: Fienberg and Rinaldo [24].*

(a)

0	+	0	+	+	+	+	+	0
+	+	+	+	0	+	+	0	0
0	+	+	0	0	+	+	+	+

(b)

+	0	0	0	0	+	0	+	0
0	+	0	+	0	0	0	0	+
0	0	+	0	+	0	+	0	0

corresponding count and this might compromise the confidentiality offered to individual respondents.

Individuals or establishments that have an uncommon combinations of attributes will show up in the contingency table in cells with small counts of "1" or "2". A count of "1" might correspond to a population unique whose identity might be at risk unless the table contains a significant number of sample uniques that are not population uniques. In this case the "true" population uniques are concealed by the counts of "1" associated with "false" uniques. Counts of "2" can lead to similar violations if the intruder is one of the two persons and, for other small counts, we have the notion of inferential or probabilistic disclosure, i.e. the possibility of determining with a high degree of certainty individuals in the database. Such small counts also have the highest disclosure risk because their upper and lower bounds are close to the true value even for modest amounts of released information. This is especially true for large sparse categorical databases (e.g., census data) in which almost all counts are zero. These counts translate into small counts in the released marginals, which in turn lead to tight upper and lower bounds.

Consider a 2×2 contingency table with cell counts n_{ij} and row and column totals n_{i+} and n_{+j} respectively, adding to the total n_{++}. If we are given the row and column totals, then the well-known Fréchet bounds for the individual cell counts are:

$$\min(n_{i+}, n_{+j}) \geq n_{ij} \geq \max(n_{i+} + n_{+j} - n, 0) \text{ for } i = 1,2, j = 1,2. \quad (2.1)$$

The extra lower bound component comes from the upper bounds on the cells complementary to (i,j). These bounds have been widely exploited in the disclosure limitation literature and have served as the basis for the development of statistical theory on copulas [33]. The link to statistical theory comes from recognizing that the minimum component $n_{i+} + n_{+j} - n$ corresponds to the MLE of the expected cell value under independence, $n_{i+}n_{+j}/n$. The bounds are also directly applicable to $I \times J$ tables and essentially a related argument can be used to derive exact sharp bounds for multi-way tables whenever the marginal totals that are fixed correspond to the minimal sufficient statistics of a log-linear model that is *decomposable*.

Next we consider a $2 \times 2 \times 2$ table with cell counts n_{ijk}, and two way marginal totals n_{ij+}, n_{i+j}, and n_{+jk}, adding to the grand total n_{+++}. Given the 2-way margins, the bounds for the count in the (i, j, k) cell for $i = 1, 2$, $j = 1, 2$,and $k = 1, 2$, are

$$\begin{aligned} &\min(n_{ij+}, n_{i+k}, n_{+jk}, n_{ijk} + n_{\bar{i}\bar{j}\bar{k}}) \\ &\geq n_{ijk} \qquad (2.2) \\ &\geq \max(n_{i++} - n_{i+k} - n_{ij+}, n_{+j+} - n_{ij+} - n_{+jk}, n_{++k} - n_{i+k} - n_{+jk}, 0) \end{aligned}$$

where $(\bar{i}, \bar{j}, \bar{k})$ is the complementary cell to (i, j, k) found by replacing 1 by 2 and 2 by 1, respectively. Equation (2.3) consists of a combination of Fréchet bounds for each of the rows, columns, and layers of the full table plus an extra upper bound component $n_{ijk} + n_{\bar{i}\bar{j}\bar{k}}$.

Fienberg [21] suggested how to use this basic construction to get bounds for an $I \times J \times K$ table by considering all possible collapsed $2 \times 2 \times 2$ versions (based on all possible permutations of the subscripts). Dobra [11] refined this construction and developed a "generalized shuttle" algorithm, extending an idea in Buzzigoli and Giusti [2] in order to obtain sharp bounds by iterating between "naive" upper bound and lower bounds. This algorithm finds the sharp bounds for decomposable models without extensive computation, which is reduced in other special cases, e.g., see Dobra and Fienberg [16]. Nonetheless, it does not scale well for large sparse tables, c.f. the fact that non-integer bound problems are NP-hard (see [6, 7]).

2.4. Link between maximum likelihood estimates and cell bounds. We now use the algebraic geometric machinery to make the link between existence of the MLE and the computation of cell bounds explicit through the following result.

PROPOSITION 2.1. *For any lattice point* $\mathbf{t}$ *on the boundary of the marginal cone, let* $\widehat{\mathbf{p}}^{\mathrm{e}}$ *be the extended MLE and let* $\mathcal{Z}_{\mathbf{t}} = \{i : \widehat{\mathbf{p}}^{\mathrm{e}}_i = 0\}$ *be the set of cells for which the extended MLE is zero. Then, each table* $\mathbf{n}$ *in the fiber* $\mathrm{P}_{\mathbf{t}}$ *is such that* $\mathbf{n}_i = 0$ *for all* $i \in \mathcal{Z}_{\mathbf{t}}$.

The set $\mathcal{Z}_{\mathbf{t}}$ is uniquely determined by the margins $\mathbf{t}$ and corresponds to one of the many patterns of sampling zeros which invalidate the existence of the MLE. For the $2 \times 2 \times 2$ table and the model of no-2nd-order interaction, "impermissible" patterns of zeros occur in pairs of cells where $i + j + k$ is odd for one and even for the other. These cells can either be in adjacent cells adding to a margin zero in one of the two-way margins or they can be of the form (i,j,k) and $(\bar{i},\bar{j},\bar{k})$ for all possible values of i,j and k, c.f., Haberman [30]. Thus in particular, the MLE does not exist when $n_{ijk} + n_{\bar{i}\bar{j}\bar{k}} = 0$. The extra component of the upper bound for this non-decomposable model in equation (2.3) is thus inextricably bound up with the existence of MLEs.

The cells *not* belonging to $\mathcal{Z}_{\mathbf{t}}$ form a (random, as it depends on the random quantity $\mathbf{t}$) *facial set* (see [29, 35]). The cells with positive entries in Tables 2 (a) and Table 3 are examples of facial sets. Proposition 2.1 then shows that the determination of the facial set associated with a given marginal table is crucial, not only for computing the extended MLE, but also for calculating individual cell bounds, as it implies that one only needs to consider the cells in the facial set for performing the tasks of counting, sampling and optimizing over the fiber.

The determination of the facial sets is an instance of what in computational geometry is known as the face-enumeration problem: the enumerations of all the faces of a given polyhedron. Unfortunately, the number of solutions of this problem is often affected by a combinatorial explosion. conducted in [20], which suggests that the number of facial sets associated to a given hierarchical log-linear model may grow super-exponentially in the dimension of the table. As a result, complete enumeration of all the facial sets is impractical. A much more efficient solution consists in finding just the facial set corresponding to the observed margins $\mathbf{t}$, using the methods developed in [23].

Table 3 shows a 4×4×4 table for which the MLE for the model of no-2nd-order interaction is nonexistent. The zero entries correspond to the complement of the facial set for the minimal sufficient statistics, which as we noted above are the sets of all two-way margins. There are 123 tables in the fiber. Table 4 provides the cell bounds given the two-way marginals computed using the shuttle algorithm [11]. Proposition 2.1 implies that the upper bounds for the entries of the zero cells, which correspond to the set $\mathcal{Z}_{\mathbf{t}}$, is zero. The integer bounds for this table are shown in Table 4. Notice that the entry range for each cell is an interval of integer points, i.e., the fiber is connected, and thus the knowledge of cell bounds is very informative for assessing the risk of disclosure.

3. More on disclosure limitation: from margins to margins and conditionals. Because data from both marginal and conditional tables are widely reported as summary data from multi-way contingency tables, we need to understand how they differ from the sets of marginals in terms of the information they convey about the entries in the tables. For example, we want to see whether or not sets of marginal and conditional distributions for a contingency table are sufficient to uniquely identify the full joint distribution. When this is not the case we can protect against

TABLE 3
A 4×4×4 table with a pattern of zeros corresponding to a non empty Z_t and, therefore, to a nonexistent MLE for the model of no-second-order interaction. Source: Fienberg and Rinaldo [24].

(:,:,1)=				(:,:,2)=				(:,:,3)=				(:,:,4)=			
0	0	0	5	0	0	1	1	0	1	2	2	4	2	3	3
4	5	5	1	0	0	6	0	0	5	5	0	2	2	2	0
1	5	0	1	5	3	2	2	0	4	0	0	2	2	0	0
1	0	0	1	5	0	2	2	3	2	4	3	2	0	0	0

TABLE 4
Sharp integer bounds of the 4×4×4 Table 3.

(:,:,1)=				(:,:,2)=			
[0, 0]	[0, 0]	[0, 0]	[5, 5]	[0, 0]	[0, 0]	[0, 2]	[0, 2]
[2, 6]	[3, 7]	[5, 5]	[1, 1]	[0, 0]	[0, 0]	[6, 6]	[0, 0]
[0, 4]	[3, 7]	[0, 0]	[0, 2]	[4, 6]	[3, 3]	[2, 2]	[1, 3]
[0, 2]	[0, 0]	[0, 0]	[0, 2]	[4, 6]	[0, 0]	[1, 3]	[0, 4]
(:,:,3)=				**(:,:,4)=**			
[0, 0]	[0, 3]	[0, 4]	[1, 3]	[4, 4]	[0, 3]	[2, 5]	[3, 3]
[0, 0]	[3, 6]	[4, 7]	[0, 0]	[0, 4]	[0, 6]	[0, 3]	[0, 0]
[0, 0]	[4, 4]	[0, 0]	[0, 0]	[0, 4]	[0, 4]	[0, 0]	[0, 0]
[3, 3]	[2, 2]	[3, 5]	[2, 4]	[2, 2]	[0, 0]	[0, 0]	[0, 0]

disclosure further by replacing a marginal table by constituent marginal and conditional components.

At first blush, one might think that there would also be a direct role for marginals and conditionals in the estimation of Bayes net models described in algebraic terms by Garcia et al. [27], especially when all variables are categorical. In such settings, we replace an omnibus log-linear model by a series of linear log-odds or *logit* models corresponding to a factorization of the joint probabilities into a product of conditional distributions and there is an interesting issue of whether we can use any reduction via minimal sufficient statistics to exploit the ideas that follow.

To extend the ideas from the preceding section, we consider a subset a of $K = \{1, ..., k\}$ and denote by $\mathbf{n}_a$ and $\mathbf{p}_a$ the vectors of marginal counts and probabilities for the variables in a, respectively, of dimension $d_a = \prod_{i \in a} d_i$. If a and b are two disjoint subsets of K, we denote by $\mathbf{n}_{ab}$ and $\mathbf{p}_{ab}$ the corresponding marginal quantities for the variables in $a \cup b$. Provided that the entries of $\mathbf{n}_b$ are strictly positive, we define the array of *observed conditional proportions* of a given b by $\mathbf{n}_{a|b} = \mathbf{n}_{ab}/\mathbf{n}_b$, and the array of *conditional probabilities* of a given b by $\mathbf{p}_{a|b} = \mathbf{p}_{ab}/\mathbf{p}_b$, where $\mathbf{p}_b > 0$.

When $k = 2$, so that $a = \{1\}$, $b = \{2\}$, any of the following sets of distributions uniquely identifies the joint distribution: (1) $\mathbf{p}_{a|b}$ and $\mathbf{p}_{b|a}$, (2) $\mathbf{p}_{a|b}$ and $\mathbf{p}_b$, or (3) $\mathbf{p}_{b|a}$ and $\mathbf{p}_a$. Cell entries can be zero as long as we do not condition on an event of zero probability. Sometimes the sets $\{\mathbf{p}_{a|b}, \mathbf{p}_a\}$ and $\{\mathbf{p}_{b|a}, \mathbf{p}_b\}$ uniquely identify the joint distribution. The following result, due to Slavkovic [37] and described in Slavkovic and Fienberg [39], characterizes this situation for a generalization to a k-way table.

THEOREM 3.1. *(Slavkovic(2004)) Consider a k-way contingency table and a pair of matrices $\mathcal{T} = \{\mathbf{p}_{a|b}, \mathbf{p}_a\}$, where $a, b \subset \{1, \ldots, k\}$ and $a \cap b = \emptyset$. If the matrix of conditional probabilities has full rank, and $d_a \geq d_b$, then $\mathcal{T}$ uniquely identifies the marginal table of probabilities $\mathbf{p}_{ab}$.*

Often, there are multiple realizations of the joint distribution for $\mathbf{n}$, i.e., there is more than one table that satisfies the constraints imposed by them. Slavkovic [37], and Slavkovic and Fienberg [39] describe the calculation of bounds given an arbitrary collection of marginals and conditionals. They use linear programming (LP) and integer programming (IP) methods and discuss potential inadequacies in treating conditional constraints via LP. These results rely on the fact that any k-way table satisfying compatible marginals and/or conditionals is a point in a convex polytope defined by a system of linear equations induced by released conditionals and marginals.

If a cell count is small and the upper bound is close to the lower bound, an intruder intent on learning about individuals represented in a table of counts knows with a high degree of certainty that there is only a small number of them possessing the characteristics corresponding to that cell. This may pose a risk of disclosure of the identity of these individuals. For example, equation (2.1) gives the bounds when all that is released are the two one-way marginals in a two-way table. When we have a single marginal or a single conditional, the cell probability is bounded below by zero and above by a corresponding marginal or a conditional value. This translates into bounds for cell counts provided we know the sample size n. When we are working with released marginals we know n, but when we work only with conditionals this is an extra piece of information that needs to be provided to rescale the proportions to infer possible values for tables of counts.

When the conditions of Theorem 3.1 are not satisfied, we can obtain bounds for cell entries, and in some two-way cases there are closed-form solutions. Slavkovic [37] and Fienberg and Slavkovic [39] derive such closed-form solutions for $2 \times J$ tables. Then the corresponding marginal and conditional cell probabilities are denoted as $p_{i+} = \sum_j p_{ij}$ for $\mathbf{p}_a, i \in a$, $p_{+j} = \sum_i p_{ij}$ for $\mathbf{p}_b, j \in b$, and $p_{i|j} = p_{ij}/p_{+j}$ for $\mathbf{p}_{a|b}$, respectively. The closed form solutions for the p_{ij}'s rely on solving a linear programming problem via the simplex method and are given in the following theorem.

THEOREM 3.2. *Consider a* $2 \times J$ *contingency table and a pair of matrices* $\mathcal{T} = \{\mathbf{p}_{a|b}, \mathbf{p}_a\}, i \in a, j \in b$. *Let*

$$UB_1 = p_{i|j} \frac{p_{i+} - \max_{r \neq j}\{p_{i|r}\}}{p_{i|j} - \max_{r \neq j}\{p_{i|r}\}},$$

and

$$UB_2 = p_{i|j} \frac{p_{i+} - \min_{r \neq j}\{p_{i|r}\}}{p_{i|j} - \min_{r \neq j}\{p_{i|r}\}}$$

Then there are sharp upper bounds (UB) and lower bounds (LB) on the cell probabilities, p_{ij} *given by*

$$UB = \begin{cases} UB_1 & \text{if} \quad p_{i+} \geq p_{i|j} \\ UB_2 & \text{if} \quad p_{i+} < p_{i|j}, \end{cases} \tag{3.1}$$

and

$$LB = \begin{cases} \max\{0, UB_2\} \text{ s.t. } UB_2 \leq UB & \text{if} \quad p_{i+} \geq p_{i|j} \\ \max\{0, UB_1\} \text{ s.t. } UB_1 \leq UB & \text{if} \quad p_{i+} < p_{i|j} \end{cases} \tag{3.2}$$

Given a set of low dimensional tables with nicely rounded conditional probability values, these bounds will be sharp. For higher dimensions, linear approximations of the bounds may be far from the true sharp bounds for the table of counts, and thus may mask the true disclosure risk. To calculate sharp IP bounds, we need either nicely rounded conditional probability values, which rarely occur in practice, or we need the observed cell counts. Thus if the observed counts are considered sensitive, the database owner is the only one who can produce the sharp IP bounds in the case of the conditionals, e.g., see Slavkovic and Smucker [38].

Using algebraic tools for determining Gröbner and Markov bases, we can find feasible solutions to the constrained maximization/minimization problem. Some advantages of this approach are that (1) we obtain sharp bounds when the linear program approach fails, and (2) we can use it to describe all possible tables satisfying costraints imposed by information given by $\mathcal{T}$. In particular, a set of minimal Markov bases allows us to build a connected Markov chain and perform a random walk over all the points in the fiber that have the same fixed marginals and/or conditionals. Thus we can either enumerate or sample from the space of tables via Sequential Importance Sampling (SIS) or Markov Chain Monte Carlo (MCMC) sampling, e.g., see Chen et al. [5]. Some disadvantages of the algebraic approach are

TABLE 5
Cell counts for the dataset analyzed by Edwards [19]. Data publicly available at http://www.hypergraph.dk/.

			1				2				D
			1		2		1		2		E
			1	2	1	2	1	2	1	2	F
1	1	1	0	0	0	0	3	0	1	0	
		2	0	1	0	0	0	1	0	0	
	2	1	1	0	1	0	7	1	4	0	
		2	0	0	0	2	1	3	0	11	
2	1	1	16	1	4	0	1	0	0	0	
		2	1	4	1	4	0	0	0	1	
	2	1	0	0	0	0	0	0	0	0	
		2	0	0	0	0	0	0	0	0	
A	B	C									

that (1) calculation of Markov bases are computationally infeasible even for tables of small dimension, and (2) for conditionals, Markov bases are extremely sensitive to rounding of cell probabilities. Slavkovic [37] provides a description of calculation and structure of Markov bases given fixed conditionals for two-way tables. In this setting, the design matrix A does not rely on a log-linear model, but is a $m \times d$ *constraint matrix* A where d is the number of cells and m the number of linear constraints induced by $\mathcal{T}$ and the row of ones induced by the fixed sample size n; the corresponding m dimensional vector $\mathbf{t}^{-1} = \begin{bmatrix} n & \mathbf{0} \end{bmatrix}$. The reported results in the examples below rely on this methodology.

4. Two examples. Here we illustrate several of the results described in the preceding sections in the context of two examples of sparse contingency tables. The examples illustrate the limits of current computational approaches. To simplify the description of log-linear models we use the common short-hand notation for marginals, referring to the variables comprising them. For example, in a 3-way table involving variables A, B, and C, we denote the 3 2-way marginals as [AB], [AC], and [BC].

4.1. Example: Genetics data. Edwards [19] reports on an analysis of genetics data in the form of a sparse 2^6 contingency table given in Table 5. The six dichotomous categorical variables, labeled with the letters A-F, record the parental alleles corresponding to six loci along a chromosome strand of a barely powder mildew fungus, for a total of 70 offspring. The original data set, described in [4], included 37 loci for 81 offsprings, with 11 missing data—a rather sparse table.

For the model implied by fixing all the 2-way margins, the MLE is nonexistent because there is one null entry in the [AB] margins. Using

TABLE 6

Example of a 2^6 sparse table with a nonexistent MLE for the model specified by fixing all 2-way margins. The '+' signs indicate cells in a facial set corresponding to one facet of the marginal cone.

			1				2				D
			1		2		1		2		E
			1	2	1	2	1	2	1	2	F
1	1	1	0	+	+	+	0	0	+	0	
		2	0	0	+	+	0	0	0	0	
	2	1	0	0	0	0	0	0	0	0	
		2	0	0	+	0	0	0	0	0	
2	1	1	0	+	0	0	+	+	+	0	
		2	0	+	+	+	0	0	+	0	
	2	1	0	0	0	0	+	0	0	0	
		2	+	+	+	0	+	0	+	0	
A	B	C									

polymake [28], we found that the marginal cone for this model has 116,764 facets, each corresponding to a different pattern of sampling zeros causing nonexistence of the MLE , but only 60 of them correspond to null margins. Table 4.1 displays the facial set associated to one of these facets. The facet of the marginal cone containing in its relative interior the null margins observed for the Table 5 is, in turn, a polyhedral cone with 11,432 facets.

Table 7 shows the set $\mathcal{Z}_t$ obtained when we release three marginals: [ABCD][CDE][ABCEF]. The cells marked with a '0' correspond to values constrained to be zero, the '+' entries are cells for which the integer lower bound is positive, and the '+0' cells indicate a zero lower integer bound. The fiber in this case consists of 30 tables.

Proposition 2.1 is about LP and not ILP. In fact, null integer upper bounds for a set of cells do not imply that the MLE does not exist. In fact, Table 8 shows a set of sharp integer upper and lower bounds for a model for which the MLE exists despite the fact that there exist strictly positive real-valued tables in the fiber determined by the prescribed margins, there are cells, highlighted in boldface, for which no positive integer entries can occur. Although the MLE is well defined, many estimated cell mean values are rather small: 28 out of 64 values were less than 0.01 and only 14 were bigger than 1. For such small estimates, which correspond mostly to the cells for which the upper and lower integer bound is zero, the standard error is clearly very large. In fact, it is reasonable to expect that cells for which the maximal integer entries compatible with the fixed margins are zero will correspond to cell estimates with large standard errors. In this sense, cell bounds and maximum likelihood inference are strongly linked.

TABLE 7
Zero patterns when the margins [CDE], [ABCD], [ABCEF] are fixed.

			1				2				D
			1		2		1		2		E
			1	2	1	2	1	2	1	2	F
1	1	1	0	0	0	0	+	0	+	0	
		2	0	+	0	0	0	+	0	0	
	2	1	+0	+0	+	0	+	+0	+	0	
		2	+0	+0	0	+	+0	+	0	+	
2	1	1	+	+0	+	0	+0	+0	+0	0	
		2	+0	+	+0	+	+0	+0	+0	+0	
	2	1	0	0	0	0	0	0	0	0	
		2	0	0	0	0	0	0	0	0	
A	B	C									

TABLE 8
Exact upper and lower bounds for the model obtained by fixing all positive 3-way margins.

			1				2				D
			1		2		1		2		E
			1	2	1	2	1	2	1	2	F
1	1	1	[0,1]	**[0,0]**	[0,2]	**[0,0]**	[1,4]	[0,1]	[0,2]	[0,1]	
		2	**[0,0]**	[0,2]	**[0,0]**	[0,2]	[0,1]	[0,2]	[0,1]	[0,1]	
	2	1	[0,1]	**[0,0]**	[0,2]	**[0,0]**	[6,9]	[0,1]	[1,4]	[0,1]	
		2	**[0,0]**	[0,1]	**[0,0]**	[0,2]	[0,1]	[1,4]	[0,1]	[9,12]	
2	1	1	[15,18]	[0,1]	[0,4]	[0,1]	[0,1]	**[0,0]**	[0,1]	**[0,0]**	
		2	[0,1]	[2,5]	[1,2]	[1,5]	**[0,0]**	[0,1]	**[0,0]**	[0,1]	
	2	1	[0,1]	**[0,0]**	[0,2]	[0,1]	[0,1]	**[0,0]**	[0,1]	**[0,0]**	
		2	**[0,0]**	[0,1]	[0,1]	[0,2]	**[0,0]**	[0,1]	**[0,0]**	[0,1]	
A	B	C									

4.2. Example 2: Data from the 1993 U.S. current population survey. Table 9 describes data extracted from the 1993 Current Population Survey. Versions of these data have been used previously to illustrate several other approaches to confidentiality protection. The resulting 8-way table contains 2,880 cells and is based on 48,842 cases; 1185 cells, approximately 41%, contain 0 count cells. This is an example of a sparse table, too often present in practice, which poses significant problems for model fitting and estimation. Almost all lower level margins (e.g., 2-way margins) contain 0 counts. Thus the existence of maximum likelihood estimates is an issue. These zeros propagate into the corresponding conditional tables.

From the disclosure risk perspective we are interested in protecting cells with small counts such as "1" and "2". There are 361 cells with count of 1 and 186 with count of 2. Our task is to reduce a potential disclosure risk

TABLE 9
Description of variables in CPS data extract.

Variable	Label	Categories
Age (in years)	A	$< 25, 25 - 55, > 55$
Employer Type (*Empolyment*)	B	Gov, Pvt, SE, Other
Education	C	<HS, HS, Bach, Bach+, Coll
Marital status (*Marital*)	D	Married, Other
Race	E	White, Non-White
Sex	F	Male, Female
Hours Worked (*HrsWorked*)	G	$< 40, 40, > 40$
Annual Salary (*Salary*)	H	$< \$50K, \$50K+$

for at least 19% of our sample, while still providing sufficient information for a "valid" statistical analysis.

To alleviate estimation problems, we recoded variables C and G from 5 and 3 categories respectively to 2 categories each yielding a reduced 8-way table with 768 cells. This table is still sparse. There are 193 zero count cells, or about 25% of the cells. About 16% of cells have high potential disclosure risk; there are 73 cells with counts of 1 and 53 with counts of 2. For this table we find two reasonable log-liner models

1: [ABCFG][ACDFG][ACDGH][ADEFG],
2: [ACDGH][ABFG][ABCG][ADFG][BEFG][DEFG],

with goodness-of-fit statistics $G^2 = 1870.64$ with 600 degrees of freedom and $G^2 = 2058.91$ with 634 degrees of freedom, respectively.

Model 1 is a decomposable graphical log-linear model whose minimal sufficient statistics are the released margins. We first evaluate if these five-way marginal tables are safe to release by analyzing the number of cells with small counts. Most of the cell counts are large and do not seem to present an immediate disclosure risk. Two of the margins are potentially problematic. The marginal table [ABCFG] has 1 cell with a count of "5", while the margin [ACDGH] has a low count of "4" and two cells with a count of "8"; e.g., see Table 10. Even without any further analysis, most agencies would not release such margins. Because we are fitting a decomposable model, this initial exploratory analysis reveals that there will be at least one cell with a tight sharp upper bound of size "4". Now we investigate if these margins are indeed safe to release accounting for the log-linear model we can fit and the estimates they provide for the reduced and full 8-way tables.

Model 1 is decomposable and thus there are closed-form solutions for the bounds given the margins. Almost all lower bounds are 0. As expected from the analysis above, the smallest upper bound is 4. There are 16 such cells, of which 4 contain counts of "1" and rest contain "0". The next smallest upper bound is 5, for 7 "0" cell counts and for 1 cell with a count of "5". The 5 cells with counts of "1" have the highest risk of disclosure.

TABLE 10

Marginal table [ACDGH] from 8-way CPS table.

		A	1		2		3	
		C	1	2	1	2	1	2
D	G	H						
1	1	1	198	139	943	567	2357	2225
		2	11	19	240	715	1009	3781
	2	1	246	144	765	294	3092	2018
		2	8	14	274	480	1040	2465
2	1	1	2327	2558	835	524	2794	3735
		2	8	14	51	105	114	770
	2	1	1411	1316	617	359	3738	3953
		2	4	15	32	68	78	372

TABLE 11

Summary of the differences between upper and lower bounds for small cell counts in the full 8-way CPS table under Model 1 and under Model 2.

	Model 1						Model 2					
Difference Cell count	0	1	2	3	4	5	0	1	2	3	4	5
0	226	112	66	52	69	62	192	94	58	40	36	26
1	-	12	15	14	13	20	-	10	8	6	2	10
2	-	-	1	3	8	4	-	-	2	2	4	4
3	-	-	-	1	4	2	-	-	-	0	0	0

The next set of cells with a considerably high disclosure risk are cells with an upper bound of size 8. There are 32 such cells (23 contain counts of "0", 4 contain counts of "1", 3 contain counts of "2", and 2 contain counts of "3"). If we focus on count cells of "1" and "2", with the release of this model we directly identified 12 out of 126 sensitive cells.

If we fit the same model to the full 8-way table with 2,880 cells, there are 660 cells with difference in bounds less than or equal to 5, with all lower bounds being 0. Most of these are "0" cell counts; however, a high disclosure risk exists for 74 cells with counts of "1", 16 cells with cell count equal to "2", and 7 cells with counts of "3"; see the summary in Table 11. Thus releasing the margins corresponding to Model 1 poses a substantial risk of disclosure.

Model 2 is a non-decomposable log-linear model and it requires an iterative algorithm for parameter estimation and extensive calculations for computing the cell bounds. This model has 6 marginals as sufficient statistics. The 5 4-way margins all appear to be safe to release with the smallest count of size "46" appearing in cell (1,4,1,1) of the margin [ABFG], but 5-way margin [ACDGH] is still problematic.

We focus our discussion only on cells with small counts, as we did for the Model 1. Since Model 2 is non-decomposable, no closed-form solutions exist for cell bounds, and we must rely on LP, which may not produce sharp bounds. In this case this was not an issue. For the reduced 8-way table, all lower bounds are 0 and the minimum upper bound again is 4. There are 16 cells with an upper bound of 4, of which four cells have count "1", and the rest are "0". The next smallest upper bound is 8, and there are 5 such cells with counts of "1", 4 cells with counts of "2", and 3 cells with counts of "3". With these margins, in comparison to the released margins under Model 1, we have eliminated the effect of the margin [ABCFG], and reduced a disclosure risk for a subset of small cell counts; however, we did not reduce the disclosure risk for the small cell counts with the highest disclosure risk. For the full 8-way table, we compare the distribution of small cell bounds for the small cell counts under the two models; see Table 11. There are no cells with counts of "3" that have very tight bounds. For the cells with counts of "2", the number of tight bounds have not substantially decreased (e.g., 16 under Model 1 vs. 12 under Model 2), but there has been a significant decrease in the number of tight bounds for the cells with count of "1" (e.g., from 74 under Model 1 to 36 under Model 2).

In theory we could enumerate the number of possible tables utilizing algebraic techniques and software such as LattE [8] or sampling techniques such as MCMC and SIS [5]. Due to the large dimension of the solution polytope for this example, however, LattE is currently unable the execute the computation because the space of possible tables is extremely large. We have also been unable to fine-tune the SIS procedure to obtain a reasonable estimate except "infinity". While it is possible to find a Markov basis corresponding to the second log-linear model, utilizing those for calculating bounds and or sampling from the space of tables is also currently computationally infeasible.

Based on Model 1, the variables B and H are conditionally independent given the remaining 6 variables. Thus we can collapse the 8-way table to a 6-way table and carry out a disclosure risk analysis on it. The collapsed table has only 96 cells, and there are only three small cell counts, two of size "2" and one of size "3", that would raise an immediate privacy concern. Furthermore, we have collapsed over the two "most" sensitive and most interesting variables for statistical analysis: Type of Employer and Income. We do not pursue this analysis here but, if other variables are of interest, we could again focus on search for the best decomposable model. With various search algorithms and criteria, out of 32,768 possible decomposable models all searches converge to [ACFG][ADEFG], a model with a likelihood ratio chi-square of $G^2 = 144.036$ and 36 degrees of freedom.

First, we recall that given only one margin, the lower bounds are all zero and the upper bound corresponds to the values of the observed margin. For example, given [ACFG], the smallest upper bound is 502 for the cell (211112), but for the small counts of "2" the upper bounds are 677 for

TABLE 12
Upper bounds for [ACFG].

		A	1	2	3
C	F	G			
1	1	1	1128	740	1893
		2	552	502	2271
	2	1	1416	1329	4381
		2	1117	1186	5677
2	1	1	1462	525	3069
		2	677	334	3039
	2	1	1268	1386	7442
		2	812	867	5769

(121112) and 1117 for (111122). Table 12 includes all of these upper bound values. We can carry out a similar analysis for the [ADEFG] margin.

For the decomposable model above, the sharp bounds are easy to calculate. The upper bound for a particular cell is the minimum of the relevant marginal counts [ACFG] and [ADEFG]. The lower bound is the maximum between the 0 count and the value equal to the ([ACFG] + [ADEFG]-[AFG]), where the marginal [AFG] is a separator in the decomposable model. The smallest bound for the whole table is on the sensitive cell (121112) with the count of 2; the bound is [0,15]. If we consider releasing the corresponding conditionals, e.g., [C|AFG] and[DE|AFG], we would obtain the same sharp bounds! In fact, any conditional that corresponds to the same marginal table and involves all variables of the marginal table will produce the same sharp bounds, e.g., [AFG|C] would have the same IP bounds as [ACFG]. The same argument holds for [DE|AFG] and [ADEFG]. Moreover, since the model is decomposable we can consider the pieces separately.

The LP relaxation bounds are typically much wider for the conditionals than for the corresponding margins, however, and the space of tables is different and often larger. For this example, due to computational complexity we were unable the obtain the counts via LattE.

5. Some open statistical problems and their geometry.

We present below a list of open problems that are pertinent to the topics introduced in this article. We purposely formulate them in rather general terms, as all of the problems pose challenges that are both of theoretical and computational nature, and we believe are relevant to the mathematical and statistical audience jointly.

5.1. Patterns for non-existence of MLEs.

PROBLEM 5.1. *Suppose that d_1 is allowed to grow, while keeping k and $d_2, \ldots, d_k$ fixed. What is the smallest integer d such that the number*

TABLE 13

An example of a table with integer gap of 1.67 for the entry $(1,1,1,1)$ *with fixed 2-way margins. For that cell the integer upper bounds is* 0. *Incidentally, we note that the MLE is defined and that the fiber contains one table only. Source: Hosten and Sturmfels [31].*

		C = 1		C = 2	
A	**B**	D = 1	D = 2	D = 1	D = 2
1	1	0	1	1	0
	2	1	0	0	0
2	1	1	0	0	0
	2	0	0	0	1

of different patterns of zeros that lead to the non-existence of MLEs is constant for $d_1 > d$?

Eriksson et al. [20] posed a related conjecture, and wondered whether some finite complexity properties of the facial structure of the marginal cone is related to the finite complexity properties of Markov bases proved in Santos and Sturmfels [36].

Carrying the algebraic statistical results on existence of MLEs to large sparse contingency tables in a fashion that allows relatively easy computational verification has proven to be difficult. Thus we pose the following challenge:

PROBLEM 5.2. *Given a marginal cone* C_A *and a vector of observed margins* $\mathbf{t} = A\mathbf{n}$, *design an algorithm for finding the facial set associated to* $\mathbf{t}$ *that is computationally feasible for very large, sparse tables.*

5.2. Sharp bounds. Linear programming relaxation methods for the problem of computing integers bounds for cell entries will often produce fractional and non-sharp bounds, e.g., for Table 13.

In recent years, researchers became aware of the seriousness of the *integer gap* problem defined as the maximum difference between the real and integer bounds—see Hoşten and Sturmfels [31] and Sullivant [43]. A relevant example is presented in Table 13. The necessary and sufficient conditions for null integer gaps given in Sullivant [42] are the geometric counterpart to similar results by Dobra and Fienberg [13] already existing in the statistical literature.

The generalized shuttle algorithm propose by Dobra [11] is based on a succeeding branch-and-bound approach to enumerate all feasible tables, thus adjusting the shuttle bounds to be sharpest, and implemented a parallel version of the enumerating procedure which permits efficient computation for large tables. Dobra and Fienberg [16] provide further details and applications. Because this algorithm substitutes for the traversal of all lattice points in the convex polytope, and this involves aspects of the *exact distribution* without the probabilities, it is not surprising that there are links with the issues of maximum likelihood estimation. When the margins correspond to decomposable graphs, the bounds have explicit representation (see [13]) and the branch and bound component is not needed. When they correspond to reducible graphs this component effectively works on the reducible components!

PROBLEM 5.3. *Can we formalize the algebraic geometric links for the bounds problem in a form that scales to large sparse tables?*

5.3. Markov Bases complexity and gaps in the fiber. By a fiber with gaps we mean a fiber in which, for some of the cells entries, the range of integer values that are compatible with the given margins is not a sequence of consecutive integers, but instead contains gaps. In the presence of such gaps, knowledge of sharp upper and lower integer bounds for the cell entries cannot be a definitive indication of the safety of a data release. By construction, Markov bases preserve connectedness in the fiber and thus they encode the maximal degree of geometric and combinatorial complexity of all the fibers associated to a given log-linear model. De Loera and Onn [9] show that the complexity of Markov bases has no bound and thus there is little hope for an efficient computation of Markov bases for problems of even moderate size, from the theoretical point of view. They also show in a constructive way that fibers can have large (in fact, arbitrarily large) gaps, a fact that can be quantified by the degree of the Markov moves.

PROBLEM 5.4. *What combinatorial and geometric tools allow us to assess and quantify gaps in a given fiber?*

These open problems have important implications for disclosure limitation methodologies. Table 14 gives an example of an integer gap for a $3\times4\times6$ with fixed 2-way margins. The fiber contains only 2 feasible tables and the range entry for the first cell is $\{0,2\}$, thus exhibit a gap, since a value of 1 cannot be observed. In principle, it is possible to generate examples of tables with arbitrarily disconnected fiber.

Markov bases are *data independent*, in the sense that they prescribe all the moves required to guarantee connectedness for *any* fiber. However, there are instances which depend on the observed table $\mathbf{n}$, when in fact some (potentially many) of the moves are not needed: for example when the observed fiber contains gaps and when the observed margins lie on the boundary of the marginal cone C_A.

PROBLEM 5.5. *Is it possible to reduce the computational burden of calculating Markov bases by computing only the moves that are relevant to the observed fiber* P_t*?*

5.4. Bounds for released margins and conditionals. The degree of the Markov moves for given conditionals is arbitrary in a sense that it

TABLE 14
Margins of a 3×4×6 table with a gap in the entry range for the (1, 1, 1) *cell. Source: De Loera and Onn [9].*

$(:, j, k) =$	$(i, :, k) =$	$(i, j, :) =$
2 1 2 0 2 0 1 0 2 0 0 2 1 0 0 2 2 0 0 1 0 2 0 2	2 1 2 3 0 0 2 1 0 0 2 1 0 0 2 1 2 3	2 2 2 2 3 1 1 1 2 2 2 2

depends on the values of the conditional probabilities, that is it depends on the smallest common divisor of the actual cell counts for a given conditional. In the disclosure limitation context, the database owner who knows the original cell counts can calculate the sharp LP and IP bounds, and the Markov bases for given conditionals only subject to the computational limitations of current optimization and algebraic software. In practice, however, the conditional values are reported as real numbers and depending on the rounding point the LP/IP bounds, the moves and thus the fibers generated for a given table will differ.

PROBLEM 5.6. *Characterize the difference between these bases, the fibers, and bounds due to rounding of the observed conditional probabilities.*

We have observed that the gap in the bounds for the cell counts, and thus the degree of gap of a given fiber, is more pronounced with conditionals than with the corresponding marginals. While the sharp bounds on the cells maybe the same, the fibers differ in their content and size resulting in different conditional distributions on the space of tables. This has important implications for exact inference, and disclosure limitation methods as certain conditionals may release less information than the corresponding margin. Consider a k-way contingency table, and two fibers; one for a matrix of conditional values $\mathbf{p}_{a|b}$ and a second for the corresponding margin $\mathbf{p}_{ab}$ where $a, b \subset \{1, ..., k\}$. The size of the first fiber will be greater than equal to the size of the second fiber. Also the Markov bases for $\mathbf{p}_{a|b}$ will include all the elements of the moves from fixed margin $\mathbf{p}_{ab}$ plus some additional ones. These observations lead to the following challenge:

PROBLEM 5.7. *Characterize the difference of two fibers, one for a conditional probability array and the other for the corresponding margin, and thus simplify the calculation of Markov bases for the conditionals by using the knowledge of the moves of the corresponding margin.*

This is related to the characterization of Theorem 3.2 when $\mathbf{p}_{a|b}$ and $\mathbf{p}_a$ do not uniquely identify the marginal table $\mathbf{p}_{ab}$.

REFERENCES

[1] Y.M.M. Bishop, S.E. Fienberg, and P.W. Holland (1975). *Discrete Multivariate Analysis: Theory and Practice*, MIT Press, Cambridge, MA. Reprinted (2007), Springer-Verlag, New York.

[2] L. Buzzigoli and A. Giusti (1999). An algorithm to calculate the lower and upper bounds of the elements of an array given its marginals, in *Proceedings of the Conference on Statistical Data Protection*. Luxemburg: Eurostat, pp. 131–147.

[3] E. Carlini and F. Rapallo (2005). *The geometry of statistical models for two-way contingency tables with fixed odds ratios*, Rendiconti dell'Istituto di Matematica dell'Università di Trieste, **37**:71–84.

[4] S.K. Christiansen and H. Giese (19991). *Genetic analysis of obligate barley powdery mildew fungus based on RFPL and virulence loci*, Theoretical and Applied Genetics, **79**:705–712.

[5] Y. Chen, I.H. Dinwoodie, and S. Sullivant (2006). *Sequential importance sampling for multiway tables*, Annals of Statistics, **34**:523–545.

[6] L.H. Cox (2002). *Bounds on entries in 3-dimensional contingency tables subject to given marginal totals*, In J. Domingo-Ferrer (Ed.), *Inference Control in Statistical Databases*, Springer-Verlag LNCS 2316, pp. 21–33.

[7] L.H. Cox (2003). *On properties of multi-dimensional statistical tables*, Journal of Statistical Planning and Inference, **117**:251–273.

[8] J.A. De Loera, R. Hemmecke, J. Tauzer, and R. Yoshida (2004). *Effective lattice point counting in rational convex polytopes*, Journal of Symbolic Computation, **38**:1273–1302.

[9] J.A. De Loera and S. Onn (2006). *Markov bases of 3-way tables are arbitrarily complicated*, Journal of Symbolic Computation, **41**:173–181.

[10] P. Diaconis and B. Sturmfels (1998). *Algebraic algorithms for sampling from conditional distribution*, Annals of Statistics, **26**:363–397.

[11] A. Dobra (2002). *Statistical Tools for Disclosure Limitation in Multi-way Contingency Tables*. Ph.D. Dissertation, Department of Statistics, Carnegie Mellon University.

[12] A. Dobra (2003). *Markov bases for decomposable graphical models*, Bernoulli, **9**(6):1–16.

[13] A. Dobra and S.E. Fienberg (2000). *Bounds for cell entries in contingency tables given marginal totals and decomposable graphs*, Proceedings of the National Academy of Sciences, **97**:11885–11892.

[14] A. Dobra and S.E. Fienberg (2001). "Bounds for cell entries in contingency tables induced by fixed marginal totals with applications to disclosure limitation," Statistical Journal of the United Nations ECE, **18**:363–371.

[15] A. Dobra and S.E. Fienberg (2003). *Bounding entries in multi-way contingency tables given a set of marginal totals*, in Y. Haitovsky, H.R. Lerche, and Y. Ritov, eds., *Foundations of Statistical Inference: Proceedings of the Shoresh Conference 2000*, Physica-Verlag, pp. 3–16.

[16] A. Dobra and S.E. Fienberg (2008). *The generalized shuttle algorithm*, in P. Gibilisco, Eva Riccomagno, Maria-Piera Rogantin (eds.) *Algebraic and Geometric Methods in Probability and Statistics*, Cambridge University Press, to appear.

[17] A. Dobra, S.E. Fienberg, and M. Trottini (2003). *Assessing the risk of disclosure of confidential categorical data*, in J. Bernardo et al., eds., *Bayesian Statistics 7*, Oxford University Press, pp. 125–144.

[18] P. Doyle, J. Lane, J. Theeuwes, and L. Zayatz (eds.) (2001). *Confidentiality, Disclosure and Data Access: Theory and Practical Applications for Statistical Agencies*. Elsevier.

[19] D. Edwards (1992). *Linkage analysis using log-linear models*, Computational Statistics and Data Analysis, **10**:281–290.

[20] N. Eriksson, S.E. Fienberg, A. Rinaldo, and S. Sullivant (2006). *Polyhedral conditions for the non-existence of the MLE for hierarchical log-linear models,* Journal of Symbolic Computation, **41**:222–233.

[21] S.E. Fienberg (1999). *Fréchet and Bonferroni bounds for multi-way tables of counts With applications to disclosure limitation,* In *Statistical Data Protection, Proceedings of the Conference, Lisbon,* Eurostat, pp. 115–131.

[22] S.E. Fienberg, U.E. Makov, M.M. Meyer, and R.J. Steele (2001). "Computing the exact distribution for a multi-way contingency table conditional on its marginal totals," in A.K.M.E. Saleh, ed., *Data Analysis from Statistical Foundations: A Festschrift in Honor of the 75th Birthday of D. A. S. Fraser,* Nova Science Publishers, Huntington, NY, pp. 145–165.

[23] S.E. Fienberg and A. Rinaldo (2006). *Computing maximum likelihood estimates in log-linear models,* Technical Report **835**, Department of Statistics, Carnegie Mellon University.

[24] S.E. Fienberg and A. Rinaldo (2007). *Three centuries of categorical data analysis: log-linear models and maximum likelihood estimation,* Journal of Statistical Planning and Inference, **137**:3430–3445.

[25] S.E. Fienberg and A.B. Slavkovic (2004a). *Making the release of confidential data from multi-way tables count,* Chance, **17**(3):5–10.

[26] S.E Fienberg and A.B. Slavkovic (2005). *Preserving the confidentiality of categorical databases when releasing information for association rules,* Data Mining and Knowledge Discovery, **11**:155–180.

[27] L. Garcia, M. Stillman, and B. Sturmfels (2005). *Algebraic geometry for Bayesian networks,* Journal of Symbolic Computation, **39**:331–355.

[28] E. Gawrilow and M. Joswig (2005). *Geometric reasoning with polymake,* Manuscript available at `arXiv:math.CO/0507273`.

[29] D. Geiger, C. Meek, and B. Sturmfels (2006). *On the toric algebra of graphical models,* Annals of Statistics, **34**:1463–1492.

[30] S.J. Haberman (1974). *The Analysis of Frequency Data,* University of Chicago Press, Chicago, Illinois.

[31] S. Hoşten and B. Sturmfels (2006). *Computing the integer programming gap,* Combinatorica, **27**:367–382.

[32] S.L. Lauritzen (1996). *Graphical Models,* Oxford University Press, New York.

[33] R.B. Nelsen (2006). *An Introduction to Copulas.* Springer-Verlag, New York.

[34] A. Rinaldo (2005). *Maximum Likelihood Estimation for Log-linear Models.* Ph.D. Dissertation, Department of Statistics, Carnegie Mellon University.

[35] A. Rinaldo (2006). *On maximum likelihood estimation for log-linear models,* submitted for publication.

[36] F. Santos and B. Sturmfels (2003). *Higher Lawrence configurations,* J. Combin. Theory Ser. A, **103**:151–164.

[37] A.B. Slavkovic (2004). *Statistical Disclosure Limitation Beyond the Margins: Characterization of Joint Distributions for Contingency Tables.* Ph.D. Dissertation, Department of Statistics, Carnegie Mellon University.

[38] A.B. Slavkovic and B. Smucker (2007). *Calculating Cell Bounds in Contingency Tables Based on Conditional Frequencies,* Technical Report, Department of Statistics, Penn State University.

[39] A.B. Slavkovic and Fienberg, S. E. (2004). *Bounds for Cell Entries in Two-way Tables Given Conditional Relative Frequencies,* In Domingo-Ferrer, J. and Torra, V. (eds.), *Privacy in Statistical Databases,* Lecture Notes in Computer Science No. 3050, pp. 30–43. New York: Springer-Verlag.

[40] A.B. Slavkovic and S.E. Fienberg (2008). *The algebraic geometry of* 2×2 *contingency tables,* forthcoming.

[41] B. Sturmfels (1995). *Gröbner Bases and Convex Polytope,* American Mathematical Society, University Lecture Series, **8**.

[42] S. Sullivant (2006). *Compressed polytopes and statistical disclosure limitation,* Tohoku Mathematical Journal, **58**(3):433–445.

[43] S. SULLIVANT (2005). *Small contingency tables with large gaps,* SIAM Journal of Discrete Mathematics, **18**(4):787–793.
[44] G.M. ZIEGLER (1998). *Lectures on Polytopes,* Springer-Verlag, New York.

USING INVARIANTS FOR PHYLOGENETIC TREE CONSTRUCTION

NICHOLAS ERIKSSON*

Abstract. Phylogenetic invariants are certain polynomials in the joint probability distribution of a Markov model on a phylogenetic tree. Such polynomials are of theoretical interest in the field of algebraic statistics and they are also of practical interest—they can be used to construct phylogenetic trees. This paper is a self-contained introduction to the algebraic, statistical, and computational challenges involved in the practical use of phylogenetic invariants. We survey the relevant literature and provide some partial answers and many open problems.

Key words. Algebraic statistics, phylogenetics, semidefinite programming, Mahalonobis norm.

AMS(MOS) subject classifications. 92B10, 92D15, 13P10, 05C05.

1. Introduction. The emerging field of algebraic statistics (cf. [42]) has at its core the belief that many statistical problems are inherently algebraic. Statistical problems are often analyzed by specifying a *model*—a family of possible probability distributions to explain the data. In particular, many statistical models are defined parametrically by polynomials and thus involve algebraic varieties. From this point of view, one would hope that the ideal of polynomials that vanish on a statistical model would give statistical information about the model. This is not a new idea in statistics, indeed, tests based on polynomials that vanish on a model include the *odds-ratio*, which is based on the determinant of a two by two matrix. The polynomials that vanish on the statistical model have come to be known as the *(algebraic) invariants* of the model.

The field of phylogenetics provides important statistical and biological models with interesting combinatorial structure. The central problem in phylogenetics is to determine the evolutionary relationships among a set of *taxa* (short for taxonomic units, which could be species, for example). To a first approximation, these relationships can be represented using rooted binary trees, where the leaves correspond to the observed taxa and the interior nodes to ancestors. For example, Figure 1 shows the relationships between a portion of a gene in seven mammalian species.

Phylogenetic invariants are polynomials in the joint probability distribution describing sequence data that vanish on distributions arising from a particular tree and model of sequence evolution. The first of the invariants for phylogenetic tree models were discovered by Lake and Cavender-Felsenstein [36, 14]. This set off a flurry of work: in mathematics, generalizing these invariants (cf. [27, 21, 52]) and in phylogenetics, using these

*Department of Statistics, University of Chicago, Chicago, IL 60637 (eriksson@galton.uchicago.edu), partially supported by the NSF (DMS-06-03448).

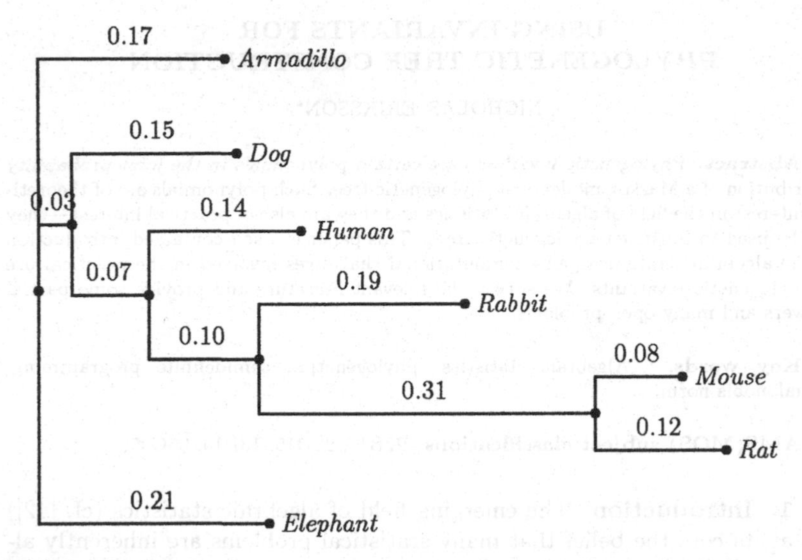

FIG. 1. *Phylogenetic tree for seven mammalian species derived from an alignment of a portion of the* HOXA *region (ENCODE region ENm010, see [17] and genome.ucsc.edu/encode). This tree was built using the dnaml maximum likelihood package from PHYLIP [24] on an alignment partially shown in Figure 2.*

invariants to construct trees (cf. [29, 38, 39, 44, 45]). However, the linear invariants didn't fare well in simulations [30] and the idea fell into disuse.

However, the study of phylogenetic invariants was revived in the field of algebraic statistics; the subsequent theoretical (cf. [2, 50, 12, 5]) and practical (cf. [13, 11, 18, 20, 34]) developments have given cause for optimism in using invariants to construct phylogenetic trees. There are benefits to these algebraic tools; however, obstacles in algebraic geometry, statistics, and computer science must be overcome if they are to live up to their potential. In this paper, we formulate and analyze some of the fundamental advantages and difficulties in using algebraic statistics to construct phylogenetic trees, describing the current research and formulating many open problems.

In geometric terms, the problem of phylogenetic tree construction can be stated as follows. We observe DNA sequences from n different taxa and wish to determine which binary tree with n leaves best describes the relationships between these sequences for a fixed model of evolution. Each of these trees corresponds to a different algebraic variety in $\mathbb{R}^{4^n}$. The DNA sequences correspond to a certain point in $\mathbb{R}^{4^n}$ as well. Picking the best tree means picking the variety that is closest to the data point in some sense. Since the data will not typically lie on the variety of any tree, we have to decide what is meant by "close".

Denote the variety (resp. ideal) associated to a tree T by $V(T)$ (resp. $I(T)$). Our main goal, then, is to understand how the polynomials in $I(T)$ can be used to select the best tree given the data. In order to answer this question, there are five fundamental obstacles.

1. Formulate an appropriate model of evolution and determine the invariants for that model, if possible in a form that can be evaluated quickly.
2. Choose a finite set of polynomials in $I(T)$ with good discriminating power between different trees.
3. Given a set of invariants for each tree, define a single score that can be used to compare different trees.
4. Since the varieties are in $\mathbb{R}^{4^n}$, each polynomial is in exponentially many unknowns. Thus even evaluating a single invariant could become difficult as n increases. This is in addition to the problem that the number of trees and the codimension of $V(T)$ increase exponentially. Phylogenetic algorithms are often used for hundreds of species. Can invariants become practical for large problems?
5. Statistical models are not complex algebraic varieties; they make sense only in the probability simplex and thus are real, semi-algebraic sets. This problem is more than theoretical—it is quite noticeable in simulated data (see Figures 6 and 7). Can semi-algebraic information be used to augment the invariants?

In the remainder of the paper, we will analyze these problems in detail, showing why they are significant and explaining some methods for dealing with them. The first problem (determining phylogenetic invariants) has been the focus of substantial research, thus we deal here with only the last four problems. We begin by introducing phylogenetics and constructing and using some phylogenetic invariants, then consider the four problems in order.

While in this paper we concentrate solely on the problem of constructing phylogenetic trees using invariants, we should note that phylogenetic invariants are interesting for many other reasons. On the theoretical side of phylogenetics, they have been used to answer questions about identifiability (e.g., [3, 37]). The study of the algebraic geometry arising from invariants has led to many interesting problems in mathematics [18, 9, 15].

2. Background. We give here a short, self-contained introduction to phylogenetics and phylogenetic invariants. For a more thorough survey of algebraic methods in phylogenetics, see [4]. Also see [23, 46] for more of the practical and combinatorial aspects of phylogenetics.

DEFINITION 2.1. Let X be a set of taxa. A *phylogenetic tree* T on X is a unrooted binary tree with $|X|$ leaves where each leaf is labelled with an element of X and each edge e of T has a weight, written t_e and called the *branch length*.

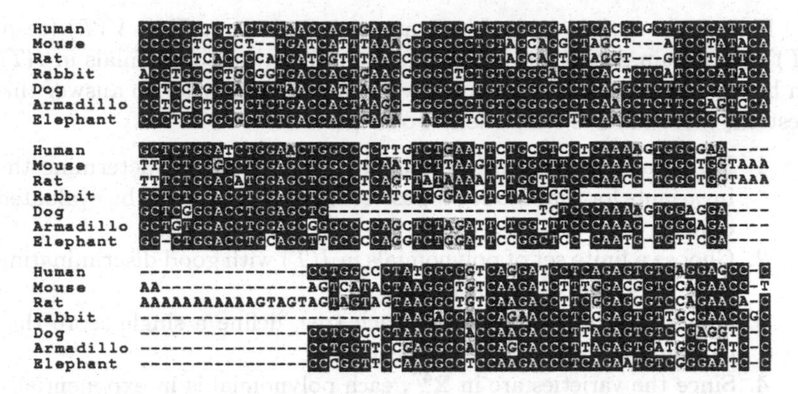

FIG. 2. *Multiple sequence alignment of length 180 from the HOXA region of seven mammalian genomes. Dashes indicate gaps; bases are colored according to their similarity across the species.*

While we include branch lengths in our definition of phylogenetic trees, our discussions about constructing trees are about only choosing the correct topology (meaning the topology of the labelled tree), not the branch lengths. While estimating branch lengths is relatively easy using maximum likelihood methods after a tree topology is fixed (e.g., with [54]), it is an interesting question whether algebraic ideas can be used to estimate branch lengths (see [48, 7] for algebraic techniques for estimating parameters in invariable-site phylogenetic models).

Phylogenetics depends on having identified *homologous characters* between the set of taxa. For example, historically, these characters might be physical characteristics of the organisms (for example, binary characters might include the following: are they unicellular or multicellular, cold-blooded or hot-blooded, egg-laying or placental mammals). In the era of genomics, the characters are typically single nucleotides or amino acids that have been inferred to be homologous (e.g., the first amino acid in a certain gene that is shared in a slightly different form among many organisms). For example, see Figure 2, which shows a multiple sequence alignment. We will throughout make the typical assumption that characters evolve independently, so that each column in Figure 2 is an independent, identically distributed (i.i.d.) sample from the model of evolution. While both DNA and amino acid data are common, we will work only with DNA and thus use the alphabet $\Sigma = \{\mathtt{A}, \mathtt{C}, \mathtt{G}, \mathtt{T}\}$.

We assume that evolution happens via a continuous-time Markov process on a phylogenetic tree (see [41] for general details about Markov chains). That is, along each edge e there is a length t_e and a rate matrix Q_e giving the instantaneous rates for evolution along edge e. Then $M_e = e^{Q_e t_e}$ is the transition matrix giving the probabilities of substitutions along the edge. In order to work with unrooted trees, we will assume that the Markov process is reversible, that is, $\pi_i M_e(i,j) = \pi_j M_e(j,i)$, where π is the stationary distribution of M_e. In order for $e^{Q_e t_e}$ to be stochastic, we must have $Q(i,i) \leq 0$, $Q(i,j) \geq 0$ for $i \neq j$, and $\sum_j Q(i,j) = 0$ for all i. Notice that since $\det(e^Q) = e^{\mathrm{tr}(Q)}$, we can recover the branch length from the transition matrix M_e as

$$t_e = \frac{1}{\mathrm{tr} Q_e} \log \det(M_e). \tag{2.1}$$

EXAMPLE 1. Let $Q_e = \begin{pmatrix} -1 & \frac{1}{3} & \frac{1}{3} & \frac{1}{3} \\ \frac{1}{3} & -1 & \frac{1}{3} & \frac{1}{3} \\ \frac{1}{3} & \frac{1}{3} & -1 & \frac{1}{3} \\ \frac{1}{3} & \frac{1}{3} & \frac{1}{3} & -1 \end{pmatrix}$ be the rate matrix for edge e, where the rows and columns are labeled by $\Sigma = \{\mathtt{A}, \mathtt{C}, \mathtt{G}, \mathtt{T}\}$. Then

$$M_e = e^{Q_e t_e} = \frac{1}{4} \begin{pmatrix} 1+3e^{-\frac{4}{3}t_e} & 1-e^{-\frac{4}{3}t_e} & 1-e^{-\frac{4}{3}t_e} & 1-e^{-\frac{4}{3}t_e} \\ 1-e^{-\frac{4}{3}t_e} & 1+3e^{-\frac{4}{3}t_e} & 1-e^{-\frac{4}{3}t_e} & 1-e^{-\frac{4}{3}t_e} \\ 1-e^{-\frac{4}{3}t_e} & 1-e^{-\frac{4}{3}t_e} & 1+3e^{-\frac{4}{3}t_e} & 1-e^{-\frac{4}{3}t_e} \\ 1-e^{-\frac{4}{3}t_e} & 1-e^{-\frac{4}{3}t_e} & 1-e^{-\frac{4}{3}t_e} & 1+3e^{-\frac{4}{3}t_e} \end{pmatrix}$$

This form of rate matrix is known as the Jukes-Cantor model [33]. For example, the probability of changing from an A to a C along edge e is given by $M_e(1,2) = \frac{1-e^{-\frac{4}{3}t_e}}{4}$.

Commonly used models that are more realistic than the Jukes-Cantor model include the Kimura 3-parameter model [35] where the rate matrices are of the form

$$\begin{pmatrix} \cdot & \gamma & \alpha & \beta \\ \gamma & \cdot & \beta & \alpha \\ \alpha & \beta & \cdot & \gamma \\ \beta & \alpha & \gamma & \cdot \end{pmatrix}$$

where $\cdot = -\gamma - \alpha - \beta$. See [42, Figure 4.7] for a description of many other possible models.

In order to obtain the joint distribution of characters at the leaves of the trees, we have to choose a root of the tree (arbitrarily, since the processes are time reversible), and run the Markov process down the edges of the tree, starting from a distribution of the characters at the root. The result is a joint probability distribution $p = (p_{\mathtt{A}\dots\mathtt{A}}, \dots, p_{\mathtt{T}\dots\mathtt{T}})$, and the important point is that the coordinates of p can be written as polynomials in the transition probabilities. That is, the model is specified parametrically by polynomials in the entries of M_e. We will forget about the specific form of the entries of $M_e = e^{Q_e t_e}$ and instead treat each entry of M_e as an unknown. Thus for the Jukes-Cantor model, we have two unknowns per edge: $\alpha_e = \frac{1+3e^{-\frac{4}{3}t_e}}{4}$ and $\beta_e = \frac{1-e^{-\frac{4}{3}t_e}}{4}$. This makes the algebraic model

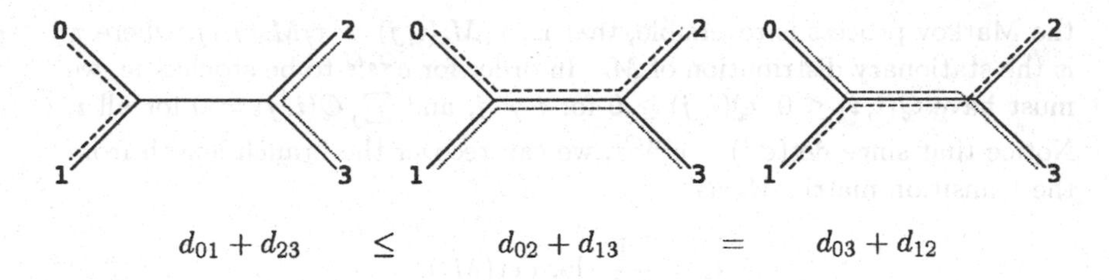

$d_{01} + d_{23} \quad \leq \quad d_{02} + d_{13} \quad = \quad d_{03} + d_{12}$

FIG. 3. *The four-point condition.*

more general than the statistical model (as it allows probabilities in the transition matrices to be negative or even complex). Although this allows algebraic tools to be used, we will see in Section 7 that it can be a disadvantage. Generally speaking, there are two types of phylogenetic models that have been thoroughly studied from the algebraic viewpoint: "group based" models such as the Jukes-Cantor and Kimura models, and variants of the general Markov model, in which no constraints are placed on the transition matrices.

In this paper, we define the *phylogenetic invariants* for a model of evolution and a tree to be the polynomials in the joint probabilities that vanish if the probabilities come from the model on the tree. For example, for a quartet tree (an unrooted binary tree with four leaves, see Figure 3) under the Jukes-Cantor model, $p_{\mathtt{AAAA}} - p_{\mathtt{CCCC}} = 0$, due to the symmetry built into the model. However, this polynomial doesn't differentiate between trees—it lies in the intersection of the ideals of the three quartet trees. Beware that there are two commonly used definitions of phylogenetic invariants. Originally, they were defined as polynomials that vanish on probability distributions arising from exactly one tree, so the above polynomial would be excluded. However, it is more algebraically convenient to take as invariants the full set of polynomials that vanish, as this forms an ideal. We spend the rest of this section deriving a particularly important polynomial.

A class of phylogenetic methods bypass working with the joint probability distribution and instead only estimate the distances between each pair of taxa. The goal then is to find a tree with branch lengths such that the distance along edges of the tree between pairs of leaves approximates the estimated pairwise distances. To use these distance methods, we first need a couple of definitions. We will concentrate in this paper on *quartet* trees, i.e., trees with four leaves. There are 3 different (unrooted, binary) trees on four leaves, we will write them (01 : 23), (02 : 13), and (03 : 12), corresponding to which pairs of leaves are joined together.

DEFINITION 2.2. *A dissimilarity map $d \in \mathbb{R}^{\binom{n}{2}}$ satisfies $d(i,j) = d(j,i) \geq 0$ and $d(i,i) = 0$. We say that d is a* tree metric *if there exists a phylogenetic tree T with non-negative branch lengths t_e such that for every* *pair i,j of taxa, $d(i,j)$ is the sum of the branch lengths t_e on the edges of T connecting i and j.*

PROPOSITION 2.1 (Four-point condition [10]). *A dissimilarity map d is a tree metric if and only if for every i,j,k, and l, the maximum of the three numbers*

$$d_{ij} + d_{kl}, \quad d_{ik} + d_{jl}, \quad \text{and} \quad d_{il} + d_{jk}$$

is attained at least twice.

EXAMPLE 2. Let us restrict our attention to a tree with four leaves, $(ij : kl)$. In this case, the four-point condition becomes (see Figure 3)

$$d_{ij} + d_{kl} \leq d_{ik} + d_{jl} = d_{il} + d_{jk}. \tag{2.2}$$

The equality in the four-point condition can be translated into a quadratic polynomial in the probabilities, however, we first have to understand how to transform the joint probabilities into distances. Distances can be estimated from data in a variety of ways (there are different formulas for the maximum likelihood estimates of the distances under different models of evolution, see [23, Chapter 13]). The formula for the general Markov model is the logdet distance, which mimics what we saw above (2.1), in that a transition matrix is estimated and the distance is taken to be the log of the determinant of this matrix.

Here we will use a simpler formula for the distance, under the Jukes-Cantor model (Example 1). The maximum likelihood estimate of the distance between two sequences under the Jukes-Cantor model is given by $d_{ij} = -\frac{1}{4}\log\left(1 - \frac{4m_{ij}}{3}\right)$ where m_{ij} is the fraction of mismatches between the two sequences, e.g.,

$$m_{12} = \sum_{i,j,k,l \in \{A,C,G,T\}, i \neq j} p_{ijkl}.$$

After plugging this distance into the four point condition, cancelling, and exponentiating, the equality in (2.2) becomes

$$\left(1 - \frac{4}{3}m_{ik}\right)\left(1 - \frac{4}{3}m_{jl}\right) - \left(1 - \frac{4}{3}m_{il}\right)\left(1 - \frac{4}{3}m_{jk}\right) = 0. \tag{2.3}$$

We will call this polynomial the four-point invariant. This construction is originally due to Cavender and Felsenstein [14].

Example 2 shows one of the first constructions on a phylogenetic invariant, in the same year as the discovery by Lake of linear invariants [36]. There is a linear change of coordinates on the probability distribution p under which $I(T)$ has a generating set of binomials. In particular, in these coordinates, a simple calculation shows that (2.3) becomes a binomial. Known as the Hadamard or Fourier transform [27, 52, 21, 50], this change of coordinates transforms the ideals of invariants for several models of evolution into toric ideals [49]. It should be emphasized, however, that this transform is only known to exist for group-based models.

The four-point invariant is a polynomial in the joint probabilities that vanishes on distributions arising from a certain quartet tree. Define the ideal $I_{\mathcal{M}}(T)$ of invariants for a model $\mathcal{M}$ of evolution on a tree T to be the set of all polynomials that are identically zero on all probability distributions arising from the model $\mathcal{M}$ on T. We will write only $I(T)$ when $\mathcal{M}$ is clear from context.

3. How to use invariants. The basic idea of using phylogenetic invariants is as follows. A multiple sequence alignment DNA alignment of n species gives rise to an empirical probability distribution $\hat{p} \in \mathbb{R}^{4^n}$. This occurs simply by counting columns of each possible type in the alignment, throwing out all columns that contain a gap (a "-" symbol). For example, Figure 2 has exactly one column that reads "CCCACCC" (the first) out of 107 gap-free columns total, so $\hat{p}_{\mathtt{CCCACCC}} = 1/107$.

Then if f is an invariant for tree T under a certain model of evolution, we expect $f(\hat{p}) \approx 0$ if (and generically only if) the alignment comes from the model on T. More precisely, where $\hat{p}_N$ is the empirical distribution after seeing N observations from the model on T, then $\lim_{N\to\infty} E(f(\hat{p}_N)) = 0$.

We thus have a rough outline of how to use phylogenetic invariants to construct trees:

1. Choose a model $\mathcal{M}$ of evolution.
2. Choose a set of invariants $\mathbf{f}_T$ for model $\mathcal{M}$ for each tree T with n leaves.
3. Evaluate each set of invariants at $\hat{p}$.
4. Pick the tree T such that $\mathbf{f}_T(\hat{p})$ is smallest (in some sense).

However, all of these steps contain difficulties: there are infinitely many polynomials to pick in exponentially many unknowns and exponentially many trees to compare. We will discuss step 2 in Section 4, step 3 in Section 6, and step 4 in Section 5. Selecting a model of evolution is difficult as well. There is, as always, a trade-off between biological realism (which could lead to hundreds of parameters per edge) and statistical usefulness of the model.

Since the rest of this paper will discuss difficulties with using invariants, we should stop and emphasize two especially promising features of invariants:

1. Invariants allow for arbitrary rate matrices. One major challenge of phylogenetics is that evolution does not always happen at one rate. But common methods for constructing trees generally assume a single rate matrix Q for all edges, leading to difficulties on data with heterogeneous rates. While methods have been developed to solve this problem (cf. [55, 25, 26]), it is a major focus of research.

In contrast, phylogenetic invariants allow for differing rate matrices within the chosen model on every edge (and in fact, even changing rate matrices along a single edge). The invariants for the Kimura 3-parameter model [35] have been shown to outperform neighbor-joining and maximum likelihood on quartet trees for heterogeneous simulated data [11]. To be fair, we should note that the invariants in this analysis were based on the correct model (i.e., the Kimura 3-parameter with heterogeneous rates, which the data was simulated from) while the maximum likelihood analysis used an incorrect model (with homogeneous rates) due to limitations in standard maximum likelihood packages.

2. Invariants perhaps can test individual features of trees. Researchers are frequently interested in the validity of a single edge in the tree. For example, we might wonder if human or dog is a closer relative to the rabbit. This amounts to wondering about how much confidence there is in the edge between the human-rabbit-mouse-rat subtree and the dog subtree in Figure 1. There are methods, most notably the bootstrap [22] and Bayesian methods (cf. [32]), which provide answers to this question, but there are concerns about their interpretation [28, 16, 40, 1].

As for phylogenetic invariants, the generators of the ideal $I(T)$ are, in many cases, built from polynomials constructed from local features of the tree. Thus invariants seem to be well suited to test individual features of a tree. For example, suppose we have n taxa. Consider a partition $\{A, B\}$ of the taxa into two sets. Construct the $|\Sigma|^{|A|} \times |\Sigma|^{|B|}$ matrix $\mathrm{Flat}_{A,B}(p)$ where the rows are indexed by assignments of Σ to the taxa in A and the columns by assignments of Σ to the taxa in B. The entry of the matrix in a given row and column is the joint probability of seeing the corresponding assignments of Σ to A and B. The following theorem is [6, Theorem 4] and deals with the general Markov model, where there are no conditions on the form of the rate matrices.

THEOREM 3.1 (Allman-Rhodes). *Let $\Sigma = \{0, 1\}$ and let T be a binary tree under the general Markov model. Then the 3×3 minors of $\mathrm{Flat}_{A,B}(p)$ generate $I(T)$ for the general Markov model, where we let A, B range over all partitions of $[n]$ that are induced by removing an edge of T.*

While the polynomials in Theorem 3.1 do not generate the ideal for the DNA alphabet, versions of these polynomials do vanish for any Markov model on a tree. A similar result also holds for the Jukes-Cantor model in Fourier coordinates; the following is part of [50, Thm 2].

THEOREM 3.2 (Sturmfels-Sullivant). *The ideal for the Jukes-Cantor DNA model is generated by polynomials of degree 1, 2, and 3 where the quadratic (resp. cubic) invariants are constructed in an explicit combinatorial manner from the edges (resp. vertices) of the tree.*

Although there are many challenges to overcome, the fact that phylogenetic invariants are associated to specific features of a tree provides hope that they can lead to a new class of statistical tests for individual features on phylogenetic trees.

4. Choosing powerful invariants. There are, of course, infinitely many polynomials in each ideal $I(T)$, and it is not clear mathematically or statistically which should be used in the set $\mathbf{f}_T$ of invariants that we test. For example, we might hope to use a generating set, or a Gröbner basis, or a set that locally defines the variety, or a set that cuts out the variety over $\mathbb{R}$. We have no actual answers to this dilemma, but we provide a few illustrative examples and suggest possible criteria for an invariant to be powerful. We will deal with the Jukes-Cantor model on a tree with four leaves; the 33 generators for this ideal can be found on the "small trees" website `www.shsu.edu/~ldg005/small-trees/` [13].

We believe that symmetry is an important factor in choosing powerful invariants. The trees with four leaves have a very large symmetry group: each tree can be written in the plane in eight different ways (for example, one tree can be written as $(01:23)$, $(10:23)$, ..., $(32:10)$), and each of these induces a different order on the probability coordinates p_{ijkl}. This symmetry group $(\mathbb{Z}_2 \times \mathbb{Z}_2 \times \mathbb{Z}_2)$ acts on the ideal $I(T)$ as well. In order that the results do not change under different orderings of the input, we should choose a set $\mathbf{f}_T$ of invariants that is closed (up to sign) under this action. After applying this action to the 33 generators, we get a set of 49 invariants. This symmetry will also play an important role in our metric learning algorithms in Section 5. See also [51] for a different perspective on symmetry in phylogenetics.

We begin by showing how different polynomials have drastically different behavior. Figure 4 shows the distribution of three of the invariants on data from simulations of alignments of length 1000 from the Jukes-Cantor model on $(01:23)$ for branch lengths ranging from 0.01 to 0.75 (similar to [30, 11, 20]). The histograms show the distributions for the simulated tree in white and the distributions for the other trees in gray and black. The four-point invariant (left) distinguishes nicely between the three trees with the correct tree tightly distributed around zero. It is correct almost all of the time. Lake's linear invariant (middle) also shows power to distinguish between all three trees, but distributions overlap much more—it is only correct about half of the time. The final polynomial seems to be biased towards selecting the wrong tree, even though it does not lie in $I(T)$ for either of the other two trees.

Figure 5 shows the performance of all the generators for this ideal on simulated data. The four-point invariant is the best, but the performance drops sharply with the other generators. Notably, the four-point invariant and several of the other powerful ones are unchanged (aside from sign) under the symmetries of the tree. While any invariant can be made symmetric by averaging, this behavior leads us to believe that invariants with a simple, symmetric form may be the best choice.

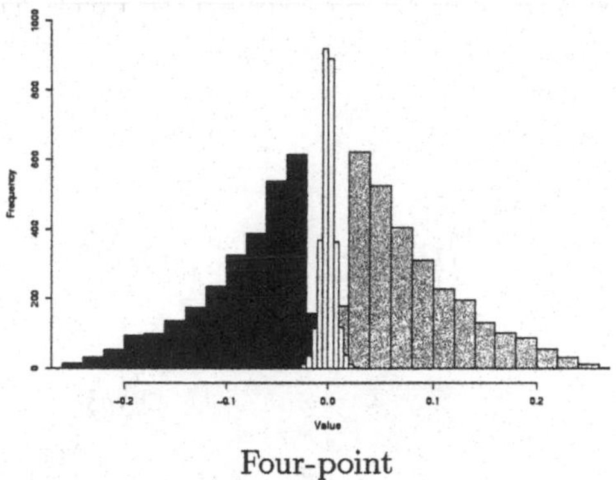

Four-point

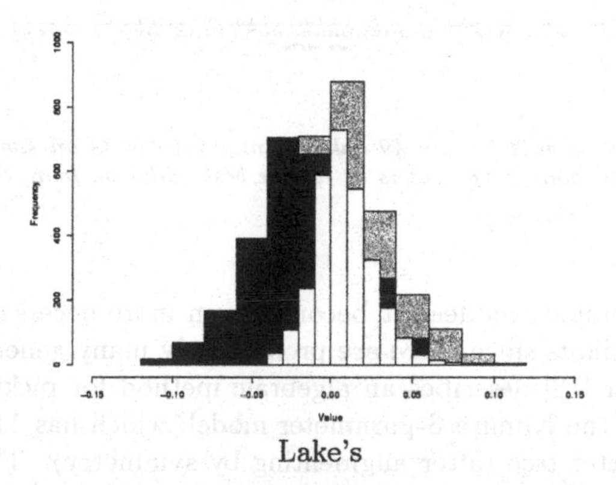

Lake's

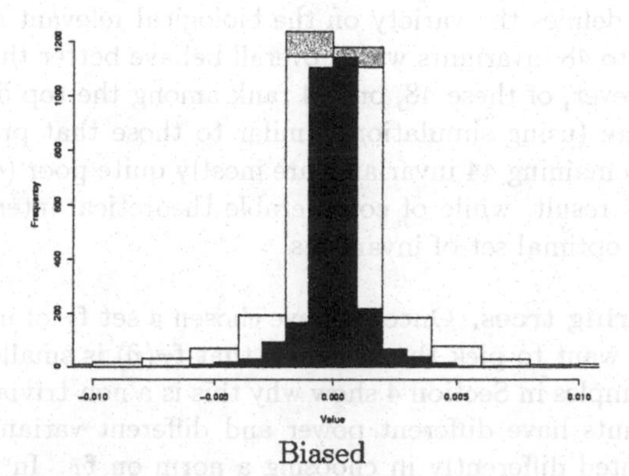

Biased

FIG. 4. *Distributions of three invariants (the four-point invariant, Lake's linear invariant, and a biased invariant) on simulated data. The white histogram corresponds to the correct tree, the black and gray are the other two trees. The invariants have quite different variances and performance.*

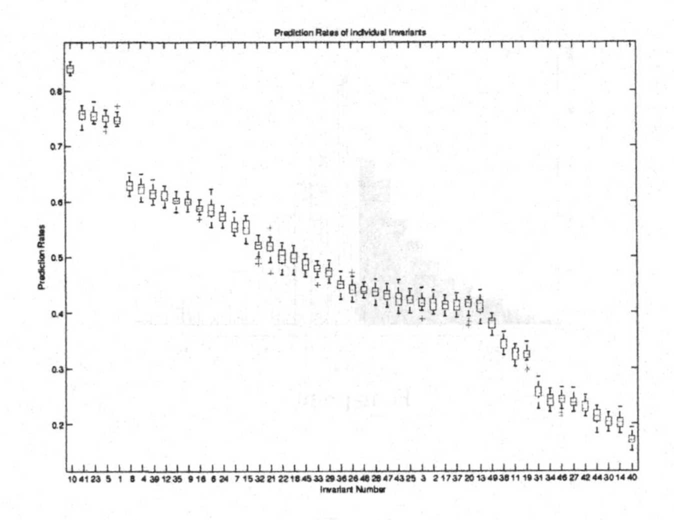

FIG. 5. *Prediction rate for the 49 Jukes-Cantor invariants on simulated data of length 100. The four-point invariant is by far the best, although four other invariants are quite good.*

For more complex models, it becomes even more necessary to pick a good set of invariants since there are prohibitively many generators of the ideal. The paper [12] describes an algebraic method for picking a subset of invariants for the Kimura 3-parameter model, which has 11612 generators for the quartet tree (after augmenting by symmetry). Their method constructs a set of invariants which is a local complete intersection, and shows that this defines the variety on the biological relevant region. This reduces the list to 48 invariants which overall behave better than all 11612 invariants. However, of these 48, only 4 rank among the top 52 invariants in prediction rate (using simulations similar to those that produced Figure 5) and the remaining 44 invariants are mostly quite poor (42% average accuracy). This result, while of considerable theoretical interest, doesn't seem to give an optimal set of invariants.

5. Comparing trees. Once we have chosen a set $\mathbf{f}_T$ of invariants for each tree T, we want to pick the tree such that $\mathbf{f}_T(\hat{p})$ is smallest (in some sense). The examples in Section 4 show why this is a non-trivial problem—different invariants have different power and different variance and thus should be weighted differently in choosing a norm on $\mathbf{f}_T$. In this section, we briefly describe an approach to normalizing the invariants to enable us to choose a tree. It is based on machine learning and was developed in [20]. It leads to large improvements over previous uses of invariants; however, it is computationally expensive and dependant on the training data. It can be thought of as finding the best single invariant which is a quadratic form in the starting set $\mathbf{f}_T$ of invariants.

There are also standard asymptotic statistical tools such as the delta method for normalizing invariants to have a common mean and variance. They have the disadvantage of depending on a linear approximation and asymptotic behavior, which might not be accurate for small datasets. Fortunately, the varieties for many phylogenetic models are smooth in the biologically significant region [12], so linear approximations may work well.

This problem is somewhat easier when we are choosing between different trees with the same (unlabelled) topology, for example, the three quartet trees. In this case the different ideals $I(T)$ are the same under a permutation of the unknowns, and thus we are comparing the same sets of polynomials (as long as the chosen set $\mathbf{f}_T$ is closed under the symmetries of T). For this reason, we will concentrate on the case of quartet trees and write $T_1 = (01:23)$, $T_2 = (02:13)$, and $T_3 = (03:12)$.

Let $\hat{p}(\theta)$ be an empirical probability distribution generated from a phylogenetic model on tree T_1 with parameters θ. Choose m invariants $\mathbf{f}_i$ $(i = 1, 2, 3)$ that are closed under the symmetries of T_1. We want a norm $\| \ \|_*$ such that

$$\|\mathbf{f}_1(\hat{p}(\theta))\|_* < \min\left(\|\mathbf{f}_2(\hat{p}(\theta))\|_*, \|\mathbf{f}_3(\hat{p}(\theta))\|_*\right) \tag{5.1}$$

is typically true, i.e., the true tree should have its associated invariants closer to zero than others on the relevant range of parameter space.

In order to scale and weigh the individual invariants, the algorithm seeks to find an optimal $\| \ \|_*$ within the class of Mahalanobis norms. Recall that given a positive (semi)definite matrix A, the Mahalanobis (semi)norm $\| \cdot \|_A$ is defined by

$$\|x\|_A = \sqrt{x^t A x}.$$

Since A is positive semidefinite, it can be written as $A = UDU^t$ where U is orthogonal and D is diagonal with non-negative entries. Thus the positive semidefinite square root $B = U\sqrt{D}U^t$ is unique. Now since $\|x\|_A^2 = x^t A x = (Bx)^t(Bx) = \|Bx\|^2$, learning such a metric is the same as finding a transformation of the space of invariants that replaces each point x with Bx under the Euclidean norm, i.e., a rotation and shrinking/stretching of the original x.

Now suppose that Θ is a finite set of parameters from which training data $\mathbf{f}_1(\hat{p}(\theta)), \mathbf{f}_2(\hat{p}(\theta)), \mathbf{f}_3(\hat{p}(\theta))$ is generated for $\theta \in \Theta$. As we saw above, each of the eight possible ways of writing each tree induces a permutation of the coordinates p_{ijkl} and thus induces a signed permutation of the coordinates of each $\mathbf{f}_i(\hat{p}(\theta))$. Write these permutations in matrix form as $\pi_1, \ldots, \pi_8$. Then the positive semidefinite matrix A must satisfy the symmetry constraints $\pi_i A = A\pi_i$ which are hyperplanes intersecting the semidefinite cone. This symmetry condition is crucial in reducing the computational cost. Given training data, the following optimization problem finds a good metric on the space of invariants.

$$\begin{aligned}
\text{Minimize:} \quad & \sum_{\theta\in\Theta} \xi(\theta) + \lambda \mathrm{tr} A \\
\text{Subject to:} \quad & \|\mathbf{f}_1(\hat{p}(\theta))\|_A^2 + \gamma \leq \|\mathbf{f}_i(\hat{p}(\theta))\|_A^2 + \xi(\theta) \quad (\text{for } i = 2, 3), \\
& \pi_i A = A\pi_i \quad (\text{for } 1 \leq i \leq 8), \qquad (5.2) \\
& \xi(\theta) \geq 0, \quad \text{and} \\
& A \succeq 0,
\end{aligned}$$

where $A \succeq 0$ denotes that A is a positive semidefinite matrix, so this is a semidefinite programming problem. There are several parameters involved in this algorithm: $\xi(\theta)$ for $\theta \in \Theta$ are slack-variables measuring the violation of (5.1), γ is a margin parameter that lets us strengthen condition (5.1), and λ is a regularization parameter to keep the trace $\mathrm{tr}A$ small while keeping A as low rank as possible. It tries to find a positive semidefinite A at a trade-off between the small violation of (5.1) and small trace A.

The metric learning problem (5.2) was inspired by some early results on metric learning algorithms [53, 47], which aim to find a Mahalanobis (semi)norm such that the mutual distances between similar examples are minimized while the distances across dissimilar examples or classes are kept large. If it becomes too computationally expensive, we can restrict A to be diagonal, which reduces the problem to a linear program. See [20] for details and simulation results. The learned metrics significantly improve on any of the individual invariants as well as on unweighted norms. The semidefinite programming algorithm is computationally feasible for approximately 100 invariants, and the choice of powerful invariants is important.

6. Efficient computation. At first glance, the problem of using invariants seems intractable for large trees for the simple reason that the number of unknowns grows exponentially with the number of leaves. However, the problem is not as bad as it may seem. Phylogenetic analyses typically use DNA sequences at most thousands of bases long, which means that the empirical distribution $\hat{p} \in \mathbb{R}^{4^n}$ will be extremely sparse even with a relatively small number of taxa.

Also the data can be sparse, this will not help unless we can write down the invariants in sparse form. If the polynomials can be written down in an effective way, they can be evaluated quickly. The determinantal form of the invariants in Theorem 3.1 provide such a form; see [18] for an algorithm to construct phylogenetic trees in polynomial time using these invariants and numerical linear algebra. Also see [2] for invariants that are (in some sense) determinantal. It seems that determinantal conditions could be particularly useful, so we suggest Problem 8.5 to computational commutative algebraists (see also [19]).

Unfortunately, for group-based models the polynomials are only sparse when written in Fourier coordinates, and the Fourier transform takes a sparse distribution p and produces a completely dense vector q. Many of the invariants are determinantal in Fourier coordinates, but since the matrices are dense, they are difficult to write down. Can these polynomials be evaluated efficiently?

7. Positivity. Recall that the four point condition (Proposition 2.1 and Figure 3) says that for a dissimilarity map d arising from the quartet tree $(01:23)$,

$$d_{01} + d_{23} \leq d_{02} + d_{13} = d_{03} + d_{12}. \tag{7.1}$$

This is true since the right two sums traverse the inner edge of the tree twice (Figure 3). We saw in Example 2 that the equality in (7.1) translates to a quadratic invariant. However, notice that if the interior branch of the tree has negative length, the equality is still satisfied, but the inequality changes so that $d_{01} + d_{23}$ is now larger than the other two sums.

The widely used neighbor-joining algorithm [43], when restricted to four taxa, reduces to finding the smallest of the three sums in the four-point condition. That is, neighbor-joining on a quartet tree involves estimating the distances as in Section 2 and then returning the tree $(ij : kl)$ that minimizes $d_{ij} + d_{kl}$. If instead we used the quadratic invariant arising from the equality in the four point condition, we would have an invariant based method that simply returns the tree $(ij : kl)$ that minimizes $|d_{ik} + d_{jl} - d_{il} - d_{jk}|$. We saw in Section 4 that this invariant performs quite well compared to the other generators of the Jukes-Cantor model. However, it compares poorly to the neighbor-joining criterion in the following way.

Figure 6 shows the difference between these two selection criteria on a projection of the six dimensional space of dissimilarity maps $\mathbb{R}^{\binom{4}{2}}$ to two dimensions. The three black lines are the projections of distances arising from the three different trees. Moving out from the center along these lines corresponds to increasing the length of the inner edge in the tree.

Geometrically, neighbor-joining can be thought of as finding the closest tree metric (a point on a black half-ray) to a dissimilarity map. The four-point invariant can't distinguish negative inner branch length (the dotted black line) and thus is much less robust than neighbor-joining. Notice that even when it picks the wrong tree, it can pick the *wrong* wrong tree—that is, the one least supported by the data. It is less robust than neighbor-joining in the "Felsenstein zone" [31], which corresponds to the region close to the center, where the inner edge is very short.

Simulations (see Figure 7) show that building trees by evaluating this quadratic invariant does not perform nearly as well as neighbor-joining. This is because many simulations with a short interior branch tend to return metrics that seem to come from trees with negative inner branch lengths.

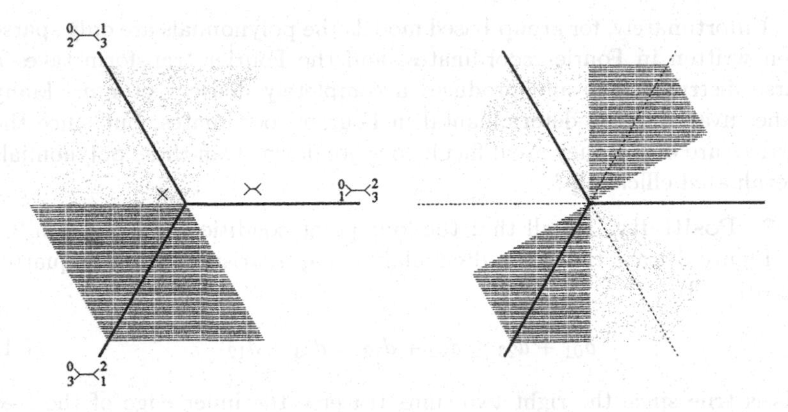

FIG. 6. *The selection criteria for neighbor-joining (left) and the four-point invariant (right) projected to two dimensions. The colored/shaded regions show which dissimilarity maps are matched to which trees by the two algorithms. The white/unshaded area corresponds to tree* (01 : 23), *the red/solid area to tree* (02 : 13) *and the blue/striped area to* (03 : 12).

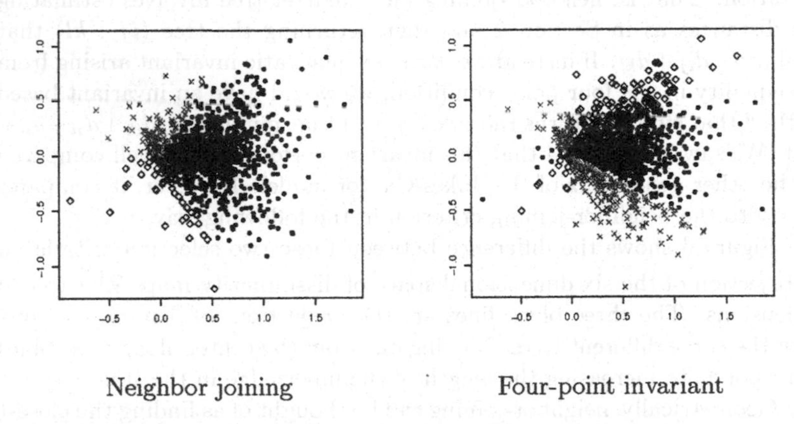

FIG. 7. *Illustration of Figure 6 on simulated data. Simulated alignments from tree* (01 : 23) *of length 100 were created for randomly chosen branch lengths between* 0.01 *and* 0.75. *Distances were estimated using the Jukes-Cantor model and projected onto two dimensions in the same way as in Figure 6. Trees were built from the distances using both neighbor-joining and the four-point invariant. Black circles correspond to distances assigned tree* (01 : 23), *red x's to tree* (02 : 13), *and blue diamonds to tree* (03 : 12).

This seems to be a large blow to the method of invariants: even the most powerful invariant on our list in Section 4 doesn't behave as well as this simple condition. However, it can be easily seen that testing the inequality is equivalent to testing the signs of the invariant instead of the absolute value, which leads us to ask if invariants can provide a way to discover conditions similar to that used in neighbor-joining (see Problem 8.7). The original paper of Cavender-Felsenstein [14] also suggested using inequalities, although no one seems to have followed up on this idea.

8. Open problems.

PROBLEM 8.1. *Can algebraic ideas be used to estimate branch lengths and other parameters in phylogenetic trees? See [48, 7] for algebraic techniques for estimating parameters in invariable-site phylogenetic models.*

PROBLEM 8.2. *Investigate the behavior of individual invariants on data from trees with heterogeneous rates. Are the best invariants the same ones that are powerful for homogeneous rates?*

PROBLEM 8.3. *Can asymptotic statistical methods be practically used to normalize invariants? Do they give any information about the power of individual invariants?*

PROBLEM 8.4. *Do the metrics constructed by the machine learning algorithm in Section 5 shed any light on the criteria for invariants to be powerful?*

PROBLEM 8.5. *Define the "determinantal closure" of an ideal I and develop algorithms to calculate it. See also [19].*

PROBLEM 8.6. *For group-based models, does Fourier analysis provide a method to efficiently evaluate polynomials in the Fourier coordinates without destroying the sparsity of the problem? Note that many of the invariants are determinental in Fourier coordinates.*

PROBLEM 8.7. *Are there other phylogenetic invariants (say for quartet trees under the Jukes-Cantor model) "similar" to the four-point invariant? We suggest the following conditions:*

1. *Be fixed (up to sign) under the* $\mathbb{Z}_2 \times \mathbb{Z}_2 \times \mathbb{Z}_2$ *symmetries of the quartet tree.*
2. *Have the following sign condition:* $\pm f(p) > 0$ *for all p from* T_2 *and* T_3 *(with perhaps a different choice of sign for* T_2 *and* T_3*). See for example, the symmetries of the left subfigure in Figure 4.*

Beware that results such as [8] on the uniqueness of the neighbor-joining criterion place some constraints on whether we can hope to find invariants mimicking this behavior.

Acknowledgments. We thank E. Allman, M. Casanellas, M. Drton, L. Pachter, J. Rhodes, and F. Sottile for enlightening discussions about these topics at the IMA. We are very appreciative of the hospitality of the IMA during our visit and of very helpful comments from two referees.

REFERENCES

[1] M. Alfaro, S. Zoller, and F. Lutzoni, *Bayes or Bootstrap? A Simulation Study Comparing the Performance of Bayesian Markov Chain Monte Carlo Sampling and Bootstrapping in Assessing Phylogenetic Confidence*, Molecular Biology and Evolution, **20** (2003), pp. 255–266.

[2] E.S. Allman and J.A. Rhodes, *Phylogenetic invariants for the general Markov model of sequence mutation*, Mathematical Biosciences, **186** (2003), pp. 113–144.

[3] ——, *The identifiability of tree topology for phylogenetic models, including covarion and mixture models*, J. Comput. Biol., **13** (2006), pp. 1101–1113 (electronic).

[4] ——, *Molecular phylogenetics from an algebraic viewpoint*, Statistica Sinica, **17** (2007), pp. 1299–1316.

[5] ——, *Phylogenetic invariants*, in Reconstructing evolution: new mathematical and computational advances, O. Gascuel and M. Steel, eds., Oxford University Press, 2007, ch. 4.

[6] ——, *Phylogenetic ideals and varieties for the general Markov model*, Advances in Applied Mathematics (2007, in press).

[7] ——, *Identifying evolutionary trees and substitution parameters for the general Markov model with invariable sites*, Mathematical Biosciences, **211** (2008), pp. 18–33.

[8] D. Bryant, *On the Uniqueness of the Selection Criterion in Neighbor-Joining*, Journal of Classification, **22**(2005), pp. 3–15.

[9] W. Buczyńska and J.A. Wiśniewski, *On geometry of binary symmetric models of phylogenetic trees*, J. Eur. Math. Soc. (JEMS), **9** (2007), pp. 609–635.

[10] P. Buneman, *A note on the metric properties of trees*, J. Combinatorial Theory Ser. B, **17** (1974), pp. 48–50.

[11] M. Casanellas and J. Fernández-Sánchez, *Performance of a new invariants method on homogeneous and nonhomogeneous quartet trees*, Mol. Biol. Evol., **24** (2007), pp. 288–293.

[12] ——, *Geometry of the Kimura 3-parameter model*, Adv Appl Math, to appear, 2008. `arxiv.org/abs/math/0702834`

[13] M. Casanellas, L.D. Garcia, and S. Sullivant, *Catalog of small trees*, in Algebraic Statistics for Computational Biology, L. Pachter and B. Sturmfels, eds., Cambridge University Press, Cambridge, UK, 2005, ch. 15, pp. 291–304.

[14] J. Cavender and J. Felsenstein, *Invariants of phylogenies in a simple case with discrete states*, Journal of Classification, **4** (1987), pp. 57–71.

[15] D. Cox and J. Sidman, *Secant varieties of toric varieties*, J. Pure Appl. Algebra, **209** (2007), pp. 651–669.

[16] B. Efron, E. Halloran, and S. Holmes, *Bootstrap confidence levels for phylogenetic trees*, Proceedings of the National Academy of Sciences, **93** (1996), pp. 13429–13429.

[17] ENCODE Project Consortium, *The ENCODE (ENCyclopedia Of DNA Elements) Project*, Science, **306** (2004), pp. 636–40.

[18] N. Eriksson, *Tree construction using singular value decomposition*, in Algebraic Statistics for Computational Biology, L. Pachter and B. Sturmfels, eds., Cambridge University Press, Cambridge, UK, 2005, ch. 19, pp. 347–358.

[19] N. Eriksson, K. Ranestad, B. Sturmfels, and S. Sullivant, *Phylogenetic algebraic geometry*, in Projective varieties with unexpected properties, C. Ciliberto, A. Geramita, B. Harbourne, R.-M. Roig, and K. Ranestad, eds., Walter de Gruyter GmbH & Co. KG, Berlin, 2005, pp. 237–255.

[20] N. Eriksson and Y. Yao, *Metric learning for phylogenetic invariants*, 2007. `arXiv.org/abs/q-bio/0703034v1`

[21] S. Evans and T. Speed, *Invariants of some probability models used in phylogenetic inference*, The Annals of Statistics, **21** (1993), pp. 355–377.

[22] J. Felsenstein, *Confidence Limits on Phylogenies: An Approach Using the Bootstrap*, Evolution, **39** (1985), pp. 783–791.

[23] ———, *Inferring Phylogenies*, Sinauer Associates, Sunderland, MA, 2003.

[24] ———, *PHYLIP (phylogeny inference package) version 3.6.* Available at `http://evolution.genetics.washington.edu/phylip.html`, 2005.

[25] N. Galtier and M. Gouy, *Inferring pattern and process: maximum-likelihood implementation of a nonhomogeneous model of DNA sequence evolution for phylogenetic analysis*, Mol. Biol. Evol, **15** (1998), pp. 871–879.

[26] O. Gascuel and S. Guindon, *Modelling the variability of evolutionary processes*, in Reconstructing evolution: new mathematical and computational advances, O. Gascuel and M. Steel, eds., Oxford University Press, 2007, ch. 3.

[27] M. Hendy and D. Penny, *A framework for the quantitative study of evolutionary trees*, Systematic Zoology, **38** (1989).

[28] D. Hillis and J. Bull, *An Empirical Test of Bootstrapping as a Method for Assessing Confidence in Phylogenetic Analysis*, Systematic Biology, **42** (1993), pp. 182–192.

[29] R. Holmquist, M. Miyamoto, and M. Goodman, *Analysis of higher-primate phylogeny from transversion differences in nuclear and mitochondrial DNA by Lake's methods of evolutionary parsimony and operator metrics*, Mol Biol Evol, **5** (1988), pp. 217–236.

[30] J. Huelsenbeck, *Performance of phylogenetic methods in simulations*, Sys Biol, **1** (1995), pp. 17–48.

[31] J. Huelsenbeck and D. Hillis, *Success of Phylogenetic Methods in the Four-Taxon Case*, Systematic Biology, **42** (1993), pp. 247–264.

[32] J. Huelsenbeck, F. Ronquist, R. Nielsen, and J. Bollback, *Bayesian Inference of Phylogeny and Its Impact on Evolutionary Biology*, Science, **294** (2001), p. 2310.

[33] T. Jukes and C. Cantor, *Evolution of protein molecules*, in Mammalian Protein Metabolism, H. Munro, ed., New York Academic Press, 1969, pp. 21–32.

[34] Y.R. Kim, O.-I. Kwon, S.-H. Paeng, and C.-J. Park, *Phylogenetic tree constructing algorithms fit for grid computing with SVD*, 2006. `arxiv.org/abs/q-bio.QM/0611015`

[35] M. Kimura, *Estimation of evolutionary sequences between homologous nucleotide sequences*, Proceedings of the National Academy of Sciences, USA, **78** (1981), pp. 454–458.

[36] J. Lake, *A rate-independent technique for analysis of nucleaic acid sequences: evolutionary parsimony*, Molecular Biology and Evolution, **4** (1987), pp. 167–191.

[37] F. Matsen, *Phylogenetic mixtures on a single tree can mimic a tree of another topology*, Systematic Biology, **56** (2007), pp. 767–775.

[38] W.C. Navidi, G.A. Churchill, and A. von Haeseler, *Methods for inferring phylogenies from nucleic acid sequence data by using maximum likelihood and linear invariants*, Mol. Biol. Evol., **8** (1991), pp. 128–143.

[39] ———, *Phylogenetic inference: linear invariants and maximum likelihood.*, Biometrics, **49** (1993), pp. 543–555.

[40] M. Newton, *Bootstrapping phytogenies: Large deviations and dispersion effects*, Biometrika, **83** (1996), p. 315.

[41] J. Norris, *Markov Chains*, Cambridge University Press, 1997.

[42] L. Pachter and B. Sturmfels, eds., *Algebraic Statistics for Computational Biology*, Cambridge University Press, Cambridge, UK, 2005.

[43] N. Saitou and M. Nei, *The neighbor joining method: a new method for reconstructing phylogenetic trees*, Molecular Biology and Evolution, **4** (1987), pp. 406–425.

[44] D. Sankoff and M. Blanchette, *Phylogenetic invariants for metazoan mitochondrial genome evolution*, Genome Informatics (1998), pp. 22–31.

[45] D. Sankoff and M. Blanchette, *Comparative genomics via phylogenetic invariants for Jukes-Cantor semigroups*, in Stochastic models (Ottawa, ON, 1998), Vol. **26** of Proceedings of the International Conference on Stochastic Models, American Mathematical Society, Providence, RI, 2000, pp. 399–418.

[46] C. Semple and M. Steel, *Phylogenetics*, vol. 24 of Oxford Lecture Series in Mathematics and its Applications, Oxford University Press, Oxford, 2003.

[47] S. Shalev-Shwartz, Y. Singer, and A.Y. Ng, *Online learning of pseudo-metrics*, in Proceedings of the Twenty-first International Conference on Machine Learning, 2004.

[48] M. Steel, D. Huson, and P. Lockhart, *Invariable sites models and their use in phylogeny reconstruction*, Systematic Biology, **49** (2000), pp. 225–232.

[49] B. Sturmfels, *Gröbner Bases and Convex Polytopes*, Vol. **8** of University Lecture Series, American Mathematical Society, 1996.

[50] B. Sturmfels and S. Sullivant, *Toric ideals of phylogenetic invariants*, J. Comput. Biol., **12** (2005), pp. 457–481.

[51] J.G. Sumner, M.A. Charleston, L.S. Jermiin, and P.D. Jarvis, *Markov invariants, plethysms, and phylogenetics*, 2007. `arxiv.org/abs/0711.3503v2`

[52] L. Székely, M. Steel, and P. Erdős, *Fourier calculus on evolutionary trees*, Advances in Applied Mathematics, **14** (1993), pp. 200–210.

[53] E. Xing, A.Y. Ng, M. Jordan, and S. Russell, *Distance metric learning, with application to clustering with side-information*, in NIPS, 2003.

[54] Z. Yang, *PAML: A program package for phylogenetic analysis by maximum likelihood*, CABIOS, **15** (1997), pp. 555–556.

[55] Z. Yang and D. Roberts, *On the use of nucleic acid sequences to infer early branchings in the tree of life*, Mol. Biol. Evol, **12** (1995), pp. 451–458.

ON THE ALGEBRAIC GEOMETRY OF POLYNOMIAL DYNAMICAL SYSTEMS

ABDUL S. JARRAH* AND REINHARD LAUBENBACHER*†

Abstract. This paper focuses on polynomial dynamical systems over finite fields. These systems appear in a variety of contexts, in computer science, engineering, and computational biology, for instance as models of intracellular biochemical networks. It is shown that several problems relating to their structure and dynamics, as well as control theory, can be formulated and solved in the language of algebraic geometry.

Key words. Polynomial dynamical system, inference of biochemical networks, control theory, computational algebra.

1. Introduction. The study of the dynamics of polynomial maps as time-discrete dynamical systems has a long tradition, in particular polynomials over the complex numbers leading to fractal structures. An example of more recent work using techniques from algebraic geometry includes the study of the algebraic and topological entropy of iterates of monomial mappings $f = (f_1, \ldots, f_n)$ over subsets of $\mathbf{C}^n$ [10]. The study of iterates of polynomial mappings (or, more generally, rational maps) over the p-adic numbers originally arose in Diophantine geometry [2]. For more recent work see, e.g., [27]. Finally, there is a long tradition of studying the iterates of polynomial maps $f : \mathbf{F}_q \longrightarrow \mathbf{F}_q$ over finite fields, primarily using techniques from combinatorics and algebraic number theory (see, e.g., [22, 23]).

The general case of *finite dynamical systems*

$$f = (f_1, \ldots, f_n) : \mathbf{F}_q^n \longrightarrow \mathbf{F}_q^n,$$

with $f_i \in \mathbf{F}_q[x_1, \ldots, x_n]$ has long been of interest in the special case $q = 2$, which includes Boolean networks and cellular automata. They are of considerable interest in engineering and computer science, for instance. Since the 1960s they are also being used increasingly as models of diverse biological and social networks.

The underlying mathematical questions in these studies for different fields are similar. They relate primarily to understanding the long-term behavior of the iterates of the mapping. In the case of monomial mappings, the matrix of exponents is usually the right mathematical object to analyze by using different methods based on the ground field. Generally, one would like to be able to use some feature of the structure of the coordinate functions f_i to deduce properties of the structure of the phase space of the system f which represents the system's dynamics. For finite systems, i.e.,

*Virginia Bioinformatics Institute, Virginia Polytechnic Institute and State University, Blacksburg, VA 24061, United States (ajarrah@vbi.vt.edu).

†reinhard@vbi.vt.edu.

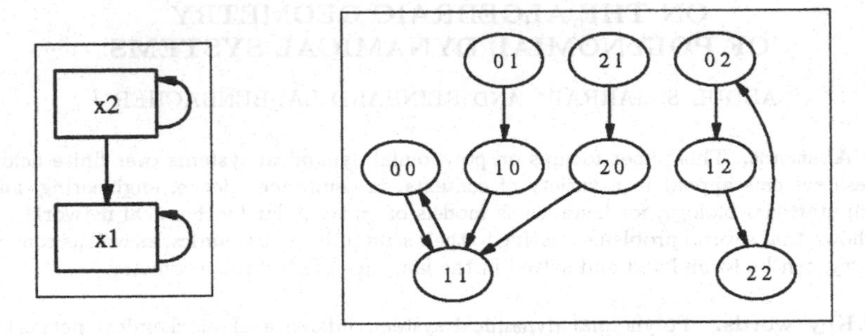

FIG. 1. *The dependency graph (left) and the phase space $\mathcal{P}(f)$ (right) of the finite dynamical system in Example 1.1.*

polynomial dynamical systems f over a finite field k, the phase space $\mathcal{P}(f)$ has the form of a directed graph with vertices the elements of k^n. There is an edge $\mathbf{v} \longrightarrow \mathbf{w}$ in $\mathcal{P}(f)$ if $f(\mathbf{v}) = \mathbf{w}$. Since the graph has finitely many vertices it is easy to see that each component has the structure of a directed *limit cycle*, with trees (*transients*) feeding into each of the nodes in the cycle.

EXAMPLE 1.1. Let $f : \mathbf{F}_3^2 \longrightarrow \mathbf{F}_3^2$ be given by $f(x_1, x_2) = (1 - x_1x_2, 1 + 2x_2)$. The phase space of f has two components, containing two limit cycles: one of length two and one of length three. See Figure 1 (right). The dependency relations among the variables are encoded in the dependency graph in Figure 1 (left).

In this case the information of interest is the number of components, the length of the limit cycles, and, possibly, the structure of the transient trees. It is also of interest to study the sequence of iterates f^r of f.

In recent years interest in polynomial dynamical systems over general finite fields has arisen also because they are useful as multistate generalizations of Boolean network models for biochemical networks, such as gene regulatory networks. In particular, the problem of network inference from experimental data can be formulated and solved in this framework. The tools used are very similar to those developed for the solution of problems in the statistical field of design of experiments and which led directly to the new field of algebraic statistics. (This connection is made explicit in [21].) Analyzing finite dynamical systems models of molecular networks requires tools that provide information about network dynamics from information about the system. Since the systems arising in applications can be quite high-dimensional it is not feasible to carry out such an analysis by exhaustive enumeration of the phase space. Thus, the studies of polynomial dynamics mentioned earlier become relevant to this type of application. Finally, an important role of models in biology and engineering is their use for the construction of controllers, be it for drug delivery to fight cancer or control of an airfoil. Here too, approaches have been developed for control systems in the context of polynomial dynamical systems.

Here we describe this circle of ideas and some existing results. Along the way we will mention open problems that arise. We believe that algebraic geometry can play an essential role as a conceptual and computational tool in this application area. Consistent with our specific research interests we will restrict ourselves to polynomial dynamical systems over finite fields.

2. Finite dynamical systems. Let k be a finite field and let

$$f = (f_1, \ldots, f_n) : k^n \longrightarrow k^n$$

be a mapping. It is well-known that the coordinate functions f_i can be represented as polynomials [24, p. 369], and this representation is unique if we require that the degree of every variable in every term of the f_i is less than $|k|$. If k is the field with two elements, then it is easily seen that f represents a Boolean network. Conversely, every Boolean network can be represented in polynomial form. Hence, polynomial dynamical systems over a finite field, which we shall call *finite dynamical systems* include *all* time-discrete dynamical systems over finite fields. Iteration of f generates dynamics which is captured by the phase space $\mathcal{P}(f)$ of f, defined above. We will first focus on the problem of inferring information about the structure of $\mathcal{P}(f)$ from the polynomials f_i.

In principle, a lot of information about $\mathcal{P}(f)$ can be gained by solving systems of polynomial equations. For instance, the fixed points of f are precisely the points on the variety given by the system

$$f_1(x_1, \ldots, x_n) - x_1 = 0, \ldots, f_n(x_1, \ldots, x_n) - x_n = 0.$$

The points of period 2 are the points on the variety given by the system $f_i^2(x_1, \ldots, x_n) - x_i = 0, i = 1, \ldots, n$, and so on. But often the most efficient way of solving these systems over a finite field, in particular a small finite field, is by exhaustive enumeration, which is not feasible for large n. A more modest goal would be to find out the number of periodic points of a given period, that is, the number of points on the corresponding variety. For finite fields this is also quite difficult without the availability of good general tools, and much work remains to be done in this direction.

The case of linear systems is the only case that has so far been treated systematically, using methods from linear algebra, as one might expect. Let M be the matrix of f. Then complete information about the number of components in $\mathcal{P}(f)$, the lengths of all the limit cycles and the structure of the transient trees can be computed from the invariant factors of A, together with the orders of their irreducible factors [11]. See [14] for an implementation of this algorithm in the computer algebra system Singular [9].

There are results available for several other special families of systems, using ad hoc combinatorial and graph-theoretical methods. For instance, we have investigated the case where all the f_i are monomials. It is of interest to characterize monomial systems all of whose periodic points are fixed points. For the field with two elements this characterization can be given in terms of the *dependency graph* of the system. This graph has as vertices the variables $x_1, \ldots, x_n$. There is a directed edge $x_i \to x_j$ if x_i appears in f_j; see Figure 1 (left) for an example. Given a directed graph it can be decomposed into strongly connected components, with each component a subgraph in which each vertex can be reached from every other vertex by a directed path. Associated to a strongly connected graph we have its *loop number* which is defined as the greatest common divisor of the lengths of all directed loops in the graph based at a fixed vertex. (It is easy to see that this number is independent of the vertex chosen.) The loop number is also known as the *index of imprimitivity* of the graph.

It is shown in [4] that the length of any cycle in the phase space of a monomial system f must divide the loop number of its dependency graph. In particular, we have the following result about fixed-point monomial systems.

THEOREM 2.1 ([4]). *All periodic points of a monomial system f are fixed points if and only if each strongly connected component of the dependency graph of f has loop number 1.*

The study of monomial systems over general finite fields can be reduced to studying linear and Boolean systems [3]. There are other interesting families of systems whose dynamics has been studied. One such family is that of Boolean networks constructed from *nested canalyzing functions*. These were introduced and studied in [18, 19]. The context is the use of Boolean networks as models for gene regulatory networks initiated by S. Kauffman [17]. We first recall the definitions of canalyzing and nested canalyzing functions from [18].

DEFINITION 2.2. *A* canalyzing function *is a Boolean function with the property that one of its inputs alone can determine the output value, for either "true" or "false" input. This input value is referred to as the* canalyzing value, *while the output value is the* canalyzed value.

EXAMPLE 2.3. The function $f(x, y) = xy$ is a canalyzing function in the variable x with canalyzing value 0 and canalyzed value 0. However, the function $f(x, y) = x + y$ is not canalyzing in either variable.

Nested canalyzing functions are a natural specialization of canalyzing functions. They arise from the question of what happens when the function does not get the canalyzing value as input but instead has to rely on its other inputs. Throughout this paper, when we refer to a function of n variables, we mean that f depends on all n variables. That is, for $1 \leq i \leq n$, there exists $(a_1, \ldots, a_n) \in \mathbf{F}_2^n$ such that $f(a_1, \ldots, a_{i-1}, a_i, a_{i+1}, \ldots, a_n) \neq f(a_1, \ldots, a_{i-1}, 1 + a_i, a_{i+1}, \ldots, a_n)$.

DEFINITION 2.4. *A Boolean function f in n variables is a* nested canalyzing function*(NCF) in the variable order $x_1, x_2, \ldots, x_n$ with canalyzing* *input values $a_1, \ldots, a_n$ and canalyzed output values $b_1, \ldots, b_n$, respectively, if it can be expressed in the form*

$$f(x_1, x_2, \ldots, x_n) = \begin{cases} b_1 & \text{if } x_1 = a_1, \\ b_2 & \text{if } x_1 \neq a_1 \text{ and } x_2 = a_2, \\ b_3 & \text{if } x_1 \neq a_1 \text{ and } x_2 \neq a_2 \text{ and } x_3 = a_3, \\ \vdots & \vdots \\ b_n & \text{if } x_1 \neq a_1 \text{ and } \cdots \text{ and } x_{n-1} \neq a_{n-1} \text{ and } x_n = a_n, \\ b_n + 1 & \text{if } x_1 \neq a_1 \text{ and } \cdots \text{ and } x_n \neq a_n. \end{cases}$$

EXAMPLE 2.5. The function $f(x, y, z) = x(y-1)z$ is nested canalyzing in the variable order x, y, z with canalyzing values 0,1,0 and canalyzed values 0,0,0, respectively. However, the function $f(x, y, z, w) = xy(z + w)$ is not a nested canalyzing function because if $x \neq 0$ and $y \neq 0$, then the value of the function is not constant for any input values for either z or w.

It is shown in [19] through extensive computer simulations that Boolean networks constructed from nested canalyzing functions show very stable dynamic behavior, with short transient trees and a small number of components in their phase space. It is this type of dynamics that gene regulatory networks are thought to exhibit. It is thus reasonable to use this class of functions preferentially in modeling such networks. To do so effectively it is necessary to have a better understanding of the properties of this class, for instance, how many nested canalyzing functions with a given number of variables there are.

In [15] a parametrization of this class is given as follows. The first step is to view Boolean functions as polynomials using the following translation:

$$x \wedge y = x \cdot y, \; x \vee y = x + y + xy, \; \neg x = x + 1.$$

It is shown in [15] that the ring of Boolean functions is isomorphic to the quotient ring $R = \mathbf{F}_2[x_1, \ldots, x_n]/I$, where $I = \langle x_i^2 - x_i : 1 \leq i \leq n \rangle$. Indexing monomials by the subsets of $[n] := \{1, \ldots, n\}$ corresponding to the variables appearing in the monomial, we can write the elements of R as

$$R = \left\{ \sum_{S \subseteq [n]} c_S \prod_{i \in S} x_i \, : \, c_S \in \mathbf{F}_2 \right\}.$$

As a vector space over $\mathbf{F}_2$, R is isomorphic to $\mathbf{F}_2^{2^n}$ via the correspondence

$$R \ni \sum_{S \subseteq [n]} c_S \prod_{i \in S} x_i \longleftrightarrow (c_\emptyset, \ldots, c_{[n]}) \in \mathbf{F}_2^{2^n},$$

for a given fixed total ordering of all square-free monomials. That is, a polynomial function corresponds to the vector of coefficients of the monomial summands. The main result in [15] is the identification of the set of nested canalyzing functions in R with a subset V^{ncf} of $\mathbf{F}_2^{2^n}$ by imposing relations on the coordinates of its elements.

DEFINITION 2.6. *Let σ be a permutation of the elements of the set $[n]$. We define a new order relation $<_\sigma$ on the elements of $[n]$ as follows: $\sigma(i) <_\sigma \sigma(j)$ if and only if $i < j$. Let r_S^σ be the maximum element of a nonempty subset S of $[n]$ with respect to the order relation $<_\sigma$. For any nonempty subset S of $[n]$, the completion of S with respect to the permutation σ, denoted by $[r_S^\sigma]$, is the set $[r_S^\sigma] = \{\sigma(1), \sigma(2), \ldots, \sigma(r_S)\}$.*

Note that, if σ is the identity permutation, then the completion is $[r_S] := \{1, 2, \ldots, r_S\}$, where r_S is the largest element of S.

THEOREM 2.7 ([12], Thm. 1). *Let $f \in R$ and let σ be a permutation of the set $[n]$. The polynomial f is a nested canalyzing function in the order $x_{\sigma(1)}, x_{\sigma(2)}, \ldots, x_{\sigma(n)}$, with input values $a_{\sigma(i)}$ and corresponding output values $b_{\sigma(i)}, 1 \leq i \leq n$, if and only if $c_{[n]} = 1$ and, for any proper subset $S \subseteq [n]$,*

$$c_S = c_{[r_S^\sigma]} \prod_{\sigma(i) \in [r_S^\sigma] \setminus S} c_{[n] \setminus \{\sigma(i)\}}.$$

COROLLARY 2.8 ([12], Cor. 1). *The set of points in $\mathbf{F}_2^{2^n}$ corresponding to nested canalyzing functions in the variable order $x_{\sigma(1)}, x_{\sigma(2)}, \ldots, x_{\sigma(n)}$, denoted by V_σ^{ncf}, is defined by*

$$V_\sigma^{ncf} = \{(c_\emptyset, \ldots, c_{[n]}) \in \mathbf{F}_2^{2^n} : c_{[n]} = 1,$$

$$c_S = c_{[r_S^\sigma]} \prod_{\sigma(i) \in [r_S^\sigma] \setminus S} c_{[n] \setminus \{\sigma(i)\}}, \text{ for } S \subseteq [n]\}.$$

It was also shown in [15] that

$$V^{ncf} = \bigcup_\sigma V_\sigma^{ncf}.$$

Counting the points on this variety for small values of n resulted in an integer sequence, which, with the help of the On-Line Encyclopedia of Integer Sequences (`http://www.research.att.com/ njas/sequences/`) led to the realization that the class of nested canalyzing functions is identical to the class of unate cascade functions that has been studied extensively in computer engineering literature. In particular, using this equality, one obtains a recursive formula for the number of nested canalyzing functions, see [15, Corollary 2.11].

It is shown in [12] that the sets V_σ^{ncf} are the irreducible components of V^{ncf}. Precisely, it is shown that for all permutations σ on $[n]$, the ideal of the variety V_σ^{ncf}, denoted by $I_\sigma := \mathbb{I}(V_\sigma^{ncf})$, is a binomial prime ideal in the polynomial ring $\overline{\mathbf{F}}_2[\{c_S : S \subseteq [n]\}]$, where $\overline{\mathbf{F}}_2$ is the algebraic closure of $\mathbf{F}_2$.

It remains to study this toric variety in more detail. In particular, a generalization of the concept of nested canalyzing function to larger finite fields remains to be worked out. Also, the approach taken here can be applied to other classes of functions important in network modeling, such as threshold functions or monotone functions.

3. Network inference. Our motivation for much of the research described in the previous section comes from our work on one of the central problems in computational systems biology. Due to the availability of so-called "omics" data sets it is now feasible to think about making large-scale mathematical models of molecular networks involving many gene transcripts (genomics), proteins (proteomics), and metabolites (metabolomics). One possible approach to this problem is to build a phenomenological model based solely or largely on the experimental data which can subsequently be refined with additional biological information about the mechanisms of interaction of the different molecular species. That is, given a data set, we are to infer a "most likely" mathematical or statistical model of the network that generated this data set. The biggest challenge is that typically the network that generated the data is high-dimensional (hundreds or thousands of molecular species/variables) and the available data sets are typically very small (tens or hundreds of data points). Also, general properties of such networks are not well-understood so that there are few general selection criteria. Therefore, it is not feasible to apply many of the existing network inference methods. In this context, two pieces of information about a molecular network are of interest to a life scientist: the "wiring diagram" of the network indicating which variables causally affect which others, and the long-term dynamic behavior of the network. Some network inference methods give only information about the wiring diagram, others provide both.

One possible modeling framework is that of polynomial dynamical systems over finite fields. In the 1960s, S. Kauffman proposed Boolean networks as good models for capturing key aspects of gene regulation [17]. In the 1990s so-called *logical models* were proposed by R. Thomas [29] as models for biochemical network, which are multi-state time-discrete dynamical systems, with both deterministic and stochastic variants. In [20] it was shown that the modeling framework of polynomial dynamical systems over finite fields is a good setting for the problem of network inference. As pointed out, these systems generalize Boolean networks and have many of the same features as logical models. The network inference problem can be formulated in this setting as follows.

Suppose that the biological system to be modeled contains n variables, e.g., genes, and we measure $r+1$ time points $\mathbf{p}_0, \ldots, \mathbf{p}_r$, using, e.g., gene chip technology, each of which can be viewed as an n-dimensional real-valued vector. The first step is to discretize the entries in the $\mathbf{p}_i$ into a prime number of states, which are viewed as entries in a finite field k. If we choose to discretize into two states by choosing a threshold, then we will obtain Boolean networks as models. The discretization step is crucial in this process as it represents the interface between the continuous and discrete worlds. Other network inference methods, such as most dynamic Bayesian network methods, also have to carry out this preprocessing step. Unfortunately, there is very little work that has been done on this problem. We have developed a new discretization method which is described in [6]. It compares favorably to other commonly used discretization methods, using different network inference methods.

Given this data set, an *admissible model*

$$f = (f_1, f_2, \ldots, f_n) : k^n \longrightarrow k^n$$

consists of a dynamical system f which satisfies the property that

$$f(\mathbf{p}_j) = (f_1(\mathbf{p}_j), \ldots, f_n(\mathbf{p}_j)) = \mathbf{p}_{j+1}.$$

The algorithm in [20] then proceeds to select such a model f, which is the most likely one based on certain specified criteria. This is done by first reducing the problem to the case of one variable, that is, to the problem of selecting the f_i separately. For this purpose, we compute the set of all functions f_i such that $f_i(\mathbf{p}_j) = \mathbf{p}^i_{j+1}$, that is, all polynomial functions f_i whose value on $\mathbf{p}_j$ is the ith coordinate of $\mathbf{p}_{j+1}$. This set can be represented as the coset $f^0 + I$, where f^0 is a particular such function and $I \subset k[x_1, \ldots, x_n]$ is the ideal of all polynomials that vanish on the given data set, also known as the *ideal of points* of $\mathbf{p}_1, \ldots, \mathbf{p}_{r-1}$. The algorithm then chooses the normal form of an interpolating polynomial f^0, based on a chosen term order. One drawback of the algorithm is that this choice of term order is typically random and can of course significantly affect the form of the model.

Modifications of the algorithm in [20] have been constructed. The algorithm in [13] starts with only data as input and computes all possible minimal wiring diagrams of polynomial models that fit the given data and outputs a most likely one, based on one of several possible model scoring methods. It does not depend on the choice of a term order. The algorithm is based on the observation that if f_i is the i-th coordinate function of a model and $\mathbf{p}, \mathbf{q}$ are data points such that $f_i(\mathbf{p}) \neq f_i(\mathbf{q})$, then the function f_i must involve at least some of the variables corresponding to coordinates in which $\mathbf{p}$ and $\mathbf{q}$ differ. This observation can be encoded in a monomial ideal M which is generated by all monomials of the form

$$\prod_{\mathbf{p}_j \neq \mathbf{q}_j} x_j$$

for all pairs of points $\mathbf{p} \neq \mathbf{q}$ such that $f_i(\mathbf{p}) \neq f_i(\mathbf{q})$. Now let $P = \langle x_{j_1}, \ldots, x_{j_t} \rangle$ be a minimal prime of M. Then it is not hard to see that the generators of P induce a minimal wiring diagram for f_i. Conversely, every such minimal diagram provides generators for a minimal prime of the ideal M. This algorithm has been implemented in Macaulay2 [8]. The algorithm comes with a collection of probability distributions on the set of minimal primes that can be used for model selection.

Another approach to the problem of dependency of the model selection process in [20] on the chosen term order is taken in [7]. The algorithm there uses the Gröbner fan of the ideal of points as a computational tool to find a most likely wiring diagram. It is clear that an upper bound for the number of different models one can obtain from the algorithm in [20] by varying the choice of term order is given by the number of cones in the Gröbner fan. The algorithm in [7] uses information about the frequency of appearance of the different variables in models built for each cone to build a consensus wiring diagram from this collection of possible models. It computes the Deegan-Packel index of power [5] to rank variables in order of significance. This index was introduced in [1], where it was computed using a Monte Carlo algorithm to generate random term orders. The use of the Gröbner fan allows a systematic computation of this index.

Note that the model space f^0+I contains *all* possible polynomial functions that fit the given data. In order to improve the performance of model selection algorithms it would be very useful to be able to select certain subspaces of functions that have favorable properties as models of particular biological systems, thereby reducing the model space. For instance, one might consider imposing certain constraints on the structure of the polynomials or on the resulting dynamics. The desire to find such constraints is what motivated the investigations described in the previous section. In order to select a class of polynomials with prescribed dynamics one needs to be able to link polynomial structure to dynamics in an algorithmic way. Similarly, in order to limit model selection to special classes of polynomials, such as nested canalyzing functions in the Boolean case, one must be able to identify efficiently the set of such functions from the model space $f^0 + I$. This problem remains open.

4. Control of finite dynamical systems. Control of biological systems is an important aspect of computational biology, ranging from the control of intracellular biochemical pathways to chemotherapy drug delivery and control of epidemiological processes. In order to apply mathematical control theory techniques it is necessary to work with a mathematical model of the system for which control theoretic tools exist. There is of course a very rich control-theoretic literature for systems of differential equations. However, the problem has also been considered in the context of polynomial dynamical systems over finite fields [25, 26, 28]. We briefly describe the general setting.

The framework developed before needs to be slightly modified to accommodate variables representing control inputs at each state transitions as well as constraints on these inputs and on the set of allowable initial conditions.

DEFINITION 4.1. *A* controlled finite dynamical system *is a function*

$$F : k^n \times k^m \longrightarrow k^n,$$

where the first set of variables $x_1, \ldots, x_n$ *represents the state variables, and the second set* $u_1, \ldots, u_m$ *represents the control variables. Furthermore, we have a system of polynomial equations*

$$Q_i(x_1, \ldots, x_n; u_1, \ldots, u_m) = 0, \; i = 1, \ldots, r,$$

which defines the variety of admissible control inputs, and another system

$$P_j(x_1, \ldots, x_n) = 0, \; j = 1, \ldots, s,$$

which defines the variety of admissible initial conditions of the system. Finally, we have another polynomial system

$$U_a(x_1, \ldots, x_n) = 0, \; a = 1, \ldots, t,$$

which defines the variety of admissible final states.

A typical optimal control problem is then stated as follows. Given a controlled system F and an admissible initialization $(x_1, \ldots, x_n)$, find a sequence of control inputs which drive the system to an admissible final state, in such a way that a suitably defined cost function is minimized. There are several ad hoc strategies of finding an optimal controller but much theoretical work remains to be done in this context.

In [16] an application of this approach to a virus competition problem was given, which we describe here in some detail in order to illustrate the definitions. In a Petri dish the center is infected with two different suitably chosen virus strains. It can be observed experimentally that, as the infection spreads to the rest of the dish a clear pattern of segmentation occurs between the two virus populations, rather than the expected mixing of the two strains. Through addition of one or the other type of virus over time the segmentation pattern can be influenced. For instance, it is possible to contain one virus strain by strategically inoculating cells in the Petri dish with the other strain. A possible application of this observation might be that one can contain the spread of a very harmful virus by the strategic introduction of another, less harmful virus strain. In this context it would be of interest to develop optimal strategies for introducing the second strain.

This problem was treated in [16] by representing the spread of the two virus populations as a polynomial dynamical system over the Galois field

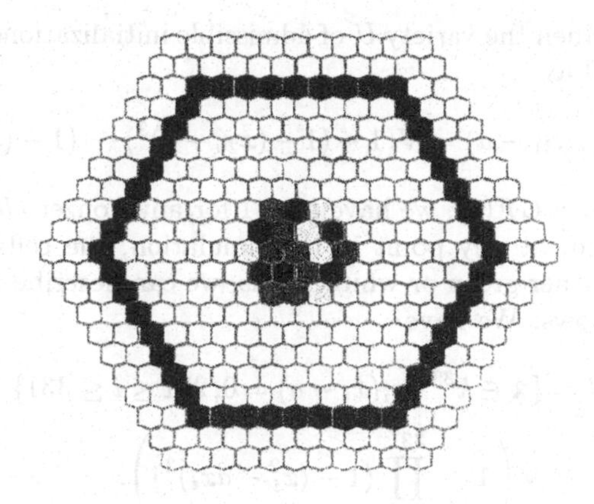

FIG. 2. *Control objective.*

$k = GF(4)$ as follows. The Petri dish is represented by 331 concentrically arranged hexagonal cells, each of which is a variable of the system. Each cell can take on 4 possible states, corresponding to being uninfected (White), infected by one of the two strains (Green or Red), or infected by both (Yellow). We begin with an arbitrary initialization of the center of the Petri dish, which is represented by the 19 innermost cells. That is, we make an arbitrary assignment of the 4 colors to these cells. All remaining cells are assigned White. The goal is to apply a series of control inputs as the infection spreads, which prevents the Red virus from spreading to the edge of the Petri dish. That is, a desirable final state of the system is any state for which the outermost ring of cells is infected only with Green virus. See Figure 2.

The rules governing the spread of infection are as follows:

1. If a cell has only one infected neighbor, then it will get the same type of infection.
2. If a cell has two infected neighbors, then one makes an assignment for the different possibilities (for details see [16]).

We use the representation $k = GF(4) := \mathbf{Z}_2[a] = \{0, a, a^2, a^3\}$. The color assignment is as follows:

Color	Field Element
Green	0
White	a
Yellow	a^2
Red	$a^3 \equiv 1$

If we represent the 331 cells by the variables $x_1, \ldots, x_{331}$, with $x_1, \ldots, x_{19}$ representing the center cells and $x_{271}, \ldots, x_{331}$ the cells in the outermost ring, then the variety U of admissible initializations of the model can be described as

$$\mathbf{V}(x_{20} - a, \ldots, x_{331} - a) = \mathbf{V}(1 - (1 - (x_{20} - a)^3) \cdots (1 - (x_{331} - a)^3)).$$

(Recall that in $k = GF(4)$, we have $b^3 = 1$ for any nonzero $b \in GF(4)$.)

As explained, at any point in the simulation, the cells in the outer ring should be either green or white. Thus, we can describe the constraint variety V as follows. We have

$$\begin{aligned}\mathbf{V} &= \{\mathbf{x} \in k^{331} | x_i(x_i - a) = 0; 271 \leq i \leq 331\} \\ &= \mathbf{V}\left(1 - \prod_{i=271}^{331} (1 - (x_i^2 - ax_i)^3)\right).\end{aligned}$$

The next step is to construct a state space model

$$f = (f_1, \ldots, f_{331}) : k^{331} \to k^{331},$$

of this experimental system, that is, a polynomial dynamical system f that approximates the dynamics observed in the laboratory. The coordinate functions f_i are polynomials in $k[x_1, \ldots, x_{331}]$ and represent the update rules of the cells x_i. Since the simulated Petri dish is homogeneous, all cells are identical, so that $f_i = f_j$ for all i, j. This is done using the algorithm in [20].

Let x represent one of the 331 cells, and let $y_1, \ldots, y_6$ represent its six immediate neighbors. We compute a symmetric polynomial $h \in k[y_1, \ldots, y_6]$ which represents the rules for the spread of the infection. That is, for a given infection status of the neighboring six cells, the polynomial takes on the appropriate value in $GF(4)$. At the same time, we compute the ideal of these points I, that is, the set of all polynomial functions in $k[y_1, \ldots, y_6]$ that vanish at these points. Since we are interested only in symmetric functions, let $\bar{I}$ be the ideal of symmetric functions inside the ideal I. The ideal $\bar{I}$ can easily be computed using computational algebra methods. Thus any possible symmetric polynomial function that can be a model of our system must be in the set $h + \bar{I}$. We choose as a model the normal form of h in the ideal $\bar{I}$, with respect to a chosen term order. In our case,

$$f = \gamma_2^2 + \gamma_2\gamma_1^3 + a^2\gamma_1^3 + a^2\gamma_1^2 + a^2\gamma_1$$

where

$$\begin{aligned}\gamma_1 &= y_1 + \cdots + y_6, \text{ and} \\ \gamma_2 &= \sum_{i \neq j} y_i y_j.\end{aligned}$$

The polynomials γ_1 and γ_2 are elementary symmetric functions in the polynomial ring $k[y_1, \ldots, y_6]$. Thus, the polynomial dynamical system $f : k^{331} \to k^{331}$ is a state space model for our system.

The next step in formulating the optimal control problem is to define a cost function for a given controller $g : GF(4)^{331} \to GF(4)^{331}$. We assume a uniform unit cost c for each cell that the controller infects with GREEN virus. Furthermore, assume that there is an "overhead" cost d attached to each intervention, independent of the number of cells transformed. Then the cost function for a controller g is given as follows. Suppose that $\{u_1, \ldots, u_r\}$ is a sequence of control inputs. Let c_i be the number of cells infected with GREEN during control input u_i. Then

$$C(g, \{u_i\}) = \sum_{i=1}^{r} (c \cdot c_i + |\delta_{c_i,0} - 1| \cdot d).$$

To find a controller g that minimizes the cost function $C(g)$ amounts to finding a way to control the system by transforming a minimal number of cells.

The paper [16] contains the construction and implementation of a suitable controller that was experimentally verified to accomplish the stated goal of containing one of the two strains. The authors were not able to show that the chosen control strategy was optimal, however.

It is well-known that optimal control of non-linear systems is difficult and few general tools are available. A common ad hoc approach is to "work backwards" from a desired end state and reconstruct control inputs in this way. One might hope that a suitable formulation of the problem in the language of algebraic geometry might allow the use of new tools from this field.

5. Discussion. It was shown in this paper that several problems about polynomial dynamical systems over finite fields can be formulated and possibly solved in the context of algebraic geometry and computational algebra. In particular, the problem of relating the structure of the defining polynomials to the resulting dynamics can be approached in this way. Likewise, the inference of a system from a given partial data set is amenable to a solution using tools from algebraic geometry. Finally, the beginnings of a control theory for such systems is expressed in the language of ideals and varieties.

However, it is clear that the results presented here barely scratch the surface of the problems and of what could be accomplished with the use of more sophisticated algebraic geometric tools.

6. Acknowledgements. The authors were partially supported by NSF Grant DMS-0511441 and NIH Grant R01 GM068947-01.

REFERENCES

[1] E. Allen, J. Fetrow, L. Daniel, S. Thomas, and D. John, Algebraic dependency models of protein signal transduction networks from time series data, Journal of Theoretical Biology, **238**:317–330, 2006.

[2] G. Call and J. Silverman, Canonical height on varieties with morphisms, Compositio Math., **89**:163–205, 1993.

[3] O. Colón-Reyes, A. Jarrah, R. Laubenbacher, and B. Sturmfels, Monomial dynamical systems over finite fields, Complex Systems, **16**(4):333–342, 2006.

[4] O. Colón-Reyes, R. Laubenbacher, and B. Pareigis, Boolean Monomial Dynamical Systems, Annals of Combinatorics, **8**:425–439, 2004.

[5] J. Deegan and E. Packel, A new index for simple n-person games, Int. J. Game Theory, **7**:113–123, 1978.

[6] E. Dimitorva, P. Vera-Licoa, J. McGee, and R. Laubnebahcer, Discretization of time series data. Submitted, 2007.

[7] E. Dimitrova, A. Jarrah, B. Stigler, and R. Laubenabcher, A Groebner Fan-based Method for Biochemical Network, ISSAC Proceedings, pp. 122–126, ACM Press, 2007.

[8] D. Grayson and M. Stillman, Macaulay 2, a software system for research in algebraic geometry, World Wide Web, `http://www.math.uiuc.edu/Macaulay2`.

[9] G.-M. Greuel, G. Pfister, and H. Schönemann, Singular 2.0, A Computer Algebra System for Polynomial Computations, Centre for Computer Algebra, University of Kaiserslautern, 2001, `http://www.singular.uni-kl.de`.

[10] B. Hasselblatt and J. Propp, Degree growth of monomial maps, `arXiv:Math.DS/0604521` v2, 2006.

[11] A. Hernández-Toledo, Linear Finite Dynamical Systems, Communications in Algebra, **33**(9):2977–2989, 2005.

[12] A. Jarrah and R. Laubenbacher, Discrete Models of Biochemical Networks: The Toric Variety of Nested Canalyzing Functions, Algebraic Biology, 2007, H. Anai and K. Horimoto and T. Kutsia, **4545**, LNCS, pp. 15–22, Springer.

[13] A. Jarrah, R. Laubenbacher, B. Stigler, and M. Stillman, Reverse-engineering of polynomial dynamical systems, Advances in Applied Mathematics, **39**(4):477–489, 2007.

[14] A. Jarrah, R. Laubenbacher, Mike Stillman, and P. Vera-Licona, An efficient algorithm for the phase space structure of linear dynamical systems over finite fields. Submitted, 2007.

[15] A. Jarrah, B. Raposa, and R. Laubenbacher, Nested canalyzing, unate cascade, and polynomial functions, Physica D, **233**:167–174, 2007.

[16] A. Jarrah, H. Vastani, K. Duca, and R. Laubenbacher, An optimal control problem for *in vitro virus competition*, 43rd IEEE Conference on Decision and Control, 2004, Invited paper, December.

[17] S. Kauffman, Metabolic stability and epigenesis in randomly constructed genetic nets, Journal of Theoretical Biology, **22**:437–467, 1969.

[18] S. Kauffman, C. Peterson, B. Samuelsson, and C. Troein, Random Boolean network models and the yeast transcriptional network, Proc. Natl. Acad. Sci. USA., **100**:14796–9, 2003.

[19] S. Kauffman, C. Peterson, B. Samuelsson, and C. Troein, Genetic networks with canalyzing Boolean rules are always stable, PNAS, **101**(49):17102–17107, 10.1073/pnas.0407783101, 2004.

[20] R. Laubenbacher and B. Stigler, A computational algebra approach to the reverse-engineering of gene regulatory networks, Journal of Theoretical Biology, **229**:523–537, 2004.

[21] ———, Design of experiments and biochemical network inference, Algebraic and Geometric Methods in Statistics, Gibilisco P., Riccomagno E., 2007, Cambridge University Press, Cambridge.

[22] L. Lidl and G. Mullen, When does a polynomial over a finite field permute the elements of the field? American Mathematical Monthly, **95**(3):243–246, 1988.

[23] ———, When does a polynomial over a finite field permute the elements of the field? II American Mathematical Monthly, **100**(1):71–74, 1993.

[24] R. Lidl and H. Niederreiter, Finite Fields, Cambridge University Press, 1997, New York.

[25] H. Marchand and M. LeBorgne, On the optimal control of polynomial dynamical systems over $\mathbf{Z}/p\mathbf{Z}$, Fourth Workshop on Discrete Event Systems, IEEE, 1998, Cagliari, Italy.

[26] ———, Partial order control of discrete event systems modeled as polynomial dynamical systems, IEEE International conference on control applications, 1998, Trieste, Italy.

[27] J. Pettigrew, J.A.G. Roberts, and F. Vivaldi, Complexity of regular invertible p-adic motions, Chaos, **11**:849–857, 2001.

[28] L. Reger and K. Schmidt, Aspects on analysis and synthesis of linear discrete systems over the finite field $GF(q)$, Proc. European Control Conference ECC2003, 2003, Cambridge University Press.

[29] R. Thomas and R. D'Ari, Biological Feedback, CRC Press, 1989.

A UNIFIED APPROACH TO COMPUTING REAL AND COMPLEX ZEROS OF ZERO-DIMENSIONAL IDEALS

JEAN BERNARD LASSERRE*, MONIQUE LAURENT†, AND PHILIPP ROSTALSKI‡

Abstract. In this paper we propose a unified methodology for computing the set $V_{\mathbb{K}}(I)$ of complex ($\mathbb{K} = \mathbb{C}$) or real ($\mathbb{K} = \mathbb{R}$) roots of an ideal $I \subseteq \mathbb{R}[x]$, assuming $V_{\mathbb{K}}(I)$ is finite. We show how moment matrices, defined in terms of a given set of generators of the ideal I, can be used to (numerically) find not only the real variety $V_{\mathbb{R}}(I)$, as shown in the authors' previous work, but also the complex variety $V_{\mathbb{C}}(I)$, thus leading to a unified treatment of the algebraic and real algebraic problem. In contrast to the real algebraic version of the algorithm, the complex analogue only uses basic numerical linear algebra because it does not require positive semidefiniteness of the moment matrix and so avoids semidefinite programming techniques. The links between these algorithms and other numerical algebraic methods are outlined and their stopping criteria are related.

Key words. Polynomial ideal, zero-dimensional ideal, complex roots, real roots, numerical linear algebra.

AMS(MOS) subject classifications. 12D10, 12E12, 12Y05, 13A15.

1. Introduction.

1.1. Motivation and contribution.

Computing all complex and/or real solutions of a system of polynomial equations is a fundamental problem in mathematics with many important practical applications. Let $I \subseteq \mathbb{R}[x] := \mathbb{R}[x_1, \ldots, x_n]$ be an ideal generated by a set of polynomials h_j ($j = 1, \ldots, m$). Fundamental problems in polynomial algebra are:

(I) The computation of the algebraic variety $V_{\mathbb{C}}(I) = \{v \in \mathbb{C}^n \mid h_j(v) = 0 \ \forall\, j = 1, \ldots, m\}$ of I,

(II) The computation of the real variety $V_{\mathbb{R}}(I) = V_{\mathbb{C}}(I) \cap \mathbb{R}^n$ of I,

as well as a set of generators for the radical ideal $J = I(V_{\mathbb{K}}(I))$ for $\mathbb{K} = \mathbb{R}$ or $\mathbb{C}$, assuming $V_{\mathbb{K}}(I)$ is finite.

One way to solve problem (II) is to first compute all complex solutions and to sort out $V_{\mathbb{R}}(I) = \mathbb{R}^n \cap V_{\mathbb{C}}(I)$ from $V_{\mathbb{C}}(I)$ afterwards. This is certainly possible when I is a zero-dimensional ideal, but even in this case one might perform many unnecessary computations, particularly if $|V_{\mathbb{R}}(I)| \ll |V_{\mathbb{C}}(I)|$, i.e. in case there are many more complex than real roots. In addition there are cases where $V_{\mathbb{R}}(I)$ is finite whereas $V_{\mathbb{C}}(I)$ is not! These two reasons

*LAAS-CNRS and Institute of Mathematics, Toulouse, France (lasserre@laas.fr).
Supported by the french national research agency ANR under grant NT05-3-41612.

†Centrum voor Wiskunde en Informatica, Kruislaan 413, 1098 SJ Amsterdam, Netherlands (M.Laurent@cwi.nl).
Supported by the Netherlands Organization for Scientific Research grant NWO 639.032.203 and by ADONET, Marie Curie Research Training Network MRTN-CT-2003-504438.

‡Automatic Control Laboratory, Physikstrasse 3, ETH Zurich, 8092 Zurich, Switzerland (rostalski@control.ee.ethz.ch).

alone provide a rationale for designing a method specialized to problem (II), that is, a method that takes into account right from the beginning the *real* algebraic nature of the problem.

In [10] we have provided a semidefinite characterization and an algorithm for approximating $V_{\mathbb{K}}(I)$ ($\mathbb{K} = \mathbb{R}, \mathbb{C}$) as well as a basis of the radical ideal $I(V_{\mathbb{K}}(I))$ (in the form of a border or Gröbner basis). The approach there utilizes well established semidefinite programming techniques and numerical linear algebra. Remarkably, all information needed to compute the above objects is contained in the so-called moment matrix (a matrix with a particular quasi-Hankel structure, indexed by a basis of $\mathbb{R}[x]$, and whose entries depend on the polynomials generating the ideal I) and the geometry behind it when this matrix is required to be positive semidefinite with maximum rank. For the task of computing the real roots and the real radical ideal $\sqrt[\mathbb{R}]{I} = I(V_{\mathbb{R}}(I))$, the method is real algebraic in nature, as we do not compute (implicitly or explicitly) any complex element of $V_{\mathbb{C}}(I)$.

The method proposed in [10] for solving problem (I) treats $\mathbb{C}^n$ as $\mathbb{R}^{2n}$ and essentially applies the same algorithm as for problem (II), but now working in $\mathbb{R}^{2n}$ instead of $\mathbb{R}^n$. Hence one has to use semidefinite matrices of much larger size since they are now indexed by a basis of $\mathbb{C}[x, \bar{x}]$ (as opposed to $\mathbb{R}[x]$ for problem (II)).

This latter remark is one of the motivations for the present paper in which we provide a method for computing $V_{\mathbb{C}}(I)$, a complex analogue of the method of [10] for computing $V_{\mathbb{R}}(I)$, which also uses a moment matrix indexed by a basis of $\mathbb{R}[x]$ (instead of $\mathbb{C}[x, \overline{x}]$ as in [10]). The algorithm is very similar to the one proposed in [10] for problem (II), except for the important fact that we now do *not* require the positivity of the moment matrix; therefore the algorithm only uses basic numerical linear algebra techniques and *no* semidefinite programming optimization. The price to pay for the reduced complexity is that our algorithm now finds a basis for an ideal J with $I \subseteq J \subseteq \sqrt{I}$ (instead of $J = \sqrt{I}$ in [10]), though with the same algebraic variety $V_{\mathbb{C}}(J) = V_{\mathbb{C}}(I)$. Note however that once a basis $\mathcal{B}$ of $\mathbb{R}[x]/J$ and the corresponding multiplication matrices are known, generators for the ideal $\sqrt{I}$ can be computed numerically e.g. via the algorithm proposed in [7].

On the other hand there is a plethora of methods and algorithms to compute the (finite) complex variety $V_{\mathbb{C}}(I)$ and certain distinguished bases as Gröbner and border bases to name just a few. This motivates the second contribution of this paper, which is to relate the proposed method based on moment matrices to existing methods and, in particular, to the method of [18] (and [17]) for the (finite) complex variety. It turns out that, by adding the positive semidefiniteness constraint, the method of [18] can be adapted and extended for computing the (finite) real variety $V_{\mathbb{R}}(I)$; this will be treated in detail in the follow-up paper [9]. Summarizing, our results provide a unified treatment of the computation of real and complex roots either by means of moment matrices or by means of a dual form characterization as in [9], [17] and [18].

1.2. Related literature. The importance and relevance to various branches of mathematics of the problem of solving systems of polynomials is reflected by the broad literature, see e.g. [6]. Various methods exist for problem (I), ranging from numerical continuation methods (see e.g. [22]), to exact symbolic methods (e.g. [19]), or more general symbolic/numeric methods (e.g. [16] or [18], see also the monograph [23]). For instance, Verschelde [24] proposes a numerical algorithm via homotopy continuation methods (cf. also [22]) whereas Rouillier [19] solves a zero-dimensional system of polynomials symbolically by giving a rational univariate representation (RUR) for its solutions, of the form $f(t) = 0$, $x_1 = \frac{g_1(t)}{g(t)}, \ldots, x_n = \frac{g_n(t)}{g(t)}$, where $f, g, g_1, \ldots, g_n \in \mathbb{K}[t]$ are univariate polynomials. The computation of the RUR relies in an essential way on the multiplication matrices in the quotient algebra $\mathbb{K}[x]/I$ which thus requires the knowledge of a corresponding linear basis of the quotient space.

The literature tailored to problem (II), i.e. to the real solving of systems of polynomials, is by far not as broad as the one for finding all (complex) solutions. Most algorithms (beside our previous work [10]) are based on real-root counting algorithms using e.g. Hermite's quadratic forms or variants of Sturm sequences (see e.g. [1] or [20] for a discussion).

1.3. Contribution. Our first contribution is a unified treatment of the cases $\mathbb{K} = \mathbb{R}$ or $\mathbb{K} = \mathbb{C}$ to obtain the 0-dimensional variety $V_{\mathbb{K}}(I)$. We will work with the space $\mathbb{R}[x]_t$ of polynomials of degree smaller or equal to t and with certain subsets of its dual space $(\mathbb{R}[x]_t)^*$, the space of linear functionals on $\mathbb{R}[x]_t$. More precisely, for an integer $t \geq D := \max_j \deg(h_j)$, set

$$\mathcal{H}_t := \{h_j x^\alpha \mid j = 1, \ldots, m \text{ and } \alpha \in \mathbb{N}^n \text{ with } |\alpha| + \deg(h_j) \leq t\} \tag{1.1}$$

and consider the two sets

$$\mathcal{K}_t := \left\{ L \in (\mathbb{R}[x]_t)^* \mid L(p) = 0 \ \forall p \in \mathcal{H}_t \right\} \tag{1.2}$$

for computing $V_{\mathbb{C}}(I)$, and

$$\mathcal{K}_{t,\succeq} := \left\{ L \in \mathcal{K}_t \,\middle|\, L(f^2) \geq 0 \ \forall f \in \mathbb{R}[x]_{\lfloor t/2 \rfloor} \right\} \tag{1.3}$$

for computing $V_{\mathbb{R}}(I)$. Obviously, the linear form associated with evaluation at $v \in V_{\mathbb{K}}(I)$, lies in the set $\mathcal{K}_{t,\succeq}$ ($\mathbb{K} = \mathbb{R}$) and its real and imaginary parts lie in the set $\mathcal{K}_t$ ($\mathbb{K} = \mathbb{C}$). Roughly speaking, by iterating on $t \in \mathbb{N}$, we will refine the description of those sets by successively adding linear conditions (and conditions stemming from SOS relations in I in the case $\mathbb{K} = \mathbb{R}$), until they contain sufficient information to enable extraction of the points $V_{\mathbb{K}}(I)$.

Sketch of the moment-matrix algorithm. Let us now give a more specific sketch of the algorithm of [10] for $V_{\mathbb{R}}(I)$ and of its extension for $V_{\mathbb{C}}(I)$ proposed in the present paper. Given $L \in (\mathbb{R}[x]_t)^*$ and $1 \leq s \leq \lfloor t/2 \rfloor$, define its moment matrix $M_s(L)$ as the matrix indexed by $\mathbb{N}^n_s = \{\alpha \in \mathbb{N}^n \mid |\alpha| = \sum_i \alpha_i \leq s\}$, with (α, β)th entry $L(x^\alpha x^\beta)$. For a matrix M, positive semidefiniteness, i.e. the property $x^T M x \geq 0$ for all vectors x, is denoted by $M \succeq 0$. Thus L satisfies the condition $L(f^2) \geq 0$ for all $f \in \mathbb{R}[x]_{\lfloor t/2 \rfloor}$, precisely when $M_{\lfloor t/2 \rfloor}(L) \succeq 0$. Consider the following rank conditions on the matrix $M_s(L)$:

$$\operatorname{rank} M_s(L) = \operatorname{rank} M_{s-1}(L), \tag{1.4}$$

$$\operatorname{rank} M_s(L) = \operatorname{rank} M_{s-d}(L), \tag{1.5}$$

after setting $d := \lceil D/2 \rceil$. Algorithm 1 is our moment-matrix algorithm for finding $V_{\mathbb{K}}(I)$.

Algorithm 1 *The moment-matrix algorithm for* $V_{\mathbb{K}}(I)$*:*

Input: $t \geq D$.

Output: A basis $\mathcal{B} \subseteq \mathbb{R}[x]_{s-1}$ of $\mathbb{K}[x]/\langle \operatorname{Ker} M_s(L) \rangle$ (which will enable the computation of $V_{\mathbb{K}}(I)$).

1: Find a generic element $L \in K_t$.
2: Check if (1.4) holds for some $D \leq s \leq \lfloor t/2 \rfloor$, or if (1.5) holds for some $d \leq s \leq \lfloor t/2 \rfloor$.
3: **if** yes **then**
4: **return** a basis $\mathcal{B} \subseteq \mathbb{R}[x]_{s-1}$ of the column space of $M_{s-1}(L)$,
5: **else**
6: Iterate (go to 1)) replacing t by $t+1$
7: **end if**

REMARK 1.1. Here $K_t = \mathcal{K}_{t,\succeq}$ for the task of computing $V_{\mathbb{R}}(I)$ as in [10], and $K_t = \mathcal{K}_t$ for the task of computing $V_{\mathbb{C}}(I)$ in the present paper. In Step 1, we say that $L \in K_t$ is generic if, for all $1 \leq s \leq \lfloor t/2 \rfloor$, $\operatorname{rank} M_s(L)$ is maximum over K_t.

Consider first the real case, treated in [10]. A first observation is that in the above definition of a generic element, it suffices to require the maximum rank property for $s = \lfloor t/2 \rfloor$. The algorithm relies on the following crucial properties. If the answer in Step 2 is 'yes' then $\langle \operatorname{Ker} M_s(L) \rangle$, the ideal generated by polynomials p of degree no more than s whose coefficient vector vec(p) lies in $\operatorname{Ker} M_s(L)$, coincides with $I(V_{\mathbb{R}}(I))$, the real radical of I. Moreover the set $\mathcal{B}$ is a basis of the quotient space $\mathbb{R}[x]/\langle \operatorname{Ker} M_s(L) \rangle$ and thus one can apply the classical eigenvalue method to compute $V_{\mathbb{R}}(I)$. Additionally, a border (or Gröbner) basis of $I(V_{\mathbb{R}}(I))$ is readily available from the kernel of the matrix $M_s(L)$ (cf. [10] for details).

We show in the present paper that the *same* algorithm works also for the task of computing finite $V_{\mathbb{C}}(I)$ (whenever finite), except we now use the set $K_t = \mathcal{K}_t$. Although the algorithms are apparently identical in the real and complex cases, the proofs of correctness are however distinct as well as the implementations. For instance, a generic element $L \in \mathcal{K}_{t,\succeq}$ is any element in the relative interior of the cone $\mathcal{K}_{t,\succeq}$ and can be found with appropriate interior-point algorithms for semidefinite programming optimization. On the other hand, a generic element in $\mathcal{K}_t$ can be found using some randomization argument (cf. details later in Section 3.1.4). Moreover, if L is a generic element of $\mathcal{K}_{t,\succeq}$, then $\operatorname{Ker} M_s(L) \subseteq \operatorname{Ker} M_s(L')$ for *all* $L' \in \mathcal{K}_{t,\succeq}$, a property which is not true in general for a generic element $L \in \mathcal{K}_t$ (namely it is not true if the algebra $\mathbb{R}[x]/I$ is not Gorenstein; cf. Section 3.2 for details). For a generic $L \in K_t$ ($K_t = \mathcal{K}_t$ or $\mathcal{K}_{t,\succeq}$), a useful property is that $\operatorname{Ker} M_s(L) \subseteq I(V_{\mathbb{K}}(I))$. This property is true in both cases $\mathbb{K} = \mathbb{R}, \mathbb{C}$. However, while this fact is fairly immediate in the real case, the proof is technically more involved in the complex case (cf. Section 3.1.2). Finally, in the complex case, if the answer is 'yes' in Step 2, we can only claim that the ideal $J := \langle \operatorname{Ker} M_s(L) \rangle$ is nested between I and $I(V_{\mathbb{C}}(I))$; as $V_{\mathbb{C}}(J) = V_{\mathbb{C}}(I)$ this property is however sufficient for the task of computing $V_{\mathbb{C}}(I)$.

Another contribution of the paper is to relate the stopping criteria (1.4) and (1.5) used in our moment based approach to the stopping criterion

$$\dim \pi_s(\mathcal{K}_t) = \dim \pi_{s-1}(\mathcal{K}_t) = \dim \pi_s(\mathcal{K}_{t+1}) \tag{1.6}$$

(where π_s denotes the projection from $(\mathbb{R}[x]_t)^*$ onto $(\mathbb{R}[x]_s)^*$) used e.g. in the method of Zhi and Reid [18].

Roughly speaking, if (1.6) holds for some $D \leq s \leq t$, then $\mathbb{R}[x]_s \cap I = \mathbb{R}[x]_s \cap \operatorname{Span}_{\mathbb{R}}(\mathcal{H}_t)$ and one can construct a basis $\mathcal{B} \subseteq \mathbb{R}[x]_{s-1}$ of $\mathbb{R}[x]/I$ (enabling computing $V_{\mathbb{C}}(I)$) (see Section 4.1 for details). Thus the condition (1.6) is a global condition on the set $\mathcal{K}_t$ while (1.4) and (1.5) are conditions on a generic element $L \in \mathcal{K}_t$. However these two types of conditions are closely related as shown in Section 4.2.

Contents of the paper. The paper is organized as follows. In Section 2 we introduce some definitions and results about polynomials and moment matrices that we need in the paper. In Section 3 we present our algorithm for computing the complex roots of a zero-dimensional ideal using moment matrices and discuss some small examples. In Section 4 we revisit the involutive base method of Zhi and Reid and compare the various stopping criteria.

2. Preliminaries.

In this section we recall some preliminaries of polynomial algebra and moment matrices used throughout the paper.

2.1. Some basics of algebraic geometry.

2.1.1. Polynomial ideals and varieties.

Let $\mathbb{R}[x] := \mathbb{R}[x_1, \ldots, x_n]$ denote the ring of multivariate polynomials in n variables. For $\alpha \in \mathbb{N}^n$, the monomial $x^\alpha := x_1^{\alpha_1} \cdots x_n^{\alpha_n}$ has degree $|\alpha| := \sum_{i=1}^n \alpha_i$. Set $\mathbb{N}_t^n := \{\alpha \in$

$\mathbb{N}^n \mid |\alpha| \leq t\}$. Then $\mathbb{T}^n := \{x^\alpha \mid \alpha \in \mathbb{N}^n\}$ denotes the set of all monomials in n variables and $\mathbb{T}^n_t := \{x^\alpha \mid \alpha \in \mathbb{N}^n_t\}$ the subset of monomials of degree smaller or equal to t. Given polynomials $h_1, \ldots, h_m \in \mathbb{R}[x]$,

$$I = \langle h_1, \ldots, h_m \rangle := \Big\{ \sum_{j=1}^m a_j h_j \mid \ a_1, \ldots, a_m \in \mathbb{R}[x] \Big\}$$

is the ideal generated by $h_1, \ldots, h_m$. The algebraic variety of I is the set

$$V_{\mathbb{C}}(I) = \{v \in \mathbb{C}^n \mid h_j(v) = 0 \ \forall j = 1, \ldots, m\}$$

of common complex zeros to all polynomials in I and its real variety is $V_{\mathbb{R}}(I) := V_{\mathbb{C}}(I) \cap \mathbb{R}^n$. The ideal I is zero-dimensional when its complex variety $V_{\mathbb{C}}(I)$ is finite. Conversely the vanishing ideal of a subset $V \subseteq \mathbb{C}^n$ is the ideal $I(V) := \{f \in \mathbb{R}[x] \mid f(v) = 0 \ \forall v \in V\}$. For an ideal $I \subseteq \mathbb{R}[x]$, we may also define the ideal

$$\sqrt{I} := \Big\{ f \in \mathbb{R}[x] \mid \ f^m \in I \text{ for some } m \in \mathbb{N} \setminus \{0\} \Big\},$$

called the radical of I, and the ideal

$$\sqrt[\mathbb{R}]{I} := \Big\{ p \in \mathbb{R}[x] \mid p^{2m} + \sum_j q_j^2 \in I \text{ for some } q_j \in \mathbb{R}[x], m \in \mathbb{N} \setminus \{0\} \Big\},$$

called the real radical ideal of I; I is radical (resp., real radical) if $I = \sqrt{I}$ (resp., $I = \sqrt[\mathbb{R}]{I}$). Obviously $I \subseteq \sqrt{I} \subseteq I(V_{\mathbb{C}}(I))$ and $I \subseteq \sqrt[\mathbb{R}]{I} \subseteq I(V_{\mathbb{R}}(I))$. The relation between vanishing and (real) radical ideals is stated in the following two famous theorems:

THEOREM 2.1. *Let $I \subseteq \mathbb{R}[x]$ be an ideal.*
(i) *Hilbert's Nullstellensatz (see, e.g., [3, §4.1])* $\sqrt{I} = I(V_{\mathbb{C}}(I))$.
(ii) *Real Nullstellensatz (see, e.g., [2, §4.1])* $\sqrt[\mathbb{R}]{I} = I(V_{\mathbb{R}}(I))$.

2.1.2. The (dual) ring of polynomials. Given a vector space A on $\mathbb{R}$, its dual space $A^* := \text{Hom}(A, \mathbb{R})$ consists of all linear functionals from A to $\mathbb{R}$. Given a subset $B \subseteq A$, set $B^\perp := \{L \in A^* \mid L(b) = 0 \ \forall b \in B\}$. Then $\text{Span}_{\mathbb{R}}(B) \subseteq (B^\perp)^\perp$, with equality when A is finite dimensional. Here $\text{Span}_{\mathbb{R}}(B) := \{\sum_{i=1}^m \lambda_i b_i \mid \lambda_i \in \mathbb{R}, b_i \in B\}$. We will mostly work here with the vector space $A = \mathbb{R}[x]$ (or subspaces). Examples of linear functionals on $\mathbb{R}[x]$ are the evaluation $p \in \mathbb{R}[x] \mapsto p(v)$ at any $v \in \mathbb{R}^n$ and, given $\alpha \in \mathbb{N}^n$, the differential functional

$$p \in \mathbb{R}[x] \mapsto \partial_\alpha[v](p) := \frac{1}{\prod_{i=1}^n \alpha_i!} \left(\frac{\partial^{|\alpha|}}{\partial x_1^{\alpha_1} \ldots x_n^{\alpha_n}} p \right)(v), \tag{2.1}$$

which evaluates at $v \in \mathbb{R}^n$ the (scaled) derivative of p; thus $\partial_0[v](p) = p(v)$ is the linear form that evaluates p at v. Note that, for $\alpha, \beta \in \mathbb{N}^n$,

$$\partial_\alpha[0] \left(\prod_{i=1}^n x_i^{\beta_i} \right) = \begin{cases} 1 \text{ if } \alpha = \beta \\ 0 \text{ otherwise} \end{cases}$$

Therefore the monomial basis $\mathbb{T}^n = \{x^\alpha \mid \alpha \in \mathbb{N}^n\}$ of $\mathbb{R}[x]$ and the basis $\{\partial_\alpha[0] \mid \alpha \in \mathbb{N}^n\}$ of $(\mathbb{R}[x])^*$ are dual bases. Throughout we will mainly use these two canonical bases. In particular, we write a polynomial $p \in \mathbb{R}[x]$ in the form $p = \sum_\alpha p_\alpha x^\alpha$, and $L \in (\mathbb{R}[x])^*$ in the form $L = \sum_\alpha y_\alpha \partial_\alpha[0]$; thus $p_\alpha = \partial_\alpha[0](p)$ and $y_\alpha = L(x^\alpha)$ are the respective coefficients of p and L in the canonical bases and $L(p) = y^T \operatorname{vec}(p) = \sum_\alpha p_\alpha y_\alpha$. Here we let $\operatorname{vec}(p) := (p_\alpha)_\alpha$ denote the vector of coefficients of the polynomial p. Finally, given $v \in \mathbb{C}^n$ and $t \in \mathbb{N}$, set $\zeta_v := (v^\alpha)_{\alpha \in \mathbb{N}^n}$ and $\zeta_{t,v} := (v^\alpha)_{\alpha \in \mathbb{N}_t^n}$; thus $\zeta_v = (\partial_0[v](x^\alpha))_\alpha$ is the coordinate sequence of the linear functional $\partial_0[v]$ in the canonical basis of $(\mathbb{R}[x])^*$.

As vector spaces, both $\mathbb{R}[x]$ and its dual $(\mathbb{R}[x])^*$ are infinite dimensional and so for practical computation it is more convenient to work with the finite dimensional subspaces $\mathbb{R}[x]_t = \{p \in \mathbb{R}[x] \mid \deg(p) \le t\}$ for $t \in \mathbb{N}$. Both vectors spaces $\mathbb{R}[x]_t$ and its dual $(\mathbb{R}[x]_t)^*$ are isomorphic to $\mathbb{R}^{\mathbb{N}_t^n}$, with canonical dual bases $\mathbb{T}_t^n$ and $\{\partial_\alpha[0] \mid \alpha \in \mathbb{N}_t^n\}$, respectively. Given an integer $s \le t$, we let π_s denote the projection from $\mathbb{R}^{\mathbb{N}_t^n}$ onto $\mathbb{R}^{\mathbb{N}_s^n}$, which can thus be interpreted as the projection from $\mathbb{R}[x]_t$ onto $\mathbb{R}[x]_s$, or from $(\mathbb{R}[x]_t)^*$ onto $(\mathbb{R}[x]_s)^*$ depending on the context.

2.1.3. The quotient algebra. Given an ideal $I \subseteq \mathbb{R}[x]$, the quotient set $\mathbb{R}[x]/I$ consists of all cosets $[f] := f + I = \{f + q \mid q \in I\}$ for $f \in \mathbb{R}[x]$, i.e. all equivalent classes of polynomials of $\mathbb{R}[x]$ modulo the ideal I. The quotient set $\mathbb{R}[x]/I$ is an algebra with addition $[f] + [g] := [f + g]$, scalar multiplication $\lambda[f] := [\lambda f]$ and with multiplication $[f][g] := [fg]$, for $\lambda \in \mathbb{R}$, $f, g \in \mathbb{R}[x]$.

A useful property is that, when I is zero-dimensional (i.e. $|V_\mathbb{C}(I)| < \infty$), then $\mathbb{R}[x]/I$ is a finite-dimensional vector space and the cardinality of $V_\mathbb{C}(I)$ is related to its dimension, as indicated in Theorem 2.2 below.

THEOREM 2.2. *Let I be an ideal in $\mathbb{R}[x]$. Then $|V_\mathbb{C}(I)| < \infty \Longleftrightarrow \dim \mathbb{R}[x]/I < \infty$. Moreover, $|V_\mathbb{C}(I)| \le \dim \mathbb{R}[x]/I$, with equality if and only if I is radical.*

A proof of this theorem and a detailed treatment of the quotient algebra $\mathbb{R}[x]/I$ can be found e.g. in [3], [23].

Assume $|V_\mathbb{C}(I)| < \infty$ and set $N := \dim \mathbb{R}[x]/I$, $|V_\mathbb{C}(I)| \le N < \infty$. Consider a set $\mathcal{B} := \{b_1, \ldots, b_N\} \subseteq \mathbb{R}[x]$ for which the cosets $[b_1], \ldots, [b_N]$ are pairwise distinct and $\{[b_1], \ldots, [b_N]\}$ is a basis of $\mathbb{R}[x]/I$; by abuse of language we also say that $\mathcal{B}$ itself is a basis of $\mathbb{R}[x]/I$. Then every $f \in \mathbb{R}[x]$ can be written in a unique way as $f = \sum_{i=1}^N c_i b_i + p$, where $c_i \in \mathbb{R}$, $p \in I$; the polynomial $\mathcal{N}_\mathcal{B}(f) := \sum_{i=1}^N c_i b_i$ is called the residue of f modulo I, or its *normal form*, with respect to the basis $\mathcal{B}$. In other words, $\operatorname{Span}_\mathbb{R}(\mathcal{B})$ and $\mathbb{R}[x]/I$ are isomorphic vector spaces.

Following Stetter [23], for an ideal $I \subseteq \mathbb{R}[x]$, define its dual space

$$\mathcal{D}[I] := I^\perp = \{L \in (\mathbb{R}[x])^* \mid L(p) = 0 \ \forall p \in I\} \qquad (2.2)$$

consisting of all linear functionals vanishing on I. Thus $\mathcal{D}[I]$ is isomorphic to $(\mathbb{R}[x]/I)^*$ and, when I is zero-dimensional,

$$\operatorname{Ker}\mathcal{D}[I] := \mathcal{D}[I]^{\perp} = \{\, p \in \mathbb{R}[x] \mid L(p) = 0 \; \forall L \in \mathcal{D}[I] \,\} = I.$$

When I is zero-dimensional and radical the sum of the real and imaginary parts of the evaluation at points $v \in V_{\mathbb{C}}(I)$ form a basis of $\mathcal{D}[I]$; that is,

$$\mathcal{D}[I] = \operatorname{Span}_{\mathbb{R}} \{ \operatorname{Re} \partial_0[v] + \operatorname{Im} \partial_0[v] \mid v \in V_{\mathbb{C}}(I) \} \,. \tag{2.3}$$

Indeed, each linear map $\operatorname{Re} \partial_0[v] + \operatorname{Im} \partial_0[v]$ ($v \in V_{\mathbb{C}}(I)$) vanishes at all $p \in I$ and thus belongs to $\mathcal{D}[I]$; moreover, they are linearly independent and $\dim \mathcal{D}[I] = \dim(\mathbb{R}[x]/I)^* = \dim \mathbb{R}[x]/I$, which is equal to $|V_{\mathbb{C}}(I)|$ since I is zero-dimensional and radical (using Theorem 2.2).

2.1.4. Multiplication operators. Given a polynomial $h \in \mathbb{R}[x]$, we can define the *multiplication (by h) operator* as

$$\begin{array}{rccl} \mathcal{X}_h : & \mathbb{R}[x]/I & \longrightarrow & \mathbb{R}[x]/I \\ & [f] & \longmapsto & \mathcal{X}_h([f]) := [hf], \end{array} \tag{2.4}$$

with adjoint operator

$$\begin{array}{rccl} \mathcal{X}_h^{\dagger} : & (\mathbb{R}[x]/I)^* & \longrightarrow & (\mathbb{R}[x]/I)^* \\ & L & \longmapsto & L \circ \mathcal{X}_h. \end{array}$$

Assume that $N := \dim \mathbb{R}[x]/I < \infty$. Then the multiplication operator $\mathcal{X}_h$ can be represented by its matrix, again denoted $\mathcal{X}_h$ for simplicity, with respect to a given basis $\mathcal{B} = \{b_1, \ldots, b_N\}$ of $\mathbb{R}[x]/I$ and then $\mathcal{X}_h^T$ represents $\mathcal{X}_h^{\dagger}$ with respect to the dual basis of $\mathcal{B}$. Namely, setting $\mathcal{N}_{\mathcal{B}}(hb_j) := \sum_{i=1}^{N} a_{ij} b_i$ for some scalars $a_{ij} \in \mathbb{R}$, the jth column of $\mathcal{X}_h$ is the vector $(a_{ij})_{i=1}^{N}$. Given $v \in \mathbb{C}^n$, define the vector $\zeta_{\mathcal{B},v} := (b_j(v))_{j=1}^{N} \in \mathbb{C}^N$, whose coordinates are the evaluations at v of the polynomials in $\mathcal{B}$. The following famous result (see e.g. [4, Chapter 2§4]) relates the eigenvalues of the multiplication operators in $\mathbb{R}[x]/I$ to the algebraic variety $V_{\mathbb{C}}(I)$. This result underlies the so-called eigenvalue method for solving polynomial equations and plays a central role in many algorithms, also in the present paper.

THEOREM 2.3. *Let I be a zero-dimensional ideal in $\mathbb{R}[x]$, $\mathcal{B}$ a basis of $\mathbb{R}[x]/I$, and $h \in \mathbb{R}[x]$. The eigenvalues of the multiplication operator $\mathcal{X}_h$ are the evaluations $h(v)$ of the polynomial h at the points $v \in V_{\mathbb{C}}(I)$. Moreover, $(\mathcal{X}_h)^T \zeta_{\mathcal{B},v} = h(v)\zeta_{\mathcal{B},v}$ for all $v \in V_{\mathbb{C}}(I)$.*

Throughout the paper we also denote by $\mathcal{X}_i := \mathcal{X}_{x_i}$ the matrix of the multiplication operator by the variable x_i. By the above theorem, the eigenvalues of the matrices $\mathcal{X}_i$ are the ith coordinate of the points $v \in V_{\mathbb{C}}(I)$. Thus the task of solving a system of polynomial equations is reduced to a task of numerical linear algebra once a basis of $\mathbb{R}[x]/I$ and a normal form algorithm are available, permitting the construction of the multiplication matrices $\mathcal{X}_i$.

2.1.5. Normal form criterion. The eigenvalue method for solving polynomial equations (recall Theorem 2.3) requires knowledge of a basis of $\mathbb{R}[x]/I$ and an algorithm to compute the normal form of a polynomial with respect to this basis. This, in turn, permits the construction of multiplication matrices $\mathcal{X}_i$ $(i = 1, \ldots, n)$ and therefore the computation of $V_{\mathbb{C}}(I)$.

A well known basis of $\mathbb{R}[x]/I$ is the set of standard monomials with respect to some monomial ordering. A classical way to obtain this basis is to compute a Gröbner basis of I from which the normal form of a polynomial can be found via a polynomial division algorithm using the given monomial ordering. (See e.g. [3, Chapter 1] for details.) Other techniques have been proposed for producing bases of the ideal I and of the vector space $\mathbb{R}[x]/I$, which do not depend on a specific monomial ordering. In particular, algorithms have been proposed for constructing border bases of I leading to general (stable by division) bases of $\mathbb{R}[x]/I$ (see [6, Chapter 4], [8] and [23]). Another normal form algorithm is proposed by Mourrain [14] (see also [15, 17]) leading to more general (namely, connected to 1) bases of $\mathbb{R}[x]/I$. The moment-matrix approach of this paper allows the computation of general polynomial bases of $\mathbb{R}[x]/I$ (or of $\mathbb{R}[x]/I(V_{\mathbb{R}}(I))$ as explained in [10]). We now recall the main notions and results about *border bases* and *rewriting families* needed for our treatment, following mainly [15, 17].

DEFINITION 2.1. *Given* $\mathcal{B} \subseteq \mathbb{T}^n$, *let* $\mathcal{B}^+ = \mathcal{B} \cup x_1\mathcal{B} \cup x_2\mathcal{B} \cup \ldots \cup x_n\mathcal{B}$ *with* $x_i\mathcal{B} := \{x_ib \mid b \in \mathcal{B},\ i = 1, \ldots, n\}$, *the expansion of* $\mathcal{B}$ *with one degree, and* $\partial\mathcal{B} := \mathcal{B}^+ \setminus \mathcal{B}$, *the border set of* $\mathcal{B}$. *The set* $\mathcal{B}$ *is said to be* connected to 1 *if each* $m \in \mathcal{B}$ *can be written as* $m = m_1 \cdots m_t$ *with* $m_1 = 1$ *and* $m_1 \cdots m_s \in \mathcal{B}$ $(s = 1, \ldots, t)$. *Moreover,* $\mathcal{B} \subseteq \mathbb{T}^n$ *is said to be* stable by division *if, for all* $m, m' \in \mathcal{B}$,

$$m \in \mathcal{B},\ m'|m \Rightarrow m' \in \mathcal{B}.$$

Obviously, $\mathcal{B}$ *is connected to 1 if it is stable by division.*

Assume $\mathcal{B} \subseteq \mathbb{T}^n$ is connected to 1. For each monomial $m \in \partial\mathcal{B}$, consider a polynomial f_m of the form

$$f_m := m - r_m \quad \text{where } r_m := \sum_{b \in \mathcal{B}} \lambda_{m,b}\, b \in \mathrm{Span}_{\mathbb{R}}(\mathcal{B}) \quad (\lambda_{m,b} \in \mathbb{R}). \tag{2.5}$$

The family

$$F := \{f_m \mid m \in \partial\mathcal{B}\}$$

is called a *rewriting family for* $\mathcal{B}$ in [15, 17] (or a $\mathcal{B}$-border prebasis in [6, Chapter 4]; note that $\mathcal{B}$ is assumed to be stable by division there). Thus a rewriting family enables expressing all monomials in $\partial\mathcal{B}$ as linear combinations of monomials in $\mathcal{B}$ modulo the ideal $\langle F\rangle$. Such a rewriting family can be used in a polynomial division algorithm to decompose any polynomial $p \in \mathbb{R}[x]$ as

$$p = r_p + \sum_{m \in \partial\mathcal{B}} u_m f_m \quad \text{where } r_p \in \mathrm{Span}_{\mathbb{R}}(\mathcal{B}),\ u_m \in \mathbb{R}[x]. \tag{2.6}$$

Therefore the set $\mathcal{B}$ spans the vector space $\mathbb{R}[x]/\langle F\rangle$ and in addition, if $\mathcal{B}$ is linearly independent in $\mathbb{R}[x]/\langle F\rangle$ then $\mathcal{B}$ is a basis of $\mathbb{R}[x]/\langle F\rangle$. This latter condition is equivalent to requiring that any polynomial can be reduced in a unique way using the rewriting family F and thus the decomposition (2.6) does not depend on the order in which the rewriting rules taken from F are applied.

Formally we can define a linear operator $\mathcal{X}_i : \mathrm{Span}_{\mathbb{R}}(\mathcal{B}) \to \mathrm{Span}_{\mathbb{R}}(\mathcal{B})$ using the rewriting family F; namely, for $b \in \mathcal{B}$, $\mathcal{X}_i(b) := x_i b$ if $x_i b \in \mathcal{B}$ and $\mathcal{X}_i(b) := N_{\mathcal{B}}(x_i b) = x_i b - f_{x_i b} = r_{x_i b}$ otherwise (recall (2.5)), and extend $\mathcal{X}_i$ to $\mathrm{Span}_{\mathbb{R}}(\mathcal{B})$ by linearity. Denote also by $\mathcal{X}_i$ the matrix of this linear operator, which can be seen as a *formal multiplication (by x_i) matrix*. The next result shows that the pairwise commutativity of the $\mathcal{X}_i$'s is sufficient to ensure the uniqueness of a decomposition (2.6). (See also [6, Chapter 4] in the case when $\mathcal{B}$ is stable by division.)

THEOREM 2.4. *[14] Let $\mathcal{B} \subseteq \mathbb{T}^n$ be a set connected to 1, let F be a rewriting family for $\mathcal{B}$, with associated formal multiplication matrices $\mathcal{X}_1, \ldots, \mathcal{X}_n$, and let $J := \langle F\rangle$ be the ideal generated by F. The following conditions are equivalent.*
(i) *The matrices $\mathcal{X}_1, \ldots, \mathcal{X}_n$ commute pairwise.*
(ii) *$\mathbb{R}[x] = \mathrm{Span}_{\mathbb{R}}(\mathcal{B}) \oplus J$, i.e. $\mathcal{B}$ is a basis of $\mathbb{R}[x]/J$.*
Then F is called a border basis *of the ideal J, and the matrix $\mathcal{X}_i$ represents the multiplication operator m_{x_i} of $\mathbb{R}[x]/J$ with respect to the basis $\mathcal{B}$.*

2.2. Bilinear forms and moment matrices.

2.2.1. Bilinear forms. Given $L \in (\mathbb{R}[x])^*$, we can define the symmetric bilinear form on $\mathbb{R}[x]$

$$\begin{aligned}(\cdot,\cdot)_L : \mathbb{R}[x] \times \mathbb{R}[x] &\to \mathbb{R}\\ (f,g) &\mapsto (f,g)_L := L(fg)\end{aligned}$$

with associated quadratic form

$$\begin{aligned}(\cdot)_L : \mathbb{R}[x] &\to \mathbb{R}\\ f &\mapsto \quad (f)_L := (f,f)_L = L(f^2).\end{aligned}$$

The kernel of this bilinear form $(\cdot,\cdot)_L$ is an ideal of $\mathbb{R}[x]$ (see e.g. [5]), which is real radical whenever the quadratic form $(\cdot)_L$ is positive semidefinite, i.e. satisfies $(f)_L = L(f^2) \geq 0$ for all $f \in \mathbb{R}[x]$ (see [12], [13]). We can define truncated analogues of $(\cdot,\cdot)_L$ and $(\cdot)_L$ on $\mathbb{R}[x]_t$ in the following way. Given $L \in (\mathbb{R}[x]_t)^*$, consider the bilinear form on $\mathbb{R}[x]_{\lfloor t/2\rfloor}$

$$\begin{aligned}(\cdot,\cdot)_L : \mathbb{R}[x]_{\lfloor t/2\rfloor} \times \mathbb{R}[x]_{\lfloor t/2\rfloor} &\to \mathbb{R}\\ (f,g) &\mapsto \quad (f,g)_L := L(fg),\end{aligned}$$

with associated quadratic form $(\cdot)_L$ on $\mathbb{R}[x]_{\lfloor t/2 \rfloor}$ defined by $(f)_L := L(f^2)$ for $f \in \mathbb{R}[x]_{\lfloor t/2 \rfloor}$.

2.2.2. Moment matrices. Fixing the canonical basis $(x^\alpha)_\alpha$ of the polynomial ring, the quadratic form $(\cdot)_L$ is positive semidefinite precisely when the matrix $(L(x^{\alpha+\beta}))_{\alpha,\beta}$ (with rows and columns indexed by $\mathbb{N}^n$ when $L \in (\mathbb{R}[x])^*$, and by $\mathbb{N}^n_{\lfloor t/2 \rfloor}$ when $L \in (\mathbb{R}[x]_t)^*$) is positive semidefinite. Note that the (α, β)-entry of this matrix depends only on the sum $\alpha + \beta$ and such a matrix is also known as the *moment matrix* associated with L. We may identify $L \in (\mathbb{R}[x])^*$ with its coordinate sequence $y := (L(x^\alpha))_{\alpha \in \mathbb{N}^n}$ in the canonical basis of $(\mathbb{R}[x])^*$, in which case we also write $L = L_y$.

Given $y \in \mathbb{R}^{\mathbb{N}^n}$, let $M(y)$ denote the matrix with rows and columns indexed by $\mathbb{N}^n$, and with (α, β)-entry $y_{\alpha+\beta}$, known as the *moment matrix* of y (or of the associated linear functional L_y). Analogously, for $L \in (\mathbb{R}[x]_t)^*$, let $y := (L(x^\alpha))_{\alpha \in \mathbb{N}^n_t}$ be the coordinate sequence of L in the canonical basis of $(\mathbb{R}[x]_t)^*$ and define the (truncated) moment matrix $M_{\lfloor t/2 \rfloor}(y)$ with rows and columns indexed by $\mathbb{N}^n_{\lfloor t/2 \rfloor}$, and with (α, β)-entry $y_{\alpha+\beta}$. Then $(\cdot)_L$ is positive semidefinite if and only if the matrix $M_{\lfloor t/2 \rfloor}(y)$ is positive semidefinite.

The kernel of $M(y)$ (resp. $M_{\lfloor t/2 \rfloor}(y)$) can be identified with the set of polynomials $p \in \mathbb{R}[x]$ (resp. $p \in \mathbb{R}[x]_{\lfloor t/2 \rfloor}$) such that $M(y)\,\mathrm{vec}(p) = 0$ (resp. $M_{\lfloor t/2 \rfloor}(y)\,\mathrm{vec}(p) = 0$). As observed above, $\mathrm{Ker}\, M(y)$ is an ideal of $\mathbb{R}[x]$ (and so $\mathrm{Ker}\, M(y) = \langle \mathrm{Ker}\, M(y) \rangle$), which is real radical when $M(y) \succeq 0$.

For $L := \partial_0[v]$, the evaluation at $v \in \mathbb{R}^n$, the quadratic form $(\cdot)_L$ is obviously positive semidefinite. Moreover, as the coordinate sequence of L in the canonical basis of $(\mathbb{R}[x])^*$ is $\zeta_v = (v^\alpha)_\alpha$, the matrix associated with $(\cdot)_L$ is just $\zeta_v \zeta_v^T$, and its kernel is the set of polynomials $p \in \mathbb{R}[x]$ that vanish at the point v. The above features explain the relevance of positive semidefinite quadratic forms and moment matrices to the problem of computing the real solutions of a system of polynomial equations. In [10] the 'real radical ideal' property of the kernel of a positive semidefinite quadratic form played a central role for finding all real roots and the real radical ideal for a zero-dimensional ideal of $\mathbb{R}[x]$. Here we will extend the method of [10] and show that, without the positive semidefiniteness assumption, bilinear forms and moment matrices can still be used for finding all complex roots of zero-dimensional systems of polynomial equations.

2.2.3. Flat extensions of moment matrices. We recall here some results about moment matrices needed throughout. We begin with recalling the following elementary property of kernels of block matrices, used for flat extensions of moment matrices in Theorems 2.5, 2.6 below.

LEMMA 2.1. *Let* $M = \begin{pmatrix} A & B \\ B^T & C \end{pmatrix}$ *be a symmetric matrix.*

(i) *Assume* $\operatorname{rank} M = \operatorname{rank} A$. *Then,* $x \in \operatorname{Ker} A \Longleftrightarrow \tilde{x} := \begin{pmatrix} x \\ 0 \end{pmatrix} \in \operatorname{Ker} M$ *and* $\operatorname{Ker} M = \operatorname{Ker} \begin{pmatrix} A & B \end{pmatrix}$.

(ii) *Assume* $M \succeq 0$. *Then,* $x \in \operatorname{Ker} A \Longleftrightarrow \tilde{x} := \begin{pmatrix} x \\ 0 \end{pmatrix} \in \operatorname{Ker} M$.

Proof. (i) As $\operatorname{rank} M = \operatorname{rank} A$, there exists a matrix U for which $B = AU$ and $C = B^T U = U^T AU$. Then, $Ax = 0 \Longrightarrow B^T x = U^T Ax = 0 \Longrightarrow M\tilde{x} = 0$. Moreover, $Ax + By = 0$ implies $By = -Ax$ and thus $B^T x + Cy = U^T Ax + U^T AUy = U^T Ax + U^T(-Ax) = 0$, showing (i).

(ii) If $Ax = 0$ then $\tilde{x}^T M \tilde{x} = x^T Ax = 0$, which implies $M\tilde{x} = 0$ when $M \succeq 0$. □

When a matrix M with the block form shown in Lemma 2.1 satisfies $\operatorname{rank} M = \operatorname{rank} A$, one says that *$M$ is a flat extension of A*. Curto and Fialkow [5] show the following result (see also [11] for a detailed exposition).

THEOREM 2.5. *[5] (Flat Extension theorem) Let* $y \in \mathbb{R}^{\mathbb{N}^n_{2t}}$ *and assume that*

$$\operatorname{rank} M_t(y) = \operatorname{rank} M_{t-1}(y).$$

Then one can extend y to $\tilde{y} \in \mathbb{R}^{\mathbb{N}^n_{2t+2}}$ *in such a way that* $\operatorname{rank} M_{t+1}(\tilde{y}) = \operatorname{rank} M_t(y)$.

Based on this, one can prove the following result which plays a central role in our moment-matrix approach (as well as in the previous paper [10]).

THEOREM 2.6. *Let* $y \in \mathbb{R}^{\mathbb{N}^n_{2t}}$ *and assume that*

$$\operatorname{rank} M_t(y) = \operatorname{rank} M_{t-1}(y).$$

Then one can extend y to $\tilde{y} \in \mathbb{R}^{\mathbb{N}^n}$ *in such a way that* $\operatorname{rank} M(\tilde{y}) = \operatorname{rank} M_t(y)$. *Moreover,* $\operatorname{Ker} M(\tilde{y}) = \langle \operatorname{Ker} M_t(y) \rangle$, *and any basis* $\mathcal{B} \subseteq \mathbb{T}^n_{t-1}$ *of the column space of* $M_t(y)$ *is a basis of* $\mathbb{R}[x]/\langle \operatorname{Ker} M(\tilde{y}) \rangle$.

Proof. The existence of $\tilde{y}$ follows applying iteratively Theorem 2.5. As $\operatorname{rank} M(\tilde{y}) = \operatorname{rank} M_t(y)$, the inclusion $\operatorname{Ker} M_t(y) \subseteq \operatorname{Ker} M(\tilde{y})$ follows from Lemma 2.1 (i). Hence $\langle \operatorname{Ker} M_t(y) \rangle \subseteq \operatorname{Ker} M(\tilde{y})$, since $\operatorname{Ker} M(\tilde{y})$ is an ideal of $\mathbb{R}[x]$. Let $\mathcal{B} \subseteq \mathbb{T}^n_{t-1}$ index a basis of the column space of $M_t(y)$. Hence $\mathcal{B}$ also indexes a basis of the column space of $M(\tilde{y})$, which implies $\operatorname{Span}_{\mathbb{R}}(\mathcal{B}) \cap \operatorname{Ker} M(\tilde{y}) = \{0\}$ and thus $\operatorname{Span}_{\mathbb{R}}(\mathcal{B}) \cap \langle \operatorname{Ker} M_t(y) \rangle = \{0\}$. We now show that

$$\mathbb{R}[x] = \operatorname{Span}_{\mathbb{R}}(\mathcal{B}) \oplus \langle \operatorname{Ker} M_t(y) \rangle. \tag{2.7}$$

For this it suffices to show that $x^\alpha \in \operatorname{Span}_{\mathbb{R}}(\mathcal{B}) + \langle \operatorname{Ker} M_t(y) \rangle$ for all $\alpha \in \mathbb{N}^n$. We use induction on $|\alpha|$. If $|\alpha| \le t$ just use the definition of $\mathcal{B}$. Next, let $|\alpha| \ge t+1$ and write $x^\alpha = x_i x^\delta$. By the induction assumption, $x^\delta = \sum_{x^\beta \in \mathcal{B}} \lambda_\beta x^\beta + q$ where $q \in \langle \operatorname{Ker} M_t(y) \rangle$. Hence, $x^\alpha = \sum_{x^\beta \in \mathcal{B}} \lambda_\beta x_i x^\beta +$

$x_i q$, with $x_i q \in \langle \operatorname{Ker} M_t(y)\rangle$. As $\deg(x_i x^\beta) \leq 1 + t - 1 = t$, each $x_i x^\beta$ lies in $\operatorname{Span}_{\mathbb{R}}(\mathcal{B}) + \langle \operatorname{Ker} M_t(y)\rangle$ and therefore x^α also lies in $\operatorname{Span}_{\mathbb{R}}(\mathcal{B}) + \langle \operatorname{Ker} M_t(y)\rangle$. Hence (2.7) holds. This implies

$$\operatorname{Ker} M(\tilde{y}) = \langle \operatorname{Ker} M_t(y)\rangle.$$

Indeed let $f \in \operatorname{Ker} M(\tilde{y})$ and write $f = r + q$ with $r \in \operatorname{Span}_{\mathbb{R}}(\mathcal{B})$ and $q \in \langle \operatorname{Ker} M_t(y)\rangle$. Thus $r = f - q \in \operatorname{Ker} M(\tilde{y}) \cap \operatorname{Span}_{\mathbb{R}}(\mathcal{B}) = \{0\}$, which implies $f = q \in \langle \operatorname{Ker} M_t(y)\rangle$. The above argument also shows that $\mathcal{B}$ is a basis of the space $\mathbb{R}[x]/\langle \operatorname{Ker} M_t(y)\rangle$. □

3. The moment-matrix approach for complex roots. In this section we show how the method from [10] can be simply adapted to find all complex roots for a zero-dimensional ideal. The method of [10] was designed to find $V_{\mathbb{R}}(I)$ (assuming it is finite) and uses the set $K_{t,\succeq}$ introduced in (1.3). We now show that only by *omitting* the positivity condition in (1.3) and working instead with the set $\mathcal{K}_t$ from (1.2), we can find the complex variety $V_{\mathbb{C}}(I)$.

3.1. Approaching I with kernels of moment matrices. Let $I = \langle h_1, \ldots, h_m\rangle$ be a zero-dimensional ideal whose associated complex variety $V_{\mathbb{C}}(I)$ has to be found. Throughout we set

$$D := \max_{j=1,\ldots,m} \deg(h_j), \quad d := \max_{j=1,\ldots,m} \lceil \deg(h_j)/2 \rceil = \lceil D/2 \rceil. \tag{3.1}$$

Recall the definition of the sets $\mathcal{H}_t$, $\mathcal{K}_t$ from (1.1), (1.2):

$$\mathcal{H}_t := \{x^\alpha h_j \in \mathbb{R}[x]_t \mid j = 1, \ldots, m, \alpha \in \mathbb{N}^n \text{ with } |\alpha| + \deg(h_j) \leq t\},$$

$$\mathcal{K}_t = \mathcal{H}_t^\perp = \{L \in (\mathbb{R}[x]_t)^* \mid L(p) = 0 \quad \forall p \in \mathcal{H}_t\}.$$

Equivalently, identifying $L \in (\mathbb{R}[x]_t)^*$ with its sequence of coefficients $y = (y_\alpha)_\alpha$ in the canonical basis of $(\mathbb{R}[x]_t)^*$ and setting $L_y := L$,

$$\mathcal{K}_t = \{y \in \mathbb{R}^{\mathbb{N}^n_t} \mid L_y(p) = y^T \operatorname{vec}(p) = 0 \quad \forall p \in \mathcal{H}_t\}.$$

For further reference, notice the following fact about the moment matrix $M_{\lfloor t/2 \rfloor}(y)$ of an arbitrary $y \in \mathbb{R}^{\mathbb{N}^n_t}$. If $\deg(fg)$, $\deg(gh) \leq \lfloor t/2 \rfloor$ then

$$\operatorname{vec}(f)^T M_{\lfloor t/2 \rfloor}(y) \operatorname{vec}(gh) = \operatorname{vec}(fg)^T M_{\lfloor t/2 \rfloor}(y) \operatorname{vec}(h) \; (= L_y(fgh)). \tag{3.2}$$

We now show several results relating the kernel of the moment matrix $M_{\lfloor t/2 \rfloor}(y)$ of $y \in \mathcal{K}_t$ to the ideal I.

3.1.1. The inclusion $I \subseteq \langle \operatorname{Ker} M_{\lfloor t/2 \rfloor}(y)\rangle$. The next two lemmas give sufficient conditions ensuring that the ideal generated by the kernel of $M_{\lfloor t/2 \rfloor}(y)$ contains the ideal I.

LEMMA 3.1. *Let $y \in \mathcal{K}_t$ and let s be an integer with $D \leq s \leq \lfloor t/2 \rfloor$. Then* $\text{vec}(h_1), \ldots, \text{vec}(h_m) \in \text{Ker}\, M_s(y)$ *and thus* $I \subseteq \langle \text{Ker}\, M_s(y) \rangle$.

Proof. For $\alpha \in \mathbb{N}^n_s$, $(M_s(y)\,\text{vec}(h_j))_\alpha = L_y(x^\alpha h_j) = 0$ since $x^\alpha h_j \in \mathcal{H}_t$ as $\deg(x^\alpha h_j) \leq s + D \leq t$. □

LEMMA 3.2. *Let $y \in \mathcal{K}_t$ and let s be an integer with $d \leq s \leq \lfloor t/2 \rfloor$. If* $\text{rank}\, M_s(y) = \text{rank}\, M_{s-d}(y)$ *then* $I \subseteq \langle \text{Ker}\, M_s(y) \rangle$.

Proof. We apply Theorem 2.6: Let $\tilde{y} \in \mathbb{R}^{\mathbb{N}^n}$ be an extension of $\pi_{2s}(y)$ such that $\text{rank}\, M(\tilde{y}) = \text{rank}\, M_s(y)$ and let $\mathcal{B} \subseteq \mathbb{T}^n_{s-d}$ index a basis of the column space of $M_s(y)$. Then $\mathcal{B}$ is a basis of $\mathbb{R}[x]/\text{Ker}\, M(\tilde{y})$ and $\text{Ker}\, M(\tilde{y}) = \langle \text{Ker}\, M_s(y) \rangle$. It suffices now to show that $h_j \in \text{Ker}\, M(\tilde{y})$ for all $j = 1, \ldots, m$. For this write $h_j = r + q$ where $r \in \text{Span}_{\mathbb{R}}(\mathcal{B})$ and $q \in \text{Ker}\, M(\tilde{y})$. Then $M(\tilde{y})\,\text{vec}(h_j) = M(\tilde{y})\,\text{vec}(r)$. As $\deg(r) \leq s-d$ and $M(\tilde{y})$ is a flat extension of $M_{s-d}(\tilde{y})$, Lemma 2.1 (i) implies that $M(\tilde{y})\,\text{vec}(r) = 0$ if and only if $M_{s-d}(\tilde{y})\,\text{vec}(r) = 0$, i.e. $\text{vec}(x^\alpha)^T M(\tilde{y})\,\text{vec}(r) = 0$ for all $\alpha \in \mathbb{N}^n_{s-d}$. Given $\alpha \in \mathbb{N}^n_{s-d}$, $\text{vec}(x^\alpha)^T M(\tilde{y})\,\text{vec}(r) = \text{vec}(x^\alpha)^T M(\tilde{y})\,\text{vec}(h_j) = L_{\tilde{y}}(x^\alpha h_j)$, which is equal to $L_y(x^\alpha h_j)$ since $\deg(x^\alpha h_j) \leq s - d + 2d \leq 2s$, and in turn is equal to 0 since $x^\alpha h_j \in \mathcal{H}_t$ and $y \in \mathcal{K}_t$. □

3.1.2. The inclusion $\langle \text{Ker}\, M_t(y) \rangle \subseteq I(V_{\mathbb{C}}(I))$ for generic y. We now show that, under some maximality assumption on the rank of the matrix $M_{\lfloor t/2 \rfloor}(y)$, the polynomial ideal $\langle \text{Ker}\, M_{\lfloor t/2 \rfloor}(y) \rangle$ is contained in $I(V_{\mathbb{C}}(I))$.

THEOREM 3.1. *Given $1 \leq s \leq \lfloor t/2 \rfloor$, let $y \in \mathcal{K}_t$ for which* $\text{rank}\, M_s(y)$ *is maximum; that is,*

$$\text{rank}\, M_s(y) = \max_{z \in \mathcal{K}_t} \text{rank}\, M_s(z). \tag{3.3}$$

Then $\langle \text{Ker}\, M_s(y) \rangle \subseteq I(V_{\mathbb{C}}(I))$.

Proof. It suffices to show that $\text{Ker}\, M_s(y) \subseteq I(V_{\mathbb{C}}(I))$. Suppose for contradiction that there exists $f \in \mathbb{R}[x]_s$ with $\text{vec}(f) \in \text{Ker}\, M_s(y)$ and $f \notin I(V_{\mathbb{C}}(I))$. Then there exists $v \in V_{\mathbb{C}}(I)$ for which $f(v) \neq 0$.

We first consider the case when $v \in \mathbb{R}^n$. Then $\zeta_{t,v} \in \mathcal{K}_t$. Set $y' := y + \zeta_{t,v}$. Then $y' \in \mathcal{K}_t$ and $\text{vec}(f) \notin \text{Ker}\, M_s(y')$ since $\text{vec}(f) \notin \text{Ker}\, M_s(\zeta_{t,v})$. Therefore, $\text{Ker}\, M_s(y') \not\subseteq \text{Ker}\, M_s(y)$, for otherwise we would have strict inclusion: $\text{Ker}\, M_s(y') \subsetneq \text{Ker}\, M_s(y)$, implying $\text{rank}\, M_s(y') > \text{rank}\, M_s(y)$ and thus contradicting the maximality assumption on $\text{rank}\, M_s(y)$. Hence there exists $f' \in \mathbb{R}[x]_s$ with $\text{vec}(f') \in \text{Ker}\, M_s(y') \setminus \text{Ker}\, M_s(y)$. We have $M_s(y)\,\text{vec}(f') = -f'(v)\zeta_{s,v}$ and $f'(v) \neq 0$. Moreover,

$$\text{vec}(f)^T M_s(y)\,\text{vec}(f') = -f(v)f'(v)\,,$$

yielding a contradiction since $f(v)f'(v) \neq 0$ and $\text{vec}(f)^T M_s(y)\,\text{vec}(f') = 0$.

We now consider the case when $v \in \mathbb{C}^n \setminus \mathbb{R}^n$. The proof is along the same lines but needs a more detailed analysis. We start with the following observation.

CLAIM 3.2. *For $v \in \mathbb{C}^n \setminus \mathbb{R}^n$ and $s \geq 1$, the two vectors $\zeta_{s,v}$ and $\zeta_{s,\overline{v}}$ are linearly independent over $\mathbb{C}$.*

Proof. Assume $\lambda\zeta_{s,v} + \mu\zeta_{s,\overline{v}} = 0$ where $\lambda, \mu \in \mathbb{C}$. Then $\lambda + \mu = 0$ (evaluating at the coordinate indexed by the constant monomial 1) and $\lambda(v_i - \overline{v}_i) = 0$ $(i = 1, \ldots, n)$, implying $\lambda = \mu = 0$ since $v_i \notin \mathbb{R}$ for some i. □

Define the two vectors

$$y' := y + f(\overline{v})\zeta_{t,v} + f(v)\zeta_{t,\overline{v}},\ y'' := y + \mathbf{i}(f(v)\zeta_{t,\overline{v}} - f(\overline{v})\zeta_{t,v}),$$

where $\mathbf{i}$ denotes the complex root of -1. Then, $y', y'' \in \mathcal{K}_t$, $\text{vec}(f) \notin \text{Ker}\, M_s(y')$ since $M_s(y')\,\text{vec}(f) = |f(v)|^2(\zeta_{s,v} + \zeta_{s,\overline{v}}) \neq 0$ and $\text{vec}(f) \notin \text{Ker}\, M_s(y'')$ since $M_s(y'')\,\text{vec}(f) = \mathbf{i}|f(v)|^2(\zeta_{s,\overline{v}} - \zeta_{s,v}) \neq 0$ (as $v \notin \mathbb{R}^n$). By the maximality assumption on $\text{rank}\, M_s(y)$, $\text{Ker}\, M_s(y') \not\subseteq \text{Ker}\, M_s(y)$ and $\text{Ker}\, M_s(y'') \not\subseteq \text{Ker}\, M_s(y)$. In what follows, $\text{Re}\, a$, $\text{Im}\, a$ denote, respectively, the real and imaginary parts of $a \in \mathbb{C}$.

CLAIM 3.3.
(i) $\text{Re}\, g(v) = 0$ *for all* $g \in \mathbb{R}[x]_s$ *with* $\text{vec}(g) \in \text{Ker}\, M_s(y') \setminus \text{Ker}\, M_s(y)$.
(ii) $\text{Im}\, g(v) = 0$ *for all* $g \in \mathbb{R}[x]_s$ *with* $\text{vec}(g) \in \text{Ker}\, M_s(y'') \setminus \text{Ker}\, M_s(y)$.

Proof. (i) For $g \in \mathbb{R}[x]_s$ with $\text{vec}(g) \in \text{Ker}\, M_s(y')$, we have:

$$\begin{aligned} 0 &= \text{vec}(f)^T M_s(y')\,\text{vec}(g) \\ &= \text{vec}(f)^T (f(\overline{v})\zeta_{s,v}\zeta_{s,v}^T + f(v)\zeta_{s,\overline{v}}\zeta_{s,\overline{v}}^T)\,\text{vec}(g) \\ &= |f(v)|^2(g(v) + g(\overline{v})) \end{aligned}$$

implying that $g(v)$ is a pure imaginary complex number, i.e., $\text{Re}\, g(v) = 0$.
(ii) Similarly, for $g \in \mathbb{R}[x]_s$ with $\text{vec}(g) \in \text{Ker}\, M_s(y'')$,

$$0 = \text{vec}(f)^T M_s(y'')\,\text{vec}(g) = \mathbf{i}|f(v)|^2(g(\overline{v}) - g(v))$$

which implies that $g(v) \in \mathbb{R}$, i.e., $\text{Im}\, g(v) = 0$. □

Fix $f' \in \mathbb{R}[x]_s$ with $\text{vec}(f') \in \text{Ker}\, M_s(y') \setminus \text{Ker}\, M_s(y)$ and fix $f'' \in \mathbb{R}[x]_s$ with $\text{vec}(f'') \in \text{Ker}\, M_s(y'') \setminus \text{Ker}\, M_s(y)$ with $f'(v) = \mathbf{i}$ and $f''(v) = 1$. Set $W_0' := \text{Ker}\, M_s(y) \cap \text{Ker}\, M_s(y')$.

CLAIM 3.4.
(i) $\text{Ker}\, M_s(y') = W_0' + \mathbb{R}\ \text{vec}(f')$.
(ii) $\text{Ker}\, M_s(y) = W_0' + \mathbb{R}\ \text{vec}(f)$.
(iii) $g(v) = 0$ *for all* $g \in \mathbb{R}[x]_s$ *with* $\text{vec}(g) \in W_0'$.

Proof. (i) Let $g \in \mathbb{R}[x]_s$ with $\text{vec}(g) \in \text{Ker}\, M_s(y') \setminus \text{Ker}\, M_s(y)$. By Claim 3.3, $g(v) = \mathbf{i}a$ for some $a \in \mathbb{R}$. As $g - af'$ vanishes at v and $\overline{v}$,

$\mathrm{vec}(g) - a\,\mathrm{vec}(f') \in \mathrm{Ker}\, M_s(y)$ and thus $\mathrm{vec}(g) - a\,\mathrm{vec}(f') \in W_0'$. This shows that $\mathrm{Ker}\, M_s(y') = W_0' + \mathbb{R}\,\mathrm{vec}(f')$.

(ii) Setting $k_0 := \dim W_0'$, we have $\dim \mathrm{Ker}\, M_s(y') = k_0 + 1$. As $W_0' + \mathbb{R}\,\mathrm{vec}(f) \subseteq \mathrm{Ker}\, M_s(y)$, we have $\dim \mathrm{Ker}\, M_s(y) \geq k_0 + 1$; moreover equality holds for otherwise one would have $\mathrm{rank}\, M_s(y) < \mathrm{rank}\, M_s(y')$. Therefore, $\mathrm{Ker}\, M_s(y) = W_0' + \mathbb{R}\,\mathrm{vec}(f)$.

(iii) Assume $\mathrm{vec}(g) \in W_0'$; then

$$0 = (f(\overline{v})\zeta_{s,v}\zeta_{s,v}^T + f(v)\zeta_{s,\overline{v}}\zeta_{s,\overline{v}}^T)\,\mathrm{vec}(g) f(\overline{v})g(v)\zeta_{s,v} + f(v)g(\overline{v})\zeta_{s,\overline{v}}$$

which, using Claim 3.2, implies that $f(\overline{v})g(v) = 0$ and thus $g(v) = 0$. □

CLAIM 3.5. *$f(v) = a(1 + \mathbf{i})$ for some $a \in \mathbb{R}$, i.e.,* $\mathrm{Re}\, f(v) = \mathrm{Im}\, f(v)$.

Proof. We first show that $\mathrm{vec}(f' + f'') \in \mathrm{Ker}\, M_s(y)$. Indeed,

$$\begin{aligned} &-M_s(y)\,\mathrm{vec}(f' + f'') \\ &= (f(\overline{v})f'(v)\zeta_{s,v} + f(v)f'(\overline{v})\zeta_{s,\overline{v}}) + \mathbf{i}(f(v)f''(\overline{v})\zeta_{s,\overline{v}} - f(\overline{v})f''(v)\zeta_{s,v}) \\ &= (\mathbf{i}f(\overline{v})\zeta_{s,v} - \mathbf{i}f(v)\zeta_{s,\overline{v}}) + \mathbf{i}(f(v)\zeta_{s,\overline{v}} - f(\overline{v})\zeta_{s,v}) = 0. \end{aligned}$$

By Claim 3.4 (ii), $\mathrm{Ker}\, M_s(y) = W_0' + \mathbb{R}\,\mathrm{vec}(f)$. Therefore, $\mathrm{vec}(f' + f'') = \mathrm{vec}(f_0) + \lambda\,\mathrm{vec}(f)$ for some $\mathrm{vec}(f_0) \in W_0'$ and $\lambda \in \mathbb{R}$. As $f_0(v) = 0$ (by Claim 3.4 (iii)), we find that $\lambda f(v) = f'(v) + f''(v) = \mathbf{i} + 1$ and thus $\mathrm{Re}\, f(v) = \mathrm{Im}\, f(v)$. □

We are now ready to conclude the proof of Theorem 3.1. We have just proven the following fact: Let $y \in \mathcal{K}_t$ for which $\mathrm{rank}\, M_s(y)$ is maximum; if $\mathrm{vec}(f) \in \mathrm{Ker}\, M_s(y)$ satisfies $f(v) \neq 0$ for some $v \in V(I) \setminus \mathbb{R}^n$, then $\mathrm{Re}\, f(v) = \mathrm{Im}\, f(v)$. On the other hand we have constructed $y' \in \mathcal{K}_t$ for which $\mathrm{rank}\, M_s(y')$ is maximum (since $\mathrm{rank}\, M_s(y') = \mathrm{rank}\, M_s(y)$ by Claim 3.4) (i) and (ii)) and $\mathrm{vec}(f') \in \mathrm{Ker}\, M_s(y')$ with $f'(v) \neq 0$ and $\mathrm{Re}\, f'(v) = 0 \neq 1 = \mathrm{Im}\, f'(v)$. Therefore we reach a contradiction. □

3.1.3. The ingredients for our algorithm for $V_{\mathbb{C}}(I)$. As a direct consequence of Lemma 3.1 and Theorem 3.1, if $y \in \mathcal{K}_t$ satisfies (3.3) for $D \leq s \leq \lfloor t/2 \rfloor$, then

$$I \subseteq \langle \mathrm{Ker}\, M_s(y) \rangle \subseteq I(V_{\mathbb{C}}(I)). \tag{3.4}$$

and thus $V_{\mathbb{C}}(I) = V_{\mathbb{C}}(\langle \mathrm{Ker}\, M_s(y) \rangle)$. This does not help to compute $V_{\mathbb{C}}(I)$ yet since we also need a basis of the quotient space $\mathbb{R}[x]/\langle \mathrm{Ker}\, M_s(y) \rangle$. A crucial feature is that, if moreover the matrix $M_s(y)$ is a flat extension of its submatrix $M_{s-1}(y)$ then, in view of Theorem 2.6, any basis $\mathcal{B}$ of the column space of $M_{s-1}(y)$ is a basis of the quotient space $\mathbb{R}[x]/\langle \mathrm{Ker}\, M_s(y) \rangle$. We now state the main result on which our algorithm is based.

THEOREM 3.6. *Let $y \in \mathcal{K}_t$, let $1 \leq s \leq \lfloor t/2 \rfloor$, assume that* $\mathrm{rank}\, M_s(y)$ *is maximum, i.e. that (3.3) holds, and consider the conditions:*

$$\mathrm{rank}\, M_s(y) = \mathrm{rank}\, M_{s-1}(y) \quad \textit{with } D \leq s \leq \lfloor t/2 \rfloor, \tag{3.5}$$

$$\mathrm{rank}\, M_s(y) = \mathrm{rank}\, M_{s-d}(y) \quad \textit{with } d \leq s \leq \lfloor t/2 \rfloor. \tag{3.6}$$

If (3.5) or (3.6) holds then (3.4) holds and any basis of the column space of $M_{s-1}(y)$ is a basis of $\mathbb{R}[x]/\langle \operatorname{Ker} M_s(y)\rangle$. Hence one can construct the multiplication matrices in $\mathbb{R}[x]/\langle \operatorname{Ker} M_s(y)\rangle$ from the matrix $M_s(y)$ and find the variety $V_{\mathbb{C}}(\langle \operatorname{Ker} M_s(y)\rangle) = V_{\mathbb{C}}(I)$ using the eigenvalue method.

Proof. Directly using Theorems 2.6, 3.1 and Lemmas 3.1, 3.2. □

The next result will be used for proving termination of our algorithm.

PROPOSITION 3.1. *Assume $1 \le |V_{\mathbb{C}}(I)| < \infty$. There exist integers t_1, t_2 such that, for any t with $\lfloor t/2 \rfloor \ge t_1 + t_2$, $\operatorname{rank} M_{t_1}(y) = \operatorname{rank} M_{t_1-d}(y)$ holds for all $y \in \mathcal{K}_t$.*

Proof. Let $y \in \mathcal{K}_t$ and assume $t \ge 2D$. Then, by Lemma 3.1, $h_1, \ldots, h_m \in \operatorname{Ker} M_{\lfloor t/2 \rfloor}(y)$. We will use the following fact which follows from (3.2): For $u \in \mathbb{R}[x]$,

$$[M_{\lfloor t/2 \rfloor}(y) \operatorname{vec}(uh_j)]_\gamma = 0 \text{ if } |\gamma| + \deg(u) \le \lfloor t/2 \rfloor. \tag{3.7}$$

Let $\mathcal{B}$ be a basis of $\mathbb{R}[x]/I$ and set $d_{\mathcal{B}} := \max_{b \in \mathcal{B}} \deg(b)$ (which is well defined as $|\mathcal{B}| < \infty$ since $|V_{\mathbb{C}}(I)| < \infty$). Write any monomial as

$$x^\alpha = r^{(\alpha)} + \sum_{j=1}^m u_j^{(\alpha)} h_j,$$

where $r^{(\alpha)} \in \operatorname{Span}_{\mathbb{R}}(\mathcal{B})$ and $u_j^{(\alpha)} \in \mathbb{R}[x]$. Set $t_1 := \max(D, d_{\mathcal{B}} + d)$,

$$t_2 := \max(\deg(u_j^{(\alpha)}) \mid j = 1, \ldots, m, |\alpha| \le t_1)$$

and let t be such that $\lfloor t/2 \rfloor \ge t_1 + t_2$. Let $y \in \mathcal{K}_t$; we show that $M_{t_1}(y)$ is a flat extension of $M_{t_1-d}(y)$. For this consider $\alpha, \gamma \in \mathbb{N}^n_{t_1}$. Then $|\gamma| + \deg(u_j^{(\alpha)}) \le t_1 + t_2 \le \lfloor t/2 \rfloor$. Hence, by (3.7), the γth component of $M_{\lfloor t/2 \rfloor}(y) \operatorname{vec}(u_j^{(\alpha)} h_j)$ is equal to 0 and thus the γth component of $M_{\lfloor t/2 \rfloor}(y) \operatorname{vec}(x^\alpha - r^{(\alpha)})$ is equal to 0. In other words, for $|\alpha| \le t_1$, the αth column of $M_{t_1}(y)$ is a linear combination of the columns of $M_{t_1}(y)$ indexed by $\mathcal{B}$ and thus $M_{t_1}(y)$ is a flat extension of $M_{t_1-d}(y)$ as $d_{\mathcal{B}} \le t_1 - d$. □

We next provide a criterion for detecting when the variety $V_{\mathbb{C}}(I)$ is empty.

PROPOSITION 3.2. *The following statements are equivalent.*

(i) $V_{\mathbb{C}}(I) = \emptyset$.

(ii) *There exist $t_1, t_2 \in \mathbb{N}$ such that, for all t with $\lfloor t/2 \rfloor \ge t_1 + t_2$ and all $y \in \mathcal{K}_t$, $y_\alpha = 0$ for all $\alpha \in \mathbb{N}^n_{t_1}$.*

Proof. If $v \in V_{\mathbb{C}}(I)$, then $y := \zeta_{t,v} + \zeta_{t,\bar{v}} \in \mathcal{K}_t$ with $y_0 = 2$; this showing (ii) $\Longrightarrow$ (i). Conversely, assume $V_{\mathbb{C}}(I) = \emptyset$. Then, by Hilbert's Nullstellensatz, $1 \in I$, i.e., $1 = \sum_{j=1}^m u_j h_j$ for some $u_j \in \mathbb{R}[x]$. Set $t_1 := D$,

$t_2 := \max_j \deg(u_j)$ and consider $y \in \mathcal{K}_t$ where $\lfloor t/2 \rfloor \geq t_1 + t_2$. Then, for each j, $[M_t(y)\,\mathrm{vec}(u_j h_j)]_\alpha = 0$ if $|\alpha| \leq t_1$ (using (3.7)). Therefore, $y_\alpha = [M_t(y)\,\mathrm{vec}(1)]_\alpha = 0$ for all $|\alpha| \leq t_1$. □

3.1.4. Sketch of the algorithm for finding $V_{\mathbb{C}}(I)$. We can now describe our algorithm for finding $V_{\mathbb{C}}(I)$. Algorithm 2 is similar to the one introduced in [10] for the task of computing real roots, except that now it only uses standard numerical linear algebra and *no* semidefinite programming.

Algorithm 2 *The moment-matrix algorithm for $V_{\mathbb{C}}(I)$:*

Input: $t \geq D$.
Output: A basis $\mathcal{B} \subseteq \mathbb{R}[x]_{s-1}$ of $\mathbb{K}[x]/\langle \mathrm{Ker}\, M_s(y)\rangle$ needed to compute $V_{\mathbb{C}}(I)$.

1: Find $y \in \mathcal{K}_t$ for which $\mathrm{rank}\, M_s(y)$ is maximum for all $1 \leq s \leq \lfloor t/2 \rfloor$.
2: Check if (3.5) holds for some $D \leq s \leq \lfloor t/2 \rfloor$, or if (3.6) holds for some $d \leq s \leq \lfloor t/2 \rfloor$.
3: **if** yes **then**
4: **return** a basis $\mathcal{B} \subseteq \mathbb{R}[x]_{s-1}$ of the column space of $M_{s-1}(y)$, and extract $V_{\mathbb{C}}(I)$ (applying the eigenvalue method to the quotient space $\mathbb{R}[x]/\langle \mathrm{Ker}\, M_s(y)\rangle$ with basis $\mathcal{B}$).
5: **else**
6: Iterate (go to 1)) replacing t by $t+1$
7: **end if**

REMARK 3.1. Proposition 3.1 guarantees the termination of this algorithm.

More details concerning Step 2 (in particular, about finding a basis of the column space and implementing the eigenvalue method) can be found in our preceding paper [10]. We now discuss the issue raised in Step 1 of Algorithm 2, that is, how to find $y \in \mathcal{K}_t$ satisfying

$$\forall s,\ 1 \leq s \leq \lfloor t/2 \rfloor,\ \mathrm{rank}\, M_s(y) = \max_{z \in \mathcal{K}_t} \mathrm{rank}\, M_s(z) =: R_{t,s}. \tag{3.8}$$

As we now show, this property is in fact a generic property of $\mathcal{K}_t$, i.e. the set of points $y \in \mathcal{K}_t$ that do not have this property has measure 0. For this, set $N_t := \dim \mathcal{K}_t$ and let $z_1, \ldots, z_{N_t} \in \mathbb{R}^{\mathbb{N}^n_t}$ be a linear basis of $\mathcal{K}_t$, so that $\mathcal{K}_t = \{\sum_{i=1}^{N_t} a_i z_i \mid a_i \in \mathbb{R}\}$. For $1 \leq s \leq \lfloor t/2 \rfloor$, set

$$\Omega_{t,s} := \left\{ a = (a_1, \ldots, a_{N_t}) \in \mathbb{R}^{N_t} \mid \mathrm{rank}\, M_s\left(\sum_{i=1}^{N_t} a_i z_i\right) < R_{t,s} \right\}.$$

LEMMA 3.3. *$\Omega_{t,s} = V_{\mathbb{R}}(\mathcal{P}_{t,s})$ for some finite set $\mathcal{P}_{t,s} \subseteq \mathbb{R}[x_1, \ldots, x_{N_t}]$ containing at least one nonzero polynomial.*

Proof. The condition $\operatorname{rank} M_s(\sum_{i=1}^{N_t} a_i z_i) < R_{t,s}$ is equivalent to requiring that all $R_{t,s} \times R_{t,s}$ submatrices of $M_s(\sum_{i=1}^{N_t} a_i z_i)$ have zero determinant. Each such determinant can be expressed as a polynomial in the variables $a_1, \ldots, a_{N_t}$. Therefore there exists a finite set $\mathcal{P}_{t,s}$ of polynomials in $\mathbb{R}[x_1, \ldots, x_{N_t}]$ for which $\Omega_{t,s} = V_{\mathbb{R}}(\mathcal{P}_{t,s})$. By definition of $R_{t,s}$, there exists $a \in \mathbb{R}^{N_t}$ for which $\operatorname{rank} M_s(\sum_i a_i z_i) = R_{t,s}$. Hence, at least one $R_{t,s} \times R_{t,s}$ minor of $M_s(\sum_i a_i z_i)$ is nonzero; that is, $p(a) \neq 0$ for some $p \in \mathcal{P}_{t,s}$ and so p is nonzero. □

Note that $\{a \in \mathbb{R}^{N_t} \mid \exists s \leq \lfloor t/2 \rfloor \text{ with } \operatorname{rank} M_s(\sum_i a_i z_i) < R_{t,s}\} = \bigcup_{s=1}^{\lfloor t/2 \rfloor} \Omega_{t,s}$; by Lemma 3.3, this set has Lebesgue measure 0, which shows that the property (3.8) is a generic property of the set $\mathcal{K}_t$. This shows:

COROLLARY 3.1. *The subset $G_t \subseteq \mathcal{K}_t$ of all generic elements (i.e. satisfying (3.8)) of $\mathcal{K}_t$ is dense in $\mathcal{K}_t$.*

Our strategy for Step 1 of Algorithm 2 is to choose $y = \sum_{i=1}^{N_t} a_i z_i$ where the scalars a_i are picked randomly according to e.g. a uniform probability distribution on $[0, 1]$. Then the maximality property (3.8) holds almost surely for y.

EXAMPLE 1. The following example

$$\begin{aligned} h_1 &= x_1^2 - 2x_1x_3 + 5\,, \\ h_2 &= x_1x_2^2 + x_2x_3 + 1\,, \\ h_3 &= 3x_2^2 - 8x_1x_3\,, \end{aligned}$$

taken from [4, Ex. 4, p.57], is used to illustrate Algorithm 2. Table 1 shows the ranks of the matrices $M_s(y)$ for generic $y \in \mathcal{K}_t$, as a function of s and t. Condition (3.5) is satisfied e.g. for $t = 8$ and $s = 3$ as we have:

$$\operatorname{rank} M_3(y) = \operatorname{rank} M_2(y), \text{ with } y \in \mathcal{K}_8.$$

TABLE 1
Rank of $M_s(y)$ for generic $y \in \mathcal{K}_t$ in Example 1.

	$t=2$	$t=4$	$t=6$	$t=8$
$s=0$	1	1	1	1
$s=1$	4	4	4	4
$s=2$		8	8	8
$s=3$			9	8
$s=4$				9

Applying Algorithm 2 we have computed the following 8 complex solutions:

$$
\begin{aligned}
v_1 &= [\,-1.101, -2.878, -2.821\,]\,,\\
v_2 &= [\,0.07665 + 2.243i, 0.461 + 0.497i, 0.0764 + 0.00834i\,]\,,\\
v_3 &= [\,0.07665 - 2.243i, 0.461 - 0.497i, 0.0764 - 0.00834i\,]\,,\\
v_4 &= [\,-0.081502 - 0.93107i, 2.350 + 0.0431i, -0.274 + 2.199i\,]\,,\\
v_5 &= [\,-0.081502 + 0.93107i, 2.350 - 0.0431i, -0.274 - 2.199i\,]\,,\\
v_6 &= [\,0.0725 + 2.237i, -0.466 - 0.464i, 0.0724 + 0.00210i\,]\,,\\
v_7 &= [\,0.0725 - 2.237i, -0.466 + 0.464i, 0.0724 - 0.00210i\,]\,,\\
v_8 &= [\,0.966, -2.813, 3.072\,]
\end{aligned}
$$

with maximum error of $\epsilon := \max_{i \leq 8, j \leq 3} |h_j(v_i)| \leq 3 \cdot 10^{-10}$. For the sake of comparison, Table 2 displays the ranks of the matrices $M_s(y)$ for generic $y \in \mathcal{K}_{t,\succeq}$; now the rank condition (3.5) is satisfied at $s = 2$ and $t = 6$; that is,

$$\operatorname{rank} M_2(y) = \operatorname{rank} M_1(y), \text{ with } y \in \mathcal{K}_{6,\succeq}.$$

TABLE 2
Rank of $M_s(y)$ for generic $y \in \mathcal{K}_{t,\succeq}$ in Example 1.

	$t=2$	$t=4$	$t=6$
$s=0$	1	1	1
$s=1$	4	4	**2**
$s=2$		8	**2**
$s=3$			3

The real roots extracted with the algorithm proposed in [10] are

$$
\begin{aligned}
v_1 &= [\,-1.101, -2.878, -2.821\,]\,,\\
v_2 &= [\,0.966, -2.813, 3.072\,]\,,
\end{aligned}
$$

with a maximum error of $\epsilon \leq 9 \cdot 10^{-11}$.

3.2. The Gorenstein case. We address here the question of when equality $I = \langle \operatorname{Ker} M_s(y) \rangle$ can be attained in (3.4). We begin with an example showing that both inclusions in (3.4) may be strict.

EXAMPLE 2. Consider the ideal $I = \langle x_1^2, x_2^2, x_1x_2 \rangle \subseteq \mathbb{R}[x_1, x_2]$. Then, $V_{\mathbb{C}}(I) = \{0\}$, $I(V_{\mathbb{C}}(I)) = \langle x_1, x_2 \rangle$, $\dim \mathbb{R}[x]/I = 3$ (with basis $\{1, x_1, x_2\}$), and $\dim \mathbb{R}[x]/I(V_{\mathbb{C}}(I)) = 1$ (with basis $\{1\}$). On the other hand, we have $\dim \mathbb{R}[x]/\langle \operatorname{Ker} M_s(y) \rangle = 2$ (with base $\{1, x_1\}$ or $\{1, x_2\}$) for any generic $y \in \mathcal{K}_t$ (i.e. satisfying the maximality property (3.8)) and $t \geq 2$, $1 \leq s \leq \lfloor t/2 \rfloor$. Indeed any such y satisfies $y_\alpha = 0$ for all $|\alpha| \geq 2$; therefore its moment matrix has the form

$$M_{\lfloor t/2 \rfloor}(y) = \begin{pmatrix} y_{00} & y_{10} & y_{01} & 0 \dots \\ y_{10} & 0 & 0 & 0 \dots \\ y_{01} & 0 & 0 & 0 \dots \\ 0 & 0 & 0 & 0 \dots \\ \vdots & \vdots & \vdots & \ddots \end{pmatrix}$$

(indexing $M_{\lfloor t/2 \rfloor}(y)$ by $1, x_1, x_2, x_1^2, x_1x_2, x_2^2, \dots$), with

$$L_y = y_{00}\partial_{00}[0] + y_{10}\partial_{10}[0] + y_{01}\partial_{01}[0].$$

Hence, $R_{t,0} = 1$ and for $s \geq 1$ $R_{t,s} = 2$ where, for generic $y_1, y_2 \in \mathcal{K}_t$, e.g. $\langle \text{Ker}\, M_s(y_1) \rangle = \langle x_1, x_1x_2, x_2^2 \rangle$ or $\langle \text{Ker}\, M_s(y_2) \rangle = \langle x_2, x_1x_2, x_1^2 \rangle$ is a strict superset of I and a strict subset of $I(V_{\mathbb{C}}(I))$.

Following Cox [6, Chapter 2], call an algebra $\mathcal{A}$ *Gorenstein* if there exists a nongenerate bilinear form

$$(\cdot, \cdot) : \mathcal{A} \times \mathcal{A} \to \mathbb{R}$$

satisfying $(fg, h) = (f, gh)$ for all $f, g, h \in \mathcal{A}$ (equivalently, if $\mathcal{A}$ and its dual space are isomorphic $\mathcal{A}$-modules). Consider the quotient algebra $\mathcal{A} = \mathbb{R}[x]/I$ where $I = \langle h_1, \dots, h_m \rangle$ is an ideal in $\mathbb{R}[x]$. Then any bilinear form on $\mathbb{R}[x]/I \times \mathbb{R}[x]/I$ is of the form

$$(f, g) \quad \mapsto \quad (f, g)_y := \text{vec}(f)^T M(y)\, \text{vec}(g)$$

for some $y \in \mathcal{K}_\infty$, after setting

$$\mathcal{K}_\infty := \{ y \in \mathbb{R}^{\mathbb{N}^n} \mid I \subseteq \text{Ker}\, M(y) \}.$$

Moreover the bilinear form $(\cdot, \cdot)_y$ is nondegenerate precisely when $I = \text{Ker}\, M(y)$. As $I = \text{Span}_{\mathbb{R}}(\cup_{t \geq 1} \mathcal{H}_t)$, we have

$$\mathcal{K}_\infty = \{ y \in \mathbb{R}^{\mathbb{N}^n} \mid L_y(p) = y^T \text{vec}(p) = 0 \;\forall p \in \cup_{t \geq 1} \mathcal{H}_t \}.$$

That is, $\mathcal{K}_\infty$ is the analogue of the sets $\mathcal{K}_t$ for $t = \infty$, and $\mathcal{K}_\infty$ is isomorphic to the dual space $\mathcal{D}[I] = I^\perp$ (recall (2.2)). Based on the above observations and Theorem 2.6 we obtain:

PROPOSITION 3.3. *Let $I = \langle h_1, \dots, h_m \rangle$ be a zero-dimensional ideal in $\mathbb{R}[x]$. The following assertions are equivalent.*
(i) *The algebra $\mathbb{R}[x]/I$ is Gorenstein.*
(ii) *There exists $y \in \mathbb{R}^{\mathbb{N}^n}$ such that $I = \text{Ker}\, M(y)$.*
(iii) *There exist $t \geq 1$ and $y \in \mathcal{K}_{2t}$ such that* $\text{rank}\, M_t(y) = \text{rank}\, M_{t-1}(y)$ *and* $I = \langle \text{Ker}\, M_t(y) \rangle$.

Proof. The equivalence of (i), (ii) follows from the definition of a Gorenstein algebra and the above observations. Assume (ii) holds. Let $\mathcal{B}$ be a basis of $\mathbb{R}[x]/I$ and suppose $\mathcal{B} \subseteq \mathbb{T}^n_{t-1}$. It is immediate to verify that $\mathcal{B}$ indexes a maximal linearly independent set of columns of $M(y)$. Hence, $\operatorname{rank} M_{t-1}(y) = \operatorname{rank} M(y) = \operatorname{rank} M_t(y)$ and $I = \langle \operatorname{Ker} M_t(y) \rangle$ (by Theorem 2.6). Thus (iii) holds. The reverse implication (iii) $\Longrightarrow$ (ii) follows directly from Theorem 2.6. □

EXAMPLE 3. Consider the ideal $I = \langle x_1^2, x_2^2 \rangle \subseteq \mathbb{R}[x_1, x_2]$. Then $\dim \mathbb{R}[x]/I = 4$ and for $y \in \mathcal{K}_{2t}$ $(t \geq 2)$,

$$M_t(y) = \begin{pmatrix} y_{00} & y_{10} & y_{01} & y_{11} & 0 \dots \\ y_{10} & 0 & y_{11} & 0 & 0 \dots \\ y_{01} & y_{11} & 0 & 0 & 0 \dots \\ y_{11} & 0 & 0 & 0 & 0 \dots \\ 0 & 0 & 0 & 0 & 0 \dots \\ \vdots & \vdots & \vdots & \vdots & \ddots \end{pmatrix}.$$

Hence the maximal rank of $M_t(y)$ is equal to 4 and for any such y, $I = \langle \operatorname{Ker} M_t(y) \rangle$. Thus $\mathbb{R}[x]/I$ is Gorenstein. Similarly, $\mathbb{R}[x]/I$ is also Gorenstein when $I = \langle x_1, x_2^3 \rangle$, but not for the ideal I of Example 2.

The next lemma illustrates how the kernels of moment matrices $M_t(y)$ for $y \in K_t$ are related to the ideal I even in the non-Gorenstein case.

LEMMA 3.4. *Assume that the ideal I is zero-dimensional. Then,*
(i) *Let $\{w_1, \dots, w_N\}$ be a linear basis of $\mathcal{K}_\infty$. Then*

$$I = \bigcap_{i=1}^{N} \langle \operatorname{Ker} M(w_i) \rangle.$$

(ii) *Let $z_1, \dots, z_{N_t}$ be a basis of $\mathcal{K}_t$. Then*

$$\langle \bigcap_{i=1}^{N_t} \operatorname{Ker} M_{\lfloor t/2 \rfloor}(z_i) \rangle \subseteq I,$$

with equality for $\lfloor t/2 \rfloor \geq D$.

Proof. (i) The inclusion $I \subseteq \cap_{i=1}^N \operatorname{Ker} M(w_i)$ is obvious. Conversely, let $q \in \mathbb{R}[x]$ with $\operatorname{vec}(q) \in \cap_{i=1}^N \operatorname{Ker} M(w_i)$; we show that $q \in I$ or, equivalently, that $L(q) = 0$ for all $L \in (\mathbb{R}[x]/I)^*$. Let $L \in (\mathbb{R}[x]/I)^*$ and so $L = L_y$ for some $y = \sum_{i=1}^N a_i w_i$ with $a_i \in \mathbb{R}$. Hence $L(q) = \sum_i a_i w_i^T \operatorname{vec}(q) = 0$ since $w_i^T \operatorname{vec}(q) = 1^T M(w_i) \operatorname{vec}(q) = 0$.
(ii) Let $q \in \mathbb{R}[x]$ with $\operatorname{vec}(q) \in \cap_{i=1}^{N_t} \operatorname{Ker} M_{\lfloor t/2 \rfloor}(z_i)$; we show that $q \in I$. Again it suffices to show that $L(q) = 0$ for every $L \in (\mathbb{R}[x]/I)^*$. As above let $L = L_y$ with $y = \sum_{i=1}^N a_i w_i$. As $w_i \in \mathcal{K}_\infty$, its projection $\pi_t(w_i)$ is an element of $\mathcal{K}_t$ and thus is of the form $\sum_{j=1}^{N_t} \lambda_{i,j} z_j$ for some scalars $\lambda_{i,j}$. Hence $L(q) = L_y(q) = \sum_i a_i \sum_j \lambda_{i,j} z_j^T \operatorname{vec}(q) = 0$ since $z_j^T \operatorname{vec}(q) = 0 \; \forall j = 1, \dots, N_t$. If $\lfloor t/2 \rfloor \geq D$ then $\operatorname{vec}(h_j) \in \operatorname{Ker} M_{\lfloor t/2 \rfloor}(y)$ for all $y \in \mathcal{K}_t$ using Lemma 3.1, which gives the desired equality $I = \langle \cap_{i=1}^{N_t} \operatorname{Ker} M_{\lfloor t/2 \rfloor}(z_i) \rangle$. □

4. Link with other symbolic-numeric methods. In this section we explore the links between the moment based approach from the preceding section with other methods in the literature, in particular the work of Zhi and Reid [18] and Mourrain et al. [15, 17]. The method we discuss here again uses the sets $\mathcal{K}_t$ introduced in (1.2) but is more global. Namely while in the moment based method we used a generic point $y \in \mathcal{K}_t$, we now use the full set $\mathcal{K}_t$ and its defining equations. More precisely, while in the moment based method the stopping criterion was a certain rank condition ((3.5) or (3.6)) on moment matrices $M_s(y)$ for a generic point $y \in \mathcal{K}_t$, the stopping criterion is now formulated in terms of the dimension of projections $\pi_s(\mathcal{K}_t)$ of the set $\mathcal{K}_t$.

The basic techniques behind the work [18] originally stem from the treatment of partial differential equations. Zharkov et al. [25, 26] were the first to apply these techniques to polynomial ideals. We will describe the idea of their algorithm (a simplified version, based on [18]) for the complex case in Section 4.1, using the language (rewriting families, multiplication matrices, etc.) presented earlier in the paper. Then, in Section 4.2 we show relations between the stopping criteria for the moment based method and the Zhi-Reid method. In a follow-up work [9] we will show that the method can be extended to the computation of real roots by adding some positivity constraints, thus working with the set $\mathcal{K}_{t,\succeq}$ in place of $\mathcal{K}_t$.

4.1. Dual space characterization of I. Again consider the sets $\mathcal{H}_t$ and $\mathcal{K}_t$ in (1.1) and (1.2). $\mathcal{K}_t$ is a linear subspace of $(\mathbb{R}[x]_t)^*$ and $\mathcal{K}_t^\perp = \text{Span}_\mathbb{R}(\mathcal{H}_t)$. For $s \leq t$, recall that π_s is the projection from $(\mathbb{R}[x]_t)^*$ onto $(\mathbb{R}[x]_s)^*$. Note that

$$(\pi_s(\mathcal{K}_t))^\perp = \mathcal{K}_t^\perp \cap \mathbb{R}[x]_s = \text{Span}_\mathbb{R}(\mathcal{H}_t) \cap \mathbb{R}[x]_s. \tag{4.1}$$

Recall that $\mathcal{D}[I] = I^\perp = \{L \in (\mathbb{R}[x])^* \mid L(p) = 0 \ \forall p \in I\}$ is isomorphic to the set $\mathcal{K}_\infty$ (by the linear mapping $y \mapsto L_y$). As $\mathcal{H}_t \subseteq I$, we have

$$\pi_s(\mathcal{D}[I]) \subseteq \pi_s(\mathcal{K}_t), \ \text{Span}_\mathbb{R}(\mathcal{H}_t) \cap \mathbb{R}[x]_s = (\pi_s(\mathcal{K}_t))^\perp \subseteq I \cap \mathbb{R}[x]_s. \tag{4.2}$$

We will show some dimension conditions on $\mathcal{K}_t$ ensuring that equality holds in (4.2), thus leading to a dual space characterization of I. The main result of this section is the following theorem, similar to some well known results from the theory of involutive bases (see [21]) and used e.g. in the algorithm of [18].

THEOREM 4.1. *Let $I = \langle h_1, \ldots, h_m \rangle$ be an ideal in $\mathbb{R}[x]$ and $D = \max_j \deg(h_j)$. Consider the following two conditions*

$$\dim \pi_s(\mathcal{K}_t) = \dim \pi_{s-1}(\mathcal{K}_t), \tag{4.3a}$$

$$\dim \pi_s(\mathcal{K}_t) = \dim \pi_s(\mathcal{K}_{t+1}). \tag{4.3b}$$

Assume that (4.3a) and (4.3b) hold for some integer s with $D \leq s \leq t$. If $\dim \pi_{s-1}(\mathcal{K}_t) = 0$ then $V_\mathbb{C}(I) = \emptyset$. Otherwise let $\mathcal{B} \subseteq \mathbb{T}^n_{s-1}$ satisfying

$$\pi_{s-1}(\mathcal{K}_t) \oplus \text{Span}_\mathbb{R}(\{\partial_\alpha[0] \mid x^\alpha \in \mathbb{T}^n_{s-1} \setminus \mathcal{B}\}) = (\mathbb{R}[x]_{s-1})^*$$

and assume that $\mathcal{B}$ is connected to 1.[1] *Then,*

- *$\mathcal{B}$ is a basis of $\mathbb{R}[x]/I$*
- *$\pi_s(\mathcal{D}[I]) = \pi_s(\mathcal{K}_t)$ and $I \cap \mathbb{R}[x]_s = \mathrm{Span}_{\mathbb{R}}(\mathcal{H}_t) \cap \mathbb{R}[x]_s$, i.e. equality holds in (4.2).*

Proof. Set $N := \dim \pi_{s-1}(\mathcal{K}_t)$. If $N = 0$, then $\pi_{s-1}(\mathcal{K}_t) = \{0\}$, $(\pi_{s-1}(\mathcal{K}_t))^\perp = \mathbb{R}[x]_{s-1}$ implying $1 \in I$ and thus $V_{\mathbb{C}}(I) = \emptyset$. Assume now $N \geq 1$. Let $\{L_1, \dots, L_N\} \subseteq \mathcal{K}_t$ such that $\mathcal{L}_1 := \{\pi_{s-1}(L_j) \mid j = 1, \dots, N\}$ forms a basis of $\pi_{s-1}(\mathcal{K}_t)$. We can complete $\mathcal{L}_1$ to a basis of $(\mathbb{R}[x]_{s-1})^*$ using members of the canonical basis. That is, let $\mathcal{B} \subseteq \mathbb{T}^n_{s-1}$, $|\mathcal{B}| = N$, $\mathcal{L}_2 := \{\partial_\alpha[0] \mid x^\alpha \in \mathbb{T}^n_{s-1} \setminus \mathcal{B}\}$ such that $\mathcal{L}_1 \cup \mathcal{L}_2$ is a basis of $(\mathbb{R}[x]_{s-1})^*$. We claim that

$$(\pi_{s-1}(\mathcal{K}_t)^\perp \oplus \mathrm{Span}_{\mathbb{R}}(\mathcal{B}) = \mathbb{R}[x]_{s-1}. \tag{4.4}$$

It suffices to verify that $\mathrm{Span}_{\mathbb{R}}(\mathcal{B}) \cap (\pi_{s-1}(\mathcal{K}_t))^\perp = \{0\}$ as the dimensions of both sides then coincide. Indeed, if $p \in \mathrm{Span}_{\mathbb{R}}(\mathcal{B})$, then $L(p) = 0$ if $L = \partial_\alpha[0]$ $(x^\alpha \in \mathbb{T}^n_{s-1} \setminus \mathcal{B})$ since p uses only monomials from $\mathcal{B}$, and if $p \in (\pi_{s-1}(\mathcal{K}_t))^\perp$ then $L(p) = 0$ if $L = L_j$ $(j \leq N)$.

As the set $\{\pi_s(L_j) \mid j \leq N\}$ is linearly independent in $\pi_s(\mathcal{K}_t)$ and $N = \dim \pi_s(\mathcal{K}_t)$ by (4.3*a*), this set is in fact a basis of $\pi_s(\mathcal{K}_t)$ and the analogue of (4.4) holds:

$$(\pi_s(\mathcal{K}_t))^\perp \oplus \mathrm{Span}_{\mathbb{R}}(\mathcal{B}) = \mathbb{R}[x]_s. \tag{4.5}$$

In particular, $\mathrm{Span}_{\mathbb{R}}(\mathcal{B}) \cap (\pi_s(\mathcal{K}_t))^\perp = \{0\}$ and any polynomial $p \in \mathbb{R}[x]_s$ can be written in a unique way as $p = r_p + f_p$, where $r_p \in \mathrm{Span}_{\mathbb{R}}(\mathcal{B})$ and $f_p \in (\pi_s(\mathcal{K}_t))^\perp = \mathbb{R}[x]_s \cap \mathrm{Span}_{\mathbb{R}}(\mathcal{H}_t)$. Thus $f_p = 0$ if $p \in \mathrm{Span}_{\mathbb{R}}(\mathcal{B})$. Set

$$F_0 := \{f_m \mid m \in \partial\mathcal{B}\} \subseteq F := \{f_m \mid m \in \mathbb{T}^n_s\} \subseteq \mathbb{R}[x]_s \cap \mathrm{Span}_{\mathbb{R}}(\mathcal{H}_t) \subseteq I. \tag{4.6}$$

Thus F_0 is a rewriting family for $\mathcal{B}$ and $F \subseteq \mathbb{R}[x]_s \cap \mathrm{Span}_{\mathbb{R}}(\mathcal{H}_t) \subseteq \mathbb{R}[x]_s \cap I$.

LEMMA 4.1. $\mathrm{Span}_{\mathbb{R}}(F) = \mathbb{R}[x]_s \cap \mathrm{Span}_{\mathbb{R}}(\mathcal{H}_t)$. *In particular,* $I = \langle F \rangle$ *if* $s \geq D = \max_j \deg(h_j)$.

Proof. Let $p \in (\pi_s(\mathcal{K}_t))^\perp = \mathbb{R}[x]_s \cap \mathrm{Span}_{\mathbb{R}}(\mathcal{H}_t)$. Write

$$p = \sum_{m \in \mathbb{T}^n_s} \lambda_m m = \sum_{m \in \mathbb{T}^n_s} \lambda_m (r_m + f_m) = r + f,$$

where $r := \sum_{m \in \mathbb{T}^n_s} \lambda_m r_m \in \mathrm{Span}_{\mathbb{R}}(\mathcal{B})$ and $f := \sum_{m \in \mathbb{T}^n_s} \lambda_m f_m \in \mathrm{Span}_{\mathbb{R}}(F)$. Thus $p - f = r \in \mathrm{Span}_{\mathbb{R}}(\mathcal{B}) \cap (\pi_s(\mathcal{K}_t))^\perp = \{0\}$, showing $p = f \in \mathrm{Span}_{\mathbb{R}}(F)$. If $s \geq D$, then each h_j lies in $\mathbb{R}[x]_s \cap \mathrm{Span}_{\mathbb{R}}(\mathcal{H}_t)$, equal to $\mathrm{Span}_{\mathbb{R}}(F)$ by the above; this thus gives $I = \langle F \rangle$. □

As $\pi_s(\mathcal{K}_{t+1}) \subseteq \pi_s(\mathcal{K}_t)$, condition (4.3*b*) implies equality of these two sets and thus of their orthogonal complements, i.e. $\mathbb{R}[x]_s \cap \mathrm{Span}_{\mathbb{R}}(\mathcal{H}_t) =$

[1] That such a basis connected to 1 exists is proved in [9].

$\mathbb{R}[x]_s \cap \mathrm{Span}_{\mathbb{R}}(\mathcal{H}_{t+1})$ (recall (4.1)). The next lemma shows that $(\pi_s(\mathcal{K}_t))^\perp = \mathbb{R}[x]_s \cap \mathrm{Span}_{\mathbb{R}}(\mathcal{H}_t)$ enjoys some ideal like properties.

LEMMA 4.2. *If* $f \in (\pi_s(\mathcal{K}_t))^\perp$ *and* $\deg(fg) \leq s$ *then* $fg \in (\pi_s(\mathcal{K}_t))^\perp$

Proof. It is sufficient to show the result for $g = x_i$. If $f \in (\pi_s(\mathcal{K}_t))^\perp = \mathbb{R}[x]_s \cap \mathrm{Span}_{\mathbb{R}}(\mathcal{H}_t)$, then $x_i f \in \mathrm{Span}_{\mathbb{R}}(\mathcal{H}_{t+1})$. As $\deg(x_i f) \leq s$, $x_i f \in \mathbb{R}[x]_s \cap \mathrm{Span}_{\mathbb{R}}(\mathcal{H}_{t+1}) = \mathbb{R}[x]_s \cap \mathrm{Span}_{\mathbb{R}}(\mathcal{H}_t)$ by (4.3*b*). □

Note that $1 \notin (\pi_s(\mathcal{K}_t))^\perp$; otherwise we would have $(\pi_s(\mathcal{K}_t))^\perp = \mathbb{R}[x]_{s-1}$ (by Lemma 4.2) and thus $\mathrm{Span}_{\mathbb{R}}(\mathcal{B}) = \{0\}$ (by (4.5)), contradicting our assumption $N \geq 1$. Hence we can choose the basis $\mathcal{B}$ satisfying (4.5) in such a way that $1 \in \mathcal{B}$. We now establish a relation between the two families F and F_0.

LEMMA 4.3. *If* $1 \in \mathcal{B}$ *then* $F \subseteq \langle F_0 \rangle$ *and so* $\langle F \rangle = \langle F_0 \rangle$.

Proof. Consider $m \in \mathbb{T}_s \setminus \mathcal{B}^+$. Write $m = x_i m_1$. Then,

$$f_m = m - r_m = x_i m_1 - r_m = x_i(r_{m_1} + f_{m_1}) - r_m = x_i r_{m_1} + x_i f_{m_1} - r_m.$$

We have $x_i r_{m_1} = r_{x_i r_{m_1}} + f_{x_i r_{m_1}}$, where $r_{x_i r_{m_1}} \in \mathrm{Span}_{\mathbb{R}}(\mathcal{B})$ and $f_{x_i r_{m_1}} \in \mathrm{Span}_{\mathbb{R}}(F_0)$ since $x_i r_{m_1} \in \mathrm{Span}_{\mathbb{R}}(\mathcal{B}^+)$. Moreover, $f_m \in F \subseteq \pi_s(\mathcal{K}_t))^\perp$, $r := r_{x_i r_{m_1}} - r_m \in \mathrm{Span}_{\mathbb{R}}(\mathcal{B})$, and $x_i f_{m_1} \in (\pi_s(\mathcal{K}_t))^\perp$ (by Lemma 4.2 since $f_{m_1} \in (\pi_s(\mathcal{K}_t))^\perp$). Therefore, $f_m - f_{x_i r_{m_1}} - x_i f_{m_1} = r \in \mathrm{Span}_{\mathbb{R}}(\mathcal{B}) \cap (\pi_s(\mathcal{K}_t))^\perp$ is thus equal to 0. This shows $f_m = f_{x_i r_{m_1}} + x_i f_{m_1}$, where $f_{x_i r_{m_1}} \in \mathrm{Span}_{\mathbb{R}}(F_0)$. Using induction on the distance of m to $\mathcal{B}$, we can conclude that $f_m \in \langle F_0 \rangle$. (The distance of m to $\mathcal{B}$ is defined as the minimum value of $|\alpha|$ for which $m = x^\alpha x^\beta$ with $x_\beta \in \mathcal{B}$; it is at most $\deg(m)$ since $1 \in \mathcal{B}$.) □

Using the rewriting family F_0 we can construct the formal multiplication matrices $\mathcal{X}_1, \ldots, \mathcal{X}_n$. Next we show that they commute pairwise.

LEMMA 4.4. *The formal multiplication matrices* $\mathcal{X}_i$ *defined using the rewriting family* F_0 *commute pairwise.*

Proof. Recall that the formal multiplication operator $\mathcal{X}_i$ is defined by $\mathcal{X}_i(m) = x_i m - f_{x_i m} = r_{x_i m}$ for any $m \in \mathcal{B}$ (and extended by linearity to $\mathrm{Span}_{\mathbb{R}}(\mathcal{B})$). We have to show that $\mathcal{X}_i(\mathcal{X}_j(m_0)) = \mathcal{X}_j(\mathcal{X}_i(m_0))$ for all $i, j \leq n$ and $m_0 \in \mathcal{B}$. Let $m_0 \in \mathcal{B}$. Assume first that $x_i m_0, x_j m_0 \notin \mathcal{B}$ and thus lie in $\partial\mathcal{B}$. We have: $\mathcal{X}_i(m_0) = r_{x_i m_0} := \sum_{b \in \mathcal{B}} a_b^i b$, implying $\mathcal{X}_j(\mathcal{X}_i(m_0)) = \mathcal{X}_j(\sum_{b \in \mathcal{B}} a_b^i b) = \sum_{b \in \mathcal{B}} a_b^i \mathcal{X}_j(b) = \sum_{b \in \mathcal{B}} a_b^i (x_j b - f_{x_j b}) = x_j(x_i m_0 - f_{x_i m_0}) - \sum_{b \in \mathcal{B}} a_b^i f_{x_j b}$. Analogously, $\mathcal{X}_i(\mathcal{X}_j(m_0)) = x_i(x_j m_0 - f_{x_j m_0}) - \sum_{b \in \mathcal{B}} a_b^j f_{x_i b}$. Therefore,

$$\begin{aligned} p := & \; \mathcal{X}_j(\mathcal{X}_i(m_0)) - \mathcal{X}_i(\mathcal{X}_j(m_0)) \\ = & \; \underbrace{x_i f_{x_j m_0} - x_j f_{x_i m_0}}_{q_1} + \underbrace{\sum_{b \in \mathcal{B}} a_b^j f_{x_i b} - \sum_{b \in \mathcal{B}} a_b^i f_{x_j b}}_{q_2}. \end{aligned}$$

Now, $\deg(p) \leq s$ (as $p \in \mathrm{Span}_{\mathbb{R}}(\mathcal{B})$), $\deg(q_2) \leq s$, implies $\deg(q_1) \leq s$. Moreover, $q_1 \in (\pi_s(\mathcal{K}_t))^{\perp}$ (by Lemma 4.2) and $q_2 \in \mathrm{Span}_{\mathbb{R}}(F)(\pi_s(\mathcal{K}_t))^{\perp}$ (by Lemma 4.1). Therefore, $p \in \mathrm{Span}_{\mathbb{R}}(\mathcal{B}) \cap (\pi_s(\mathcal{K}_t))^{\perp} = \{0\}$, which shows the desired identity $\mathcal{X}_i(\mathcal{X}_j(m_0)) = \mathcal{X}_j(\mathcal{X}_i(m_0))$. The proof is analogous in the other cases; say, $x_i m_0 \in \mathcal{B}$, $x_j m_0 \in \partial\mathcal{B}$. □

COROLLARY 4.1. *Assume $\mathcal{B}$ is connected to 1. Then,*

- *$\mathcal{B}$ is a basis of $\mathbb{R}[x]/\langle F_0\rangle = \mathbb{R}[x]/\langle F\rangle = \mathbb{R}[x]/I$.*
- *$(\pi_s(\mathcal{K}_t))^{\perp} = \mathbb{R}[x]_s \cap \mathrm{Span}_{\mathbb{R}}(\mathcal{H}_t)$ and $\pi_s(\mathcal{K}_t) = \pi_s(\mathcal{D}[I])$.*

Proof. As $\mathcal{B}$ is connected to 1, F_0 is a rewriting family for $\mathcal{B}$, and the associated multiplication matrices commute pairwise (by Lemma 4.4), we can conclude using Theorem 2.4 that the set $\mathcal{B}$ is a basis of $\mathbb{R}[x]/\langle F_0\rangle$. Now $\langle F_0\rangle = \langle F\rangle$ (by Lemma 4.3) and $I = \langle F\rangle$ since $s \geq D$ (by Lemma 4.1). Finally, write $p \in I \cap \mathbb{R}[x]_s$ as $p = r + q$, where $r \in \mathrm{Span}_{\mathbb{R}}(\mathcal{B})$ and $q \in (\pi_s(\mathcal{K}_t))^{\perp} \subseteq I$ (by (4.1)). Thus $p - q = r \in \mathrm{Span}_{\mathbb{R}}(\mathcal{B}) \cap I = \{0\}$, which shows the identity $I \cap \mathbb{R}[x]_s = (\pi_s(\mathcal{K}_t))^{\perp}$ and thus $\pi_s(\mathcal{D}[I]) = \pi_s(\mathcal{K}_t)$. □

This concludes the proof of Theorem 4.1. □

EXAMPLE 4. We continue with Example 1 to illustrate the condition on the dimension of $\mathcal{K}_t$ and its projections $\pi_s(\mathcal{K}_t)$. Table 3 shows the dimension of the set $\pi_s(\mathcal{K}_t)$ for various orders t and projection orders s. Note that the conditions (4.3*a*) and (4.3*b*) are satisfied at $(t, s) = (7, 4)$, i.e.

$$\dim \pi_4(\mathcal{K}_7) = \dim \pi_3(\mathcal{K}_7)$$
$$\dim \pi_4(\mathcal{K}_7) = \dim \pi_4(\mathcal{K}_8)\,.$$

TABLE 3
Dimension of the set $\pi_s(\mathcal{K}_t)$ in Example 4.

	$t=3$	$t=4$	$t=5$	$t=6$	$t=7$	$t=8$	$t=9$
$s=0$	1	1	1	1	1	1	1
$s=1$	4	4	4	4	4	4	4
$s=2$	7	7	7	7	7	7	7
$s=3$	11	10	9	8	**8**	8	8
$s=4$		12	10	9	**8**	**8**	8
$s=5$			12	10	9	8	8
$s=6$				12	10	9	8
$s=7$					12	10	9
$s=8$						12	10
$s=9$							12

4.2. Links between the stopping criteria of both methods. We show some connections between the stopping criteria (3.5) and (3.6) for the moment based method and the stopping criteria (4.3a), (4.3b) for the Zhi-Reid method [18].

First we show that the rank condition (3.5) for a generic element $y \in \mathcal{K}_t$ at a pair (t, s) implies the conditions (4.3a) – (4.3b) at the pair $(t, 2s)$.

PROPOSITION 4.1. *Assume* $\operatorname{rank} M_s(y) = \operatorname{rank} M_{s-1}(y)$ *for some generic* $y \in \mathcal{K}_t$ *and* $D \leq s \leq \lfloor t/2 \rfloor$. *Then,* $\dim \pi_{2s}(\mathcal{K}_t) = \dim \pi_{2s-1}(\mathcal{K}_t) = \dim \pi_{2s}(\mathcal{K}_{t+1})$, *i.e. (4.3a) and (4.3b) hold at the pair* $(t, 2s)$.

Proof. Consider the linear mapping

$$\begin{aligned} \varphi: \ \pi_{2s}(\mathcal{K}_t) &\to \pi_{2s-1}(\mathcal{K}_t) \\ \pi_{2s}(z) &\mapsto \pi_{2s-1}(z). \end{aligned}$$

As φ is onto, if we can show that φ is one-to-one, then this will imply $\dim \pi_{2s}(\mathcal{K}_t) = \dim \pi_{2s-1}(\mathcal{K}_t)$. Let $z \in \mathcal{K}_t$ for which $\pi_{2s-1}(z) = 0$; we show that $\pi_{2s}(z) = 0$. Set $R := \operatorname{rank} M_s(y) = \operatorname{rank} M_{s-1}(y)$ and let $\mathcal{B} \subseteq \mathbb{T}^n_{s-1}$ index a maximum linearly independent set of columns of $M_{s-1}(y)$, thus also of $M_s(y)$. Consider the element $y' := y + z$. Thus $\pi_{2s-1}(y') = \pi_{2s-1}(y)$ and the matrices $M_s(y)$ and $M_s(y')$ differ only at their entries indexed by $\mathbb{T}^n_s \setminus \mathbb{T}^n_{s-1}$. As $M_{s-1}(y') = M_{s-1}(y)$, $R = \operatorname{rank} M_{s-1}(y') \leq \operatorname{rank} M_s(y') \leq R$, where the latter inequality follows from the fact that y is generic, implying $\operatorname{rank} M_s(y') = R$. Hence $\mathcal{B}$ also indexes a maximal linearly independent set of columns of $M_s(y')$. Pick $x^\gamma \in \mathbb{T}^n_s \setminus \mathbb{T}^n_{s-1}$. Then $\operatorname{vec}(x^\gamma - q) \in \operatorname{Ker} M_s(y)$ and $\operatorname{vec}(x^\gamma - q') \in \operatorname{Ker} M_s(y')$ for some $q, q' \in \operatorname{Span}_{\mathbb{R}}(\mathcal{B})$. For any $x^\alpha \in \mathbb{T}^n_{s-1}$, we have $(M_s(y)\operatorname{vec}(q - q'))_\alpha = (M_s(y)\operatorname{vec}(q))_\alpha - (M_s(y')\operatorname{vec}(q'))_\alpha = (M_s(y)\operatorname{vec}(x^\gamma))_\alpha - (M_s(y')\operatorname{vec}(x^\gamma))_\alpha = y_{\alpha+\gamma} - y'_{\alpha+\gamma} = 0$ as $\pi_{2s-1}(y) = \pi_{2s-1}(y')$. Therefore $M_{s-1}(y)\operatorname{vec}(q - q') = 0$, implying $q = q'$ as $q - q' \in \operatorname{Span}_{\mathbb{R}}(\mathcal{B})$ and $\mathcal{B}$ indexes linearly independent columns of $M_{s-1}(y)$. This now implies $M_s(y)\operatorname{vec}(x^\gamma) = M_s(y)\operatorname{vec}(q) = M_s(y')\operatorname{vec}(q') = M_s(y')\operatorname{vec}(x^\gamma)$, i.e. $\pi_{2s}(y) = \pi_{2s}(y')$, giving $\pi_{2s}(z) = 0$. Thus we have shown $\dim \pi_{2s}(\mathcal{K}_t) = \dim \pi_{2s-1}(\mathcal{K}_t)$.

We now show $\pi_{2s}(\mathcal{K}_t) = \pi_{2s}(\mathcal{K}_{t+1})$; it suffices to show the inclusion $\pi_{2s}(\mathcal{K}_t) \subseteq \pi_{2s}(\mathcal{K}_{t+1})$. Let $z \in \mathcal{K}_t$ be generic; we show that $\pi_{2s}(z) \in \pi_{2s}(\mathcal{K}_{t+1})$. Note that $\operatorname{rank} M_s(z) = \operatorname{rank} M_{s-1}(z)$ since both z and y are generic. By Theorem 2.6 there exists an extension $z^* \in \mathbb{R}^{\mathbb{N}^n}$ of $\pi_{2s}(z)$ satisfying $\operatorname{rank} M(z^*) = \operatorname{rank} M_s(z)$ and $\operatorname{Ker} M(z^*) = \langle \operatorname{Ker} M_s(z) \rangle$. From $s \geq D$, it follows that $h_j \in \operatorname{Ker} M_s(z) \subseteq \operatorname{Ker} M(z^*)$, which implies $L_{z^*}(h_j x^\alpha) = 0$ for all α, i.e. $z^* \in \mathcal{K}_\infty$. Hence $\pi_{2s}(z) = \pi_{2s}(z^*)$ with $\pi_{2s}(z^*) \in \pi_{2s}(\mathcal{K}_{t+1})$. With G_t denoting the set of generic elements of $\mathcal{K}_t$, we have just shown that $\pi_{2s}(G_t) \subseteq \pi_{2s}(\mathcal{K}_{t+1})$. This implies that $\pi_{2s}(\mathcal{K}_t) \subseteq \pi_{2s}(\mathcal{K}_{t+1})$ since G_t is dense in $\mathcal{K}_t$ (by Corollary 3.1), which concludes the proof. □

REMARK 4.1. The above result combined with the result of Proposition 3.1, which shows that (3.5) holds for *all* $y \in \mathcal{K}_t$ and some (t, s) provides an alternative proof for the termination of the method proposed in [18].

We now prove some converse relation: If (4.3a) holds then (3.5) and (3.6) eventually hold too.

PROPOSITION 4.2. *Assume* $\dim \pi_s(\mathcal{K}_t) = \dim \pi_{s-1}(\mathcal{K}_t)$ *with* $D \leq s \leq \lfloor t/2 \rfloor$, *i.e. (4.3a) holds.*
(i) *For all* $y \in \mathcal{K}_{t'}$ *with* $t' := t+s+2d-2$, $\operatorname{rank} M_{s-1+d}(y) = \operatorname{rank} M_{s-1}(y)$.
(ii) *For all* $y \in \mathcal{K}_{t''}$ *with* $t'' := t+s$, $\operatorname{rank} M_s(y) = \operatorname{rank} M_{s-1}(y)$.

Proof. (i) Recall from the proof of Theorem 4.1 that any $m \in \mathbb{T}^n_s$ can be written as $m = r_m + f_m$, $r_m \in \operatorname{Span}_{\mathbb{R}}(\mathcal{B})$, $f_m \in \operatorname{Span}_{\mathbb{R}}(\mathcal{H}_t)$, $\mathcal{B} \subseteq \mathbb{T}^n_{s-1}$. (Indeed these facts could be proved without using (4.3b).) An easy induction on $k \geq 0$ permits to show that any $m \in \mathbb{T}^n_{s+k}$ can be written as $m = r_m + f_m$ where $r_m \in \operatorname{Span}_{\mathbb{R}}(\mathcal{B})$ and $f_m \in \operatorname{Span}_{\mathbb{R}}(\mathcal{H}_{t+k})$. Let $y \in \mathcal{K}_{t'}$. We show that $M_{s+d-1}(y)$ is a flat extension of $M_{s-1}(y)$. For this pick $m, m' \in \mathbb{T}^n_{s+d-1}$. Write $m = r_m + f_m$ with $f_m \in \operatorname{Span}_{\mathbb{R}}(\mathcal{H}_{t+d-1})$ and $r_m = \sum_{b \in \mathcal{B}} \lambda_b b$ ($\lambda_b \in \mathbb{R}$). Then the (m', m)th entry of $M_{s+d-1}(y)$ is equal to $L_y(mm') = L_y(m' f_m) + \sum_{b \in \mathcal{B}} \lambda_b L_y(m'b)$. Now $L_y(m' f_m) = 0$ since $m' f_m \in \mathcal{H}_{t'=t+s+2d-2}$ as $f_m \in \mathcal{H}_{t+d-1}$ and $\deg(m') \leq s+d-1$. Therefore $(M_{s+d-1}(y))_{m',m} = \sum_{b \in \mathcal{B}} \lambda_b (M_{s+d-1}(y))_{m',b}$, showing that the mth column of $M_{s+d-1}(y)$ can be written as a linear combination of its columns indexed by $\mathcal{B}$, i.e. $\operatorname{rank} M_{s-1+d}(y) = \operatorname{rank} M_{s-1}(y)$. The proof for (ii) is analogous. □

PROPOSITION 4.3. *Assume the assumptions of Theorem 4.1 hold. Then for any* $z \in \mathcal{K}_t$, *there exists* $y \in K_\infty$ *with* $\pi_{s-1}(y) = \pi_{s-1}(z)$ *and* $\operatorname{rank} M(y) = \operatorname{rank} M_{s-1}(y)$.

Proof. Let $z \in \mathcal{K}_t$. By Theorem 4.1, $\pi_s(z) \in \pi_s(\mathcal{K}_t) = \pi_s(\mathcal{D}[I]) \sim \pi_s(\mathcal{K}_\infty)$. Hence there exists $y \in \mathcal{K}_\infty$ with $\pi_s(z) = \pi_s(y)$. Let $\mathcal{B} \subseteq \mathbb{T}^n_{s-1}$ be a basis of $\mathbb{R}[x]/I$ (use Theorem 4.1). As $I \subseteq \operatorname{Ker} M(y)$, it follows that $\operatorname{rank} M(y) = \operatorname{rank} M_{s-1}(y)$. □

EXAMPLE 5. We consider again Example 2 with $I = \langle x_1^2, x_2^2, x_1 x_2 \rangle \subseteq \mathbb{R}[x_1, x_2]$, introduced earlier as an example with a non-Gorenstein quotient algebra $\mathbb{R}[x]/I$. The dimension of the space $\pi_s(\mathcal{K}_t)$ for different t and s is shown in Table 4 and the ranks of $M_s(y)$ for generic $y \in \mathcal{K}_t$ in Table 5.

TABLE 4
Dimension of the set $\pi_s(\mathcal{K}_t)$ *in Example 5.*

	$t=0$	$t=1$	$t=2$	$t=3$
$s=0$	1	1	1	1
$s=1$		3	**3**	3
$s=2$			**3**	**3**
$s=3$				3

Observe that

$$\dim \pi_2(\mathcal{K}_2) = \dim \pi_1(\mathcal{K}_2) = \dim \pi_2(\mathcal{K}_3)$$

TABLE 5
Rank of $M_s(y)$ for generic $y \in \mathcal{K}_t$ in Example 5.

	$t = 0$	$t = 2$	$t = 4$	$t = 6$
$s = 0$	1	1	1	1
$s = 1$		2	2	2
$s = 2$			2	2
$s = 3$				2

and

$$\operatorname{rank} M_2(y) = \operatorname{rank} M_1(y) \quad \forall\, y \in \mathcal{K}_4$$

as predicted by Proposition 4.2 (ii).

5. Conclusion. In this paper we have presented a new method for computing the complex variety of a zero-dimensional ideal I given by its generators. The method is a complex analogue of the moment-matrix algorithm proposed in [10] for the task of computing the *real* variety of an ideal and its real radical ideal. In contrast to the latter algorithm, the newly proposed method does not use semidefinite optimization and is based purely on numerical linear algebra.

The two methods allow a unified treatment of the algebraic and real-algebraic root finding problems. Remarkably, all information needed to compute the roots is contained in the moment matrix of a single generic linear form associated to the problem. The moment matrix can be computed numerically and, simply by adding or neglecting a positive semidefiniteness constraint, one can move from one problem to the other. While the methods are almost identical in the real and complex cases, substantially different proofs were needed for the complex case.

Furthermore, we have shown how this algorithm is related to other methods in the field, particularly to border basis methods and the Zhi-Reid algorithm based on involutive bases. Indeed, simple relationships between their stopping criteria and the rank conditions used in the moment-matrix method have been established.

In a follow-up paper [9] we show how these other numerical-algebraic methods for complex roots can be adapted to design real-algebraic variants for computing real roots directly by incorporating sums of squares conditions extracted from moment matrices.

REFERENCES

[1] S. BASU, R. POLLACK, AND M.-F. ROY, *Algorithms in real algebraic geometry*, Vol. **10** of Algorithms and Computations in Mathematics, Springer-Verlag, 2003.

[2] J. BOCHNAK, M. COSTE, AND M.-F. ROY, *Real Algebraic Geometry*, Springer, 1998.

[3] D. Cox, J. Little, and D. O'Shea, *Ideals, Varieties, and Algorithms: An Introduction to Computational Algebraic Geometry and Commutative Algebra*, Springer, 2005.

[4] ———, *Using Algebraic Geometry*, Springer, 1998.

[5] R. Curto and L. Fialkow, *Solution of the truncated complex moment problem for flat data*, Memoirs of the American Mathematical Society, **119** (1996), pp. 1–62.

[6] A. Dickenstein and I.Z. Emiris, eds., *Solving Polynomial Equations: Foundations, Algorithms, and Applications*, Vol. **14** of Algorithms and Computation in Mathematics, Springer, 2005.

[7] I. Janovitz-Freireich, L. Rónyai, and Ágnes Szántó, *Approximate radical of ideals with clusters of roots*, in ISSAC '06: Proceedings of the 2006 International Symposium on Symbolic and Algebraic Computation, New York, NY, USA, 2006, ACM Press, pp. 146–153.

[8] A. Kehrein and M. Kreuzer, *Characterizations of border bases*, J. of Pure and Applied Algebra, **196** (2005), pp. 251–270.

[9] J. Lasserre, M. Laurent, and P. Rostalski, *Computing the real variety of an ideal: A real algebraic and symbolic-numeric algorithm.* (Research report, LAAS Toulouse, France, 2007.) Short version in Proceedings of the Conference SAC08, Fortaleza, Brasil, March 16–20, 2008.

[10] ———, *Semidefinite characterization and computation of zero-dimensional real radical ideals.* To appear in Found. Comp. Math., Published online October 2007.

[11] M. Laurent, *Sums of squares, moment matrices and optimization over polynomials.* This IMA volume, Emerging Applications of Algebraic Geometry, M. Putinar and S. Sullivant, eds.

[12] ———, *Revisiting two theorems of Curto and Fialkow*, Proc. Amer. Math. Soc., **133** (2005), pp. 2965–2976.

[13] H. Möller, *An inverse problem for cubature formulae*, Computat. Technol., **9** (2004), pp. 13–20.

[14] B. Mourrain, *A new criterion for normal form algorithms*, in AAECC, 1999, pp. 430–443.

[15] ———, *Symbolic-Numeric Computation*, Trends in Mathematics, Birkhäuser, 2007, ch. Pythagore's Dilemma, Symbolic-Numeric Computation, and the Border Basis Method, pp. 223–243.

[16] B. Mourrain, F. Rouillier, and M.-F. Roy, *Bernstein's basis and real root isolation*, in Combinatorial and Computational Geometry, Mathematical Sciences Research Institute Publications, Cambridge University Press, 2005, pp. 459–478.

[17] B. Mourrain and P. Trebuchet, *Generalized normal forms and polynomial system solving*, ISSAC, 2005: 253–260.

[18] G. Reid and L. Zhi, *Solving nonlinear polynomial system via symbolic-numeric elimination method*, in Proceedings of the International Conference on Polynomial System Solving, J. Faugère and F. Rouillier, eds., 2004, pp. 50–53.

[19] F. Rouillier, *Solving zero-dimensional systems through the rational univariate representation*, Journal of Applicable Algebra in Engineering, Communication and Computing, **9** (1999), pp. 433–461.

[20] F. Rouillier and P. Zimmermann, *Efficient isolation of polynomial real roots*, Journal of Computational and Applied Mathematics, **162** (2003), pp. 33–50.

[21] W. Seiler, *Involution - The Formal Theory of Differential Equations and its Applications in Computer Algebra and Numerical Analysis*, habilitation thesis, Computer Science, University of Mannheim, 2001.

[22] A. Sommese and C. Wampler, *The Numerical Solution of Systems of Polynomials Arising in Engineering and Science*, World Scientific Press, Singapore, 2005.

[23] H.J. STETTER, *Numerical Polynomial Algebra*, Society for Industrial and Applied Mathematics, Philadelphia, PA, USA, 2004.

[24] J. VERSCHELDE, *PHCPACK: A general-purpose solver for polynomial systems by homotopy continuation*, ACM Transactions on Mathematical Software, **25** (1999), pp. 251–276.

[25] A. ZHARKOV AND Y. BLINKOV, *Involutive bases of zero-dimensional ideals*, Preprint E5-94-318, Joint Institute for Nuclear Research, Dubna, 1994.

[26] ———, *Involutive approach to investigating polynomial systems*, in Proceedings of SC 93, International IMACS Symposium on Symbolic Computation: New Trends and Developments, Lille, June 14–17, 1993, Math. Comp. Simul., Vol. **42**, 1996, pp. 323–332.

SUMS OF SQUARES, MOMENT MATRICES AND OPTIMIZATION OVER POLYNOMIALS

MONIQUE LAURENT*

April 30, 2008

Abstract. We consider the problem of minimizing a polynomial over a semialgebraic set defined by polynomial equations and inequalities, which is NP-hard in general. Hierarchies of semidefinite relaxations have been proposed in the literature, involving positive semidefinite moment matrices and the dual theory of sums of squares of polynomials. We present these hierarchies of approximations and their main properties: asymptotic/finite convergence, optimality certificate, and extraction of global optimum solutions. We review the mathematical tools underlying these properties, in particular, some sums of squares representation results for positive polynomials, some results about moment matrices (in particular, of Curto and Fialkow), and the algebraic eigenvalue method for solving zero-dimensional systems of polynomial equations. We try whenever possible to provide detailed proofs and background.

Key words. positive polynomial, sum of squares of polynomials, moment problem, polynomial optimization, semidefinite programming.

AMS(MOS) subject classifications. 13P10, 13J25, 13J30, 14P10, 15A99, 44A60, 90C22, 90C30.

Contents

*CWI, Kruislaan 413, 1098 SJ Amsterdam, The Netherlands. Supported by the Netherlands Organisation for Scientific Research grant NWO 639.032.203 and by ADONET, Marie Curie Research Training Network MRTN-CT-2003-504438.

1. Introduction. This survey focuses on the following *polynomial optimization problem*: Given polynomials $p, g_1, \ldots, g_m \in \mathbb{R}[\mathbf{x}]$, find

$$p^{\min} := \inf_{x \in \mathbb{R}^n} p(x) \text{ subject to } g_1(x) \geq 0, \ldots, g_m(x) \geq 0, \tag{1.1}$$

the infimum of p over the basic closed semialgebraic set

$$K := \{x \in \mathbb{R}^n \mid g_1(x) \geq 0, \ldots, g_m(x) \geq 0\}. \tag{1.2}$$

Here $\mathbb{R}[\mathbf{x}] = \mathbb{R}[\mathbf{x}_1, \ldots, \mathbf{x}_n]$ denotes the ring of multivariate polynomials in the n-tuple of variables $\mathbf{x} = (\mathbf{x}_1, \ldots, \mathbf{x}_n)$. This is a hard, in general nonconvex, optimization problem. The objective of this paper is to survey relaxations methods for this problem, that are based on relaxing positivity over K by sums of squares decompositions, and the dual theory of moments. The polynomial optimization problem arises in numerous applications. In the rest of the Introduction, we present several instances of this problem, discuss the scope of the paper, and give some preliminaries about polynomials and semidefinite programming.

1.1. The polynomial optimization problem. We introduce several instances of problem (1.1).

The unconstrained polynomial minimization problem. This is the problem

$$p^{\min} = \inf_{x \in \mathbb{R}^n} p(x), \tag{1.3}$$

of minimizing a polynomial p over the full space $K = \mathbb{R}^n$. We now mention several problems which can be cast as instances of the unconstrained polynomial minimization problem.

Testing matrix copositivity. An $n \times n$ symmetric matrix M is said to be *copositive* if $x^T Mx \geq 0$ for all $x \in \mathbb{R}^n_+$; equivalently, M is copositive if and only if $p^{\min} = 0$ in (1.3) for the polynomial $p := \sum_{i,j=1}^n \mathbf{x}_i^2 \mathbf{x}_j^2 M_{ij}$. Testing whether a matrix is not copositive is an NP-complete problem [94].

The partition problem. The partition problem asks whether a given sequence $a_1, \ldots, a_n$ of positive integer numbers can be partitioned, i.e., whether $x^T a = 0$ for some $x \in \{\pm 1\}^n$. Equivalently, the sequence can be partitioned if $p^{\min} = 0$ in (1.3) for the polynomial $p := (\sum_{i=1}^n a_i \mathbf{x}_i)^2 + \sum_{i=1}^n (\mathbf{x}_i^2 - 1)^2$. The partition problem is an NP-complete problem [40].

The distance realization problem. Let $d = (d_{ij})_{ij \in E} \in \mathbb{R}^E$ be a given set of scalars (distances) where E is a given set of pairs ij with $1 \leq i < j \leq n$. Given an integer $k \geq 1$ one says that d is realizable in $\mathbb{R}^k$ if there exist vectors $v_1, \ldots, v_n \in \mathbb{R}^k$ such that $d_{ij} = \|v_i - v_j\|$ for all $ij \in E$. Equivalently, d is realizable in $\mathbb{R}^k$ if $p^{\min} = 0$ for the polynomial $p := \sum_{ij \in E} (d_{ij}^2 - \sum_{h=1}^k (\mathbf{x}_{ih} - \mathbf{x}_{jh})^2)^2$ in the variables $\mathbf{x}_{ih}$ $(i = 1, \ldots, n, h = 1, \ldots, k)$. Checking whether d is realizable in $\mathbb{R}^k$ is an NP-complete problem, already for dimension $k = 1$ (Saxe [123]).

Note that the polynomials involved in the above three instances have degree 4. Hence the unconstrained polynomial minimization problem is a hard problem, already for degree 4 polynomials, while it is polynomial time solvable for degree 2 polynomials (cf. Section 3.2). The problem (1.1) also contains (0/1) linear programming.

(0/1) Linear programming. Given a matrix $A \in \mathbb{R}^{m \times n}$ and vectors $b \in \mathbb{R}^m$, $c \in \mathbb{R}^n$, the linear programming problem can be formulated as

$$\min c^T x \text{ s.t. } Ax \leq b,$$

thus it is of the form (1.1) where the objective function and the constraints are all linear (degree at most 1) polynomials. As is well known it can be solved in polynomial time (cf. e.g. [128]). If we add the quadratic constraints $x_i^2 = x_i$ $(i = 1, \ldots, n)$ we obtain the 0/1 linear programming problem:

$$\min c^T x \text{ s.t. } Ax \leq b,\ x_i^2 = x_i \ \forall i = 1, \ldots, n,$$

well known to be NP-hard.

The stable set problem. Given a graph $G = (V, E)$, a set $S \subseteq V$ is said to be stable if $ij \notin E$ for all $i, j \in S$. The stable set problem asks for the maximum cardinality $\alpha(G)$ of a stable set in G. Thus it can be formulated as

$$\alpha(G) = \max_{x \in \mathbb{R}^V} \sum_{i \in V} x_i \text{ s.t. } x_i + x_j \leq 1 \ (ij \in E), \ x_i^2 = x_i \ (i \in V) \quad (1.4)$$

$$= \max_{x \in \mathbb{R}^V} \sum_{i \in V} x_i \text{ s.t. } x_i x_j = 0 \ (ij \in E), \ x_i^2 - x_i = 0 \ (i \in V). \quad (1.5)$$

Alternatively, using the theorem of Motzkin-Straus [93], the stability number $\alpha(G)$ can be formulated via the program

$$\frac{1}{\alpha(G)} = \min \ x^T (I + A_G) x \ \text{ s.t. } \sum_{i \in V} x_i = 1, \ x_i \geq 0 \ (i \in V). \quad (1.6)$$

Using the characterization mentioned above for copositive matrices, one can derive the following further formulation for $\alpha(G)$

$$\alpha(G) = \inf \ t \ \text{ s.t. } t(I + A_G) - J \ \text{ is copositive,} \quad (1.7)$$

which was introduced in [32] and further studied e.g. in [46] and references therein. Here, J is the all ones matrix, and A_G is the adjacency matrix of G, defined as the $V \times V$ 0/1 symmetric matrix whose (i, j)th entry is 1 precisely when $i \neq j \in V$ and $ij \in E$. As computing $\alpha(G)$ is an NP-hard problem (see, e.g., [40]), we see that problem (1.1) is NP-hard already in the following two instances: the objective function is linear and the constraints are quadratic polynomials (cf. (1.5)), or the objective function is quadratic and the constraints are linear polynomials (cf. (1.6)). We will use the stable set problem and the following max-cut problem in Section 8.2 to illustrate the relaxation methods for polynomial problems in the 0/1 (or ± 1) case.

The max-cut problem. Let $G = (V, E)$ be a graph and $w_{ij} \in \mathbb{R}$ $(ij \in E)$ be weights assigned to its edges. A *cut* in G is the set of edges $\{ij \in E \mid i \in S, j \in V \setminus S\}$ for some $S \subseteq V$ and its weight is the sum of the weights of its edges. The max-cut problem, which asks for a cut of maximum weight, is NP-hard [40]. Note that a cut can be encoded by $x \in \{\pm 1\}^V$ by assigning $x_i = 1$ to nodes $i \in S$ and $x_i = -1$ to nodes $i \in V \setminus S$ and the weight of the cut is encoded by the function $\sum_{ij \in E} (w_{ij}/2)(1 - x_i x_j)$. Therefore the max-cut problem can be formulated as the polynomial optimization problem

$$\text{mc}(G, w) := \max \sum_{ij \in E} (w_{ij}/2)(1 - x_i x_j) \ \text{ s.t. } x_1^2 = 1, \ldots, x_n^2 = 1. \quad (1.8)$$

1.2. The scope of this paper. As the polynomial optimization problem (1.1) is NP-hard, several authors, in particular Lasserre [65–67], Nesterov [95], Parrilo [103, 104], Parrilo and Sturmfels [107], Shor [138–141], have proposed to approximate the problem (1.1) by a hierarchy of convex (in fact, semidefinite) relaxations. Such relaxations can be constructed using representations of nonnegative polynomials as sums of squares of polynomials and the dual theory of moments. The paradigm underlying this approach is that, while testing whether a polynomial is nonnegative is a hard problem, testing whether a polynomial is a sum of squares of polynomials can be formulated as a semidefinite problem. Now, efficient algorithms exist for solving semidefinite programs (to any arbitrary precision). Thus approximations for the infimum of p over a semialgebraic set K can be computed efficiently. Moreover, under some assumptions on the set K, asymptotic (sometimes even finite) convergence to $p^{\min}$ can be proved and one may be able to compute global minimizers of p over K. For these tasks the interplay between positive polynomials and sums of squares of polynomials on the one hand, and the dual objects, moment sequences and matrices on the other hand, plays a significant role. The above is a rough sketch of the theme of this survey paper. Our objective is to introduce the main theoretical tools and results needed for proving the various properties of the approximation scheme, in particular about convergence and extraction of global minimizers. Whenever possible we try to provide detailed proofs and background.

The link between positive (nonnegative) polynomials and sums of squares of polynomials is a classic question which goes back to work of Hilbert at the end of the nineteenth century. As Hilbert himself already realized not every nonnegative polynomial can be written as a sum of squares; he in fact characterized the cases when this happens (cf. Theorem 3.4). This was the motivation for Hilbert's 17th problem, posed in 1900 at the International Congress of Mathematicians in Paris, asking whether every nonnegative polynomial can be written as a sum of squares of *rational* functions. This was later in 1927 answered in the affirmative by E. Artin whose work lay the foundations for the field of *real algebraic geometry*. Some of the milestone results include the Real Nullstellensatz which is the real analogue of Hilbert's Nullstellensatz for the complex field, the Positivstellensatz and its refinements by Schmüdgen and by Putinar, which are most relevant to our optimization problem. We will present a brief exposition on this topic in Section 3 where, besides some simple basic results about positive polynomials and sums of squares, we present a proof for Putinar's Positivstellensatz.

The study of positive polynomials is intimately linked to the theory of moments, via the following duality relation: A sequence $y \in \mathbb{R}^{\mathbb{N}^n}$ is the sequence of moments of a nonnegative measure μ on $\mathbb{R}^n$ (i.e. $y_\alpha = \int x^\alpha \mu(dx)$ $\forall \alpha \in \mathbb{N}^n$) if and only if $y^T p := \sum_\alpha y_\alpha p_\alpha \geq 0$ for any nonnegative polynomial $p = \sum_\alpha p_\alpha \mathbf{x}^\alpha \in \mathbb{R}[\mathbf{x}]$. Characterizing moment sequences is a classical problem, relevant to operator theory and several other areas in mathematics (see e.g. [1, 64] and references therein). Indeed, sequences of moments of nonnegative measures correspond to positive linear functionals on $\mathbb{R}[\mathbf{x}]$; moreover, the linear functionals that are positive on the cone of sums of squares correspond to the sequences y whose moment matrix $M(y) := (y_{\alpha+\beta})_{\alpha,\beta\in\mathbb{N}^n}$ is positive semidefinite. Curto and Fialkow have accomplished a systematic study of the *truncated* moment problem, dealing with sequences of moments up to a given order. We will discuss some of their results that are most relevant to polynomial optimization in Section 5 and refer to [23–26, 38] and further references therein for detailed information.

Our goal in this survey is to provide a tutorial on the real algebraic tools and the results from moment theory needed to understand their application to polynomial optimization, mostly on an elementary level to make the topic accessible to non-specialists. We obviously do not pretend to offer a comprehensive treatment of these areas for which excellent accounts can be found in the literature and we apologize for all omissions and imprecisions. For a more advanced exposition on positivity and sums of squares and links to the moment problem, we refer in particular to the article by Scheiderer [125] and Schmüdgen [127] in this volume, to the survey article by Helton and Putinar [52], and to the monographs by Prestel and Delzell [114] and by Marshall [87, 90].

1.3. Preliminaries on polynomials and semidefinite programs.

We introduce here some notation and preliminaries about polynomials, matrices and semidefinite programs. We will introduce further notation and preliminaries later on in the text when needed.

Polynomials. Throughout, $\mathbb{N}$ denotes the set of nonnegative integers and we set $\mathbb{N}^n_t := \{\alpha \in \mathbb{N}^n \mid |\alpha| := \sum_{i=1}^n \alpha_i \leq t\}$ for $t \in \mathbb{N}$. $\mathbb{R}[\mathbf{x}_1, \ldots, \mathbf{x}_n]$ denotes the ring of multivariate polynomials in n variables, often abbreviated as $\mathbb{R}[\mathbf{x}]$ where $\mathbf{x}$ stands for the n-tuple $(\mathbf{x}_1, \ldots, \mathbf{x}_n)$. Throughout we use the boldfaced letters $\mathbf{x}_i, \mathbf{x}, \mathbf{y}, \mathbf{z}$, etc., to denote *variables*, while the letters $x_i, x, y, z, \ldots$ stand for real valued scalars or vectors. For $\alpha \in \mathbb{N}^n$, $\mathbf{x}^\alpha$ denotes the monomial $\mathbf{x}_1^{\alpha_1} \cdots \mathbf{x}_n^{\alpha_n}$ whose degree is $|\alpha| := \sum_{i=1}^n \alpha_i$. $\mathbb{T}^n := \{\mathbf{x}^\alpha \mid \alpha \in \mathbb{N}^n\}$ is the set of all monomials and, for $t \in \mathbb{N}$, $\mathbb{T}^n_t := \{\mathbf{x}^\alpha \mid \alpha \in \mathbb{N}^n_t\}$ is the set of monomials of degree $\leq t$. Consider a polynomial $p \in \mathbb{R}[\mathbf{x}]$, $p = \sum_{\alpha\in\mathbb{N}^n} p_\alpha \mathbf{x}^\alpha$, where there are only finitely many nonzero p_α's. When $p_\alpha \neq 0$, $p_\alpha \mathbf{x}^\alpha$ is called a term of p. The degree of p is $\deg(p) := \max(t \mid p_\alpha \neq 0 \text{ for some } \alpha \in \mathbb{N}^n_t)$ and throughout we set

$$d_p := \lceil \deg(p)/2 \rceil \quad \text{for } p \in \mathbb{R}[\mathbf{x}]. \tag{1.9}$$

For the set $K = \{x \in \mathbb{R}^n \mid g_1(x) \geq 0, \ldots, g_m(x) \geq 0\}$ from (1.2), we set

$$d_K := \max(d_{g_1}, \ldots, d_{g_m}). \tag{1.10}$$

We let $\mathbb{R}[\mathbf{x}]_t$ denote the set of polynomials of degree $\leq t$.

A polynomial $p \in \mathbb{R}[\mathbf{x}]$ is said to be homogeneous (or a form) if all its terms have the same degree. For a polynomial $p \in \mathbb{R}[\mathbf{x}]$ of degree d, $p = \sum_{|\alpha|\leq d} p_\alpha \mathbf{x}^\alpha$, its homogenization is the polynomial $\tilde{p} \in \mathbb{R}[\mathbf{x}, \mathbf{x}_{n+1}]$ defined by $\tilde{p} := \sum_{|\alpha|\leq d} p_\alpha \mathbf{x}^\alpha \mathbf{x}_{n+1}^{d-|\alpha|}$.

For a polynomial $p \in \mathbb{R}[\mathbf{x}]$, $p = \sum_\alpha p_\alpha \mathbf{x}^\alpha$, $\text{vec}(p) := (p_\alpha)_{\alpha\in\mathbb{N}^n}$ denotes its sequence of coefficients in the monomial basis of $\mathbb{R}[\mathbf{x}]$; thus $\text{vec}(p) \in \mathbb{R}^\infty$, the subspace of $\mathbb{R}^{\mathbb{N}^n}$ consisting of the sequences with finitely many nonzero coordinates. Throughout the paper we often identify a polynomial p with its coordinate sequence $\text{vec}(p)$ and, for the sake of compactness in the notation, we often use the letter p instead of $\text{vec}(p)$; that is, we use the same letter p to denote the polynomial $p \in \mathbb{R}[\mathbf{x}]$ and its sequence of coefficients $(p_\alpha)_\alpha$. We will often deal with matrices indexed by $\mathbb{N}^n$ or $\mathbb{N}^n_t$. If M is such a matrix, indexed say by $\mathbb{N}^n$, and $f, g \in \mathbb{R}[\mathbf{x}]$, the notation $f^T M g$ stands for $\text{vec}(f)^T M \text{vec}(g) = \sum_{\alpha,\beta} f_\alpha g_\beta M_{\alpha,\beta}$. In particular, we say that a polynomial f lies in the kernel of M if $Mf := M\text{vec}(f) = 0$, and $\text{Ker}\, M$ can thus be seen as a subset of $\mathbb{R}[\mathbf{x}]$. When $\deg(p) \leq t$, $\text{vec}(p)$ can also be seen be seen as a vector of $\mathbb{R}^{\mathbb{N}^n_t}$, as $p_\alpha = 0$ whenever $|\alpha| \geq t+1$.

For a subset $A \subseteq \mathbb{R}^n$, $\text{Span}_{\mathbb{R}}(A) := \{\sum_{j=1}^m \lambda_j a_j \mid a_j \in A, \lambda_j \in \mathbb{R}\}$ denotes the linear span of A, and $\text{conv}(A) := \{\sum_{j=1}^m \lambda_j a_j \mid a_j \in A, \lambda_j \in \mathbb{R}_+, \sum_j \lambda_j = 1\}$ denotes the convex hull of A. Throughout $e_1, \ldots, e_n$ denote the standard unit vectors in $\mathbb{R}^n$, i.e. $e_i = (0, \ldots, 0, 1, 0, \ldots, 0)$ with 1 at the ith position. Moreover $\overline{z}$ denotes the complex conjugate of $z \in \mathbb{C}$.

Positive semidefinite matrices. For an $n\times n$ real symmetric matrix M, the notation $M \succeq 0$ means that M is positive semidefinite, i.e. $x^T M x \geq 0$ for all $x \in \mathbb{R}^n$. Here are several further equivalent characterizations: $M \succeq 0$ if and only if any of the equivalent properties (1)-(3) holds.

(1) $M = VV^T$ for some $V \in \mathbb{R}^{n\times n}$; such a decomposition is sometimes known as a Gram decomposition of M. Here V can be chosen in $\mathbb{R}^{n\times r}$ where $r = \text{rank}\, M$.

(2) $M = (v_i^T v_j)_{i,j=1}^n$ for some vectors $v_1, \ldots, v_n \in \mathbb{R}^n$. Here the v_i's may be chosen in $\mathbb{R}^r$ where $r = \text{rank}\, M$.

(3) All eigenvalues of M are nonnegative.

The notation $M \succ 0$ means that M is positive definite, i.e. $M \succeq 0$ and $\text{rank}\, M = n$ (equivalently, all eigenvalues are positive). When M is an infinite matrix, the notation $M \succeq 0$ means that every finite principal submatrix of M is positive semidefinite. Sym_n denotes the set of symmetric $n\times n$ matrices and PSD_n the subset of positive semidefinite matrices; PSD_n is a convex cone in Sym_n. $\mathbb{R}^{n\times n}$ is endowed with the usual inner product

$$\langle A, B\rangle = Tr(A^T B) = \sum_{i,j=1}^n a_{ij} b_{ij}$$

for two matrices $A = (a_{ij}), B = (b_{ij}) \in \mathbb{R}^{n\times n}$. As is well known, the cone PSD_n is self-dual, since PSD_n coincides with its dual cone $(\mathrm{PSD}_n)^* := \{A \in \mathrm{Sym}_n \mid \langle A, B\rangle \geq 0\ \forall B \in \mathrm{PSD}_n\}$.

Flat extensions of matrices. The following notion of *flat extension* of a matrix will play a central role in the study of moment matrices with finite atomic measures, in particular, in Section 5.

DEFINITION 1.1. *Let X be a symmetric matrix with block form*

$$X = \begin{pmatrix} A & B \\ B^T & C \end{pmatrix}. \tag{1.11}$$

One says that X is a flat extension *of A if* $\mathrm{rank}\, X = \mathrm{rank}\, A$ *or, equivalently, if $B = AW$ and $C = B^TW = W^TAW$ for some matrix W. Obviously, if X is a flat extension of A, then $X \succeq 0 \Longleftrightarrow A \succeq 0$.*

We recall for further reference the following basic properties of the kernel of a positive semidefinite matrix. Recall first that, for $M \in \mathrm{PSD}_n$ and $x \in \mathbb{R}^n$, $x \in \mathrm{Ker}\, M$ (i.e. $Mx = 0$) $\Longleftrightarrow x^TMx = 0$.

LEMMA 1.2. *Let X be a symmetric matrix with block form (1.11).*

(i) *If $X \succeq 0$ or if* $\mathrm{rank}\, X = \mathrm{rank}\, A$, *then* $x \in \mathrm{Ker}\, A \Longrightarrow \begin{pmatrix} x \\ 0\end{pmatrix} \in \mathrm{Ker}\, X$.

(ii) *If* $\mathrm{rank}\, X = \mathrm{rank}\, A$, *then* $\mathrm{Ker}\, X = \mathrm{Ker}\,(A\ B)$.

Proof. (i) $Ax = 0 \Longrightarrow 0 = x^TAx = (x^T\ 0)\, X \begin{pmatrix} x \\ 0\end{pmatrix}$, which implies $X\begin{pmatrix} x \\ 0\end{pmatrix} = 0$ if $X \succeq 0$. If $\mathrm{rank}\, X = \mathrm{rank}\, A$, then $B = AW$ for some matrix W and thus $B^Tx = 0$, giving $X\begin{pmatrix} x \\ 0\end{pmatrix} = 0$.
(ii) Obviously, $\mathrm{rank}\, X \geq \mathrm{rank}\,(A\ B) \geq \mathrm{rank}\, A$. If $\mathrm{rank}\, X = \mathrm{rank}\, A$, equality holds throughout, which implies $\mathrm{Ker}\, X = \mathrm{Ker}\,(A\ B)$. □

Semidefinite programs. Consider the program

$$p^* := \sup_{X \in \mathrm{Sym}_n} \langle C, A\rangle \text{ s.t. } X \succeq 0,\ \langle A_j, X\rangle = b_j\ (j = 1, \ldots, m) \tag{1.12}$$

in the matrix variable X, where we are given $C, A_1, \ldots, A_m \in \mathrm{Sym}_n$ and $b \in \mathbb{R}^m$. This is the standard (primal) form of a semidefinite program; its dual semidefinite program reads:

$$d^* := \inf_{y \in \mathbb{R}^m} b^Ty \text{ s.t. } \sum_{j=1}^m y_jA_j - C \succeq 0 \tag{1.13}$$

in the variable $y \in \mathbb{R}^m$. Obviously,

$$p^* \leq d^*, \tag{1.14}$$

known as *weak duality*. Indeed, if X is feasible for (1.12) and y is feasible for (1.13), then $0 \leq \langle X, \sum_{j=1}^m y_j A_j - C\rangle = b^T y - \langle C, X\rangle$. One crucial issue in duality theory is to identify sufficient conditions that ensure equality in (1.14), i.e. a *zero duality gap*, in which case one speaks of *strong duality*. We say that (1.12) is *strictly feasible* when there exists $X \succ 0$ which is feasible for (1.12); analogously (1.13) is strictly feasible when there exists y feasible for (1.13) with $\sum_{j=1}^m y_j A_j - C \succ 0$.

THEOREM 1.3. *If the primal program (1.12) is strictly feasible and its dual (1.13) is feasible, then $p^* = d^*$ and (1.13) attains its supremum. Analogously, if (1.13) is strictly feasible and (1.12) is feasible, then $p^* = d^*$ and (1.12) attains its infimum.*

Semidefinite programs are convex programs. As one can test in polynomial time whether a given rational matrix is positive semidefinite (using e.g. Gaussian elimination), semidefinite programs can be solved in polynomial time to any fixed precision using the ellipsoid method (cf. [45]). Algorithms based on the ellipsoid method are however not practical since their running time is prohibitively high. Interior-point methods turn out to be the method of choice for solving semidefinite programs in practice; they can find an approximate solution (to any given precision) in polynomially many iterations and their running time is efficient in practice for medium sized problems. There is a vast literature devoted to semidefinite programming and interior-point algorithms; cf. e.g. [96, 117, 147, 150, 156].

We will use (later in Section 6.6) the following geometric property of semidefinite programs. We formulate the property for the program (1.12), but the analogous property holds for (1.13) as well.

LEMMA 1.4. *Let $\mathcal{R} := \{X \in PSD_n \mid \langle A_j, X\rangle = b_j \ (j = 1, \ldots, m)\}$ denote the feasible region of the semidefinite program (1.12). If $X^* \in \mathcal{R}$ has maximum rank, i.e.* $\operatorname{rank} X^* = \max_{X \in \mathcal{R}} \operatorname{rank} X$, *then* $\operatorname{Ker} X^* \subseteq \operatorname{Ker} X$ *for all $X \in \mathcal{R}$. In particular, if X^* is an optimum solution to (1.12) for which* $\operatorname{rank} X^*$ *is maximum, then* $\operatorname{Ker} X^* \subseteq \operatorname{Ker} X$ *for any other optimum solution X.*

Proof. Let $X^* \in \mathcal{R}$ for which $\operatorname{rank} X^*$ is maximum and let $X \in \mathcal{R}$. Then $X' := \frac{1}{2}(X^* + X) \in \mathcal{R}$, with $\operatorname{Ker} X' = \operatorname{Ker} X^* \cap \operatorname{Ker} X \subseteq \operatorname{Ker} X^*$. Thus equality $\operatorname{Ker} X' = \operatorname{Ker} X^*$ holds by the maximality assumption on $\operatorname{rank} X^*$, which implies $\operatorname{Ker} X^* \subseteq \operatorname{Ker} X$. The last statement follows simply by adding the constraint $\langle C, X\rangle = p^*$ to the description of the set $\mathcal{R}$. □

Geometrically, what the above lemma says is that the maximum rank matrices in $\mathcal{R}$ correspond to the matrices lying in the relative interior of the convex set $\mathcal{R}$. And the maximum rank optimum solutions to the program (1.12) are those lying in the relative interior of the optimum face $\langle C, X\rangle = p^*$ of the feasible region $\mathcal{R}$. As a matter of fact primal-dual interior-point algorithms that follow the so-called central path to solve a semidefinite program return a solution lying in the relative interior of the optimum face (cf. [156] for details). Thus (under certain conditions) it is easy to return an optimum solution of maximum rank; this feature will be useful for the extraction of global minimizers to polynomial optimization problems (cf. Section 6.6). In contrast it is *hard* to find optimum solutions of *minimum* rank. Indeed it is easy to formulate hard problems as semidefinite programs with a rank condition. For instance, given a sequence $a \in \mathbb{N}^n$, the program

$$p^* := \min \langle aa^T, X\rangle \text{ s.t. } X \succeq 0,\ X_{ii} = 1\ (i = 1, \ldots, n), \operatorname{rank} X = 1$$

solves the partition problem introduced in Section 1.1. Indeed any $X \succeq 0$ with diagonal entries all equal to 1 and with rank 1 is of the form $X = xx^T$ for some $x \in \{\pm 1\}^n$. Therefore, the sequence $a = (a_1, \ldots, a_n)$ can be partitioned precisely when $p^* = 0$, in which case any optimum solution $X = xx^T$ gives a partition of a, as $a^T x = \sum_{i=1}^n a_i x_i = 0$.

1.4. Contents of the paper. We provide in Section 2 more detailed algebraic preliminaries about polynomial ideals and varieties and the resolution of systems of polynomial equations. This is relevant to the problem of extracting global minimizers for the polynomial optimization problem (1.1) and can be read separately. Then the rest of the paper is divided into two parts. Part 1 contains some background results about positive polynomials and sums of squares (Section 3) and about the theory of moments (Section 4), and more detailed results about (truncated) moment matrices, in particular, from Curto and Fialkow (Section 5). Part 2 presents the application to polynomial optimization; namely, the main properties of the moment/SOS relaxations (Section 6), some further selected topics dealing in particular with approximations of positive polynomials by sums of squares and various approaches to unconstrained polynomial minimization (Section 7), and exploiting algebraic structure to reduce the problem size (Section 8).

2. Algebraic preliminaries.

We group here some preliminaries on polynomial ideals and varieties, and on the eigenvalue method for solving systems of polynomial equations. For more information, see, e.g., [6, 19, 21, 22, 144].

2.1. Polynomial ideals and varieties. Let $\mathcal{I}$ be an ideal in $\mathbb{R}[\mathbf{x}]$; that is, $\mathcal{I}$ is an additive subgroup of $\mathbb{R}[\mathbf{x}]$ satisfying $fg \in \mathcal{I}$ whenever $f \in \mathcal{I}$ and $g \in \mathbb{R}[\mathbf{x}]$. Given $h_1, \ldots, h_m \in \mathbb{R}[\mathbf{x}]$,

$$(h_1, \ldots, h_m) := \Big\{ \sum_{j=1}^m u_j h_j \mid u_1, \ldots, u_m \in \mathbb{R}[\mathbf{x}] \Big\}$$

denotes the *ideal generated by* $h_1, \ldots, h_m$. By the finite basis theorem, any ideal in $\mathbb{R}[\mathbf{x}]$ admits a finite set of generators. Given an ideal $\mathcal{I} \subseteq \mathbb{R}[\mathbf{x}]$, define

$$V_{\mathbb{C}}(\mathcal{I}) := \{x \in \mathbb{C}^n \mid f(x) = 0\ \forall f \in \mathcal{I}\},\ V_{\mathbb{R}}(\mathcal{I}) := V_{\mathbb{C}}(\mathcal{I}) \cap \mathbb{R}^n;$$

$V_{\mathbb{C}}(\mathcal{I})$ is the *(complex) variety* associated to $\mathcal{I}$ and $V_{\mathbb{R}}(\mathcal{I})$ is its *real variety*. Thus, if $\mathcal{I}$ is generated by $h_1, \ldots, h_m$, then $V_{\mathbb{C}}(\mathcal{I})$ (resp., $V_{\mathbb{R}}(\mathcal{I})$) is the set of common complex (resp., real) zeros of $h_1, \ldots, h_m$. Observe that $V_{\mathbb{C}}(\mathcal{I})$ is closed under complex conjugation, i.e., $\overline{v} \in V_{\mathbb{C}}(\mathcal{I})$ for all $v \in V_{\mathbb{C}}(\mathcal{I})$, since $\mathcal{I}$ consists of polynomials with real coefficients. When $V_{\mathbb{C}}(\mathcal{I})$ is finite, the ideal $\mathcal{I}$ is said to be *zero-dimensional.* Given $V \subseteq \mathbb{C}^n$,

$$\mathcal{I}(V) := \{f \in \mathbb{R}[\mathbf{x}] \mid f(v) = 0 \ \forall v \in V\}$$

is the *vanishing ideal* of V. Moreover,

$$\sqrt{\mathcal{I}} := \{f \in \mathbb{R}[\mathbf{x}] \mid f^k \in \mathcal{I} \text{ for some integer } k \geq 1\}$$

is the *radical* of $\mathcal{I}$ and

$$\sqrt[\mathbb{R}]{\mathcal{I}} := \left\{ f \in \mathbb{R}[\mathbf{x}] \mid f^{2k} + \sum_{j=1}^{m} p_j^2 \in \mathcal{I} \ \text{ for some } k \geq 1, \ p_1, \ldots, p_m \in \mathbb{R}[\mathbf{x}] \right\}$$

is the *real radical* of $\mathcal{I}$. The sets $\mathcal{I}(V)$, $\sqrt{\mathcal{I}}$ and $\sqrt[\mathbb{R}]{\mathcal{I}}$ are again ideals in $\mathbb{R}[\mathbf{x}]$. Obviously, for an ideal $\mathcal{I} \subseteq \mathbb{R}[\mathbf{x}]$,

$$\mathcal{I} \subseteq \sqrt{\mathcal{I}} \subseteq \mathcal{I}(V_{\mathbb{C}}(\mathcal{I})), \ \mathcal{I} \subseteq \sqrt[\mathbb{R}]{\mathcal{I}} \subseteq \mathcal{I}(V_{\mathbb{R}}(\mathcal{I})).$$

The following celebrated results relate (real) radical and vanishing ideals.

THEOREM 2.1. *Let $\mathcal{I}$ be an ideal in $\mathbb{R}[\mathbf{x}]$.*

(i) **(Hilbert's Nullstellensatz)** *(see, e.g., [21, §4.1])* $\sqrt{\mathcal{I}} = \mathcal{I}(V_{\mathbb{C}}(\mathcal{I}))$.

(ii) **(The Real Nullstellensatz)** *(see, e.g., [10, §4.1])* $\sqrt[\mathbb{R}]{\mathcal{I}} = \mathcal{I}(V_{\mathbb{R}}(\mathcal{I}))$.

The ideal $\mathcal{I}$ is said to be *radical* when $\mathcal{I} = \sqrt{\mathcal{I}}$, and *real radical* when $\mathcal{I} = \sqrt[\mathbb{R}]{\mathcal{I}}$. Roughly speaking, the ideal $\mathcal{I}$ is radical if all points of $V_{\mathbb{C}}(\mathcal{I})$ have single multiplicity. For instance, the ideal $\mathcal{I} := (\mathbf{x}^2)$ is not radical since $V_{\mathbb{C}}(\mathcal{I}) = \{0\}$ and $\mathbf{x} \in \mathcal{I}(V_{\mathbb{C}}(\mathcal{I})) \setminus \mathcal{I}$. Obviously, $\mathcal{I} \subseteq \mathcal{I}(V_{\mathbb{C}}(\mathcal{I})) \subseteq \mathcal{I}(V_{\mathbb{R}}(\mathcal{I}))$. Hence, $\mathcal{I}$ real radical $\Longrightarrow \mathcal{I}$ radical. Moreover,

$$\mathcal{I} \text{ real radical with } |V_{\mathbb{R}}(\mathcal{I})| < \infty \Longrightarrow V_{\mathbb{C}}(\mathcal{I}) = V_{\mathbb{R}}(\mathcal{I}) \subseteq \mathbb{R}^n. \tag{2.1}$$

Indeed, $\mathcal{I}(V_{\mathbb{C}}(\mathcal{I})) = \mathcal{I}(V_{\mathbb{R}}(\mathcal{I}))$ implies $V_{\mathbb{C}}(\mathcal{I}(V_{\mathbb{C}}(\mathcal{I}))) = V_{\mathbb{C}}(\mathcal{I}(V_{\mathbb{R}}(\mathcal{I})))$. Now, $V_{\mathbb{C}}(\mathcal{I}(V_{\mathbb{C}}(\mathcal{I}))) = V_{\mathbb{C}}(\mathcal{I})$, and $V_{\mathbb{C}}(\mathcal{I}(V_{\mathbb{R}}(\mathcal{I}))) = V_{\mathbb{R}}(\mathcal{I})$ since $V_{\mathbb{R}}(\mathcal{I})$ is an algebraic subset of $\mathbb{C}^n$ as it is finite. We will often use the following characterization of (real) radical ideals which follows directly from the (Real) Nullstellensatz:

$$\begin{array}{c} \mathcal{I} \text{ is radical (resp., real radical)} \\ \Longleftrightarrow \\ \text{The only polynomials vanishing at all points of } V_{\mathbb{C}}(\mathcal{I}) \\ \text{(resp., all points of } V_{\mathbb{R}}(\mathcal{I})) \text{ are the polynomials in } \mathcal{I}. \end{array} \tag{2.2}$$

The following lemma gives a useful criterion for checking whether an ideal is (real) radical.

LEMMA 2.2. *Let $\mathcal{I}$ be an ideal in $\mathbb{R}[\mathbf{x}]$.*

(i) *$\mathcal{I}$ is radical if and only if*

$$\forall f \in \mathbb{R}[\mathbf{x}] \quad f^2 \in \mathcal{I} \Longrightarrow f \in \mathcal{I}. \tag{2.3}$$

(ii) *$\mathcal{I}$ is real radical if and only if*

$$\forall p_1, \ldots, p_m \in \mathbb{R}[\mathbf{x}] \quad \sum_{j=1}^{m} p_j^2 \in \mathcal{I} \Longrightarrow p_1, \ldots, p_m \in \mathcal{I}. \tag{2.4}$$

Proof. The 'only if' part is obvious in (i), (ii); we prove the 'if part'.

(i) Assume that (2.3) holds. Let $f \in \mathbb{R}[\mathbf{x}]$. We show $f^k \in \mathcal{I} \Longrightarrow f \in \mathcal{I}$ using induction on $k \geq 1$. Let $k \geq 2$. Using (2.3), we deduce $f^{\lceil k/2 \rceil} \in \mathcal{I}$. As $\lceil k/2 \rceil \leq k-1$, we deduce $f \in \mathcal{I}$ using the induction assumption.

(ii) Assume that (2.4) holds. Let $f, p_1, \ldots, p_m \in \mathbb{R}[\mathbf{x}]$ such that $f^{2k} + \sum_{j=1}^m p_j^2 \in \mathcal{I}$; we show that $f \in \mathcal{I}$. First we deduce from (2.4) that $f^k, p_1, \ldots, p_m \in \mathcal{I}$. As (2.4) implies (2.3), we next deduce from the case (i) that $f \in \mathcal{I}$. □

We now recall the following simple fact about interpolation polynomials, which we will need at several occasions in the paper.

LEMMA 2.3. *Let $V \subseteq \mathbb{C}^n$ with $|V| < \infty$. There exist polynomials $p_v \in \mathbb{C}[\mathbf{x}]$ (for $v \in V$) satisfying $p_v(v) = 1$ and $p_v(u) = 0$ for all $u \in V \setminus \{v\}$; they are known as* Lagrange interpolation polynomials *at the points of V. Assume moreover that V is closed under complex conjugation, i.e., $V = \overline{V} := \{\overline{v} \mid v \in V\}$. Then we may choose the interpolation polynomials in such a way that they satisfy $p_{\overline{v}} = \overline{p_v}$ for all $v \in V$ and, given scalars a_v ($v \in V$) satisfying $a_{\overline{v}} = \overline{a_v}$ for all $v \in V$, there exists $p \in \mathbb{R}[\mathbf{x}]$ taking the prescribed values $p(v) = a_v$ at the points $v \in V$.*

Proof. Fix $v \in V$. For $u \in V$, $u \neq v$, pick an index $i_u \in \{1, \ldots, n\}$ for which $u_{i_u} \neq v_{i_u}$ and define the polynomial $p_v := \prod_{u \in V \setminus \{v\}} \frac{\mathbf{x}_{i_u} - u_{i_u}}{v_{i_u} - u_{i_u}}$. Then the polynomials p_v ($v \in V$) satisfy the lemma. If $\overline{V} = V$, then we can choose the interpolation polynomials in such a way that $p_{\overline{v}} = \overline{p_v}$ for all $v \in V$. Indeed, for $v \in V \cap \mathbb{R}^n$, simply replace p_v by its real part and, for $v \in V \setminus \mathbb{R}^n$, pick p_v as before and choose $p_{\overline{v}} := \overline{p_v}$. Finally, if $a_{\overline{v}} = \overline{a_v}$ for all $v \in V$, then the polynomial $p := \sum_{v \in V} a_v p_v$ has real coefficients and satisfies $p(v) = a_v$ for $v \in V$. □

The algebraic tools just introduced here permit to show the following result of Parrilo [105], giving a sum of squares decomposition for every polynomial nonnegative on a finite variety assuming radicality of the associated ideal.

THEOREM 2.4. [105] *Consider the semialgebraic set*

$$K := \{x \in \mathbb{R}^n \mid h_1(x) = 0, \ldots, h_{m_0}(x) = 0, g_1(x) \geq 0, \ldots, g_m(x) \geq 0\}, \tag{2.5}$$

where $h_1,\ldots,h_{m_0},g_1,\ldots,g_m\in\mathbb{R}[\mathbf{x}]$ *and* $m_0\geq 1$, $m\geq 0$. *Assume that the ideal* $\mathcal{I}:=(h_1,\ldots,h_{m_0})$ *is zero-dimensional and radical. Then every nonnegative polynomial on* K *is of the form* $u_0+\sum_{j=1}^m u_jg_j+q$, *where* $u_0,u_1,\ldots,u_m$ *are sums of squares of polynomials and* $q\in\mathcal{I}$.

Proof. Partition $V:=V_{\mathbb{C}}(\mathcal{I})$ into $S\cup T\cup\overline{T}$, where $S=V\cap\mathbb{R}^n$, $T\cup\overline{T}=V\setminus\mathbb{R}^n$. Let p_v ($v\in V_{\mathbb{C}}(\mathcal{I})$) be interpolation polynomials at the points of V, satisfying $p_{\overline{v}}=\overline{p_v}$ for $v\in T$ (as in Lemma 2.3). We first show the following fact: If $f\in\mathbb{R}[\mathbf{x}]$ is nonnegative on the set S, then $f=\sigma+q$ where σ is a sum of squares of polynomials and $q\in\mathcal{I}$. For this, for $v\in S\cup T$, let $\gamma_v=\sqrt{f(v)}$ be a square root of $f(v)$ (thus, $\gamma_v\in\mathbb{R}$ if $v\in S$) and define the polynomials $q_v\in\mathbb{R}[\mathbf{x}]$ by $q_v:=\gamma_vp_v$ for $v\in S$ and $q_v:=\gamma_vp_v+\overline{\gamma_vp_v}$ for $v\in T$. The polynomial $f-\sum_{v\in S\cup T}(q_v)^2$ vanishes at all points of V; hence it belongs to $\mathcal{I}$, since $\mathcal{I}$ is radical. This shows that $f=\sigma+q$, where σ is a sum of squares and $q\in\mathcal{I}$.

Suppose now that $f\in\mathbb{R}[\mathbf{x}]$ is nonnegative on the set K. In view of Lemma 2.3, we can construct polynomials $s_0,s_1,\ldots,s_m\in\mathbb{R}[\mathbf{x}]$ taking the following prescribed values at the points in V: If $v\in V\setminus S$, or if $v\in S$ and $f(v)\geq 0$, $s_0(v):=f(v)$ and $s_j(v):=0$ ($j=1,\ldots,m$). Otherwise, $v\notin K$ and thus $g_{j_v}(v)<0$ for some $j_v\in\{1,\ldots,m\}$; then $s_{j_v}(v):=\frac{f(v)}{g_{j_v}(v)}$ and $s_0(v)=s_j(v):=0$ for $j\in\{1,\ldots,m\}\setminus\{j_v\}$. By construction, each of the polynomials $s_0,s_1,\ldots,s_m$ is nonnegative on S. Using the above result, we can conclude that $s_j=\sigma_j+q_j$, where σ_j is a sum of squares and $q_j\in\mathcal{I}$, for $j=0,1,\ldots,m$. Now the polynomial $q:=f-s_0-\sum_{j=1}^m s_jg_j$ vanishes at all points of V and thus belongs to $\mathcal{I}$. Therefore, $f=s_0+\sum_{j=1}^m s_jg_j+q=\sigma_0+\sum_{j=1}^m\sigma_jg_j+q'$, where $q':=q+q_0+\sum_{j=1}^m q_jg_j\in\mathcal{I}$ and σ_0,σ_j are sums of squares of polynomials. □

2.2. The quotient algebra $\mathbb{R}[\mathbf{x}]/\mathcal{I}$. Given an ideal $\mathcal{I}$ in $\mathbb{R}[\mathbf{x}]$, the elements of the quotient space $\mathbb{R}[\mathbf{x}]/\mathcal{I}$ are the cosets $[f]:=f+\mathcal{I}=\{f+q\mid q\in\mathcal{I}\}$. $\mathbb{R}[\mathbf{x}]/\mathcal{I}$ is a $\mathbb{R}$-vector space with addition $[f]+[g]=[f+g]$ and scalar multiplication $\lambda[f]=[\lambda f]$, and an algebra with multiplication $[f][g]=[fg]$, for $\lambda\in\mathbb{R}$, $f,g\in\mathbb{R}[\mathbf{x}]$. Given $h\in\mathbb{R}[\mathbf{x}]$, the '*multiplication by h operator*'

$$\begin{array}{cccc} m_h: & \mathbb{R}[\mathbf{x}]/\mathcal{I} & \longrightarrow & \mathbb{R}[\mathbf{x}]/\mathcal{I} \\ & f \mod \mathcal{I} & \longmapsto & fh \mod \mathcal{I} \end{array} \tag{2.6}$$

is well defined. As we see later in Section 2.4, multiplication operators play a central role in the computation of the variety $V_{\mathbb{C}}(\mathcal{I})$. In what follows we often identify a subset of $\mathbb{R}[\mathbf{x}]$ with the corresponding subset of $\mathbb{R}[\mathbf{x}]/\mathcal{I}$ consisting of the cosets of its elements. For instance, given $\mathcal{B}=\{b_1,\ldots,b_N\}\subseteq\mathbb{R}[\mathbf{x}]$, if the cosets $[b_1],\ldots,[b_N]$ generate $\mathbb{R}[\mathbf{x}]/\mathcal{I}$, i.e., if any $f\in\mathbb{R}[\mathbf{x}]$ can be written as $\sum_{j=1}^N\lambda_jb_j+q$ for some $\lambda\in\mathbb{R}^N$ and $q\in\mathcal{I}$, then we also say by abuse of language that the set $\mathcal{B}$ itself is generating in $\mathbb{R}[\mathbf{x}]/\mathcal{I}$. Analogously, if the cosets $[b_1],\ldots,[b_N]$ are pairwise distinct and form a linearly independent subset of $\mathbb{R}[\mathbf{x}]/\mathcal{I}$, i.e., if $\sum_{j=1}^{N} \lambda_j b_j \in \mathcal{I}$ implies $\lambda = 0$, then we say that $\mathcal{B}$ is linearly independent in $\mathbb{R}[\mathbf{x}]/\mathcal{I}$.

Theorem 2.6 below relates the cardinality of $V_{\mathbb{C}}(\mathcal{I})$ and the dimension of the quotient vector space $\mathbb{R}[\mathbf{x}]/\mathcal{I}$. This is a classical result (see, e.g., [21]), which we will use repeatedly in our treatment. The following simple fact will be used in the proof.

LEMMA 2.5. *Let $\mathcal{I} \subseteq \mathbb{R}[\mathbf{x}]$ with $|V_{\mathbb{C}}(\mathcal{I})| < \infty$. Partition $V_{\mathbb{C}}(\mathcal{I})$ into $V_{\mathbb{C}}(\mathcal{I}) = S \cup T \cup \overline{T}$ where $S = V_{\mathbb{C}}(\mathcal{I}) \cap \mathbb{R}^n$, and let p_v be interpolation polynomials at the points of $V_{\mathbb{C}}(\mathcal{I})$ satisfying $p_{\overline{v}} = \overline{p_v}$ for all $v \in V_{\mathbb{C}}(\mathcal{I})$. The set $\mathcal{L} := \{p_v \ (v \in S), \mathrm{Re}(p_v), \mathrm{Im}(p_v) \ (v \in T)\}$ is linearly independent in $\mathbb{R}[\mathbf{x}]/\mathcal{I}$ and generates $\mathbb{R}[\mathbf{x}]/\mathcal{I}(V_{\mathbb{C}}(\mathcal{I}))$.*

Proof. Assume $\sum_{v \in S} \lambda_v p_v + \sum_{v \in T} \lambda_v \mathrm{Re}(p_v) + \lambda'_v \mathrm{Im}(p_v) \in \mathcal{I}$. Evaluating this polynomial at $v \in V_{\mathbb{C}}(\mathcal{I})$ yields that all scalars λ_v, λ'_v are 0. Thus $\mathcal{L}$ is linearly independent in $\mathbb{R}[\mathbf{x}]/\mathcal{I}$. Given $f \in \mathbb{R}[\mathbf{x}]$, the polynomial $f - \sum_{v \in V_{\mathbb{C}}(\mathcal{I})} f(v) p_v$ lies in $\mathcal{I}(V_{\mathbb{C}}(\mathcal{I}))$. Now, $\sum_{v \in V_{\mathbb{C}}(\mathcal{I})} f(v) p_v = \sum_{v \in S} f(v) p_v + \sum_{v \in T} 2\,\mathrm{Re}(f(v) p_v)$ can be written as a linear combination of $\mathrm{Re}(p_v)$ and $\mathrm{Im}(p_v)$. This implies that $\mathcal{L}$ generates $\mathbb{R}[\mathbf{x}]/\mathcal{I}(V_{\mathbb{C}}(\mathcal{I}))$. $\square$

THEOREM 2.6. *An ideal $\mathcal{I} \subseteq \mathbb{R}[\mathbf{x}]$ is zero-dimensional (i.e., $|V_{\mathbb{C}}(\mathcal{I})| < \infty$) if and only if the vector space $\mathbb{R}[\mathbf{x}]/\mathcal{I}$ is finite dimensional. Moreover, $|V_{\mathbb{C}}(\mathcal{I})| \leq \dim \ \mathbb{R}[\mathbf{x}]/\mathcal{I}$, with equality if and only if the ideal $\mathcal{I}$ is radical.*

Proof. Assume $k := \dim \mathbb{R}[\mathbf{x}]/\mathcal{I} < \infty$. Then, the set $\{1, \mathbf{x}_1, \ldots, \mathbf{x}_1^k\}$ is linearly dependent in $\mathbb{R}[\mathbf{x}]/\mathcal{I}$. Thus there exist scalars $\lambda_0, \ldots, \lambda_k$ (not all zero) for which the polynomial $f := \sum_{h=0}^{k} \lambda_h \mathbf{x}_1^h$ belongs to $\mathcal{I}$. Thus, for $v \in V_{\mathbb{C}}(\mathcal{I})$, $f(v) = 0$, which implies that v_1 takes only finitely many values. Applying the same reasoning to the other coordinates, we deduce that $V_{\mathbb{C}}(\mathcal{I})$ is finite.

Assume now $|V_{\mathbb{C}}(\mathcal{I})| < \infty$. Say, $\{v_1 \mid v \in V_{\mathbb{C}}(\mathcal{I})\} = \{a_1, \ldots, a_k\}$. Then the polynomial $f := \prod_{h=1}^{k} (\mathbf{x}_1 - a_h)$ belongs to $\mathcal{I}(V_{\mathbb{C}}(\mathcal{I}))$. By Theorem 2.1, $f \in \sqrt{\mathcal{I}}$, i.e., $f^{m_1} \in \mathcal{I}$ for some integer $m_1 \geq 1$. Hence the set $\{[1], [\mathbf{x}_1], \ldots, [\mathbf{x}_1^{km_1}]\}$ is linearly dependent in $\mathbb{R}[\mathbf{x}]/\mathcal{I}$ and thus, for some integer $n_1 \geq 1$, $[\mathbf{x}_1^{n_1}]$ lies in $\mathrm{Span}_{\mathbb{R}}([1], \ldots, [\mathbf{x}_1^{n_1-1}])$. Similarly, for any other coordinate $\mathbf{x}_i$, $[\mathbf{x}_i^{n_i}] \in \mathrm{Span}_{\mathbb{R}}([1], \ldots, [\mathbf{x}_i^{n_i-1}])$ for some integer $n_i \geq 1$. From this one can easily derive that the set $\{[\mathbf{x}^\alpha] \mid 0 \leq \alpha_i \leq n_i - 1 \ (1 \leq i \leq n)\}$ generates $\mathbb{R}[\mathbf{x}]/\mathcal{I}$, which shows that $\dim \mathbb{R}[\mathbf{x}]/\mathcal{I} < \infty$.

Assume $V_{\mathbb{C}}(\mathcal{I})$ is finite and let $\mathcal{L}$ be as in Lemma 2.5. As $\mathcal{L}$ is linearly independent in $\mathbb{R}[\mathbf{x}]/\mathcal{I}$ with $|\mathcal{L}| = |V_{\mathbb{C}}(\mathcal{I})|$ we deduce that $\dim \mathbb{R}[\mathbf{x}]/\mathcal{I} \geq |V_{\mathbb{C}}(\mathcal{I})|$. Moreover, if $\mathcal{I}$ is radical then $\mathcal{I} = \mathcal{I}(V_{\mathbb{C}}(\mathcal{I}))$ and thus $\mathcal{L}$ is also generating in $\mathbb{R}[\mathbf{x}]/\mathcal{I}$, which implies $\dim \mathbb{R}[\mathbf{x}]/\mathcal{I} = |V_{\mathbb{C}}(\mathcal{I})|$. Finally, if $\mathcal{I}$ is not radical, there exists a polynomial $f \in \mathcal{I}(V_{\mathbb{C}}(\mathcal{I})) \setminus \mathcal{I}$ and it is easy to verify that the set $\mathcal{L} \cup \{f\}$ is linearly independent in $\mathbb{R}[\mathbf{x}]/\mathcal{I}$. $\square$

For instance, the ideal $\mathcal{I} := (\mathbf{x}_i^2 - \mathbf{x}_i \mid i = 1, \ldots, n)$ is radical and zero-dimensional, since $V_{\mathbb{C}}(\mathcal{I}) = \{0,1\}^n$, and the set $\{\prod_{l \in L} \mathbf{x}_l \mid L \subseteq \{1, \ldots, n\}\}$ is a linear basis of $\mathbb{R}[\mathbf{x}]/\mathcal{I}$.

Assume $N := \dim \mathbb{R}[\mathbf{x}]/\mathcal{I} < \infty$ and let $\mathcal{B} = \{b_1, \ldots, b_N\} \subseteq \mathbb{R}[\mathbf{x}]$ be a basis of $\mathbb{R}[\mathbf{x}]/\mathcal{I}$; that is, any polynomial $f \in \mathbb{R}[\mathbf{x}]$ can be written in a unique way as

$$f = \underbrace{\sum_{j=1}^{N} \lambda_j b_j}_{\text{res}_{\mathcal{B}}(f)} + q, \text{ where } q \in \mathcal{I} \text{ and } \lambda \in \mathbb{R}^N;$$

in short, $p \equiv \sum_{j=1}^{N} \lambda_j b_j \mod \mathcal{I}$. The polynomial $\text{res}_{\mathcal{B}}(f) := \sum_{j=1}^{N} \lambda_j b_j$ is called the *residue of f modulo $\mathcal{I}$ with respect to the basis $\mathcal{B}$*. In other words, the vector space $\text{Span}_{\mathbb{R}}(\mathcal{B}) := \{\sum_{j=1}^{N} \lambda_j b_j \mid \lambda \in \mathbb{R}^N\}$ is isomorphic to $\mathbb{R}[\mathbf{x}]/\mathcal{I}$. As recalled in the next section, the set $\mathcal{B}_{\succ}$ of standard monomials with respect to any monomial ordering is a basis of $\mathbb{R}[\mathbf{x}]/\mathcal{I}$; then the residue of a polynomial f w.r.t. $\mathcal{B}_{\succ}$ is also known as the *normal form* of f w.r.t. the given monomial ordering. Let us mention for further reference the following variation of Lemma 2.3.

LEMMA 2.7. *Let $\mathcal{I}$ be a zero-dimensional ideal in $\mathbb{R}[\mathbf{x}]$ and let $\mathcal{B}$ be a basis of $\mathbb{R}[\mathbf{x}]/\mathcal{I}$. There exist interpolation polynomials p_v at the points of $V_{\mathbb{C}}(\mathcal{I})$, where each p_v is a linear combination of members of $\mathcal{B}$.*

Proof. Given a set of interpolation polynomials p_v, replace p_v by its residue modulo $\mathcal{I}$ with respect to $\mathcal{B}$. □

2.3. Gröbner bases and standard monomial bases. A classical method for constructing a linear basis of the quotient vector space $\mathbb{R}[\mathbf{x}]/\mathcal{I}$ is to determine a Gröbner basis of the ideal $\mathcal{I}$ with respect to some given monomial ordering; then the corresponding set of standard monomials provides a basis of $\mathbb{R}[\mathbf{x}]/\mathcal{I}$. We recall here a few basic definitions about monomial orderings, Gröbner bases, and standard monomials. A *monomial ordering* '$\succ$' is a total ordering of the set $\mathbb{T}_n = \{\mathbf{x}^\alpha \mid \alpha \in \mathbb{N}^n\}$ of monomials, which is a well-ordering and satisfies the condition: $\mathbf{x}^\alpha \succ \mathbf{x}^\beta \Longrightarrow \mathbf{x}^{\alpha+\gamma} \succ \mathbf{x}^{\beta+\gamma}$. We also write $a\mathbf{x}^\alpha \succ b\mathbf{x}^\beta$ if $\mathbf{x}^\alpha \succ \mathbf{x}^\beta$ and $a, b \in \mathbb{R} \setminus \{0\}$. Examples of monomial orderings are the *lexicographic order* '$\succ_{lex}$', where $\mathbf{x}^\alpha \succ_{lex} \mathbf{x}^\beta$ if $\alpha > \beta$ for a lexicographic order on $\mathbb{N}^n$, or the *graded lexicographic order* '$\succ_{grlex}$', where $\mathbf{x}^\alpha \succ_{grlex} \mathbf{x}^\beta$ if $|\alpha| > |\beta|$, or $|\alpha| = |\beta|$ and $\mathbf{x}^\alpha \succ_{lex} \mathbf{x}^\beta$. The latter is an example of a *total degree monomial ordering*, i.e., a monomial ordering $\succ$ such that $\mathbf{x}^\alpha \succ \mathbf{x}^\beta$ whenever $|\alpha| > |\beta|$.

Fix a monomial ordering $\succ$ on $\mathbb{R}[\mathbf{x}]$. For a nonzero polynomial $f = \sum_\alpha f_\alpha \mathbf{x}^\alpha$, its *terms* are the quantities $f_\alpha \mathbf{x}^\alpha$ with $f_\alpha \neq 0$ and its *leading term* $\text{LT}(f)$ is defined as the maximum $f_\alpha \mathbf{x}^\alpha$ with respect to the given ordering for which $f_\alpha \neq 0$. Let $\mathcal{I}$ be an ideal in $\mathbb{R}[\mathbf{x}]$. Its *leading term ideal* is $\text{LT}(\mathcal{I}) := (\text{LT}(f) \mid f \in \mathcal{I})$ and the set

$$\mathcal{B}_{\succ} := \mathbb{T}_n \setminus \text{LT}(\mathcal{I}) = \{\mathbf{x}^\alpha \mid \text{LT}(f) \text{ does not divide } \mathbf{x}^\alpha \quad \forall f \in \mathcal{I}\}$$

is the set of *standard monomials.* A finite subset $G \subseteq \mathcal{I}$ is called a *Gröbner basis* of $\mathcal{I}$ if $\mathrm{LT}(\mathcal{I}) = \mathrm{LT}(G)$; that is, if the leading term of every nonzero polynomial in $\mathcal{I}$ is divisible by the leading term of some polynomial in G. Hence $\mathbf{x}^\alpha \in \mathcal{B}_\succ$ if and only if $\mathbf{x}^\alpha$ is not divisible by the leading term of any polynomial in G. A Gröbner basis always exists and it can be constructed, e.g., using the algorithm of Buchberger.

Once a monomial ordering $\succ$ is fixed, one can apply the division algorithm. Given nonzero polynomials $f, g_1, \ldots, g_m$, the division algorithm applied to dividing f by $g_1, \ldots, g_m$ produces polynomials $u_1, \ldots, u_m$ and r satisfying $f = \sum_{j=1}^m u_j g_j + r$, no term of r is divisible by $\mathrm{LT}(g_j)$ $(j = 1, \ldots, m)$ if $r \neq 0$, and $\mathrm{LT}(f) \succ \mathrm{LT}(u_j g_j)$ if $u_j \neq 0$. Hence $\deg(f) \geq \deg(u_j g_j)$ if $u_j \neq 0$, when the monomial ordering is a graded lexicographic order. When the polynomials $g_1, \ldots, g_m$ form a Gröbner basis of the ideal $\mathcal{I} := (g_1, \ldots, g_m)$, the remainder r is uniquely determined and r is a linear combination of the set of standard monomials, i.e., $r \in \mathrm{Span}_{\mathbb{R}}(\mathcal{B}_\succ)$; in particular, $f \in \mathcal{I}$ if and only if $r = 0$. In other words, the set $\mathcal{B}_\succ$ of standard monomials is a basis of the quotient vector space $\mathbb{R}[\mathbf{x}]/\mathcal{I}$.

EXAMPLE 2.8. Consider the polynomial $f = \mathbf{x}^2\mathbf{y} + \mathbf{x}\mathbf{y}^2 + \mathbf{y}^2$ to be divided by the polynomials $h_1 = \mathbf{x}\mathbf{y} - 1$, $h_2 = \mathbf{y}^2 - 1$. Fix the lex order with $\mathbf{x} > \mathbf{y}$. Then $\mathrm{LT}(f) = \mathbf{x}^2\mathbf{y}$, $\mathrm{LT}(h_1) = \mathbf{x}\mathbf{y}$, $\mathrm{LT}(h_2) = \mathbf{y}^2$. As $\mathrm{LT}(h_1) | \mathrm{LT}(f)$, we write

$$f = \mathbf{x}^2\mathbf{y} + \mathbf{x}\mathbf{y}^2 + \mathbf{y}^2 = \underbrace{(\mathbf{x}\mathbf{y} - 1)}_{h_1}(\mathbf{x} + \mathbf{y}) + \underbrace{\mathbf{x} + \mathbf{y}^2 + \mathbf{y}}_{q}.$$

Now $\mathrm{LT}(q) = \mathbf{x}$ is not divisible by $\mathrm{LT}(h_1), \mathrm{LT}(h_2)$, but $\mathrm{LT}(h_2)$ divides the term $\mathbf{y}^2$ of q. Thus write

$$q = \underbrace{(\mathbf{y}^2 - 1)}_{h_2} + \mathbf{x} + \mathbf{y} + 1.$$

This gives

$$f = h_1(\mathbf{x} + \mathbf{y}) + h_2 + \mathbf{x} + \mathbf{y} + 1. \tag{2.7}$$

No term of the polynomial $r := \mathbf{x} + \mathbf{y} + 1$ is divisible by $\mathrm{LT}(h_1), \mathrm{LT}(h_2)$, thus r is the remainder of the division of f by h_1, h_2 (in that order). If we do the division by h_2, h_1 then we get the following decomposition:

$$f = (\mathbf{x} + 1)h_2 + \mathbf{x}h_1 + 2\mathbf{x} + 1. \tag{2.8}$$

Thus (2.7), (2.8) are two disctinct decompositions of f of the form

$$f = \sum_{i=1}^2 u_i h_i + r$$

where no term of r is divisible by $\mathrm{LT}(h_1), \mathrm{LT}(h_2)$. Hence the remainder is not uniquely defined. This is because the set $\{h_1, h_2\}$ is not a Gröbner basis of the ideal $\mathcal{I} := (h_1, h_2)$. Indeed the polynomial

$$h_3 := \mathbf{y}h_1 - \mathbf{x}h_2 = \mathbf{y}(\mathbf{x}\mathbf{y} - 1) - \mathbf{x}(\mathbf{y}^2 - 1) = \mathbf{x} - \mathbf{y} \in \mathcal{I}$$

and $\mathrm{LT}(h_3) = \mathbf{x}$ is not divisible by $\mathrm{LT}(h_1), \mathrm{LT}(h_2)$. For the given monomial ordering, the set of standard monomials is $\mathcal{B} = \{1, \mathbf{y}\}$, the set $\{h_2, h_3\}$ is a Gröbner basis of $\mathcal{I}$, and $\dim \mathbb{R}[\mathbf{x}]/\mathcal{I} = 2 = |V_{\mathbb{C}}(\mathcal{I})|$ with $V_{\mathbb{C}}(\mathcal{I}) = \{(1,1), (-1,-1)\}$.

2.4. Solving systems of polynomial equations. One of the attractive features of Lasserre's method for minimizing a polynomial over a semialgebraic set is that, when some technical rank condition holds for the optimum solution of the given relaxation, then this relaxation is in fact exact and moreover one can extract global minimizers for the original problem. This extraction procedure requires to solve a system of polynomial equations

$$h_1(x) = 0, \ldots, h_{m_0}(x) = 0,$$

where the ideal $\mathcal{I} := (h_1, \ldots, h_{m_0})$ is zero-dimensional (and in fact radical). This problem has received considerable attention in the literature. We present the so-called eigenvalue method (also known as the Stetter-Möller method [92]) which relates the points of $V_{\mathbb{C}}(\mathcal{I})$ to the eigenvalues of the multiplication operators in the quotient space $\mathbb{R}[\mathbf{x}]/\mathcal{I}$. See, e.g., [19, 37, 144] for a detailed account on this method and various other methods for solving systems of polynomial equations.

Fix a basis $\mathcal{B} = \{b_1, \ldots, b_N\}$ of $\mathbb{R}[\mathbf{x}]/\mathcal{I}$ and let M_h denote the matrix of the multiplication operator operator m_h from (2.6) with respect to the basis $\mathcal{B}$. Namely, for $j = 1, \ldots, N$, let $\mathrm{res}_{\mathcal{B}}(hb_j) = \sum_{i=1}^N a_{ij} b_i$ denote the residue of hb_j modulo $\mathcal{I}$ w.r.t. $\mathcal{B}$, i.e.,

$$hb_j - \sum_{i=1}^N a_{ij} b_i \in \mathcal{I}; \tag{2.9}$$

then the jth column of M_h is equal to the vector $(a_{ij})_{i=1}^N$. When $h = \mathbf{x}_i$, the multiplication matrices $M_{\mathbf{x}_i}$ $(i = 1, \ldots, n)$ are also known as the *companion matrices* of the ideal $\mathcal{I}$. Theorem 2.9 below shows that the coordinates of the points $v \in V$ can be obtained from the eigenvalues of the companion matrices. As a motivation we first treat the univariate case.

Motivation: The univariate case. Given a univariate polynomial

$$p = \mathbf{x}^d - p_{d-1}\mathbf{x}^{d-1} - \ldots - p_0$$

consider the ideal $\mathcal{I} = (p)$ (obviously zero-dimensional). The set $\mathcal{B} = \{1, \mathbf{x}, \ldots, \mathbf{x}^{d-1}\}$ is a basis of $\mathbb{R}[\mathbf{x}]/(p)$. With respect to $\mathcal{B}$, the multiplication matrix $M_{\mathbf{x}}$ has the form

$$M_{\mathbf{x}} = \begin{pmatrix} 0 & \dots & 0 & p_0 \\ & & & p_1 \\ & I & & \vdots \\ & & & p_{d-1} \end{pmatrix}$$

where I is the identity matrix of size $(d-1) \times (d-1)$. One can verify that $\det(M_{\mathbf{x}} - tI) = (-1)^d p(t)$. Therefore, the eigenvalues of the companion matrix $M_{\mathbf{x}}$ are precisely the roots of the polynomial p. We now see how this fact extends to the multivariate case.

The multivariate case. The multiplication operators $m_{\mathbf{x}_1}, \dots, m_{\mathbf{x}_n}$ commute pairwise. Therefore the set $\{M_f \mid f \in \mathbb{R}[\mathbf{x}]\}$ is a commutative algebra of $N \times N$ matrices. For a polynomial $h \in \mathbb{R}[\mathbf{x}]$, $h = \sum_\alpha h_\alpha \mathbf{x}^\alpha$, note

$$M_h = h(M_{\mathbf{x}_1}, \dots, M_{\mathbf{x}_n}) = \sum_\alpha h_\alpha (M_{\mathbf{x}_1})^{\alpha_1} \cdots (M_{\mathbf{x}_n})^{\alpha_n} =: h(M),$$
$$M_h = 0 \Longleftrightarrow h \in \mathcal{I}.$$

Based on this, one can easily find the *minimal polynomial* of M_h (i.e. the monic polynomial $p \in \mathbb{R}[\mathbf{t}]$ of smallest degree for which $p(M_h) = 0$). Indeed, for $p = \sum_{i=0}^d p_i \mathbf{t}^i \in \mathbb{R}[\mathbf{t}]$, $p(M_h) = \sum_i p_i (M_h)^i = M_{p(h)} = 0$ if and only if $p(h) \in \mathcal{I}$. Thus one can find the minimal polynomial of M_h by computing the smallest integer d for which the set $\{[1], [h], \dots, [h^d]\}$ is linearly dependent in $\mathbb{R}[\mathbf{x}]/\mathcal{I}$. In particular, the minimal polynomial of $M_{\mathbf{x}_i}$ is the monic generator of the elimination ideal $\mathcal{I} \cap \mathbb{R}[\mathbf{x}_i]$.

Let $p_v \in \mathbb{R}[\mathbf{x}]$ be Lagrange interpolation polynomials at the points of $V_\mathbb{C}(\mathcal{I})$. As observed in Lemma 2.7, we may assume that $p_v \in \text{Span}_\mathbb{R}(\mathcal{B})$ for all $v \in V_\mathbb{C}(\mathcal{I})$. For a polynomial $p \in \text{Span}_\mathbb{R}(\mathcal{B})$, $p = \sum_{i=1}^N a_i b_i$ with $a_i \in \mathbb{R}$, let $\text{vec}_\mathcal{B}(p) := (a_i)_{i=1}^N$ denote the vector of its coefficients in $\mathcal{B}$. Set $\zeta_{\mathcal{B},v} := (b_i(v))_{i=1}^N \in \mathbb{C}^N$, the vector of evaluations at v of the polynomials in the basis $\mathcal{B}$. Observe that

$$\{\zeta_{\mathcal{B},v} \mid v \in V_\mathbb{C}(\mathcal{I})\} \text{ is linearly independent in } \mathbb{C}^N. \tag{2.10}$$

Indeed assume $\sum_{v \in V_\mathbb{C}(\mathcal{I})} \lambda_v \zeta_{\mathcal{B},v} = 0$, i.e., $\sum_{v \in V_\mathbb{C}(\mathcal{I})} \lambda_v b_i(v) = 0$ for $i = 1, \dots, N$. As $\mathcal{B}$ is a basis of $\mathbb{R}[\mathbf{x}]/\mathcal{I}$, this implies that $\sum_{v \in V_\mathbb{C}(\mathcal{I})} \lambda_v f(v) = 0$ for any $f \in \mathbb{R}[\mathbf{x}]$. Applying this to $f := \text{Re}(p_v), \text{Im}(p_v)$ we find $\lambda_v = 0\ \forall v$.

THEOREM 2.9. **(Stickelberger eigenvalue theorem)** *Let $h \in \mathbb{R}[\mathbf{x}]$. The set $\{h(v) \mid v \in V_\mathbb{C}(\mathcal{I})\}$ is the set of eigenvalues of M_h. More precisely,*

$$M_h^T \zeta_{\mathcal{B},v} = h(v) \zeta_{\mathcal{B},v} \quad \forall v \in V_\mathbb{C}(\mathcal{I}) \tag{2.11}$$

and, if $\mathcal{I}$ is radical, then

$$M_h \text{vec}_\mathcal{B}(p_v) = h(v) \text{vec}_\mathcal{B}(p_v) \quad \forall v \in V_\mathbb{C}(\mathcal{I}). \tag{2.12}$$

Proof. We first show (2.11). Indeed, $(M_h^T \zeta_{\mathcal{B},v})_j = \sum_{i=1}^N b_i(v) a_{ij}$ is equal to $h(v)b_j(v)$ (using (2.9)). Thus $h(v)$ is an eigenvalue of M_h^T with eigenvector $\zeta_{\mathcal{B},v}$. Note that $\zeta_{\mathcal{B},v} \neq 0$ by (2.10).

We now show (2.12) if $\mathcal{I}$ is radical. Say, $p_v = \sum_{j=1}^N c_j b_j$, i.e., $\mathrm{vec}_{\mathcal{B}}(p_v) = (c_j)_{j=1}^N$. The i-th component of $q := M_h \mathrm{vec}_{\mathcal{B}}(p_v)$ is $q_i = \sum_{j=1}^N a_{ij} c_j$. In order to show $q_i = h(v)c_i$ for all i, it suffices to show that the polynomial $f := \sum_{i=1}^N (q_i - h(v)c_i)b_i$ belongs to $\mathcal{I}$ and, as $\mathcal{I}$ is radical, this holds if we can show that f vanishes at $V_{\mathbb{C}}(\mathcal{I})$. Now,

$$\begin{aligned} f &= \sum_{i=1}^N \Big(\sum_{j=1}^N a_{ij}c_j - h(v)c_i\Big)b_i = \sum_{j=1}^N c_j\Big(\sum_{i=1}^N a_{ij}b_i\Big) - h(v)\sum_{i=1}^N c_i b_i \\ &= \sum_{j=1}^N c_j\Big(\sum_{i=1}^N a_{ij}b_i - hb_j + hb_j\Big) - h(v)p_v \\ &\equiv \sum_{j=1}^N c_j h b_j - h(v)p_v = (h - h(v))p_v \mod \mathcal{I} \end{aligned}$$

(using (2.9)). Thus, f vanishes at $V_{\mathbb{C}}(\mathcal{I})$.

Remains to show that any eigenvalue λ of M_h belongs to the set $h(V_{\mathbb{C}}(\mathcal{I})) := \{h(v) \mid v \in V_{\mathbb{C}}(\mathcal{I})\}$. If $\mathcal{I}$ is radical, this is clear since we have already found $|V_{\mathbb{C}}(\mathcal{I})| = N$ linearly independent eigenvectors $\zeta_{\mathcal{B},v}$ ($v \in V_{\mathbb{C}}(\mathcal{I})$) (by (2.10)). Otherwise, assume $\lambda \notin h(V_{\mathbb{C}}(\mathcal{I}))$. Then the system $h_1(x) = 0, \ldots, h_{m_0}(x) = 0, h(x) - \lambda = 0$ has no solution. By Hilbert's Nullstellensatz (Theorem 2.1), $1 \in (h_1, \ldots, h_{m_0}, h - \lambda)$. That is, $1 = \sum_{j=1}^{m_0} f_j h_j + f(h - \lambda)$ for some polynomials f_j, f. Hence,

$$I = M_1 = M_{\sum_{j=1}^{m_0} f_j h_j + f(h-\lambda)} = \sum_{j=1}^{m_0} M_{f_j h_j} + M_f(M_h - \lambda I) = M_f(M_h - \lambda I)$$

since $M_{f_j h_j} = 0$ as $f_j h_j \in \mathcal{I}$. Thus $M_h - \lambda I$ is nonsingular which means that λ is not an eigenvalue of M_h. □

EXAMPLE 2.10. Consider the ideal $\mathcal{I} = (h_1, h_2, h_3) \subseteq \mathbb{R}[\mathbf{x}, \mathbf{y}]$ where

$$\begin{aligned} h_1 &= \mathbf{x}^2 + 2\mathbf{y}^2 - 2\mathbf{y} \\ h_2 &= \mathbf{x}\mathbf{y}^2 - \mathbf{x}\mathbf{y} \\ h_3 &= \mathbf{y}^3 - 2\mathbf{y}^2 + \mathbf{y}. \end{aligned}$$

Obviously, $V_{\mathbb{C}}(\mathcal{I}) = \{(0,0), (0,1)\}$. One can show that, with respect to the lexicographic order with $\mathbf{x} > \mathbf{y}$, the set $\{h_1, h_2, h_3\}$ is a Gröbner basis of $\mathcal{I}$. As the leading terms of h_1, h_2, h_3 are $\mathbf{x}^2, \mathbf{x}\mathbf{y}^2, \mathbf{y}^3$, the corresponding set of standard monomials is $\mathcal{B} = \{1, \mathbf{y}, \mathbf{y}^2, \mathbf{x}, \mathbf{x}\mathbf{y}\}$ and $\dim \mathbb{R}[\mathbf{x}, \mathbf{y}]/\mathcal{I} = 5$. As $\mathbf{x}^2\mathbf{y} \equiv -2\mathbf{y}^2 + 2\mathbf{y} \mod \mathcal{I}$, the multiplication matrices read:

$$M_{\mathbf{x}} = \begin{pmatrix} 0 & 0 & 0 & 0 & 0 \\ 0 & 0 & 0 & 2 & 2 \\ 0 & 0 & 0 & -2 & -2 \\ 1 & 0 & 0 & 0 & 0 \\ 0 & 1 & 1 & 0 & 0 \end{pmatrix}, \; M_{\mathbf{y}} = \begin{pmatrix} 0 & 0 & 0 & 0 & 0 \\ 1 & 0 & -1 & 0 & 0 \\ 0 & 1 & 2 & 0 & 0 \\ 0 & 0 & 0 & 0 & 0 \\ 0 & 0 & 0 & 1 & 1 \end{pmatrix}$$

and their characteristic polynomials are $\det(M_{\mathbf{x}} - tI) = t^5$, $\det(M_{\mathbf{y}} - tI) = t^2(t-1)^3$.

EXAMPLE 2.11. Consider now the ideal $\mathcal{I} = (\mathbf{x}^2, \mathbf{y}^2)$ in $\mathbb{R}[\mathbf{x}, \mathbf{y}]$. Obviously, $V_{\mathbb{C}}(\mathcal{I}) = \{(0,0)\}$, $\{\mathbf{x}^2, \mathbf{y}^2\}$ is a Gröbner basis w.r.t. any monomial ordering, with corresponding set $\mathcal{B} = \{1, \mathbf{x}, \mathbf{y}, \mathbf{xy}\}$ of standard monomials. Thus $\dim \mathbb{R}[\mathbf{x}, \mathbf{y}]/\mathcal{I} = 4$,

$$M_{\mathbf{x}} = \begin{pmatrix} 0 & 0 & 0 & 0 \\ 1 & 0 & 0 & 0 \\ 0 & 0 & 0 & 0 \\ 0 & 0 & 1 & 0 \end{pmatrix}, \; M_{\mathbf{y}} = \begin{pmatrix} 0 & 0 & 0 & 0 \\ 0 & 0 & 0 & 0 \\ 1 & 0 & 0 & 0 \\ 0 & 1 & 0 & 0 \end{pmatrix},$$

both with characteristic polynomial t^4.

By Theorem 2.9, the eigenvalues of the companion matrices $M_{\mathbf{x}_i}$ are the coordinates v_i of the points $v \in V_{\mathbb{C}}(\mathcal{I})$. It is however not clear how to put these coordinates together for recovering the full vectors v. For this it is better to use the eigenvectors $\zeta_{\mathcal{B},v}$ of the transpose multiplication matrices. Recall that a square matrix M is *non-derogatory* if all its eigenspaces have dimension 1; that is, if $\dim \mathrm{Ker}(M - \lambda I) = 1$ for each eigenvalue λ of M. The next result follows directly from Theorem 2.9.

LEMMA 2.12. *The following holds for a multiplication matrix M_h.*
(i) *If M_h^T is non-derogatory then $h(v)$ ($v \in V_{\mathbb{C}}(\mathcal{I})$) are pairwise distinct.*
(ii) *If $\mathcal{I}$ is radical and $h(v)$ ($v \in V_{\mathbb{C}}(\mathcal{I})$) are pairwise distinct, then M_h^T is non-derogatory.*

Finding $V_{\mathbb{C}}(\mathcal{I})$ via the eigenvectors of a non-derogatory multiplication matrix M_h. Assume we can find $h \in \mathbb{R}[\mathbf{x}]$ for which the matrix M_h^T is non-derogatory. We can assume without loss of generality that the chosen basis $\mathcal{B}$ of $\mathbb{R}[\mathbf{x}]/\mathcal{I}$ contains the constant polynomial $b_1 = 1$. Let λ be an eigenvalue of M_h^T with eigenvector u. By Theorem 2.9, $\lambda = h(v)$ and u is a scalar multiple of $\zeta_{\mathcal{B},v}$ for some $v \in V_{\mathbb{C}}(\mathcal{I})$; by rescaling (i.e. replace u by u/u_1 where u_1 is the component of u indexed by $b_1 = 1$), we may assume $u = \zeta_{\mathcal{B},v}$. If $\mathbf{x}_1, \ldots, \mathbf{x}_n \in \mathcal{B}$, one can read the coordinates of v directly from the eigenvector u. Otherwise, express $\mathbf{x}_i$ as a linear combination modulo $\mathcal{I}$ of the members of $\mathcal{B}$, say, $\mathbf{x}_i = \sum_{j=1}^N c_j b_j \mod \mathcal{I}$. Then, $v_i = \sum_{j=1}^N c_j b_j(v)$ can be computed from the coordinates of the eigenvector u.

One can show that if there exists some $h \in \mathbb{R}[\mathbf{x}]$ for which M_h^T is non-derogatory, then there exists a linear such polynomial $h = \sum_{i=1}^n c_i \mathbf{x}_i$. Following the strategy of Corless, Gianni and Trager [20], one can find such h by choosing the c_i's at random. Then, with high probability, $h(v)$ ($v \in V_{\mathbb{C}}(\mathcal{I})$) are pairwise distinct. If $\mathcal{I}$ is radical then M_h^T is non-derogatory (by Lemma 2.12). If we succeed to find a non-derogatory matrix after a few trials, we can proceed to compute $V_{\mathbb{C}}(\mathcal{I})$; otherwise we are either unlucky or there exists no non-derogatory matrix. Then one possibility is to compute the radical $\sqrt{\mathcal{I}}$ of $\mathcal{I}$ using, for instance, the following characterization:

$$\sqrt{\mathcal{I}} = (h_1, \ldots, h_m, (p_1)_{red}, \ldots, (p_n)_{red})$$

where p_i is the monic generator of $\mathcal{I} \cap \mathbb{R}[\mathbf{x}_i]$ and $(p_i)_{red}$ is its square-free part. The polynomial p_i can be found in the following way: Let k_i be the smallest integer for which the set $\{[1], [\mathbf{x}_i], \ldots, [\mathbf{x}_i^{k_i}]\}$ is linearly dependent in $\mathbb{R}[\mathbf{x}]/\mathcal{I}$. Then the polynomial $\mathbf{x}_i^{k_i} + \sum_{j=0}^{k_i-1} c_j \mathbf{x}_i^j$ lies in $\mathcal{I}$ for some scalars c_j and, by the minimality of k_i, it generates $\mathcal{I} \cap \mathbb{R}[\mathbf{x}_i]$.

EXAMPLE 2.10 (continued). None of $M_{\mathbf{x}}^T$, $M_{\mathbf{y}}^T$ is non-derogatory. Indeed, 0 is the only eigenvalue of $M_{\mathbf{x}}^T$ whose corresponding eigenspace is $\operatorname{Ker} M_{\mathbf{x}}^T = \{u \in \mathbb{R}^5 \mid u_2 = u_3, u_4 = u_5 = 0\}$ with dimension 2 and spanned by $\zeta_{\mathcal{B},(0,0)}$ and $\zeta_{\mathcal{B},(0,1)}$. The eigenspace of $M_{\mathbf{y}}^T$ for eigenvalue 0 is $\operatorname{Ker} M_{\mathbf{y}}^T = \{u \in \mathbb{R}^5 \mid u_2 = u_3 = u_4 = 0\}$ with dimension 2 and spanned by $\zeta_{\mathcal{B},(0,0)}$ and $(0,0,0,1,0)^T$. The eigenspace with eigenvalue 1 is $\operatorname{Ker}(M_{\mathbf{y}}^T - I) = \{u \in \mathbb{R}^5 \mid u_1 = u_2 = u_3, u_4 = u_5\}$ also with dimension 2 and spanned by $\zeta_{\mathcal{B},(0,1)}$ and $(0,0,0,1,1)^T$. Thus, for $h = \mathbf{y}$, this gives an example where $h(v)$ ($v \in V_{\mathbb{C}}(\mathcal{I})$) are pairwise distinct, yet the matrix M_h^T is not non-derogatory. On the other hand, for $h = 2\mathbf{x} + 3\mathbf{y}$,

$$M_h = 2M_{\mathbf{x}} + 3M_{\mathbf{y}} = \begin{pmatrix} 0 & 0 & 0 & 0 & 0 \\ 3 & 0 & -3 & 4 & 4 \\ 0 & 3 & 6 & -4 & -4 \\ 2 & 0 & 0 & 0 & 0 \\ 0 & 2 & 2 & 3 & 3 \end{pmatrix}$$

and M_h^T is non-derogatory. Indeed, M_h^T has two eigenvalues 0, 3. The eigenspace for the eigenvalue 0 is spanned by $\zeta_{\mathcal{B},(0,0)}$, permitting to extract the root $v = (0,0)$, and the eigenspace for the eigenvalue 3 is spanned by $\zeta_{\mathcal{B},(0,1)}$, permitting to extract the root $v = (0,1)$. □

EXAMPLE 2.11 (continued). In this example *every* matrix M_h^T is derogatory. Indeed, say $h = a + b\mathbf{x} + c\mathbf{y} + d\mathbf{x}^2 + e\mathbf{x}\mathbf{y} + \ldots$. Then,

$$M_h = \begin{pmatrix} a & b & c & e \\ 0 & a & 0 & c \\ 0 & 0 & a & b \\ 0 & 0 & 0 & a \end{pmatrix}.$$

Thus a is the only eigenvalue of M_h^T with eigenvector space of dimension at least 2. □

Part 1: Sums of Squares and Moments

3. Positive polynomials and sums of squares.

3.1. Some basic facts. A concept which will play a central role in the paper is the following notion of *sum of squares.* A polynomial p is said to be a *sum of squares of polynomials,* sometimes abbreviated as '*p is SOS*', if p can be written as $p = \sum_{j=1}^m u_j^2$ for some $u_1, \dots, u_m \in \mathbb{R}[\mathbf{x}]$. Given $p \in \mathbb{R}[\mathbf{x}]$ and $S \subseteq \mathbb{R}^n$, the notation '$p \geq 0$ on S' means '$p(x) \geq 0$ for all $x \in S$', in which case we say that p is nonnegative on S; analogously, $p > 0$ on S means that p is positive on S. We begin with some simple properties of sums of squares.

LEMMA 3.1. *If $p \in \mathbb{R}[\mathbf{x}]$ is a sum of squares then* $\deg(p)$ *is even and any decomposition $p = \sum_{j=1}^m u_j^2$ where $u_j \in \mathbb{R}[\mathbf{x}]$ satisfies* $\deg(u_j) \leq \deg(p)/2$ *for all j.*

Proof. Assume p is SOS. Then $p(x) \geq 0$ for all $x \in \mathbb{R}^n$ and thus $\deg(p)$ must be even, say $\deg(p) = 2d$. Write $p = \sum_{j=1}^m u_j^2$ and let $k := \max_j \deg(u_j)$. Assume $k \geq d+1$. Write each $u_j = \sum_\alpha u_{j,\alpha}\mathbf{x}^\alpha$ as $u_j = a_j + b_j$, where $b_j := \sum_{\alpha \mid |\alpha|=k} u_{j,\alpha}\mathbf{x}^\alpha$ and $a_j := u_j - b_j$. Then $p - \sum_j a_j^2 - 2a_jb_j = \sum_j b_j^2$. Here $\sum_j b_j^2$ is a homogeneous polynomial of degree $2k \geq 2d+2$, while $p - \sum_j a_j^2 - 2a_jb_j$ is a polynomial of degree $\leq 2k-1$, which yields a contradiction. This shows $\deg(u_j) \leq d$ for all j. □

LEMMA 3.2. *Let p be a homogeneous polynomial of degree $2d$. If p is SOS, then p is a sum of squares of homogeneous polynomials (each of degree d).*

Proof. Assume $p = \sum_{j=1}^m u_j^2$ where $u_j \in \mathbb{R}[\mathbf{x}]$. Write $u_j = a_j + b_j$ where a_j is the sum of the terms of degree d of u_j and thus $\deg(b_j) \leq d-1$. Then, $p - \sum_{j=1}^m a_j^2 = \sum_{j=1}^m b_j^2 + 2a_jb_j$ is equal to 0, since otherwise the right hand side has degree $\leq 2d-1$ and the left hand side is homogeneous of degree $2d$. □

LEMMA 3.3. *Consider a polynomial $p \in \mathbb{R}[\mathbf{x}]$ and its homogenization $\tilde{p} \in \mathbb{R}[\mathbf{x}, \mathbf{x}_{n+1}]$. Then, $p \geq 0$ on $\mathbb{R}^n$ (resp., p SOS) $\Longleftrightarrow$ $\tilde{p} \geq 0$ on $\mathbb{R}^{n+1}$ (resp., $\tilde{p}$ SOS).*

Proof. The 'if part' follows from the fact that $p(x) = \tilde{p}(x, 1)$ for all $x \in \mathbb{R}^n$. Conversely, if $p \geq 0$ on $\mathbb{R}^n$ then $d := \deg(p)$ is even and $\tilde{p}(x, x_{n+1}) = x_{n+1}^d \tilde{p}(x/x_{n+1}, 1) = x_{n+1}^d p(x) \geq 0$ whenever $x_{n+1} \neq 0$. Thus $\tilde{p} \geq 0$ by continuity. An analogous argument shows that, if $p = \sum_j u_j^2$ with $u_j \in \mathbb{R}[\mathbf{x}]$, then $\tilde{p} = \sum_j \tilde{u}_j^2$, where $\tilde{u}_j$ is the homogenization of u_j. □

3.2. Sums of squares and positive polynomials: Hilbert's result. Throughout the paper,

$$\mathcal{P}_n := \{p \in \mathbb{R}[\mathbf{x}] \mid p(x) \geq 0 \ \forall x \in \mathbb{R}^n\} \tag{3.1}$$

denotes the set of nonnegative polynomials on $\mathbb{R}^n$ (also called *positive semidefinite* polynomials in the literature) and

$$\Sigma_n := \{p \in \mathbb{R}[\mathbf{x}] \mid p \text{ SOS}\} \tag{3.2}$$

is the set of polynomials that are sums of squares; we sometimes omit the index n and simply write $\mathcal{P} = \mathcal{P}_n$ and $\Sigma = \Sigma_n$ when there is no danger of confusion on the number of variables. We also set

$$\mathcal{P}_{n,d} := \mathcal{P}_n \cap \mathbb{R}[\mathbf{x}]_d, \ \Sigma_{n,d} := \Sigma_n \cap \mathbb{R}[\mathbf{x}]_d.$$

Obviously any polynomial which is SOS is nonnegative on $\mathbb{R}^n$; that is,

$$\Sigma_n \subseteq \mathcal{P}_n, \ \Sigma_{n,d} \subseteq \mathcal{P}_{n,d}. \tag{3.3}$$

As is well known (cf. Lemma 3.5), equality holds in (3.3) for $n = 1$ (i.e. for univariate polynomials), but the inclusion $\Sigma_n \subseteq \mathcal{P}_n$ is strict for $n \geq 2$. The following celebrated result of Hilbert [56] classifies the pairs (n, d) for which equality $\Sigma_{n,d} = \mathcal{P}_{n,d}$ holds.

THEOREM 3.4. *$\Sigma_{n,d} = \mathcal{P}_{n,d} \iff n = 1$, or $d = 2$, or $(n, d) = (2, 4)$.*

We give below the arguments for the equality $\Sigma_{n,d} = \mathcal{P}_{n,d}$ in the two cases $n = 1$, or $d = 2$, which are simple and which were already well known in the late 19th century. In his paper [56] David Hilbert proved that $\mathcal{P}_{2,4} = \Sigma_{2,4}$; moreover he proved that any nonnegative polynomial in $n = 2$ variables with degree 4 is a sum of *three* squares; equivalently, any nonnegative ternary quartic form is a sum of *three* squares. Choi and Lam [17] gave a relatively simple proof for the equality $\mathcal{P}_{2,4} = \Sigma_{2,4}$, based on geometric arguments about the cone $\Sigma_{2,4}$; their proof shows a decomposition into *five* squares. Powers et al. [110] found a new approach to Hilbert's theorem and gave a proof of the *three* squares result in the nonsingular case.

LEMMA 3.5. *Any nonnegative univariate polynomial is a sum of two squares.*

Proof. Assume p is a univariate polynomial and $p \geq 0$ on $\mathbb{R}$. Then the roots of p are either real with even multiplicity, or appear in complex conjugate pairs. Thus $p = c\prod_{i=1}^{r}(\mathbf{x} - a_i)^{2r_i} \cdot \prod_{j=1}^{s}((\mathbf{x} - b_j)^2 + c_j^2)^{s_j}$ for some scalars $a_i, b_j, c_j, c \in \mathbb{R}$, $c > 0$, and $r, s, r_i, s_j \in \mathbb{N}$. This shows that p is SOS. To see that p can be written as a sum of two squares, use the identity $(a^2 + b^2)(c^2 + d^2) = (ac + bd)^2 + (ad - bc)^2$ (for $a, b, c, d \in \mathbb{R}$). □

LEMMA 3.6. *Any nonnegative quadratic polynomial is a sum of squares.*

Proof. Let $p \in \mathbb{R}[\mathbf{x}]_2$ of the form $p = \mathbf{x}^T Q\mathbf{x} + 2c^T\mathbf{x} + b$, where Q is a symmetric $n \times n$ matrix, $c \in \mathbb{R}^n$ and $b \in \mathbb{R}$. Its homogenization is $\tilde{p} = \mathbf{x}^T Q\mathbf{x} + 2\mathbf{x}_{n+1}c^T\mathbf{x} + b\mathbf{x}_{n+1}^2$, thus of the form $\tilde{p} = \tilde{\mathbf{x}}^T\tilde{Q}\tilde{\mathbf{x}}$, after setting

$\tilde{\mathbf{x}} := (\mathbf{x}, \mathbf{x}_{n+1})$ and $\tilde{Q} := \begin{pmatrix} Q & c^T \\ c & b \end{pmatrix}$. By Lemma 3.3, $\tilde{p} \geq 0$ on $\mathbb{R}^{n+1}$ and thus the matrix $\tilde{Q}$ is positive semidefinite. Therefore, $\tilde{Q} = \sum_j u^{(j)}(u^{(j)})^T$ for some $u^{(j)} \in \mathbb{R}^{n+1}$, which gives $\tilde{p} = \sum_j (\sum_{i=1}^{n+1} u_i^{(j)} \mathbf{x}_i)^2$ is SOS and thus p too is SOS (by Lemma 3.3 again). □

According to Hilbert's result (Theorem 3.4), for any pair $(n, d) \neq (2, 4)$ with $n \geq 2$, $d \geq 4$ even, there exists a polynomial in $\mathcal{P}_{n,d} \setminus \Sigma_{n,d}$. Some well known such examples include the Motzkin and Robinson polynomials described below.

EXAMPLE 3.7. The polynomial $p := \mathbf{x}_1^2\mathbf{x}_2^2(\mathbf{x}_1^2 + \mathbf{x}_2^2 - 3) + 1$, known as the ***Motzkin polynomial***, belongs to $\mathcal{P}_{2,6} \setminus \Sigma_{2,6}$. Indeed, $p(x_1, x_2) \geq 0$ if $x_1^2 + x_2^2 \geq 3$. Otherwise, set $x_3^2 := 3 - x_1^2 - x_2^2$. By the arithmetic geometric mean inequality, we have $\frac{x_1^2+x_2^2+x_3^2}{3} \geq \sqrt[3]{x_1^2x_2^2x_3^2}$, giving again $p(x_1, x_2) \geq 0$. One can verify directly that p cannot be written as a sum of squares of polynomials. Indeed, assume $p = \sum_k u_k^2$, where $u_k = a_k\mathbf{x}_1^3 + b_k\mathbf{x}_1^2\mathbf{x}_2 + c_k\mathbf{x}_1\mathbf{x}_2^2 + d_k\mathbf{x}_2^3 + e_k\mathbf{x}_1^2 + f_k\mathbf{x}_1\mathbf{x}_2 + g_k\mathbf{x}_2^2 + h_k\mathbf{x}_1 + i_k\mathbf{x}_2 + j_k$ for some scalars $a_k, \ldots, j_k \in \mathbb{R}$. Looking at the coefficient of $\mathbf{x}_1^6$ in p, we find $0 = \sum_k a_k^2$, giving $a_k = 0$ for all k; analogously $d_k = 0$ for all k. Next, looking at the coefficient of $\mathbf{x}_1^4$ and $\mathbf{x}_2^4$ yields $e_k = g_k = 0$ for all k; then looking at the coefficient of $\mathbf{x}_1^2, \mathbf{x}_2^2$ yields $h_k = i_k = 0$ for all k; finally the coefficient of $\mathbf{x}_1^2\mathbf{x}_2^2$ in p is equal to $-3 = \sum_k f_k^2$, yielding a contradiction. Note that this argument shows in fact that $p - \rho$ is not a sum of squares for any scalar $\rho \in \mathbb{R}$.

Therefore the ***homogeneous Motzkin form*** $M := \mathbf{x}_1^2\mathbf{x}_2^2(\mathbf{x}_1^2 + \mathbf{x}_2^2 - 3\mathbf{x}_3^2) + \mathbf{x}_3^6$ is nonnegative but not a sum of squares.

The polynomial $p := \mathbf{x}_1^6 + \mathbf{x}_2^6 + \mathbf{x}_3^6 - \sum_{1 \leq i < j \leq 3}(\mathbf{x}_i^2\mathbf{x}_j^2(\mathbf{x}_i^2 + \mathbf{x}_j^2)) + 3\mathbf{x}_1^2\mathbf{x}_2^2\mathbf{x}_3^2$, known as the ***Robinson form***, is nonnegative but not a sum of squares. See e.g. [120] for details.

We refer to Reznick [120] for a nice overview and historic discussion of Hilbert's results. More examples of positive polynomials that are not sums of squares can be found e.g. in the recent papers [16], [121] and references therein.

3.3. Recognizing sums of squares of polynomials.

We now indicate how to recognize whether a polynomial can be written as a sum of squares via semidefinite programming. The next result was discovered independently by several authors; cf. e.g. [18], [112].

LEMMA 3.8. **Recognizing sums of squares.**
Let $p \in \mathbb{R}[\mathbf{x}]$, $p = \sum_{\alpha \in \mathbb{N}^n_{2d}} p_\alpha \mathbf{x}^\alpha$, be a polynomial of degree $\leq 2d$. The following assertions are equivalent.

(i) *p is a sum of squares.*
(ii) *The following system in the matrix variable $X = (X_{\alpha,\beta})_{\alpha,\beta \in \mathbb{N}^n_d}$ is feasible:*

$$\begin{cases} X \succeq 0 \\ \sum_{\beta,\gamma \in \mathbb{N}^n_d \mid \beta+\gamma=\alpha} X_{\beta,\gamma} = p_\alpha \quad (|\alpha| \leq 2d). \end{cases} \tag{3.4}$$

Proof. Let $\mathbf{z}_d := (\mathbf{x}^\alpha \mid |\alpha| \leq d)$ denote the vector containing all monomials of degree at most d. Then for polynomials $u_j \in \mathbb{R}[\mathbf{x}]_d$, we have $u_j = \text{vec}(u_j)^T \mathbf{z}_d$ and thus $\sum_j u_j^2 = \mathbf{z}_d^T(\sum_j \text{vec}(u_j)\text{vec}(u_j)^T)\mathbf{z}_d$. Therefore, p is a sum of squares of polynomials if and only if $p = \mathbf{z}_d^T X \mathbf{z}_d$ for some positive semidefinite matrix X. Equating the coefficients of the two polynomials p and $\mathbf{z}_d^T X \mathbf{z}_d$, we find the system (3.4). □

Thus to decide whether the polynomial p can be written as a sum of squares one has to verify existence of a positive semidefinite matrix X satisfying the linear equations in (3.4) and any Gram decomposition of X gives a sum of square decomposition for p. For this reason this method is often called the *Gram-matrix method* in the literature (e.g. [18]). The system (3.4) is a system in the matrix variable X, which is indexed by $\mathbb{N}^n_d$ and thus has size $\binom{n+d}{d}$, and with $\binom{n+2d}{2d}$ equations. Therefore, this system has polynomial size if either n is fixed, or d is fixed. The system (3.4) is an instance of semidefinite program. Thus finding a sum of square decomposition of a polynomial can be done using semidefinite programming. Note also that if p has a sum of squares decomposition then it has one involving at most $|\mathbb{N}^n_d| = \binom{n+d}{d}$ squares.

We now illustrate the method on a small example.

EXAMPLE 3.9. Suppose we want to find a sum of squares decomposition for the polynomial $p = \mathbf{x}^4 + 2\mathbf{x}^3\mathbf{y} + 3\mathbf{x}^2\mathbf{y}^2 + 2\mathbf{x}\mathbf{y}^3 + 2\mathbf{y}^4 \in \mathbb{R}[\mathbf{x},\mathbf{y}]_4$. As p is a form of degree 4, we want to find $X \succeq 0$ indexed by $\mathbf{x}^2, \mathbf{xy}, \mathbf{y}^2$ satisfying

$$p = (\mathbf{x}^2\ \mathbf{xy}\ \mathbf{y}^2) \underbrace{\begin{pmatrix} a & b & c \\ b & d & e \\ c & e & f \end{pmatrix}}_{X} \begin{pmatrix} \mathbf{x}^2 \\ \mathbf{xy} \\ \mathbf{y}^2 \end{pmatrix}.$$

Equating coefficients:

$$\begin{array}{ll} \mathbf{x}^4 = \mathbf{x}^2 \cdot \mathbf{x}^2 & 1 = a \\ \mathbf{x}^3\mathbf{y} = \mathbf{x}^2 \cdot \mathbf{xy} & 2 = 2b \\ \mathbf{x}^2\mathbf{y}^2 = \mathbf{xy} \cdot \mathbf{xy} = \mathbf{x}^2 \cdot \mathbf{y}^2 & 3 = d + 2c \\ \mathbf{xy}^3 = \mathbf{xy} \cdot \mathbf{y}^2 & 2 = 2e \\ \mathbf{y}^4 = \mathbf{y}^2 \cdot \mathbf{y}^2 & 2 = f \end{array}$$

we find $X = \begin{pmatrix} 1 & 1 & c \\ 1 & 3-2c & 1 \\ c & 1 & 2 \end{pmatrix}$. Therefore $X \succeq 0 \iff -1 \leq c \leq 1$. E.g. for $c = -1$, $c = 0$, we find, respectively, the matrix

$$X = \begin{pmatrix} 1 & 0 \\ 1 & 2 \\ -1 & 1 \end{pmatrix}\begin{pmatrix} 1 & 1 & -1 \\ 0 & 2 & 1 \end{pmatrix}, \begin{pmatrix} 1 & 0 & 0 \\ 1 & \sqrt{\frac{3}{2}} & \sqrt{\frac{1}{2}} \\ 0 & \sqrt{\frac{3}{2}} & -\sqrt{\frac{1}{2}} \end{pmatrix}\begin{pmatrix} 1 & 1 & 0 \\ 0 & \sqrt{\frac{3}{2}} & \sqrt{\frac{3}{2}} \\ 0 & \sqrt{\frac{1}{2}} & -\sqrt{\frac{1}{2}} \end{pmatrix}$$

giving, respectively, the decompositions $p = (\mathbf{x}^2 + \mathbf{xy} - \mathbf{y}^2)^2 + (\mathbf{y}^2 + 2\mathbf{xy})^2$ and $p = (\mathbf{x}^2 + \mathbf{xy})^2 + \frac{3}{2}(\mathbf{xy} + \mathbf{y}^2)^2 + \frac{1}{2}(\mathbf{xy} - \mathbf{y}^2)^2$.

3.4. SOS relaxations for polynomial optimization. Although we will come back to it in detail in Section 6, we already introduce here the SOS relaxations for the polynomial optimization problem (1.1) as this will motivate our exposition later in this section of several representation results for positive polynomials. Note first that problem (1.1) can obviously be reformulated as

$$p^{\min} = \sup \rho \text{ s.t. } p - \rho \geq 0 \text{ on } K = \sup \rho \text{ s.t. } p - \rho > 0 \text{ on } K. \tag{3.5}$$

That is, computing $p^{\min}$ amounts to finding the supremum of the scalars ρ for which $p - \rho$ is nonnegative (or positive) on the set K. To tackle this hard problem it is a natural idea (going back to work of Shor [138–140], Nesterov [95], Lasserre [65], Parrilo [103, 104]) to replace the nonnegativity condition by some simpler condition, involving sums of squares, which can then be tackled using semidefinite programming. For instance, in the unconstrained case when $K = \mathbb{R}^n$, consider the parameter

$$p^{\text{sos}} := \sup \rho \text{ s.t. } p - \rho \text{ is SOS.} \tag{3.6}$$

As explained in the previous section, the parameter p^{sos} can be computed via a semidefinite program involving a matrix of size $|\mathbb{N}^n_d|$ if p has degree $2d$. Obviously, $p^{\text{sos}} \leq p^{\min}$, but as follows from Hilbert's result (Theorem 3.4), the inequality may be strict. For instance, when p is the Motzkin polynomial considered in Example 3.7, then $p^{\text{sos}} = -\infty < p^{\min} = 0$ as p vanishes at $(\pm 1, \pm 1)$. In the constrained case, one way to relax the condition '$p - \rho \geq 0$ on K' is by considering a sum of square decomposition of the form $p - \rho = s_0 + \sum_{j=1}^m s_j g_j$ where s_0, s_j are SOS. This yields the parameter:

$$p^{\text{sos}} := \sup \rho \text{ s.t. } p - \rho = s_0 + \sum_{j=1}^{m} s_j g_j \text{ with } s_0, s_j \text{ SOS.} \tag{3.7}$$

Again $p^{\text{sos}} \leq p^{\min}$ and, under certain assumption on the polynomials g_j describing the set K (cf. Theorem 3.20 below), equality holds. The above formulation does not lead yet to a semidefinite program, since it is not obvious how to bound the degrees of the polynomials s_0, s_j as cancellation of terms may occur in $s_0 + \sum_j s_j g_j$. To get a semidefinite program one may consider for any integer t with $2t \geq \max(\deg p, \deg(g_1), \ldots, \deg(g_m))$ the parameter

$$p_t^{\text{sos}} := \sup \rho \ \text{ s.t. } \ \begin{array}{l} p - \rho = s_0 + \sum_{j=1}^m s_j g_j \text{ with } s_0, s_j \in \Sigma, \\ \deg(s_0), \deg(s_j g_j) \leq 2t. \end{array} \tag{3.8}$$

Hence each p_t^{sos} can be computed via a semidefinite program involving matrices of size $|\mathbb{N}_t^n|$, $p_t^{\text{sos}} \leq p_{t+1}^{\text{sos}} \leq p^{\text{sos}} \leq p^{\min}$, and $\lim_{t\to\infty} p_t^{\text{sos}} = p^{\text{sos}}$.

3.5. Convex quadratic polynomial optimization. Here we consider problem (1.1) in the convex quadratic case, i.e. when $p, -g_1, \ldots, -g_m$ are convex quadratic polynomials. Say $p = \mathbf{x}^T Q \mathbf{x} + 2c^T \mathbf{x} + b$, $g_j = \mathbf{x}^T Q_j \mathbf{x} + 2c_j^T \mathbf{x} + b_j$, where Q, Q_j are symmetric $n \times n$ matrices satisfying $Q, -Q_j \succeq 0$, $c, c_j \in \mathbb{R}^n$, and $b, b_j \in \mathbb{R}$. First observe that, in the convex quadratic case (i.e. when $-Q_1, \ldots, -Q_m \succeq 0$), the semialgebraic set K from (1.2) admits the following semidefinite representation:

$$K = \{x \in \mathbb{R}^n \mid \quad \begin{array}{l} \exists X \in \mathbb{R}^{n\times n} \text{ such that } \begin{pmatrix} 1 & x^T \\ x & X \end{pmatrix} \succeq 0, \\ \left\langle \begin{pmatrix} b_j & c_j^T \\ c_j & Q_j \end{pmatrix}, \begin{pmatrix} 1 & x^T \\ x & X \end{pmatrix} \right\rangle \geq 0 \ (j = 1, \ldots, m)\}, \end{array} \tag{3.9}$$

(direct verification). Therefore, when minimizing a linear polynomial p over K, the infimum $p^{\min}$ is already given by its first order moment relaxation, i.e. $p^{\min} = p_1^{\text{mom}}$ (see Section 4.2 for the definition of p_1^{mom}).

We now observe that when p is a convex quadratic polynomial then, under some technical condition, the infimum $p^{\min}$ is given by the first order sum of squares bound p_1^{sos}. Let $J(x^*) := \{j \in \{1, \ldots, m\} \mid g_j(x^*) = 0\}$ for $x^* \in K$, and consider the following (MFCQ) constraint qualification:

$$\exists w \in \mathbb{R}^n \quad \text{for which} \quad w^T \nabla g_j(x^*) > 0 \ \forall j \in J(x^*); \tag{3.10}$$

equivalently, $\sum_{j\in J(x^*)} \lambda_j \nabla g_j(x^*) = 0$ with $\lambda_j \geq 0 \ \forall j$ implies $\lambda_j = 0 \ \forall j$.

LEMMA 3.10. *[65] Consider problem (1.1) where $p, -g_1, \ldots, -g_m$ are quadratic convex polynomials and assume that the set K from (1.2) is compact. If there exists a local (thus global) minimizer x^* satisfying (3.10), then $p_1^{sos} = p^{min}$.*

Proof. The bound p_1^{sos} is defined by

$$\begin{aligned} p_1^{\text{sos}} &= \sup_{\rho,\lambda_j\in\mathbb{R}} \rho \ \text{ s.t. } p - \rho - \textstyle\sum_{j=1}^m \lambda_j g_j \in \Sigma, \ \lambda_1, \ldots, \lambda_m \geq 0 \\ &= \sup_{\rho,\lambda_j\in\mathbb{R}} \rho \ \text{ s.t. } p - \rho - \textstyle\sum_{j=1}^m \lambda_j g_j \in \mathcal{P}, \ \lambda_1, \ldots, \lambda_m \geq 0, \end{aligned}$$

where the last equality follows using Lemma 3.6, It suffices now to show that $p^{\min}$ is feasible for the program defining p_1^{sos}. For this let $x^* \in K$ be a local minimizer of p over the set K satisfying (3.10). Then there exist scalars $\lambda_1, \ldots, \lambda_m \geq 0$ for which the first order Karush-Kuhn-Tucker conditions hold (cf. e.g. [101, §12.5]). That is, $\lambda_j g_j(x^*) = 0 \ \ \forall j$ and $\nabla p(x^*) = \sum_j \lambda_j \nabla g_j(x^*)$, implying

$$Qx^* + c = \sum_{j=1}^{m} \lambda_j (Q_j x^* + c_j). \tag{3.11}$$

We claim that

$$p - p^{\min} - \sum_{j=1}^{m} \lambda_j g_j = (\mathbf{x} - x^*)^T (Q - \sum_{j=1}^{m} \lambda_j Q_j)(\mathbf{x} - x^*). \tag{3.12}$$

Indeed, $p - p^{\min} - \sum_{j=1}^{m} \lambda_j g_j = p - p^{\min} + \sum_j \lambda_j (g_j(x^*) - g_j)$ is equal to $\mathbf{x}^T(Q - \sum_j \lambda_j Q_j)\mathbf{x} + 2(c - \sum_j \lambda_j c_j)^T \mathbf{x} - (x^*)^T(Q - \sum_j \lambda_j Q_j)x^* + 2(\sum_j \lambda_j c_j - c)^T x^*$ which, using (3.11), gives the desired identity (3.12). As $Q - \sum_j \lambda_j Q_j \succeq 0$, (3.12) implies that $p - p^{\min} - \sum_{j=1}^{m} \lambda_j g_j$ is nonnegative over $\mathbb{R}^n$, which concludes the proof. □

3.6. Some representation results for positive polynomials.

Certificates for positivity via the Positivstellensatz. A classical result about polynomials is Hilbert's Nullstellensatz which characterizes when a system of polynomials in $\mathbb{C}[\mathbf{x}]$ has a common root in $\mathbb{C}^n$. The next result is sometimes called the *weak* Nullstellensatz, while the result of Theorem 2.1 (i) is Hilbert's *strong* Nullstellensatz.

THEOREM 3.11. *(cf. e.g. [21])* **Hilbert's (weak) Nullstellensatz.** *Given polynomials $h_1, \ldots, h_m \in \mathbb{C}[\mathbf{x}]$, the system $h_1(x) = 0, \ldots, h_m(x) = 0$ does not have a common root in $\mathbb{C}^n$ if and only if $1 \in (h_1, \ldots, h_m)$, i.e. $1 = \sum_{j=1}^{m} u_j h_j$ for some polynomials $u_j \in \mathbb{C}[\mathbf{x}]$.*

As a trivial example, the system $h_1 := \mathbf{x} + 1 = 0$, $h_2 := \mathbf{x}^2 + 1 = 0$ has no common root, which is certified by the identity $1 = (1 - \mathbf{x})/2 h_1 + h_2/2$. The above result works only in the case of an algebraically closed field (like $\mathbb{C}$). For instance, $\mathbf{x}^2 + 1 = 0$ has no solution in $\mathbb{R}$, but 1 does not belong to the ideal generated by $\mathbf{x}^2 + 1$. A basic property of the real field $\mathbb{R}$ is that $\sum_{i=1}^{n} a_i^2 = 0 \Longrightarrow a_1 = \ldots = a_n = 0$, i.e. -1 is not a sum of squares in $\mathbb{R}$. These properties are formalized in the theory of *formally real* fields (cf. [14, 114]) and one of the objectives of real algebraic geometry is to understand when systems of real polynomial equations and inequalities have solutions in $\mathbb{R}^n$. An answer is given in Theorem 3.12 below, known as the Positivstellensatz, due to Stengle [142]. A detailed exposition can be found e.g. in [87, 114]. We need some definitions. Given polynomials $g_1, \ldots, g_m \in \mathbb{R}[\mathbf{x}]$, set $g_J := \prod_{j \in J} g_j$ for $J \subseteq \{1, \ldots, m\}$, $g_0 := 1$. The set

$$T(g_1, \ldots, g_m) := \Big\{ \sum_{J \subseteq \{1, \ldots, m\}} u_J g_J \mid u_J \in \Sigma \Big\}, \tag{3.13}$$

is called the *preordering* on $\mathbb{R}[\mathbf{x}]$ generated by $g_1, \ldots, g_m$. As in (1.2), let $K = \{x \in \mathbb{R}^n \mid g_1(x) \geq 0 \ldots g_m(x) \geq 0\}$.

THEOREM 3.12. **Positivstellensatz.** *Given a polynomial* $p \in \mathbb{R}[\mathbf{x}]$,

(i) $p > 0$ *on* $K \iff pf = 1 + g$ *for some* $f, g \in T(g_1, \ldots, g_m)$.

(ii) $p \geq 0$ *on* $K \iff pf = p^{2k} + g$ *for some* $f, g \in T(g_1, \ldots, g_m)$ *and* $k \in \mathbb{N}$.

(iii) $p = 0$ *on* $K \iff -p^{2k} \in T(g_1, \ldots, g_m)$ *for some* $k \in \mathbb{N}$.

COROLLARY 3.13. **Real Nullstellensatz.** *Given* $p, h_1, \ldots, h_m \in \mathbb{R}[\mathbf{x}]$, p *vanishes on* $\{x \in \mathbb{R}^n \mid h_j(x) = 0 \ (j = 1, \ldots, m)\}$ *if and only if* $p^{2k} + s = \sum_{j=1}^m u_j h_j$ *for some* $u_j \in \mathbb{R}[\mathbf{x}]$, $s \in \Sigma$, $k \in \mathbb{N}$.

COROLLARY 3.14. **Solution to Hilbert's 17th problem.** *Given* $p \in \mathbb{R}[\mathbf{x}]$, *if* $p \geq 0$ *on* $\mathbb{R}^n$, *then* $p = \sum_j \left(\frac{a_j}{b_j}\right)^2$ *for some* $a_j, b_j \in \mathbb{R}[\mathbf{x}]$.

Following Parrilo [103, 104], one may interpret the above results in terms of certificates of infeasiblity of certain systems of polynomial systems of equations and inequalities. First, observe that Hilbert's Nullstellensatz can be interpreted as follows: Either a system of polynomial equations is feasible, which can be certified by giving a common solution x; or it is infeasible, which can be certified by giving a Nullstellensatz certificate of the form $1 = \sum_{j=1}^m u_j h_j$. Parrilo makes the analogy with Farkas' lemma for linear programming; indeed, given $A \in \mathbb{R}^{m \times n}$, $b \in \mathbb{R}^m$, Farkas' lemma asserts that, either the linear system $Ax \leq b, x \geq 0$ has a solution, or it is infeasible, which can be certified by giving a solution y to the alternative system $A^T y \geq 0$, $y \geq 0$, $y^T b < 0$. (Cf. e.g. [128, §7.3]). This paradigm extends to the real solutions of systems of polynomial inequalities and equations, as the following reformulation of the Positivstellensatz (cf e.g. [14]) shows.

THEOREM 3.15. *Let* f_r $(r = 1, \ldots, s)$, g_l $(l = 1, \ldots, t)$, h_j $(j = 1, \ldots, m)$ *be polynomials in* $\mathbb{R}[\mathbf{x}]$. *Then one of the following holds.*

(i) *Either the system* $f_r(x) \neq 0$ $(r = 1, \ldots, s)$, $g_l(x) \geq 0$ $(l = 1, \ldots, t)$, $h_j(x) = 0$ $(j = 1, \ldots, m)$ *has a solution in* $\mathbb{R}^n$.

(ii) *Or* $\prod_{r=1}^s f_r^{2d_r} + \sum_{J \subseteq \{1,\ldots,t\}} s_J g_J + \sum_{j=1}^m u_j h_j = 0$ *for some* $d_r \in \mathbb{N}$, $s_J \in \Sigma$ *and* $u_j \in \mathbb{R}[\mathbf{x}]$.

Thus the Positivstellensatz can be seen as a generalization of Hilbert's Nullstellensatz and of Farkas' lemma (for linear programming) and one can search for bounded degree certificates that the system in Theorem 3.15 (i) has no real solution, using semidefinite programming. See [103, 104] for further discussion and references.

One may try to use the Positivstellensatz to approximate the optimization problem (1.1). Namely, in view of Theorem 3.12 (i), one can replace the condition '$p - \rho > 0$ on K' in (3.5) by the condition '$(p - \rho)f = 1 + g$ for some $f, g \in T(g_1, \ldots, g_m)$' and this remains a formulation for $p^{\min}$. However, although membership in $T(g_1, \ldots, g_m)$ with bounded degrees can be formulated via a semidefinite program, this does not lead to a semidefinite programming formulation for $p^{\min}$ because of the presence of the product ρf where both ρ and f are variables. In the case when the semialgebraic set K is compact one may instead use the following refinement of the Positivstellensatz of Schmüdgen. (See [127, 130] for a more elementary exposition of Schmüdgen's result and [131] for degree bounds.)

THEOREM 3.16 (**Schmüdgen's Positivstellensatz [126]**). *Let K be as in (1.2) and assume that the semialgebraic set K in (1.2) is compact. Given $p \in \mathbb{R}[\mathbf{x}]$, if $p > 0$ on K, then $p \in T(g_1, \ldots, g_m)$.*

This now leads to a hierarchy of semidefinite relaxations for $p^{\min}$, as the programs

$$\sup \rho \text{ s.t. } p - \rho = \sum_{J \subseteq \{1,\ldots,m\}} u_J g_J \text{ with } u_J \in \Sigma,\ \deg(u_J g_J) \leq t$$

are semidefinite programs whose optimum values converge to $p^{\min}$ as t goes to ∞. However, a drawback is that Schmüdgen's representation involves 2^m sums of squares, thus leading to possibly quite large semidefinite programs. As proposed by Lasserre [65], one may use the further refinement of Schmüdgen's Positivstellensatz proposed by Putinar [115], which holds under some condition on the compact set K.

Putinar's Positivstellensatz. Given polynomials $g_1, \ldots, g_m \in \mathbb{R}[\mathbf{x}]$, the set

$$\mathbf{M}(g_1, \ldots, g_m) := \{u_0 + \sum_{j=1}^{m} u_j g_j \mid u_0, u_j \in \Sigma\}, \tag{3.14}$$

is called the *quadratic module generated by* $g_1, \ldots, g_m$. (We use the boldface letter $\mathbf{M}$ for a quadratic module $\mathbf{M}(g_1, \ldots, g_m)$ to avoid confusion with a moment matrix $M(y)$.) Consider the condition

$$\exists f \in \mathbf{M}(g_1, \ldots, g_m) \text{ s.t. } \{x \in \mathbb{R}^n \mid f(x) \geq 0\} \text{ is a compact set.} \tag{3.15}$$

Obviously, (3.15) implies that K is compact, since $K \subseteq \{x \mid f(x) \geq 0\}$ for any $f \in \mathbf{M}(g_1, \ldots, g_m)$. Note also that (3.15) is an assumption on the description of K, rather than on the set K itself. Condition (3.15) holds, e.g., if the set $\{x \in \mathbb{R}^n \mid g_j(x) \geq 0\}$ is compact for one of the constraints defining K. It also holds when the description of K contains a set of polynomial equations $h_1 = 0, \ldots, h_{m_0} = 0$ with a compact set of common real roots. If an explicit ball of radius R is known containing K, then it suffices to add the (redundant) constraint $R^2 - \sum_{i=1}^n \mathbf{x}_i^2 \geq 0$ in order to obtain a description of K satisfying (3.15). More detailed information can be found in [57, 114]; e.g. it is shown there that condition (3.15) holds when $m \leq 2$.

As we now see the condition (3.15) admits several equivalent reformulations. Consider the following conditions

$$\exists N \in \mathbb{N} \text{ for which } N - \sum_{i=1}^{n} \mathbf{x}_i^2 \in \mathbf{M}(g_1, \ldots, g_m), \tag{3.16}$$

$$\forall p \in \mathbb{R}[\mathbf{x}]\ \exists N \in \mathbb{N} \text{ for which } N \pm p \in \mathbf{M}(g_1, \ldots, g_m), \tag{3.17}$$

$$\begin{aligned} &\exists p_1, \ldots, p_s \in \mathbb{R}[\mathbf{x}] \text{ s.t. } p_I \in \mathbf{M}(g_1, \ldots, g_m)\ \forall I \subseteq \{1, \ldots, s\} \\ &\text{and } \{x \in \mathbb{R}^n \mid p_1(x) \geq 0, \ldots, p_s(x) \geq 0\} \text{ is compact.} \end{aligned} \tag{3.18}$$

Here we set $p_I := \prod_{i \in I} p_i$ for $I \subseteq \{1, \ldots, s\}$. One can easily verify the equivalence of (3.16) and (3.17), and the equivalence with (3.18) follows directly using Schmüdgen's theorem (Theorem 3.16) as we now observe.

LEMMA 3.17. *The conditions (3.15), (3.16), (3.17), (3.18) are all equivalent.*

Proof. The implications (3.17) $\Longrightarrow$ (3.16) $\Longrightarrow$ (3.15) $\Longrightarrow$ (3.18) are obvious. To derive the implication (3.18) $\Longrightarrow$ (3.17), apply Schmüdgen's theorem (Theorem 3.16) to the compact set $K_0 := \{x \in \mathbb{R}^n \mid p_1(x) \geq 0, \ldots, p_s(x) \geq 0\}$. Given $p \in \mathbb{R}[\mathbf{x}]$, there exists $N > 0$ for which $N \pm p > 0$ on K_0. Now Theorem 3.16 implies that $N \pm p = \sum_{I \subseteq \{1,\ldots,s\}} s_I p_I$ for some $s_I \in \Sigma$. As each $p_I \in \mathbf{M}(g_1, \ldots, g_m)$, this shows that $N \pm p \in \mathbf{M}(g_1, \ldots, g_m)$. □

DEFINITION 3.18. *Given $g_1, \ldots, g_m \in \mathbb{R}[\mathbf{x}]$, the quadratic module $\mathbf{M}(g_1, \ldots, g_m)$ is said to be* Archimedean *when the condition (3.17) holds.*

EXAMPLE 3.19. *For the polynomials $g_i := \mathbf{x}_i - 1/2$ $(i = 1, \ldots, n)$ and $g_{n+1} := 1 - \prod_{i=1}^{n} \mathbf{x}_i$, the module $\mathbf{M}(g_1, \ldots, g_{n+1})$ is not Archimedean [114, Ex. 6.3.1]. To see it, consider a lexicographic monomial ordering on $\mathbb{R}[\mathbf{x}]$ and define the set M of polynomials $p \in \mathbb{R}[\mathbf{x}]$ satisfying $p = 0$, or $p \neq 0$ whose leading term $p_\alpha \mathbf{x}^\alpha$ satisfies either $p_\alpha > 0$ and $\alpha \neq (1, \ldots, 1) \mod 2$, or $p_\alpha < 0$ and $\alpha = (1, \ldots, 1) \mod 2$. Then M is a quadratic module (cf. Definition 3.27) and $g_1, \ldots, g_{n+1} \in M$, implying $\mathbf{M}(g_1, \ldots, g_{n+1}) \subseteq M$. For any $N \in \mathbb{R}$, the polynomial $N - \sum_{i=1}^{n} \mathbf{x}_i^2$ does not lie in M, which implies that it also does not lie in $\mathbf{M}(g_1, \ldots, g_{n+1})$. This shows that $\mathbf{M}(g_1, \ldots, g_{n+1})$ is not Archimedean.*

THEOREM 3.20. *(Putinar [115]; see also Jacobi and Prestel [57]). Let K be as in (1.2) and assume that the quadratic modume $\mathbf{M}(g_1, \ldots, g_m)$ is Archimedean. For $p \in \mathbb{R}[\mathbf{x}]$, if $p > 0$ on K then $p \in \mathbf{M}(g_1, \ldots, g_m)$.*

As noted by Lasserre [65], this implies directly the asymptotic convergence to $p^{\min}$ of the hierarchy of bounds from (3.8). We will come back to this hierarchy in Section 6. We refer to Nie and Schweighofer [100] for degree bounds in representations in $\mathbf{M}(g_1, \ldots, g_m)$. We present a proof of Theorem 3.20 in Section 3.7.

Some other representation results. Several other representation results for positive polynomials exist in the literature. Let us just briefly mention a few.

THEOREM 3.21. *(Pólya [108]; see [109] for a proof). Let $p \in \mathbb{R}[\mathbf{x}]$ be a homogeneous polynomial. If $p > 0$ on the simplex $\{x \in \mathbb{R}^n_+ \mid \sum_{i=1}^n x_i = 1\}$, then there exists $r \in \mathbb{N}$ for which the polynomial $(\sum_{i=1}^n \mathbf{x}_i)^r p$ has all its coefficients nonnegative.*

THEOREM 3.22. *(Reznick [119]) Let $p \in \mathbb{R}[\mathbf{x}]$ be a homogeneous polynomial. If $p > 0$ on $\mathbb{R}^n \setminus \{0\}$, then there exists $r \in \mathbb{N}$ for which the polynomial $(\sum_{i=1}^n \mathbf{x}_i^2)^r p$ is a sum of squares.*

EXAMPLE 3.23. Consider the 5×5 symmetric matrix whose entries are all equal to 1 except $M_{1,2} = M_{2,3} = M_{3,4} = M_{4,5} = M_{5,1} = -1$ and let $p_M := \sum_{i,j=1}^5 M_{i,j}\mathbf{x}_i^2\mathbf{x}_j^2$. Recall from Section 1.1 that M is copositive precisely when p_M is nonnegative. Parrilo [103] proved that, while p_M is not a SOS, $(\sum_{i=1}^5 \mathbf{x}_i^2)p_M$ is a SOS, which shows that p_M is nonnegative and thus M is copositive.

As an illustration let us briefly sketch how Pólya's theorem can be used to derive a hierarchy of SOS approximations for the stable set problem. See e.g. [31] for further applications, and [30] for a comparison of the hierarchies based on Putinar's and Pólya's theorems.

EXAMPLE 3.24. Consider the stable set problem introduced in Section 1.1 and the formulation (1.7) for $\alpha(G)$. For $t \in \mathbb{N}$ define the polynomial $p_{G,t} := \sum_{i,j \in V} \mathbf{x}_i^2\mathbf{x}_j^2(t(I + A_G) - J)_{i,j}$. For $r \in \mathbb{N}$, the parameters

$$\inf\ t \text{ s.t. } \Big(\sum_{i \in V} \mathbf{x}_i^2\Big)^r p_{G,t} \text{ is SOS}$$

provide a hierarchy of upper bounds for $\alpha(G)$. Based on an analysis of Pólya's theorem, de Klerk and Pasechnik [32] proved the *finite* convergence to $\alpha(G)$ and they conjecture that finite convergence takes place at $r = \alpha(G) - 1$. (See [46] for partial results, also for a comparison of the above parameter with the approximation of $\alpha(G)$ derived via Putinar's theorem, mentioned in Example 8.16).

One can also search for a different type of certificate for positivity of a polynomial p over K defined by polynomial inequalities $g_1 \geq 0, \ldots, g_m \geq 0$; namely of the form

$$p = \sum_{\beta \in \mathbb{N}^m} c_\beta \prod_{j=1}^m g_j^{\beta_j} \text{ with finitely many nonzero } c_\beta \in \mathbb{R}_+. \tag{3.19}$$

On the moment side this corresponds to Hausdorff-type moment conditions, and this yields hierarchies of *linear programming* relaxations for polynomial optimization. Sherali and Adams [137] use this type of representation for

0/1 polynomial optimization problems. As an example let us mention Handelman's characterization for positivity over a polytope.

THEOREM 3.25. *(Handelman [49]) let $p \in \mathbb{R}[\mathbf{x}]$ and let $K = \{x \in \mathbb{R}^n \mid g_1(x) \geq 0, \ldots, g_m(x) \geq 0\}$ be a polytope, i.e. the g_j's are linear polynomials and K is bounded. If $p > 0$ on K then p has a decomposition (3.19).*

The following result holds for a general compact semialgebraic set K, leading to a hierarchy of LP relaxations for problem (1.1). We refer to Lasserre [68, 70] for a detailed discussion and comparison with the SDP based approach.

THEOREM 3.26. *[35, 36] Assume K is compact and the polynomials $g_1, \ldots, g_m$ satisfy $0 \leq g_j \leq 1$ on K $\forall j$ and, together with the constant polynomial 1, they generate the algebra $\mathbb{R}[\mathbf{x}]$, i.e. $\mathbb{R}[\mathbf{x}] = \mathbb{R}[1, g_1, \ldots, g_m]$. Then any $p \in \mathbb{R}[\mathbf{x}]$ positive on K has a representation of the form*

$$p = \sum_{\alpha,\beta \in \mathbb{N}^n} c_{\alpha\beta} \prod_{j=1}^{m} g_j^{\alpha_j} \prod_{j=1}^{m} (1 - g_j)^{\beta_j}$$

for finitely many nonnegative scalars $c_{\alpha\beta}$.

3.7. Proof of Putinar's theorem. Schweighofer [132] gave a proof for Theorem 3.20 which is more elementary than Putinar's original proof and uses only Pólya's theorem (Theorem 3.21). Later M. Schweighofer communicated to us an even shorter and simpler proof which is just a combination of classical ideas from real algebraic geometry with an ingeneous argument of Marshall (in Claim 3.32).

DEFINITION 3.27. *Call a set $M \subseteq \mathbb{R}[\mathbf{x}]$ a* quadratic module *if it contains 1 and is closed under addition and multiplication with squares, i.e., $1 \in M$, $M + M \subseteq M$ and $\Sigma M \subseteq M$. Call a quadratic module M* proper *if $-1 \notin M$ (i.e. $M \neq \mathbb{R}[\mathbf{x}]$).*

Given $g_1, \ldots, g_m \in \mathbb{R}[\mathbf{x}]$ the set $\mathbf{M}(g_1, \ldots, g_m)$ introduced in (3.14) is obviously a quadratic module. We begin with some preliminary results about quadratic modules.

LEMMA 3.28. *If $M \subseteq \mathbb{R}[\mathbf{x}]$ is a quadratic module, then $\mathcal{I} := M \cap -M$ is an ideal.*

Proof. For $f \in \mathbb{R}[\mathbf{x}]$ and $g \in \mathcal{I}$, $fg = \left(\frac{f+1}{2}\right)^2 g - \left(\frac{f-1}{2}\right)^2 g \in \mathcal{I}$. □

LEMMA 3.29. *Let $M \subseteq \mathbb{R}[\mathbf{x}]$ be a maximal proper quadratic module. Then $M \cup -M = \mathbb{R}[\mathbf{x}]$.*

Proof. Assume $f \in \mathbb{R}[\mathbf{x}] \setminus (M \cup -M)$. By maximality of M, the quadratic modules $M + f\Sigma$ and $M - f\Sigma$ are not proper, i.e., we find $g_1, g_2 \in M$ and $s_1, s_2 \in \Sigma$ such that $-1 = g_1 + s_1 f$ and $-1 = g_2 - s_2 f$. Multiplying the first equation by s_2 and the second one by s_1, we get $s_1 + s_2 + s_1 g_2 + s_2 g_1 = 0$. This implies $s_1, s_2 \in \mathcal{I} := M \cap -M$. Since

$\mathcal{I}$ is an ideal, we get $s_1 f \in \mathcal{I} \subseteq M$ and therefore $-1 = g_1 + s_1 f \in M$, a contradiction. □

LEMMA 3.30. *Let $M \subseteq \mathbb{R}[\mathbf{x}]$ be a maximal proper quadratic module which is Archimedean, set $\mathcal{I} := M \cap -M$ and let $f \in \mathbb{R}[\mathbf{x}]$. Then there is exactly one $a \in \mathbb{R}$ such that $f - a \in \mathcal{I}$.*

Proof. Consider the sets

$$A := \{a \in \mathbb{R} \mid f - a \in M\} \quad \text{and} \quad B := \{b \in \mathbb{R} \mid b - f \in M\}.$$

As M is Archimedean, the sets A and B are not empty. We have to show that $A \cap B$ is a singleton. Since M is proper, it does not contain any negative real number. Therefore $a \leq b$ for all $a \in A$, $b \in B$. Set $a_0 := \sup A$ and $b_0 := \inf B$. Thus $a_0 \leq b_0$. Moreover, $a_0 = b_0$. Indeed if $a_0 < c < b_0$, then $f - c \notin M \cup -M$, which contradicts the fact that $\mathbb{R}[\mathbf{x}] = M \cup -M$ (by Lemma 3.29). It suffices now to show that $a_0 \in A$ and $b_0 \in B$, since this will imply that $A \cap B = \{a_0\}$ and thus conclude the proof. We show that $a_0 = \sup A \in A$. For this assume that $a_0 \notin A$, i.e., $f - a_0 \notin M$. Then $M' := M + (f - a_0)\Sigma$ is a quadratic module that cannot be proper by the maximality of M; that is, $-1 = g + (f - a_0)s$ for some $g \in M$ and $s \in \Sigma$. As M is Archimedean we can choose $N \in \mathbb{N}$ such that $N - s \in M$ and $\epsilon \in \mathbb{R}$ such that $0 < \epsilon < \frac{1}{N}$. As $a_0 - \epsilon \in A$, we have $f - (a_0 - \epsilon) \in M$. Then we have $-1 + \epsilon s = g + (f - a_0 + \epsilon)s \in M$ and $\epsilon N - \epsilon s \in M$. Adding these two equations, we get $\epsilon N - 1 \in M$ which is impossible since $\epsilon N - 1 < 0$ and M is proper. One can prove that $b_0 \in B$ in the same way. □

We now prove Theorem 3.20. Assume $p \in \mathbb{R}[\mathbf{x}]$ is positive on K; we show that $p \in \mathbf{M}(g_1, \ldots, g_m)$. We state two intermediary results.

CLAIM 3.31. *There exists $s \in \Sigma$ such that $sp \in 1 + \mathbf{M}(g_1, \ldots, g_m)$.*

Proof. We have to prove that the quadratic module $M_0 := \mathbf{M}(g_1, \ldots, g_m) - p\Sigma$ is not proper. For this assume that M_0 is proper; we show the existence of $a \in K$ for which $p(a) \leq 0$, thus contradicting the assumption $p > 0$ on K. By Zorn's lemma, we can extend M_0 to a maximal proper quadratic module $M \supseteq M_0$. As $M \supseteq \mathbf{M}(g_1, \ldots, g_m)$, M is Archimedean. Applying Lemma 3.30, there exists $a \in \mathbb{R}^n$ such that $\mathbf{x}_i - a_i \in \mathcal{I} := M \cap -M$ for all $i \in \{1, \ldots, n\}$. Since $\mathcal{I}$ is an ideal (by Lemma 3.28), $f - f(a) \in \mathcal{I}$ for any $f \in \mathbb{R}[\mathbf{x}]$. In particular, for $f = g_j$, we find that $g_j(a) = g_j - (g_j - g_j(a)) \in M$ since $g_j \in \mathbf{M}(g_1, \ldots, g_m) \subseteq M$ and $-(g_j - g_j(a)) \in M$, which implies $g_j(a) \geq 0$. Therefore, $a \in K$. Finally, $-p(a) = (p - p(a)) - p \in M$ since $p - p(a) \in \mathcal{I} \subseteq M$ and $-p \in M_0 \subseteq M$, which implies $-p(a) \geq 0$. □

CLAIM 3.32. *There exist $g \in \mathbf{M}(g_1, \ldots, g_m)$ and $N \in \mathbb{N}$ such that $N - g \in \Sigma$ and $gp \in 1 + \mathbf{M}(g_1, \ldots, g_m)$.*

Proof. (Marshall [90, 5.4.4]). Choose s as in Claim 3.31, i.e. $s \in \Sigma$ and $sp - 1 \in \mathbf{M}(g_1, \ldots, g_m)$. Using (3.17), there exists $k \in \mathbb{N}$ such that

$2k - s$, $2k - s^2p - 1 \in \mathbf{M}(g_1, \ldots, g_m)$. Set $g := s(2k - s)$ and $N := k^2$. Then $g \in \mathbf{M}(g_1, \ldots, g_m)$, $N - g = k^2 - 2ks + s^2 = (k - s)^2 \in \Sigma$. Moreover, $gp - 1 = s(2k - s)p - 1 = 2k(sp - 1) + (2k - s^2p - 1) \in \mathbf{M}(g_1, \ldots, g_m)$, since $sp - 1, 2k - s^2p - 1 \in \mathbf{M}(g_1, \ldots, g_m)$. □

We can now conclude the proof. Choose g, N as in Claim 3.32 and $k \in \mathbb{N}$ such that $k + p \in \mathbf{M}(g_1, \ldots, g_m)$. We may assume $N > 0$. Note that

$$\left(k - \frac{1}{N}\right) + p = \frac{1}{N}\Big((N - g)(k + p) + (gp - 1) + kg\Big) \in \mathbf{M}(g_1, \ldots, g_m).$$

Applying this iteratively we can make $k = (kN)\frac{1}{N}$ smaller and smaller until reaching 0 and thus obtain $p \in \mathbf{M}(g_1, \ldots, g_m)$. This concludes the proof of Theorem 3.20.

3.8. The cone $\Sigma_{n,d}$ of sums of squares is closed. As we saw earlier, for d even, the inclusion $\Sigma_{n,d} \subseteq \mathcal{P}_{n,d}$ is strict except in the three special cases $(n, d) = (1, d)$, $(n, 2)$, $(2, 4)$. One may wonder how much the two cones differ in the other cases. This question will be addressed in Section 7.1, where we will mention a result of Blekherman (Theorem 7.1) showing that, when the degree d is fixed and the number n of variables grows, there are much more nonnegative polynomials than sums of squares. On the other hand, one can show that any nonnegative polynomial is the limit (coordinate-wise) of a sequence of SOS polynomials (cf. Theorem 7.3). However, as the cone $\Sigma_{n,d}$ is a closed set, the degrees of the polynomials occurring in such sequence cannot be bounded. We now prove a more general result which will imply that $\Sigma_{n,d}$ is closed. Given polynomials $g_1, \ldots, g_m \in \mathbb{R}[\mathbf{x}]$ and an integer t, set $g_0 := 1$ and define the set

$$\mathbf{M}_t(g_1, \ldots, g_m) := \left\{ \sum_{j=0}^{m} s_j g_j \mid s_j \in \Sigma, \deg(s_j g_j) \leq t \ (0 \leq j \leq m) \right\}, \quad (3.20)$$

which can be seen as the "truncation at degree t" of the quadratic module $\mathbf{M}(g_1, \ldots, g_m)$. Let K be as in (1.2). Its interior $\mathrm{int}(K)$ (for the Euclidean topology) consists of the points $x \in K$ for which there exists a (full dimensional) ball centered at x and contained in K. Obviously

$$K' := \{x \in \mathbb{R}^n \mid g_j(x) > 0 \ \forall j = 1, \ldots, m\} \subseteq \mathrm{int}(K).$$

The inclusion may be strict (e.g. $0 \in \mathrm{int}(K) \setminus K'$ for $K = \{x \in \mathbb{R} \mid x^2 \geq 0\}$). However,

$$\mathrm{int}(K) \neq \emptyset \Longleftrightarrow K' \neq \emptyset \quad (3.21)$$

assuming no g_j is the zero polynomial. Indeed if $K' = \emptyset$ and B is a ball contained in K, then the polynomial $\prod_{j=1}^{m} g_j$ vanishes on K and thus on B, hence it must be the zero polynomial, a contradiction. The next result will also be used later in Section 6 to show the absence of duality gap between the moment/SOS relaxations.

THEOREM 3.33. *[111, 132] If K has a nonempty interior then $\mathbf{M}_t(g_1, \ldots, g_m)$ is closed in $\mathbb{R}[\mathbf{x}]_t$ for any $t \in \mathbb{N}$.*

Proof. Note that $\deg(s_j g_j) \le t$ is equivalent to $\deg s_j \le 2k_j$, setting $k_j := \lfloor (t - \deg(g_j))/2 \rfloor$. Set $\Lambda_j := \dim \mathbb{R}[\mathbf{x}]_{k_j} = |\mathbb{N}^n_{k_j}|$. Then any polynomial $f \in \mathbf{M}_t(g_1, \ldots, g_m)$ is of the form $f = \sum_{j=0}^m s_j g_j$ with $s_j = \sum_{l_j=1}^{\Lambda_j} (u_{l_j}^{(j)})^2$ for some $u_1^{(j)}, \ldots, u_{\Lambda_j}^{(j)} \in \mathbb{R}[\mathbf{x}]_{k_j}$. In other words, $\mathbf{M}_t(g_1, \ldots, g_m)$ is the image of the following map

$$\varphi: \quad \mathcal{D} := (\mathbb{R}[\mathbf{x}]_{k_0})^{\Lambda_0} \times \ldots \times (\mathbb{R}[\mathbf{x}]_{k_m})^{\Lambda_m} \rightarrow \mathbb{R}[\mathbf{x}]_t$$

$$u = ((u_{l_0}^{(0)})_{l_0=1}^{\Lambda_0}, \ldots, (u_{l_m}^{(m)})_{l_m=1}^{\Lambda_m}) \mapsto \varphi(u) = \sum_{j=0}^m \sum_{l_j=1}^{\Lambda_j} (u_{l_j}^{(j)})^2 g_j.$$

We may identify the domain $\mathcal{D}$ of φ with the space $\mathbb{R}^\Lambda$ (of suitable dimension Λ); choose a norm on this space and let S denote the unit sphere in $\mathcal{D}$. Then $V := \varphi(S)$ is a compact set in the space $\mathbb{R}[\mathbf{x}]_t$, which is also equipped with a norm. Note that any $f \in \mathbf{M}_t(g_1, \ldots, g_m)$ is of the form $f = \lambda v$ for some $\lambda \in \mathbb{R}_+$ and $v \in V$. We claim that $0 \notin V$. Indeed, by assumption, $\mathrm{int}(K) \neq \emptyset$ and thus, by (3.21), there exists a full dimensional ball $B \subseteq K$ such that each polynomial g_j ($j = 1, \ldots, m$) is positive on B. Hence, for any $u \in S$, if $\varphi(u)$ vanishes on B then each polynomial arising as component of u vanishes on B, implying $u = 0$. This shows that $\varphi(u) \neq 0$ if $u \in S$, i.e. $0 \notin V$. We now show that $\mathbf{M}_t(g_1, \ldots, g_m)$ is closed. For this consider a sequence $f_k \in \mathbf{M}_t(g_1, \ldots, g_m)$ ($k \ge 0$) converging to a polynomial f; we show that $f \in \mathbf{M}_t(g_1, \ldots, g_m)$. Write $f_k = \lambda_k v_k$ where $v_k \in V$ and $\lambda_k \in \mathbb{R}_+$. As V is compact there exists a subsequence of $(v_k)_{k \ge 0}$, again denoted as $(v_k)_{k \ge 0}$ for simplicity, converging say to $v \in V$. As $0 \notin V$, $v \neq 0$ and thus $\lambda_k = \frac{\|f_k\|}{\|v_k\|}$ converges to $\frac{\|f\|}{\|v\|}$ as $k \to \infty$. Therefore, $f_k = \lambda_k v_k$ converges to $\frac{\|f\|}{\|v\|} v$ as $k \to \infty$, which implies $f = \frac{\|f\|}{\|v\|} v \in \mathbf{M}_t(g_1, \ldots, g_m)$. □

COROLLARY 3.34. *The cone $\Sigma_{n,d}$ is a closed cone.*

Proof. Apply Theorem 3.33 to $\mathbf{M}_d(g_1) = \Sigma_{n,d}$ for $g_1 := 1$. □

When the set K has an empty interior one can prove an analogue of Theorem 3.33 after factoring out through the vanishing ideal of K. More precisely, with $\mathcal{I}(K) = \{f \in \mathbb{R}[\mathbf{x}] \mid f(x) = 0 \ \forall x \in K\}$, consider the mapping $p \in \mathbb{R}[\mathbf{x}] \mapsto p' = p + \mathcal{I}(K) \in \mathbb{R}[\mathbf{x}]/\mathcal{I}(K)$ mapping any polynomial to its coset in $\mathbb{R}[\mathbf{x}]/\mathcal{I}(K)$. Define the image under this mapping of the quadratic module $\mathbf{M}_t(g_1, \ldots, g_m)$

$$\mathbf{M}'_t(g_1, \ldots, g_m) = \{p' \mid p \in \mathbf{M}_t(g_1, \ldots, g_m)\} \subseteq \mathbb{R}[\mathbf{x}]_t/\mathcal{I}(K). \tag{3.22}$$

Marshall [88] proved the following extension of Theorem 3.33. Note indeed that if K has a nonempty interior, then $\mathcal{I}(K) = \{0\}$ and thus $\mathbf{M}'_t(g_1, \ldots, g_m) = \mathbf{M}_t(g_1, \ldots, g_m)$.

THEOREM 3.35. *[88] The set $\mathbf{M}'_t(g_1,\ldots,g_m)$ is closed in $\mathbb{R}[\mathbf{x}]_t/\mathcal{I}(K)$.*

Proof. The proof is along the same lines as that of Theorem 3.33, except one must now factor out through the ideal $\mathcal{I}(K)$. Set $g_0 := 1$, $J := \{j \in \{0,1,\ldots,m\} \mid g_j \notin \mathcal{I}(K)\}$ and $\text{Ann}(g) := \{p \in \mathbb{R}[\mathbf{x}] \mid pg \in \mathcal{I}(K)\}$ for $g \in \mathbb{R}[\mathbf{x}]$. For $j = 0,\ldots,m$, set $k_j := \lfloor (t - \deg(g_j))/2 \rfloor$ (as in the proof of Theorem 3.33), $\mathcal{A}_j := \mathbb{R}[\mathbf{x}]_{k_j} \cap \text{Ann}(g_j)$. Let $\mathcal{B}_j \subseteq \mathbb{R}[\mathbf{x}]_{k_j}$ be a set of monomials forming a basis of $\mathbb{R}[\mathbf{x}]_{k_j}/\mathcal{A}_j$; that is, any polynomial $f \in \mathbb{R}[\mathbf{x}]_{k_j}$ can be written in a unique way as $p = r+q$ where $r \in \text{Span}_{\mathbb{R}}(\mathcal{B}_j)$ and $q \in \mathcal{A}_j$. Let $\Lambda_j := |\mathcal{B}_j| = \dim \mathbb{R}[\mathbf{x}]_{k_j}/\mathcal{A}_j$. Consider the mapping

$$\varphi :\ \mathcal{D} := \prod_{j\in J} (\mathbb{R}[\mathbf{x}]_{k_j}/\mathcal{A}_j)^{\Lambda_j} \to \mathbb{R}[\mathbf{x}]_t/\mathcal{I}(K)$$

$$u = ((u_{l_j}^{(j)} \mod \mathcal{A}_j)_{l_j=1}^{\Lambda_j})_{j\in J} \mapsto \varphi(u) = \sum_{j\in J}\sum_{l_j=1}^{\Lambda_j} (u_{l_j}^{(j)})^2 g_j \mod \mathcal{I}(K).$$

Note first that φ is well defined; indeed, $u-v \in \mathcal{A}_j$ implies $(u-v)g_j \in \mathcal{I}(K)$ and thus $u^2g_j - v^2g_j = (u+v)(u-v)g_j \in \mathcal{I}(K)$. Next we claim that the image of the domain $\mathcal{D}$ under φ is precisely the set $\mathbf{M}'_t(g_1,\ldots,g_m)$. That is,

$$\forall f \in \mathbf{M}_t(g_1,\ldots,g_m),\ \exists u \in \mathcal{D} \text{ s.t. } f - \sum_{j\in J}\sum_{l_j=1}^{\Lambda_j} (u_{l_j}^{(j)})^2 g_j \in \mathcal{I}(K). \tag{3.23}$$

For this write $f = \sum_{j=0}^m s_j g_j$ where $s_j \in \Sigma$ and $\deg(s_j) \le t - \deg(g_j)$. Say $s_j = \sum_{h_j} (a_{h_j}^{(j)})^2$ where $a_{h_j}^{(j)} \in \mathbb{R}[\mathbf{x}]_{k_j}$. Write $a_{h_j}^{(j)} = r_{h_j}^{(j)} + q_{h_j}^{(j)}$, where $r_{h_j}^{(j)} \in \text{Span}_{\mathbb{R}}(\mathcal{B}_j)$ and $q_{h_j}^{(j)} \in \mathcal{A}_j$. Then, $s_jg_j = \sum_{j\in J}(\sum_{h_j}(r_{h_j}^{(j)})^2)g_j \mod \mathcal{I}(K)$ since $q_{h_j}^{(j)} g_j \in \mathcal{I}(K)$ as $q_{h_j}^{(j)} \in \mathcal{A}_j \subseteq \text{Ann}(g_j)$. Moreover, as each $r_{h_j}^{(j)}$ lies in $\text{Span}_{\mathbb{R}}(\mathcal{B}_j)$ with $|\mathcal{B}_j| = \Lambda_j$, by the Gram-matrix method (recall Section 3.3), we deduce that $\sum_{h_j}(r_{h_j}^{(j)})^2$ can be written as another sum of squares involving only Λ_j squares, i.e. $\sum_{h_j}(r_{h_j}^{(j)})^2 = \sum_{l_j=1}^{\Lambda_j}(u_{l_j}^{(j)})^2$ with $u_{l_j}^{(j)} \in \text{Span}_{\mathbb{R}}(\mathcal{B}_j)$; this shows (3.23). We now show that $\varphi^{-1}(0) = 0$. For this assume that $\varphi(u) = 0$, i.e. $f := \sum_{l_j=1}^{\Lambda_j}(u_{l_j}^{(j)})^2 g_j \in \mathcal{I}(K)$; we show that $u_{l_j}^{(j)} \in \mathcal{A}_j$ for all j, l_j. Fix $x \in K$. Then $f(x) = 0$ and, as $g_j(x) \ge 0\ \ \forall j$, $(u_{l_j}^{(j)}(x))^2 g_j(x) = 0$, i.e. $u_{l_j}^{(j)}(x)g_j(x) = 0\ \ \forall j, l_j$. This shows that each polynomial $u_{l_j}^{(j)} g_j$ lies in $\mathcal{I}(K)$, that is, $u_{l_j}^{(j)} \in \mathcal{A}_j$. We can now proceed as in the proof of Theorem 3.33 to conclude that $M'_t(g_1,\ldots,g_m)$ is a closed set in $\mathbb{R}[\mathbf{x}]_t/\mathcal{I}(K)$. □

4. Moment sequences and moment matrices.

4.1. Some basic facts. We introduce here some basic definitions and facts about measures, moment sequences and moment matrices.

Moment sequences. Throughout we consider nonnegative Borel measures on $\mathbb{R}^n$; thus, when speaking of 'a measure', we implicitly assume that it is nonnegative. A probability measure is a measure with total mass $\mu(\mathbb{R}^n) = 1$.

Given a measure μ on $\mathbb{R}^n$, its *support* $\mathrm{supp}(\mu)$ is the smallest closed set $S \subseteq \mathbb{R}^n$ for which $\mu(\mathbb{R}^n \setminus S) = 0$. We say that μ is a measure *on* K or a measure *supported by* $K \subseteq \mathbb{R}^n$ if $\mathrm{supp}(\mu) \subseteq K$.

Given $x \in \mathbb{R}^n$, δ_x denotes the *Dirac measure* at x, with support $\{x\}$ and having mass 1 at x and mass 0 elsewhere.

When the support of a measure μ is finite, say, $\mathrm{supp}(\mu) = \{x_1, \ldots, x_r\}$, then μ is of the form $\mu = \sum_{i=1}^r \lambda_i \delta_{x_i}$ for some $\lambda_1, \ldots, \lambda_r > 0$; the x_i are called the *atoms* of μ and one also says that μ is a *r-atomic* measure.

Given a measure μ on $\mathbb{R}^n$, the quantity $y_\alpha := \int x^\alpha \mu(dx)$ is called its *moment of order* α. Then, the sequence $(y_\alpha)_{\alpha \in \mathbb{N}^n}$ is called the sequence of moments of the measure μ and, given $t \in \mathbb{N}$, the (truncated) sequence $(y_\alpha)_{\alpha \in \mathbb{N}^n_t}$ is called the sequence of moments of μ up to order t. When y is the sequence of moments of a measure we also say that μ is a *representing measure* for y. The sequence of moments of the Dirac measure δ_x is the vector $\zeta_x := (x^\alpha)_{\alpha \in \mathbb{N}^n}$, called the *Zeta vector* of x (see the footnote on page 255 for a motivation). Given an integer $t \geq 1$, $\zeta_{t,x} := (x^\alpha)_{\alpha \in \mathbb{N}^n_t}$ denotes the *truncated Zeta vector*.

A basic problem in the theory of moments concerns the characterization of (infinite or truncated) moment sequences, i.e., the characterization of those sequences $y = (y_\alpha)_\alpha$ that are the sequences of moments of some measure. Given a subset $K \subseteq \mathbb{R}^n$, the *K-moment problem* asks for the characterization of those sequences that are sequences of moments of some measure supported by the set K. This problem has received considerable attention in the literature, especially in the case when $K = \mathbb{R}^n$ (the basic moment problem) or when K is a compact semialgebraic set, and it turns out to be related to our polynomial optimization problem, as we see below in this section. For more information on the moment problem see e.g. [8, 9, 23–26, 39, 63, 64, 115, 126] and references therein.

Moment matrices. The following notions of moment matrix and shift operator play a central role in the moment problem. Given a sequence $y = (y_\alpha)_{\alpha \in \mathbb{N}^n} \in \mathbb{R}^{\mathbb{N}^n}$, its *moment matrix* is the (infinite) matrix $M(y)$ indexed by $\mathbb{N}^n$, with (α, β)th entry $y_{\alpha+\beta}$, for $\alpha, \beta \in \mathbb{N}^n$. Similarly, given an integer $t \geq 1$ and a (truncated) sequence $y = (y_\alpha)_{\alpha \in \mathbb{N}^n_{2t}} \in \mathbb{R}^{\mathbb{N}^n_{2t}}$, its *moment matrix of order* t is the matrix $M_t(y)$ indexed by $\mathbb{N}^n_t$, with (α, β)th entry $y_{\alpha+\beta}$, for $\alpha, \beta \in \mathbb{N}^n_t$.

Given $g \in \mathbb{R}[\mathbf{x}]$ and $y \in \mathbb{R}^{\mathbb{N}^n}$, define the new sequence

$$gy := M(y)g \in \mathbb{R}^{\mathbb{N}^n}, \tag{4.1}$$

called *shifted vector*, with αth entry $(gy)_\alpha := \sum_\beta g_\beta y_{\alpha+\beta}$ for $\alpha \in \mathbb{N}^n$. The notation gy will also be used for denoting the truncated vector $((gy)_\alpha)_{\alpha \in \mathbb{N}^n_t}$

of $\mathbb{R}^{\mathbb{N}^n_t}$ for an integer $t \geq 1$. The moment matrices of gy are also known as the *localizing matrices*, since they can be used to "localize" the support of a representing measure for y.

Moment matrices and bilinear forms on $\mathbb{R}[\mathbf{x}]$. Given $y \in \mathbb{R}^{\mathbb{N}^n}$, define the linear form $L_y \in (\mathbb{R}[\mathbf{x}])^*$ by

$$L_y(f) := y^T \text{vec}(f) = \sum_\alpha y_\alpha f_\alpha \quad \text{for } f = \sum_\alpha f_\alpha \mathbf{x}^\alpha \in \mathbb{R}[\mathbf{x}]. \tag{4.2}$$

We will often use the following simple 'calculus' involving moment matrices.

LEMMA 4.1. *Let $y \in \mathbb{R}^{\mathbb{N}^n}$, $L_y \in (\mathbb{R}[\mathbf{x}])^*$ the associated linear form from (4.2), and let $f, g, h \in \mathbb{R}[\mathbf{x}]$.*
(i) $L_y(fg) = \text{vec}(f)^T M(y) \text{vec}(g)$; in particular, $L_y(f^2) = \text{vec}(f)^T M(y) \text{vec}(f)$, $L_y(f) = \text{vec}(1)^T M(y) \text{vec}(f)$.
(ii) $L_y(fgh) = \text{vec}(f)^T M(y) \text{vec}(gh) = \text{vec}(fg)^T M(y) \text{vec}(h) = \text{vec}(f)^T M(hy) \text{vec}(g)$.

Proof. (i) Setting $f = \sum_\alpha f_\alpha \mathbf{x}^\alpha$, $g = \sum_\beta g_\beta \mathbf{x}^\beta$, we have $fg = \sum_\gamma (\sum_{\alpha,\beta|\alpha+\beta=\gamma} f_\alpha g_\beta) \mathbf{x}^\gamma$. Then $L_y(fg) = \sum_\gamma y_\gamma (\sum_{\alpha,\beta|\alpha+\beta=\gamma} f_\alpha g_\beta)$, while $\text{vec}(f)^T M(y) \text{vec}(g) = \sum_\alpha \sum_\beta f_\alpha g_\beta y_{\alpha+\beta}$, thus equal to $L_y(fg)$. The last part of (i) follows directly.
(ii) By (i) $\text{vec}(f)^T M(y) \text{vec}(gh) = \text{vec}(fg)^T M(y) \text{vec}(h) = L_y(fgh)$, in turn equal to $\sum_{\alpha,\beta,\gamma} f_\alpha g_\beta h_\gamma y_{\alpha+\beta+\gamma}$. Finally, $\text{vec}(f)^T M(hy) \text{vec}(g) = \sum_\delta (hy)_\delta (fg)_\delta = \sum_\delta (\sum_\gamma h_\gamma y_{\gamma+\delta})(\sum_{\alpha,\beta|\alpha+\beta=\delta} f_\alpha g_\beta)$ which, by exchanging the summations, is equal to $\sum_{\alpha,\beta,\gamma} f_\alpha g_\beta h_\gamma y_{\alpha+\beta+\gamma} = L_y(fgh)$. □

Given $y \in \mathbb{R}^{\mathbb{N}^n}$, we can also define the bilinear form on $\mathbb{R}[\mathbf{x}]$

$$(f, g) \in \mathbb{R}[\mathbf{x}] \times \mathbb{R}[\mathbf{x}] \mapsto L_y(fg) = \text{vec}(f)^T M(y) \text{vec}(g),$$

whose associated quadratic form

$$f \in \mathbb{R}[\mathbf{x}] \mapsto L_y(f^2) = \text{vec}(f)^T M(y) \text{vec}(f)$$

is positive semidefinite, i.e. $L_y(f^2) \geq 0$ for all $f \in \mathbb{R}[\mathbf{x}]$, precisely when the moment matrix $M(y)$ is positive semidefinite.

Necessary conditions for moment sequences. The next lemma gives some easy well known necessary conditions for moment sequences.

LEMMA 4.2. *Let $g \in \mathbb{R}[\mathbf{x}]$ and $d_g = \lceil \deg(g)/2 \rceil$.*
(i) *If $y \in \mathbb{R}^{\mathbb{N}^n_{2t}}$ is the sequence of moments (up to order $2t$) of a measure μ, then $M_t(y) \succeq 0$ and $\text{rank}\, M_t(y) \leq |\text{supp}(\mu)|$. Moreover, for $p \in \mathbb{R}[\mathbf{x}]_t$, $M_t(y)p = 0$ implies $\text{supp}(\mu) \subseteq V_\mathbb{R}(p) = \{x \in \mathbb{R}^n \mid p(x) = 0\}$. Therefore, $\text{supp}(\mu) \subseteq V_\mathbb{R}(\text{Ker}\, M_t(y))$.*
(ii) *If $y \in \mathbb{R}^{\mathbb{N}^n_{2t}}$ ($t \geq d_g$) is the sequence of moments of a measure μ supported by the set $\{x \in \mathbb{R}^n \mid g(x) \geq 0\}$, then $M_{t-d_g}(gy) \succeq 0$.*

(iii) *If* $y \in \mathbb{R}^{\mathbb{N}^n}$ *is the sequence of moments of a measure* μ, *then* $M(y) \succeq 0$. *Moreover, if* $\text{supp}(\mu) \subseteq \{x \in \mathbb{R}^n \mid g(x) \geq 0\}$, *then* $M(gy) \succeq 0$ *and, if* μ *is* r*-atomic, then* rank $M(y) = r$.

Proof. (i) For $p \in \mathbb{R}[\mathbf{x}]_t$, $p^T M_t(y) p$ is equal to

$$\sum_{\alpha,\beta \in \mathbb{N}^n_t} p_\alpha p_\beta y_{\alpha+\beta} = \sum_{\alpha,\beta \in \mathbb{N}^n_t} p_\alpha p_\beta \int x^{\alpha+\beta} \mu(dx) = \int p(x)^2 \mu(dx) \geq 0,$$

which shows that $M_t(y) \succeq 0$. If $M_t(y)p = 0$, then $0 = p^T M_t(y) p = \int p(x)^2 \mu(dx)$. This implies that the support of μ is contained in the set $V_\mathbb{R}(p)$ of real zeros of p. [To see it, note that, as $V_\mathbb{R}(p)$ is a closed set, $\text{supp}(\mu) \subseteq V_\mathbb{R}(p)$ holds if we can show that $\mu(\mathbb{R}^n \setminus V_\mathbb{R}(p)) = 0$. Indeed, $\mathbb{R}^n \setminus V_\mathbb{R}(p) = \bigcup_{k \geq 0} U_k$, setting $U_k := \{x \in \mathbb{R}^n \mid p(x)^2 \geq \frac{1}{k}\}$ for positive $k \in \mathbb{N}$. As $0 = \int p(x)^2 \mu(dx) = \int_{\mathbb{R}^n \setminus V_\mathbb{R}(p)} p(x)^2 \mu(dx) \geq \int_{U_k} p(x)^2 \mu(dx) \geq \frac{1}{k} \mu(U_k)$, this implies $\mu(U_k) = 0$ for all k and thus $\mu(\mathbb{R}^n \setminus V_\mathbb{R}(p)) = 0$.] The inequality rank $M_t(y) \leq |\text{supp}(\mu)|$ is trivial if μ has an infinite support. So assume that μ is r-atomic, say, $\mu = \sum_{i=1}^r \lambda_i \delta_{x_i}$ where $\lambda_1, \ldots, \lambda_r > 0$ and $x_1, \ldots, x_r \in \mathbb{R}^n$. Then, $M_t(y) = \sum_{i=1}^r \lambda_i \zeta_{t,x_i} (\zeta_{t,x_i})^T$, which shows that rank $M_t(y) \leq r$.
(ii) For $p \in \mathbb{R}[\mathbf{x}]_{t-d_g}$, $p^T M_{t-d_g}(gy) p$ is equal to

$$\sum_{\alpha,\beta \in \mathbb{N}^n_{t-d_g}} \sum_{\gamma \in \mathbb{N}^n} p_\alpha p_\beta g_\gamma y_{\alpha+\beta+\gamma} = \int_K g(x) p(x)^2 \mu(dx) \geq 0,$$

which shows that $M_{t-d_g}(gy) \succeq 0$.
(iii) The first two claims follow directly from (i), (ii). Assume now $\mu = \sum_{i=1}^r \lambda_i \delta_{x_i}$ where $\lambda_i > 0$ and $x_1, \ldots, x_r$ are distinct points of $\mathbb{R}^n$. One can easily verify that the vectors ζ_{x_i} $(i = 1, \ldots, r)$ are linearly independent (using, e.g., the existence of interpolation polynomials at $x_1, \ldots, x_r$; see Lemma 2.3). Then, as $M(y) = \sum_{i=1}^r \lambda_i \zeta_{x_i} \zeta_{x_i}^T$, rank $M(y) = r$. □

Note that the inclusion $\text{supp}(\mu) \subseteq V_\mathbb{R}(\text{Ker}\, M_t(y))$ from Lemma 4.2 (i) may be strict in general; see Fialkow [38] for such an example. On the other hand, we will show in Theorem 5.19 that when $M_t(y) \succeq 0$ with $\text{rank}\, M_t(y) = \text{rank}\, M_{t-1}(y)$, then equality $\text{supp}(\mu) = V_\mathbb{R}(\text{Ker}\, M_t(y))$ holds. The next result follows directly from Lemma 4.2; d_K was defined in (1.10).

COROLLARY 4.3. *If* $y \in \mathbb{R}^{\mathbb{N}^n_{2t}}$ *is the sequence of moments (up to order* $2t$*) of a measure supported by the set* K *then, for any* $t \geq d_K$,

$$M_t(y) \succeq 0, \quad M_{t-d_{g_j}}(g_j y) \succeq 0 \; (j = 1, \ldots, m). \tag{4.3}$$

We will discuss in Section 5 several results of Curto and Fialkow showing that, under certain restrictions on the rank of the matrix $M_t(y)$, the condition (4.3) is sufficient for ensuring that y is the sequence of moments of a measure supported by K. Next we indicate how the above results lead to the moment relaxations for the polynomial optimization problem.

4.2. Moment relaxations for polynomial optimization. Lasserre [65] proposed the following strategy to approximate the problem (1.1). First observe that

$$p^{\min} := \inf_{x \in K} p(x) = \inf_{\mu} \int_K p(x)\mu(dx)$$

where the second infimum is taken over all probability measures μ on $\mathbb{R}^n$ supported by the set K. Indeed, for any $x_0 \in K$, $p(x_0) = \int p(x)\mu(dx)$ for the Dirac measure $\mu := \delta_{x_0}$, showing $p^{\min} \geq \inf_\mu \int p(x)\mu(dx)$. Conversely, as $p(x) \geq p^{\min}$ for all $x \in K$, $\int_K p(x)\mu(dx) \geq \int_K p^{\min}\mu(dx) = p^{\min}$, since μ is a probability measure. Next note that $\int p(x)\mu(dx) = \sum_\alpha p_\alpha \int x^\alpha \mu(dx) = p^T y$, where $y = (\int x^\alpha \mu(dx))_\alpha$ denotes the sequence of moments of μ. Therefore, $p^{\min}$ can be reformulated as

$$p^{\min} = \inf\ p^T y \ \text{ s.t. } y_0 = 1,\ y \text{ has a representing measure on } K. \tag{4.4}$$

Following Lemma 4.2 one may instead require in (4.4) the weaker conditions $M(y) \succeq 0$ and $M(g_j y) \succeq 0\ \forall j$, which leads to the following lower bound for $p^{\min}$

$$\begin{aligned} p^{\mathrm{mom}} &:= \inf_{y \in \mathbb{R}^{\mathbb{N}^n}} p^T y \ \text{ s.t. } y_0 = 1, M(y) \succeq 0, M(g_j y) \succeq 0\ (1 \leq j \leq m) \\ &= \inf_{L \in (\mathbb{R}[\mathbf{x}])^*} L(p) \ \text{ s.t. } L(1) = 1, L(f) \geq 0\ \forall f \in \mathbf{M}(g_1, \ldots, g_m). \end{aligned} \tag{4.5}$$

The equivalence between the two formulations in (4.5) follows directly using the correspondence (4.2) between $\mathbb{R}^{\mathbb{N}^n}$ and linear functionals on $\mathbb{R}[\mathbf{x}]$, and Lemma 4.1 which implies

$$M(y) \succeq 0, M(g_j y) \succeq 0\ \forall j \leq m \iff L_y(f) \geq 0\ \forall f \in \mathbf{M}(g_1, \ldots, g_m). \tag{4.6}$$

It is not clear how to compute the bound p^{mom} since the program (4.5) involves infinite matrices. To obtain a semidefinite program we consider instead truncated moment matrices in (4.5), which leads to the following hierarchy of lower bounds for $p^{\min}$

$$\begin{aligned} p_t^{\mathrm{mom}} &= \inf_{L \in (\mathbb{R}[\mathbf{x}]_{2t})^*} L(p) \ \text{ s.t. } && L(1) = 1, \\ & && L(f) \geq 0\ \forall f \in \mathbf{M}_{2t}(g_1, \ldots, g_m) \\ &= \inf_{y \in \mathbb{R}^{\mathbb{N}^n_{2t}}} p^T y \ \text{ s.t. } && y_0 = 1,\ M_t(y) \succeq 0, \\ & && M_{t-d_{g_j}}(g_j y) \succeq 0\ (j = 1, \ldots, m) \end{aligned} \tag{4.7}$$

for $t \geq \max(d_p, d_K)$. The equivalence between the two formulations in (4.7) follows from the truncated analogue of (4.6):

$$M_t(y) \succeq 0, M_{t-d_j}(g_j y) \succeq 0\ \forall j \leq m \iff L_y(f) \geq 0\ \forall f \in \mathbf{M}_{2t}(g_1, \ldots, g_m).$$

Thus p_t^{mom} can be computed via a semidefinite program involving matrices of size $|\mathbb{N}_t^n|$. Obviously, $p_t^{\text{mom}} \leq p_{t+1}^{\text{mom}} \leq p^{\text{mom}} \leq p^{\min}$. Moreover,

$$p_t^{\text{sos}} \leq p_t^{\text{mom}}; \tag{4.8}$$

indeed if $p - \rho \in \mathbf{M}_{2t}(g_1, \ldots, g_m)$ and L is feasible for (4.7) then $L(p) - \rho = L(p - \rho) \geq 0$. Therefore, $p^{\text{sos}} \leq p^{\text{mom}}$.

LEMMA 4.4. *If the set K has a nonempty interior, then the program (4.7) is strictly feasible.*

Proof. Let μ be a measure with $\text{supp}(\mu) = B$ where B is a ball contained in K. (For instance, define μ by $\mu(A) := \lambda(A \cap B)$ for any Borel set A, where $\lambda(\cdot)$ is the Lebesgue measure on $\mathbb{R}^n$.) Let y be the sequence of moments of μ. Then, $M(g_j y) \succ 0$ for all $j = 0, \ldots, m$, setting $g_0 := 1$. Positive semidefiniteness is obvious. If $p \in \text{Ker}\, M(g_j y)$ then $\int_B p(x)^2 g_j(x) \mu(dx) = 0$, which implies $B = \text{supp}(\mu) \subseteq V_{\mathbb{R}}(g_j p)$ and thus $p = 0$. □

In the next section we discuss in more detail the duality relationship between sums of squares of polynomials and moment sequences. We will come back to both SOS/moment hierarchies and their application to the optimization problem (1.1) in Section 6.

4.3. The moment problem. The moment problem asks for the characterization of the sequences $y \in \mathbb{R}^{\mathbb{N}^n}$ having a representing measure; the analogous problem can be posed for truncated sequences $y \in \mathbb{R}^{\mathbb{N}_t^n}$ ($t \geq 1$ integer). This problem is intimately linked to the characterization of the duals of the cone $\mathcal{P}$ of nonnegative polynomials (from (3.1)) and of the cone Σ of sums of squares (from (3.2)).

Duality between sums of squares and moment sequences. For an $\mathbb{R}$-vector space A, A^* denotes its *dual* vector space consisting of all linear maps $L : A \to \mathbb{R}$. Any $a \in A$ induces an element $\Lambda_a \in (A^*)^*$ by setting $\Lambda_a(L) := L(a)$ for $L \in A^*$; hence there is a natural homomorphism from A to $(A^*)^*$, which is an isomorphism when A is finite dimensional. Given a cone $B \subseteq A$, its dual cone is $B^* := \{L \in A^* \mid L(b) \geq 0 \ \forall b \in B\}$. There is again a natural homomorphism from B to $(B^*)^*$, which is an isomorphism when A is finite dimensional and B is a closed convex cone. Here we consider $A = \mathbb{R}[\mathbf{x}]$ and the convex cones $\mathcal{P}, \Sigma \subseteq \mathbb{R}[\mathbf{x}]$, with dual cones $\mathcal{P}^* = \{L \in (\mathbb{R}[\mathbf{x}])^* \mid L(p) \geq 0 \ \forall p \in \mathcal{P}\}$, $\Sigma^* = \{L \in (\mathbb{R}[\mathbf{x}])^* \mid L(p^2) \geq 0 \ \forall p \in \mathbb{R}[\mathbf{x}]\}$. As mentioned earlier, we may identify a polynomial $p = \sum_\alpha p_\alpha \mathbf{x}^\alpha$ with its sequence of coefficients $\text{vec}(p) = (p_\alpha)_\alpha \in \mathbb{R}^\infty$, the set of sequences in $\mathbb{R}^{\mathbb{N}^n}$ with finitely many nonzero components; analogously we may identify a linear form $L \in (\mathbb{R}[\mathbf{x}])^*$ with the sequence $y := (L(\mathbf{x}^\alpha))_\alpha \in \mathbb{R}^{\mathbb{N}^n}$, so that $L = L_y$ (recall (4.2)), i.e. $L(p) = \sum_\alpha p_\alpha y_\alpha = y^T \text{vec}(p)$. In other words, we identify $\mathbb{R}[\mathbf{x}]$ with $\mathbb{R}^\infty$ via $p \mapsto \text{vec}(p)$ and $\mathbb{R}^{\mathbb{N}^n}$ with $(\mathbb{R}[\mathbf{x}])^*$ via $y \mapsto L_y$. We now describe the duals of the cones $\mathcal{P}, \Sigma \subseteq \mathbb{R}[\mathbf{x}]$. For this consider the cones in $\mathbb{R}^{\mathbb{N}^n}$

$$\mathcal{M} := \{y \in \mathbb{R}^{\mathbb{N}^n} \mid y \text{ has a representing measure}\}, \tag{4.9}$$

$$\mathcal{M}_{\succeq} := \{y \in \mathbb{R}^{\mathbb{N}^n} \mid M(y) \succeq 0\}. \tag{4.10}$$

PROPOSITION 4.5. *The cones $\mathcal{M}$ and $\mathcal{P}$ (resp., $\mathcal{M}_{\succeq}$ and Σ) are dual of each other. That is, $\mathcal{P} = \mathcal{M}^*$, $\mathcal{M}_{\succeq} = \Sigma^*$, $\mathcal{M} = \mathcal{P}^*$, $\Sigma = (\mathcal{M}_{\succeq})^*$.*

Proof. The first two equalities are easy. Indeed, if $p \in \mathcal{P}$ and $y \in \mathcal{M}$ has a representing measure μ, then $y^T\text{vec}(p) = \sum_\alpha p_\alpha y_\alpha = \sum_\alpha p_\alpha \int_K x^\alpha \mu(dx) = \int p(x)\mu(dx) \geq 0$, which shows the inclusions $\mathcal{P} \subseteq \mathcal{M}^*$ and $\mathcal{M} \subseteq \mathcal{P}^*$. The inclusion $\mathcal{M}^* \subseteq \mathcal{P}$ follows from the fact that, if $p \in \mathcal{M}^*$ then, for any $x \in \mathbb{R}^n$, $p(x) = \text{vec}(p)^T\zeta_x \geq 0$ (since $\zeta_x = (x^\alpha)_\alpha \in \mathcal{M}$ as it admits the Dirac measure δ_x as representing measure) and thus $p \in \mathcal{P}$. Given $y \in \mathbb{R}^{\mathbb{N}^n}$, $M(y) \succeq 0$ if and only if $\text{vec}(p)^T M(y)\text{vec}(p) = y^T\text{vec}(p^2) \geq 0$ for all $p \in \mathbb{R}[\mathbf{x}]$ (use Lemma 4.1), i.e. $y^T\text{vec}(f) \geq 0$ for all $f \in \Sigma$; this shows $\mathcal{M}_{\succeq} = \Sigma^*$ and thus the inclusion $\Sigma \subseteq (\mathcal{M}_{\succeq})^*$. The remaining two inclusions $\mathcal{P}^* \subseteq \mathcal{M}$ and $(\mathcal{M}_{\succeq})^* \subseteq \Sigma$ are proved, respectively, by Haviland [51] and by Berg, Christensen and Jensen [9]. □

Obviously, $\mathcal{M} \subseteq \mathcal{M}_{\succeq}$ (by Lemma 4.2) and $\Sigma \subseteq \mathcal{P}$. As we saw earlier, the inclusion $\Sigma \subseteq \mathcal{P}$ holds at equality when $n = 1$ and it is strict for $n \geq 2$. Therefore, $\mathcal{M} = \mathcal{M}_{\succeq}$ when $n = 1$ (this is Hamburger's theorem) and the inclusion $\mathcal{M} \subseteq \mathcal{M}_{\succeq}$ is strict when $n \geq 2$. There are however some classes of sequences y for which the reverse implication

$$y \in \mathcal{M}_{\succeq} \Longrightarrow y \in \mathcal{M} \tag{4.11}$$

holds. Curto and Fialkow [23] show that this is the case when the matrix $M(y)$ has finite rank.

THEOREM 4.6. [23] *If $M(y) \succeq 0$ and $M(y)$ has finite rank r, then y has a (unique) r-atomic representing measure.*

We will come back to this result in Theorem 5.1 below. This result plays in fact a crucial role in the application to polynomial optimization, since it permits to give an optimality certificate for the semidefinite hierarchy based on moment matrices; see Section 6 for details. We next discuss another class of sequences for which the implication (4.11) holds.

Bounded moment sequences. Berg, Christensen, and Ressel [10] show that the implication (4.11) holds when the sequence y is bounded, i.e., when there is a constant $C > 0$ for which $|y_\alpha| \leq C$ for all $\alpha \in \mathbb{N}^n$. More generally, Berg and Maserick [11] show that (4.11) holds when y is *exponentially bounded*[1], i.e. when $|y_\alpha| \leq C_0 C^{|\alpha|}$ for all $\alpha \in \mathbb{N}^n$, for some

[1]Our definition is equivalent to that of Berg and Maserick [11] who say that y is exponentially bounded when $|y_\alpha| \leq C_0\sigma(\alpha)$ $\forall\alpha$, for some $C_0 > 0$ and some function, called an absolute value, $\sigma : \mathbb{N}^n \to \mathbb{R}_+$ satisfying $\sigma(0) = 1$ and $\sigma(\alpha + \beta) \leq \sigma(\alpha)\sigma(\beta)$ $\forall\alpha, \beta \in \mathbb{N}^n$. Indeed, setting $C := \max_{i=1,\dots,n} \sigma(e_i)$ we have $\sigma(\alpha) \leq C^{|\alpha|}$ and, conversely, the function $\alpha \mapsto \sigma(\alpha) := C^{|\alpha|}$ is an absolute value.

constants $C_0, C > 0$. The next result shows that a sequence $y \in \mathbb{R}^{\mathbb{N}^n}$ has a representing measure supported by a compact set if and only if it is exponentially bounded with $M(y) \succeq 0$.

THEOREM 4.7. [11] *Let $y \in \mathbb{R}^{\mathbb{N}^n}$ and $C > 0$. Then y has a representing measure supported by the set $K := [-C, C]^n$ if and only if $M(y) \succeq 0$ and there is a constant $C_0 > 0$ such that $|y_\alpha| \leq C_0 C^{|\alpha|}$ for all $\alpha \in \mathbb{N}^n$.*

The proof uses the following intermediary results.

LEMMA 4.8. *Assume $M(y) \succeq 0$ and $|y_\alpha| \leq C_0 C^{|\alpha|}$ for all $\alpha \in \mathbb{N}^n$, for some constants $C_0, C > 0$. Then $|y_\alpha| \leq y_0 C^{|\alpha|}$ for all $\alpha \in \mathbb{N}^n$.*

Proof. If $y_0 = 0$ then $y = 0$ since $M(y) \succeq 0$ and the lemma holds. Assume $y_0 > 0$. Rescaling y we may assume $y_0 = 1$; we show $|y_\alpha| \leq C^{|\alpha|}$ for all α. As $M(y) \succeq 0$, we have $y_\alpha^2 \leq y_{2\alpha}$ for all α. Then, $|y_\alpha| \leq (y_{2^k\alpha})^{1/2^k}$ for any integer $k \geq 1$ (easy induction) and thus $|y_\alpha| \leq (C_0 C^{2^k|\alpha|})^{1/2^k} = C_0^{1/2^k} C^{|\alpha|}$. Letting k go to ∞, we find $|y_\alpha| \leq C^{|\alpha|}$. □

LEMMA 4.9. *Given $C > 0$ and $K = [-C, C]^n$, the set*

$$S := \{y \in \mathbb{R}^{\mathbb{N}^n} \mid y_0 = 1,\ M(y) \succeq 0,\ |y_\alpha| \leq C^{|\alpha|}\ \forall \alpha \in \mathbb{N}^n\}$$

is a convex set whose extreme points are the Zeta vectors $\zeta_x = (x^\alpha)_{\alpha \in \mathbb{N}^n}$ for $x \in K$.

Proof. S is obviously convex. Let y be an extreme point of S. Fix $\alpha_0 \in \mathbb{N}^n$. Our first goal is to show

$$y_{\alpha+\alpha_0} = y_\alpha y_{\alpha_0}\ \forall \alpha \in \mathbb{N}^n. \tag{4.12}$$

For this, define the sequence $y^{(\epsilon)} \in \mathbb{R}^{\mathbb{N}^n}$ by $y_\alpha^{(\epsilon)} := C^{|\alpha_0|} y_\alpha + \epsilon y_{\alpha+\alpha_0}$ for $\alpha \in \mathbb{N}^n$, for $\epsilon \in \{\pm 1\}$. Therefore, $|y_\alpha^{(\epsilon)}| \leq C^{|\alpha_0|}(1+\epsilon) C^{|\alpha|}\ \ \forall \alpha$. We now show that $M(y^{(\epsilon)}) \succeq 0$. Fix $p \in \mathbb{R}[\mathbf{x}]$; we have to show that

$$p^T M(y^{(\epsilon)}) p = \sum_{\gamma,\gamma'} p_\gamma p_{\gamma'} y_{\gamma+\gamma'}^{(\epsilon)} \geq 0. \tag{4.13}$$

For this, define the new sequence $z := M(y)\text{vec}(p^2) \in \mathbb{R}^{\mathbb{N}^n}$ with $z_\alpha = \sum_{\gamma,\gamma'} p_\gamma p_{\gamma'} y_{\alpha+\gamma+\gamma'}$ for $\alpha \in \mathbb{N}^n$. Then, $|z_\alpha| \leq (\sum_{\gamma,\gamma'} |p_\gamma p_{\gamma'}| C^{|\gamma|+|\gamma'|}) C^{|\alpha|}$ $\forall \alpha$. Moreover, $M(z) \succeq 0$. Indeed, using the fact that $z = M(y)\text{vec}(p^2) = gy$ for $g := p^2$ (recall (4.1)) combined with Lemma 4.1, we find that $q^T M(z) q = q^T M(gy) q = \text{vec}(pq)^T M(y) \text{vec}(pq) \geq 0$ for all $q \in \mathbb{R}[\mathbf{x}]$. Hence Lemma 4.8 implies $-z_0 C^{|\alpha|} \leq z_\alpha \leq z_0 C^{|\alpha|}\ \ \forall \alpha$; applying this to $\alpha = \alpha_0$, we get immediately relation (4.13). Therefore, $M(y^{(\epsilon)}) \succeq 0$. Applying again Lemma 4.8, we deduce that $|y_\alpha^{(\epsilon)}| \leq y_0^{(\epsilon)} C^{|\alpha|}\ \ \forall \alpha$.

If $y_0^{(\epsilon)} = 0$ for some $\epsilon \in \{\pm 1\}$, then $y^{(\epsilon)} = 0$, which implies directly (4.12). Assume now $y_0^{(\epsilon)} > 0$ for both $\epsilon = 1, -1$. Then each $\frac{y^{(\epsilon)}}{y_0^{(\epsilon)}}$ belongs to

S and $y = \frac{y_0^{(1)}}{2C^{|\alpha_0|}} \frac{y^{(1)}}{y_0^{(1)}} + \frac{y_0^{(-1)}}{2C^{|\alpha_0|}} \frac{y^{(-1)}}{y_0^{(-1)}}$ is a convex combination of two points of S. As y is an extreme point of S, $y \in \left\{ \frac{y^{(1)}}{y_0^{(1)}}, \frac{y^{(-1)}}{y_0^{(-1)}} \right\}$, which implies again (4.12).

As relation (4.12) holds for all $\alpha_0 \in \mathbb{N}^n$, setting $x := (y_{e_i})_{i=1}^n$, we find that $x \in K$ and $y_\alpha = x^\alpha$ for all α, i.e. $y = \zeta_x$. □

Proof of Theorem 4.7. Assume that $M(y) \succeq 0$ and $|y_\alpha| \leq C_0 C^{|\alpha|}$ for all α; we show that y has a representing measure supported by K. By Lemma 4.8, $|y_\alpha| \leq y_0 C^{|\alpha|}\ \forall \alpha$. If $y_0 = 0$, $y = 0$ and we are done. Assume $y_0 = 1$ (else rescale y). Then y belongs to the convex set S introduced in Lemma 4.9. By the Krein-Milman theorem, y is a convex combination of extreme points of S. That is, $y = \sum_{j=1}^m \lambda_j \zeta_{x_j}$ where $\lambda_j > 0$ and $x_j \in K$. In other words, $\mu := \sum_{j=1}^m \lambda_j \delta_{x_j}$ is a representing measure for y supported by K. Conversely, assume that y has a representing measure μ supported by K. Set $C := \max(|x_i| \mid x \in K, i = 1, \ldots, n)$. Then $|y_\alpha| \leq \int_K |x^\alpha| \mu(dx) \leq \max_{x \in K} |x^\alpha| \mu(K) \leq \mu(K) C^{|\alpha|}$, which concludes the proof of Theorem 4.7. □

4.4. The K-moment problem. We now consider the K-moment problem where, as in (1.2), $K = \{x \in \mathbb{R}^n \mid g_1(x) \geq 0, \ldots, g_m(x) \geq 0\}$ is a semialgebraic set. Define the cones

$$\mathcal{M}_K := \{y \in \mathbb{R}^{\mathbb{N}^n} \mid y \text{ has a representing measure supported by } K\} \quad (4.14)$$

$$\mathcal{M}_{\succeq}^{sch}(g_1, \ldots, g_m) := \{y \in \mathbb{R}^{\mathbb{N}^n} \mid M(g_J y) \succeq 0\ \forall J \subseteq \{1, \ldots, m\}\}, \quad (4.15)$$

$$\mathcal{M}_{\succeq}^{put}(g_1, \ldots, g_m) := \{y \in \mathbb{R}^{\mathbb{N}^n} \mid M(y) \succeq 0,\ M(g_j y) \succeq 0\ (1 \leq j \leq m)\}, \quad (4.16)$$

setting $g_\emptyset := 1$, $g_J := \prod_{j \in J} g_j$ for $J \subseteq \{1, \ldots, m\}$. (The indices 'sch' and 'put' refer respectively to Schmüdgen and to Putinar; see Theorems 4.10 and 4.11 below.) Define also the cone

$$\mathcal{P}_K := \{p \in \mathbb{R}[\mathbf{x}] \mid p(x) \geq 0\ \forall x \in K\}$$

and recall the definition of $T(g_1, \ldots, g_m)$ from (3.13) and $\mathbf{M}(g_1, \ldots, g_m)$ from (3.14). Obviously,

$$\mathcal{M}_K \subseteq \mathcal{M}_{\succeq}^{sch}(g_1, \ldots, g_m) \subseteq \mathcal{M}_{\succeq}^{put}(g_1, \ldots, g_m),$$
$$\mathbf{M}(g_1, \ldots, g_m) \subseteq T(g_1, \ldots, g_m) \subseteq \mathcal{P}_K.$$

One can verify that

$$\mathcal{P}_K = (\mathcal{M}_K)^*, \quad \mathcal{M}_{\succeq}^{sch}(g_1, \ldots, g_m) = (T(g_1, \ldots, g_m))^*,$$
$$\mathcal{M}_{\succeq}^{put}(g_1, \ldots, g_m) = (\mathbf{M}(g_1, \ldots, g_m))^*$$

(the details are analogous to those for Proposition 4.5, using Lemma 4.1). Moreover, $\mathcal{M}_K = (\mathcal{P}_K)^*$ (Haviland [51]). The following results give the counterparts of Theorems 3.16 and 3.20, respectively, for the 'moment side'. See e.g. [114] for a detailed treatment and background.

THEOREM 4.10. *(Schmüdgen [126]) If K is compact, then $\mathcal{M}_K = \mathcal{M}_{\succeq}^{sch}(g_1, \ldots, g_m)$. Moreover, every positive polynomial p on K (i.e. $p > 0$ on K) belongs to $T(g_1, \ldots, g_m)$.*

THEOREM 4.11. *(Putinar [115]) Assume $\mathbf{M}(g_1, \ldots, g_m)$ is Archimedean. Then $\mathcal{M}_K = \mathcal{M}_{\succeq}^{put}(g_1, \ldots, g_m)$. Moreover, every positive polynomial on K belongs to $\mathbf{M}(g_1, \ldots, g_m)$.*

Let us conclude this section with a few words about the proof technique for Theorems 4.6, 4.10, 4.11. Assume $y \in \mathbb{R}^{\mathbb{N}^n}$ satisfies $M(y) \succeq 0$ and let L_y be the corresponding linear map as in (4.2). The assumption $M(y) \succeq 0$ means that L_y is nonnegative on the cone Σ. The kernel of $M(y)$ can be identified with the set $\mathcal{I} := \{p \in \mathbb{R}[\mathbf{x}] \mid M(y)p = 0\}$ which is an ideal in $\mathbb{R}[\mathbf{x}]$ (see Lemma 5.2) and the quotient space $A := \mathbb{R}[\mathbf{x}]/\mathcal{I}$ has the structure of an algebra. One can define an inner product on A by setting $\langle p, q \rangle := p^T M(y) q = L_y(pq)$. In this way, A is a Hilbert space. For $i = 1, \ldots, n$, consider the multiplication operator $m_{\mathbf{x}_i} : A \longrightarrow A$ defined by $m_{\mathbf{x}_i}(p) = \mathbf{x}_i p \mod \mathcal{I}$. Obviously, the operators $m_{\mathbf{x}_1}, \ldots, m_{\mathbf{x}_n}$ commute pairwise. (See also Section 2.4.)

Under the assumption of Theorem 4.6, $M(y)$ has a finite rank and thus the Hilbert space A has a finite dimension. Curto and Fialkow [23] use the spectral theorem and the Riesz representation theorem for proving the existence of a representing measure for y. Consider now the case when the assumptions of Schmüdgen's theorem hold; that is, the operator L_y is nonnegative on the cone $T(g_1, \ldots, g_m)$. As K is compact, there exists $\rho > 0$ for which the polynomial $\rho^2 - \sum_{i=1}^n \mathbf{x}_i^2$ is positive on K. Using the Positivstellensatz, this implies that $(\rho^2 - \sum_{i=1}^n \mathbf{x}_i^2)g = 1 + h$ for some $g, h \in T(g_1, \ldots, g_m)$. Then, the main step in Schmüdgen's proof consist of showing that the operators $m_{\mathbf{x}_i}$ are bounded; namely, $\langle \mathbf{x}_i p, \mathbf{x}_i p \rangle \leq \rho^2 \langle p, p \rangle$ for all $p \in \mathbb{R}[\mathbf{x}]$. Then the existence of a representing measure μ for y follows using the spectral theorem and Schmüdgen uses Weierstrass theorem for proving that the support of μ is contained in K. This proof uses in an essential way functional analytic methods. Schweighofer [130] gives an alternative more elementary proof for Schmüdgen's theorem, which uses only the Positivstellensatz and Pólya's theorem (Theorem 3.21); moreover, starting from a certificate: $(\rho^2 - \sum_{i=1}^n \mathbf{x}_i^2)g = 1 + h$ with $g, h \in T(g_1, \ldots, g_m)$ given by the Positivstellensatz, he constructs explicitly a representation of a positive polynomial on K proving its membership in $T(g_1, \ldots, g_m)$. Recently Schmüdgen [127] gives another proof for the 'sum of squares' part of his theorem; after proving that the preordering $T(g_1, \ldots, g_m)$ is Archimedean (using Stengle's Positivstellensataz), his proof is short and quite elementary (it uses the one-dimensional Hamburger moment problem and the approximation of the square root function by polynomials). Schweighofer [132] also gives an alternative elementary proof for Putinar's theorem relying on

Pólya's theorem; we have presented a proof for the 'sum of squares' part of Putinar's theorem in Section 3.7. We will give in Section 5.1 an alternative elementary proof for Theorem 4.6, based on the fact that $\operatorname{Ker} M(y)$ is a real radical ideal and using the Real Nullstellensatz.

5. More about moment matrices.

We group here several results about moment matrices, mostly from Curto and Fialkow [23–25], which will have important applications to the optimization problem (1.1).

5.1. Finite rank moment matrices.

We have seen in Lemma 4.2 (iii) that, if a sequence $y \in \mathbb{R}^{\mathbb{N}^n}$ has a r-atomic representing measure, then its moment matrix $M(y)$ is positive semidefinite and its rank is equal to r. Curto and Fialkow [23] show that the reverse implication holds. More precisely, they show the following result, thus implying Theorem 4.6.

THEOREM 5.1. [23] *Let* $y \in \mathbb{R}^{\mathbb{N}^n}$.

(i) *If* $M(y) \succeq 0$ *and* $M(y)$ *has finite rank* r, *then* y *has a unique representing measure* μ. *Moreover,* μ *is* r*-atomic and* $\operatorname{supp}(\mu) = V_{\mathbb{C}}(\operatorname{Ker} M(y))$ $(\subseteq \mathbb{R}^n)$.

(ii) *If* y *has a* r*-atomic representing measure, then* $M(y) \succeq 0$ *and* $M(y)$ *has rank* r.

Assertion (ii) is just Lemma 4.2 (iii). We now give a simple proof for Theorem 5.1 (i) (taken from [80]), which uses an algebraic tool (the Real Nullstellensatz) in place of the tools from functional analysis (the spectral theorem and the Riesz representation theorem) used in the original proof of Curto and Fialkow [23].

Recall that one says that 'a polynomial f lies in the kernel of $M(y)$' when $M(y)f := M(y)\operatorname{vec}(f) = 0$, which permits to identify the kernel of $M(y)$ with a subset of $\mathbb{R}[\mathbf{x}]$. Making this identification enables us to claim that 'the kernel of $M(y)$ is an ideal in $\mathbb{R}[\mathbf{x}]$' (as observed by Curto and Fialkow [23]) or, when $M(y) \succeq 0$, that 'the kernel is a radical ideal' (as observed by Laurent [80]) or even 'a real radical ideal' (as observed by Möller [91], or Scheiderer [124]). Moreover, linearly independent sets of columns of $M(y)$ correspond to linearly independent sets in the quotient vector space $\mathbb{R}[\mathbf{x}]/\operatorname{Ker} M(y)$. These properties, which play a crucial role in the proof, are reported in the next two lemmas.

LEMMA 5.2. *The kernel* $\mathcal{I} := \{p \in \mathbb{R}[\mathbf{x}] \mid M(y)p = 0\}$ *of a moment matrix* $M(y)$ *is an ideal in* $\mathbb{R}[\mathbf{x}]$. *Moreover, if* $M(y) \succeq 0$, *then* $\mathcal{I}$ *is a real radical ideal.*

Proof. We apply Lemma 4.1. Assume $f \in \mathcal{I}$ and let $g \in \mathbb{R}[\mathbf{x}]$. For any $h \in \mathbb{R}[\mathbf{x}]$, $\operatorname{vec}(h)^T M(y)\operatorname{vec}(fg) = \operatorname{vec}(hg)^T M(y)\operatorname{vec}(f) = 0$, implying that $fg \in \mathcal{I}$. Assume now $M(y) \succeq 0$. We show that $\mathcal{I}$ is real radical. In view of Lemma 2.2, it suffices to show that, for any $g_1, \ldots, g_m \in \mathbb{R}[\mathbf{x}]$,

$$\sum_{j=1}^{m} g_j^2 \in \mathcal{I} \Longrightarrow g_1, \ldots, g_m \in \mathcal{I}.$$

Indeed, if $\sum_{j=1}^m g_j^2 \in \mathcal{I}$ then $0 = \text{vec}(1)^T M(y) \text{vec}(\sum_{j=1}^m g_j^2) = \sum_{j=1}^m g_j^T M(y) g_j$. As $g_j^T M(y) g_j \geq 0$ since $M(y) \succeq 0$, this implies $0 = g_j^T M(y) g_j$ and thus $g_j \in \mathcal{I}$ for all j. □

LEMMA 5.3. *For $\mathcal{B} \subseteq \mathbb{T}_n$, $\mathcal{B}$ indexes a maximum linearly independent set of columns of $M(y)$ if and only if $\mathcal{B}$ is a basis of the quotient vector space $\mathbb{R}[\mathbf{x}]/\operatorname{Ker} M(y)$.*

Proof. Immediate verification. □

Proof of Theorem 5.1(i). Assume $M(y) \succeq 0$ and $r := \text{rank } M(y) < \infty$. By Lemmas 5.2 and 5.3, the set $\mathcal{I} := \operatorname{Ker} M(y)$ is a real radical zero-dimensional ideal in $\mathbb{R}[\mathbf{x}]$. Hence, using (2.1) and Theorem 2.6, $V_{\mathbb{C}}(\mathcal{I}) \subseteq \mathbb{R}^n$ and $|V_{\mathbb{C}}(\mathcal{I})| = \dim \mathbb{R}[\mathbf{x}]/\mathcal{I} = r$. Let $p_v \in \mathbb{R}[\mathbf{x}]$ $(v \in V_{\mathbb{C}}(\mathcal{I}))$ be interpolation polynomials at the points of $V_{\mathbb{C}}(\mathcal{I})$. Setting $\lambda_v := p_v^T M(y) p_v$, we now claim that the measure $\mu := \sum_{v \in V_{\mathbb{C}}(\mathcal{I})} \lambda_v \delta_v$ is the unique representing measure for y.

LEMMA 5.4. $M(y) = \sum_{v \in V_{\mathbb{C}}(\mathcal{I})} \lambda_v \zeta_v \zeta_v^T$.

Proof. Set $N := \sum_{v \in V_{\mathbb{C}}(\mathcal{I})} \lambda_v \zeta_v \zeta_v^T$. We first show that $p_u^T M(y) p_v = p_u^T N p_v$ for all $u, v \in V_{\mathbb{C}}(\mathcal{I})$. This identity is obvious if $u = v$. If $u \neq v$ then $p_u^T N p_v = 0$; on the other hand, $p_u^T M(y) p_v = \text{vec}(1)^T M(y) \text{vec}(p_u p_v) = 0$ where we use Lemma 4.1 for the first equality and the fact that $p_u p_v \in \mathcal{I}(V_{\mathbb{C}}(\mathcal{I})) = \mathcal{I}$ for the second equality. As the set $\{p_v \mid v \in V_{\mathbb{C}}(\mathcal{I})\}$ is a basis of $\mathbb{R}[\mathbf{x}]/\mathcal{I}$ (by Lemma 2.5), we deduce that $f^T M(y) g = f^T N g$ for all $f, g \in \mathbb{R}[\mathbf{x}]$, implying $M(y) = N$. □

LEMMA 5.5. *The measure $\mu = \sum_{v \in V_{\mathbb{C}}(\mathcal{I})} \lambda_v \delta_v$ is r-atomic and it is the unique representing measure for y.*

Proof. μ is a representing measure for y by Lemma 5.4 and μ is r-atomic since $p_v^T M(y) p_v > 0$ as $p_v \notin \mathcal{I}$ for $v \in V_{\mathbb{C}}(\mathcal{I})$. We now verify the unicity of such measure. Say, μ' is another representing measure for y. By Lemma 4.2, $r = \text{rank } M(y) \leq r' := |\text{supp}(\mu')|$; moreover, $\text{supp}(\mu') \subseteq V_{\mathbb{C}}(\mathcal{I})$, implying $r' \leq |V_{\mathbb{C}}(\mathcal{I})| = r$. Thus, $r = r'$, $\text{supp}(\mu') = \text{supp}(\mu) = V_{\mathbb{C}}(\mathcal{I})$ and $\mu = \mu'$. □

This concludes the proof of Theorem 5.1. □

We now make an observation, which will be useful for the proof of Theorem 5.23 below.

LEMMA 5.6. *Assume $M(y) \succeq 0$ and $r := \text{rank } M(y) < \infty$. Set $\mathcal{I} := \operatorname{Ker} M(y)$. If, for some integer $t \geq 1$, $\text{rank } M_t(y) = r$, then there exist interpolation polynomials p_v $(v \in V_{\mathbb{C}}(\mathcal{I}))$ having degree at most t.*

Proof. As $\text{rank } M_t(y) = \text{rank } M(y)$, one can choose a basis $\mathcal{B}$ of $\mathbb{R}[\mathbf{x}]/\mathcal{I}$ where $\mathcal{B} \subseteq \mathbb{T}_t^n$. (Recall Lemma 5.3.) Let q_v $(v \in V_{\mathbb{C}}(\mathcal{I}))$ be interpolation polynomials at the points of $V_{\mathbb{C}}(\mathcal{I})$. Replacing each q_v by its residue p_v modulo $\mathcal{I}$ w.r.t. the basis $\mathcal{B}$, we obtain a new set of interpolation polynomials p_v ($v \in V_{\mathbb{C}}(\mathcal{I})$) with $\deg p_v \leq t$. □

We saw in Lemma 5.2 that the kernel of an infinite moment matrix is an ideal in $\mathbb{R}[\mathbf{x}]$. We now observe that, although the kernel of a *truncated* moment matrix cannot be claimed to be an ideal, it yet enjoys some 'truncated ideal like' properties. We use the notion of *flat extension* of a matrix, introduced earlier in Definition 1.1, as well as Lemma 1.2.

LEMMA 5.7. *Let* $f, g \in \mathbb{R}[\mathbf{x}]$.
(i) *If* $\deg(fg) \leq t-1$ *and* $M_t(y) \succeq 0$, *then*

$$f \in \operatorname{Ker} M_t(y) \Longrightarrow fg \in \operatorname{Ker} M_t(y). \tag{5.1}$$

(ii) *If* $\deg(fg) \leq t$ *and* $\operatorname{rank} M_t(y) = \operatorname{rank} M_{t-1}(y)$, *then (5.1) holds.*

Proof. It suffices to show the result for $g = \mathbf{x}_i$ since the general result follows from repeated applications of this special case. Then, $h := f\mathbf{x}_i = \sum_\alpha f_\alpha \mathbf{x}^{\alpha+e_i} = \sum_{\alpha|\alpha\geq e_i} f_{\alpha-e_i}\mathbf{x}^\alpha$. For $\alpha \in \mathbb{N}^n_{t-1}$, we have:

$$\begin{aligned}(M_t(y)h)_\alpha &= \sum_\gamma h_\gamma y_{\alpha+\gamma} = \sum_{\gamma|\gamma\geq e_i} f_{\gamma-e_i} y_{\alpha+\gamma}\\ &= \sum_\gamma f_\gamma y_{\alpha+\gamma+e_i} = (M_t(y)f)_{\alpha+e_i} = 0.\end{aligned}$$

In view of Lemma 1.2, this implies $M_t(y)h = 0$ in both cases (i), (ii). □

5.2. Finite atomic measures for truncated moment sequences. Theorem 5.1 characterizes the infinite sequences having a finite atomic representing measure. The next question is to characterize the *truncated* sequences $y \in \mathbb{R}^{\mathbb{N}^n_{2t}}$ having a finite atomic representing measure μ. It turns out that, for a truncated sequence, the existence of a representing measure implies the existence of another one with a *finite* support (this is not true for infinite sequences). This result, due to Bayer and Teichmann [7], strengthens an earlier result of Putinar [116] which assumed the existence of a measure with a compact support. We thank M. Schweighofer for suggestions about Theorem 5.8 and its proof.

THEOREM 5.8. [7] *If a truncated sequence* $y \in \mathbb{R}^{\mathbb{N}^n_t}$ *has a representing measure* μ, *then it has another representing measure* ν *which is finitely atomic with at most* $\binom{n+t}{t}$ *atoms. Moreover, if* $S \subseteq \mathbb{R}^n$ *is measurable with* $\mu(\mathbb{R}^n \setminus S) = 0$, *then one can choose* ν *such that* $\operatorname{supp}(\nu) \subseteq S$. *In particular, one can choose* ν *with* $\operatorname{supp}(\nu) \subseteq \operatorname{supp}(\mu)$.

Proof. Let $S \subseteq \mathbb{R}^n$ be measurable with $\mu(\mathbb{R}^n \setminus S) = 0$. Let $\mathcal{I} \subseteq \mathbb{R}[\mathbf{x}]$ denote an ideal that is maximal with respect to the property that $\mu(\mathbb{R}^n \setminus (V_{\mathbb{R}}(\mathcal{I}) \cap S)) = 0$ (such an ideal exists by assumption since $\mathbb{R}[\mathbf{x}]$ is Noetherian). Set $S' := V_{\mathbb{R}}(\mathcal{I}) \cap S$; thus $\mu(\mathbb{R}^n \setminus S') = 0$. We will in fact construct ν with $\operatorname{supp}(\nu) \subseteq S'$. For this, let $C \subseteq \mathbb{R}^{\mathbb{N}^n_t}$ denote the convex cone generated by the vectors $\zeta_{t,x} = (x^\alpha)_{\alpha\in\mathbb{N}^n_t}$ for $x \in S'$. Then its closure $\overline{\mathcal{C}}$ is a closed convex cone in $\mathbb{R}^{\mathbb{N}^n_t}$ and therefore it is equal to the intersection of its supporting halfspaces. That is,

$$\overline{\mathcal{C}} = \{z \in \mathbb{R}^{\mathbb{N}^n_t} \mid c^T z \geq 0 \ \forall c \in H\}$$

for some $H \subseteq \mathbb{R}^{\mathbb{N}^n_t}$. Obviously, $y \in \overline{\mathcal{C}}$ since, for any $c \in H$,

$$c^T y = \sum_\alpha c_\alpha y_\alpha = \int_{S'} (\sum_\alpha c_\alpha x^\alpha)\mu(dx) \geq 0$$

as $\sum_\alpha c_\alpha x^\alpha = c^T \zeta_{t,x} \geq 0$ for all $x \in S'$. Moreover,

$$y \text{ belongs to the relative interior of } \overline{\mathcal{C}}. \tag{5.2}$$

Indeed, consider a supporting hyperplane $\{z \mid c^T z = 0\}$ ($c \in H$) that does not contain $\overline{\mathcal{C}}$. We show that $c^T y > 0$.

For this, assume $c^T y = 0$ and set $X := \{x \in S' \mid c^T \zeta_{t,x} > 0\}$, $X_k := \{x \in S' \mid c^T \zeta_{t,x} \geq \frac{1}{k}\}$ for $k \geq 1$ integer. Then, $X \neq \emptyset$ and $X = \bigcup_{k\geq 1} X_k$. We have

$$0 = c^T y = \int_X c^T \zeta_{t,x}\mu(dx) \geq \int_{X_k} c^T \zeta_{t,x}\mu(dx) \geq \frac{1}{k}\mu(X_k) \geq 0,$$

implying $\mu(X_k) = 0$. This shows that $\mu(X) = 0$. Now consider the polynomial $f := \sum_\alpha c_\alpha \mathbf{x}^\alpha \in \mathbb{R}[\mathbf{x}]_t$ and the ideal $\mathcal{J} := \mathcal{I} + (f) \subseteq \mathbb{R}[\mathbf{x}]$. Then, $V_\mathbb{R}(\mathcal{J}) = V_\mathbb{R}(\mathcal{I}) \cap V_\mathbb{R}(f)$, $V_\mathbb{R}(\mathcal{J}) \cap S = S' \cap V_\mathbb{R}(f)$, $X = S' \setminus V_\mathbb{R}(f)$ and thus $\mathbb{R}^n \setminus (V_\mathbb{R}(\mathcal{J}) \cap S) = (\mathbb{R}^n \setminus S') \cup X$ has measure 0 since $\mu(\mathbb{R}^n \setminus S') = \mu(X) = 0$. This implies $\mathcal{J} = \mathcal{I}$ by our maximality assumption on $\mathcal{I}$, i.e., $f \in \mathcal{I}$. Hence f vanishes on $V_\mathbb{R}(\mathcal{I})$ and thus on S', contradicting the fact that $X \neq \emptyset$.

Therefore, (5.2) holds and thus y belongs to the cone $\mathcal{C}$, since the two cones $\mathcal{C}$ and its closure $\overline{\mathcal{C}}$ have the same relative interior. Using Carathéodory's theorem, we deduce that y can be written as a conic combination of at most $|\mathbb{N}^n_t| = \binom{n+t}{t}$ vectors $\zeta_{t,x}$ ($x \in S'$); that is, y has an atomic representing measure on $S' \subseteq S$ with at most $\binom{n+t}{t}$ atoms. □

As an illustration, consider e.g. the case $n = t = 1$ and the uniform measure μ on $[-1, 1]$ with $y = (2, 0) \in \mathbb{R}^{\mathbb{N}^1_1}$. Theorem 5.8 tells us that there is another representing measure ν for y with at most two atoms. Indeed, the Dirac measure at the origin represents y, but if we exclude the origin then we need two atoms to represent y. Finding alternative measures with a small number of atoms is also known as the problem of finding cubature (or quadrature) rules for measures. The next result is a direct consequence of Theorem 5.8.

COROLLARY 5.9. *For $K \subseteq \mathbb{R}^n$ and $y \in \mathbb{R}^{\mathbb{N}^n_t}$, the following assertions (i)-(iii) are equivalent: (i) y has a representing measure on K; (ii) y has an* *atomic representing measure on* K*; (iii)* $y = \sum_{i=1}^{N} \lambda_i \delta_{x_i}$ *for some* $\lambda_i > 0$, $x_i \in K$.

We mentioned earlier in Section 4.3 the Riesz-Haviland theorem which claims that an infinite sequence $y \in \mathbb{R}^{\mathbb{N}^n}$ has a representing measure on a closed subset $K \subseteq \mathbb{R}^n$ if and only if $y^T p \geq 0$ for all polynomials p nonnegative on K; that is, $\mathcal{M}_K = (\mathcal{P}_K)^*$ in terms of conic duality. One may naturally wonder whether there is an analogue of this result for the truncated moment problem. For this, define

$$\mathcal{P}_{K,t} := \{p \in \mathbb{R}[\mathbf{x}]_t \mid p(x) \geq 0 \ \forall x \in K\}.$$

Obviously, $\mathcal{M}_{K,t} \subseteq (\mathcal{P}_{K,t})^*$; Tchakaloff [145] proved that equality holds when K is compact.

Here is an example (taken from [26]) showing that the inclusion $\mathcal{M}_{K,t} \subseteq (\mathcal{P}_{K,t})^*$ can be strict.

EXAMPLE 5.10. Consider the sequence $y := (1, 1, 1, 1, 2) \in \mathbb{R}^{\mathbb{N}^1_4}$ (here $n = 1$). Thus,

$$M_2(y) = \begin{pmatrix} 1 & 1 & 1 \\ 1 & 1 & 1 \\ 1 & 1 & 2 \end{pmatrix} \succeq 0.$$

Hence $y \in (\mathcal{P}_4)^*$ (since any univariate nonnegative polynomial is a sum of squares). However y does not have a representing measure. Indeed, if μ is a representing measure for y, then its support is contained in $V_{\mathbb{C}}(\operatorname{Ker} M_2(y)) \subseteq \{1\}$ since the polynomial $1 - x$ lies in $\operatorname{Ker} M_2(y)$. But then μ would be the Dirac measure δ_1 which would imply $y_4 = 1$, a contradiction.

Curto and Fialkow [26] can however prove the following results. We omit the proofs which use the Riesz representation theorem and a technical result of [143] about limits of measures.

THEOREM 5.11. [26, Th. 2.4] *Let* $y \in \mathbb{R}^{\mathbb{N}^n_{2t}}$ *and let* K *be a closed subset of* $\mathbb{R}^n$*. If* $y \in (\mathcal{P}_{K,2t})^*$*, then the subsequence* $(y_\alpha)_{\alpha \in \mathbb{N}^n_{2t-1}}$ *has a representing measure on* K*.*

THEOREM 5.12. [26, Th. 2.2] *Let* $y \in \mathbb{R}^{\mathbb{N}^n_{2t}}$ *and let* K *be a closed subset of* $\mathbb{R}^n$*. Then* y *has a representing measure on* K *if and only if* y *admits an extension* $\tilde{y} \in \mathbb{R}^{\mathbb{N}^n_{2t+2}}$ *such that* $\tilde{y} \in (\mathcal{P}_{K,2t+2})^*$*.*

Note that Theorem 5.12 implies in fact the Riesz-Haviland theorem $\mathcal{M}_K = (\mathcal{P}_K)^*$; to see it, use the following result of Stochel, which shows that the truncated moment problem is in fact more general than the (infinite) moment problem.

THEOREM 5.13. [143, Th. 4] *Let* $y \in \mathbb{R}^{\mathbb{N}^n}$ *and let* $K \subseteq \mathbb{R}^n$ *be a closed set. Then* y *has a representing measure on* K *if and only if, for each integer* $t \geq 1$*, the subsequence* $(y_\alpha)_{\alpha \in \mathbb{N}^n_t}$ *has a representing measure on* K*.*

5.3. Flat extensions of moment matrices. The main result in this section is Theorem 5.14 below, which provides a key result about flat extensions of moment matrices. Indeed it permits to extend a truncated sequence $y \in \mathbb{R}^{\mathbb{N}^n_{2t}}$, satisfying some 'flat extension' assumption, to an infinite sequence $\tilde{y} \in \mathbb{R}^{\mathbb{N}^n}$, satisfying $\operatorname{rank} M(\tilde{y}) = \operatorname{rank} M_t(y)$. In this way one can then apply the tools developed for infinite moment sequences (e.g., Theorem 5.1) to truncated sequences. Recall the notion of 'flat extension' from Definition 1.1.

THEOREM 5.14. (**Flat extension theorem** [23]) *Let $y \in \mathbb{R}^{\mathbb{N}^n_{2t}}$. If $M_t(y)$ is a flat extension of $M_{t-1}(y)$, then one can extend y to a (unique) vector $\tilde{y} \in \mathbb{R}^{\mathbb{N}^n_{2t+2}}$ in such a way that $M_{t+1}(\tilde{y})$ is a flat extension of $M_t(y)$.*

The rest of the section is devoted to the proof of Theorem 5.14. We give in fact two proofs. While the first one is completely elementary with some more technical details, the second one is less technical but uses some properties of Gröbner bases.

First proof. We begin with a characterization of moment matrices, which we will use in the proof.

LEMMA 5.15. *Let M be a symmetric matrix indexed by $\mathbb{N}^n_t$. Then, M is a moment matrix, i.e., $M = M_t(y)$ for some sequence $y \in \mathbb{R}^{\mathbb{N}^n_{2t}}$, if and only if the following holds:*

(i) *$M_{\alpha,\beta} = M_{\alpha-e_i,\beta+e_i}$ for all $\alpha, \beta \in \mathbb{N}^n_t$, $i \in \{1, \ldots, n\}$ such that $\alpha_i \geq 1$, $|\beta| \leq t-1$.*

(ii) *$M_{\alpha,\beta} = M_{\alpha-e_i+e_j,\beta+e_i-e_j}$ for all $\alpha, \beta \in \mathbb{N}^n_t$, $i, j \in \{1, \ldots, n\}$ such that $\alpha_i, \beta_j \geq 1$, $|\alpha| = |\beta| = t$.*

Proof. The 'if part' being obvious, we show the 'only if' part. That is, we assume that (i), (ii) hold and we show that $M(\alpha, \beta) = M(\alpha', \beta')$ whenever $\alpha + \beta = \alpha' + \beta'$. For this we use induction on the parameter $\delta_{\alpha\beta,\alpha'\beta'} := \min(\|\alpha - \alpha'\|_1, \|\beta - \beta'\|_1)$. If $\delta_{\alpha\beta,\alpha'\beta'} = 0$, then $(\alpha, \beta) = (\alpha', \beta')$ and there is nothing to prove. If $\delta_{\alpha\beta,\alpha'\beta'} = 1$, then the result holds by assumption (i). Assume now that $\delta_{\alpha\beta,\alpha'\beta'} \geq 2$.

Consider first the case when $|\alpha| + |\beta| \leq 2t - 1$. As $\alpha \neq \alpha'$ we may assume without loss of generality that $\alpha'_i \geq \alpha_i + 1$ for some i, implying $\beta'_i \leq \beta_i - 1$. Define $(\alpha'', \beta'') := (\alpha - e_i, \beta + e_i)$. Then, $\delta_{\alpha\beta,\alpha''\beta''} = \delta_{\alpha\beta,\alpha'\beta'} - 1$. If $|\beta'| \leq t - 1$, then $M_{\alpha,\beta} = M_{\alpha'',\beta''}$ by the induction assumption and $M_{\alpha'',\beta''} = M_{\alpha',\beta'}$ by (i), implying the desired result. Assume now that $|\beta'| = t$ and thus $|\alpha'| \leq t - 1$. Then, $|\alpha| - |\alpha'| = t - |\beta| \geq 0$ and thus $\alpha_i \geq \alpha'_i + 1$ for some i, yielding $\beta'_i \geq \beta_i + 1$. Define $(\alpha'', \beta'') := (\alpha' + e_i, \beta' - e_i)$. Then $\delta_{\alpha\beta,\alpha''\beta''} = \delta_{\alpha\beta,\alpha'\beta'} - 1$. Therefore, $M_{\alpha,\beta} = M_{\alpha'',\beta''}$ by the induction assumption and $M_{\alpha'',\beta''} = M_{\alpha',\beta'}$ by (i), implying the desired result.

We can now suppose that $|\alpha| = |\beta| = |\alpha'| = |\beta'| = t$. Hence, $\alpha'_i \geq \alpha_i + 1$ for some i and $\beta'_j \geq \beta_j + 1$ for some j; moreover $i \neq j$. Define $(\alpha'', \beta'') := (\alpha' - e_i + e_j, \beta' + e_i - e_j)$. Then, $\delta_{\alpha\beta,\alpha''\beta''} = \delta_{\alpha\beta,\alpha'\beta'} - 2$. Therefore, $M_{\alpha,\beta} = M_{\alpha'',\beta''}$ by the induction assumption and $M_{\alpha'',\beta''} = M_{\alpha',\beta'}$ by (ii), implying the desired result. □

Set $M_t := M_t(y) = \begin{pmatrix} A & B \\ B^T & C \end{pmatrix}$, where $A := M_{t-1}(y)$. By assumption, M_t is a flat extension of A. Our objective is to construct a flat extension $N := \begin{pmatrix} M_t & D \\ D^T & E \end{pmatrix}$ of M_t, which is a moment matrix. As M_t is a flat extension of A, we can choose a subset $\mathcal{B} \subseteq \mathbb{N}^n_{t-1}$ indexing a maximum set of linearly independent columns of M_t. Then any column of M_t can be expressed (in a unique way) as a linear combination of columns indexed by $\mathcal{B}$. In other words, for any polynomial $p \in \mathbb{R}[\mathbf{x}]_t$, there exists a unique polynomial $r \in \mathrm{Span}_{\mathbb{R}}(\mathcal{B})$ for which $p - r \in \mathrm{Ker}\, M_t$.

Lemma 5.7 (ii) plays a central role in the construction of the matrix N, i.e., of the matrices D and E. Take $\gamma \in \mathbb{N}^n_{t+1}$ with $|\gamma| = t+1$. Say, $\gamma_i \geq 1$ for some $i = 1, \ldots, n$ and $\mathbf{x}^{\gamma - e_i} - r \in \mathrm{Ker}\, M_t$, where $r \in \mathrm{Span}_{\mathbb{R}}(\mathcal{B})$. Then it follows from Lemma 5.7 (ii) that $\mathbf{x}_i(\mathbf{x}^{\gamma-e_i} - r)$ belongs to the kernel of N, the desired flat extension of M_t. In other words, $N\mathrm{vec}(\mathbf{x}^\gamma) = N\mathrm{vec}(\mathbf{x}_i r)$, which tells us how to define the γth column of N, namely, by $D\mathrm{vec}(\mathbf{x}^\gamma) = M_t\mathrm{vec}(\mathbf{x}_i r)$ and $E\mathrm{vec}(\mathbf{x}^\gamma) = D^T\mathrm{vec}(\mathbf{x}_i r)$. We now verify that these definitions are *good*, i.e., that they do not depend on the choice of the index i for which $\gamma_i \geq 1$.

LEMMA 5.16. *Let $\gamma \in \mathbb{N}^n$ with $|\gamma| = t+1$, $\gamma_i, \gamma_j \geq 1$ and let $r, s \in \mathrm{Span}_{\mathbb{R}}(\mathcal{B})$ for which $\mathbf{x}^{\gamma-e_i} - r$, $\mathbf{x}^{\gamma-e_j} - s \in \mathrm{Ker}\, M_t$. Then we have $M_t\mathrm{vec}(\mathbf{x}_i r - \mathbf{x}_j s) = 0$ (implying that D is well defined) and $D^T\mathrm{vec}(\mathbf{x}_i r - \mathbf{x}_j s) = 0$ (implying that E is well defined).*

Proof. We first show that $M_t\mathrm{vec}(\mathbf{x}_i r - \mathbf{x}_j s) = 0$. In view of Lemma 1.2 (ii), it suffices to show that $\mathrm{vec}(\mathbf{x}^\alpha)^T M_t\mathrm{vec}(\mathbf{x}_i r - \mathbf{x}_j s) = 0$ for all $\alpha \in \mathbb{N}^n_{t-1}$. Fix $\alpha \in \mathbb{N}^n_{t-1}$. Then,

$$\begin{aligned}\mathrm{vec}(\mathbf{x}^\alpha)^T M_t\mathrm{vec}(\mathbf{x}_i r - \mathbf{x}_j s) &= \mathrm{vec}(\mathbf{x}_i\mathbf{x}^\alpha)^T M_t r - \mathrm{vec}(\mathbf{x}_j\mathbf{x}^\alpha)^T M_t s \\ &= \mathrm{vec}(\mathbf{x}_i\mathbf{x}^\alpha)^T M_t\mathrm{vec}(\mathbf{x}^{\gamma-e_i}) - \mathrm{vec}(\mathbf{x}_j\mathbf{x}^\alpha)^T M_t\mathrm{vec}(\mathbf{x}^{\gamma-e_j}) \\ &= y^T\mathrm{vec}(\mathbf{x}_i\mathbf{x}^\alpha\mathbf{x}^{\gamma-e_i}) - y^T\mathrm{vec}(\mathbf{x}_j\mathbf{x}^\alpha\mathbf{x}^{\gamma-e_j}) = 0,\end{aligned}$$

where we have used the fact that $r - \mathbf{x}^{\gamma-e_i}, s - \mathbf{x}^{\gamma-e_j} \in \mathrm{Ker}\, M_t$ for the second equality, and Lemma 4.1 for the third equality. We now show that $D^T\mathrm{vec}(\mathbf{x}_i r - \mathbf{x}_j s) = 0$, i.e., $\mathrm{vec}(\mathbf{x}_i r - \mathbf{x}_j s)^T D\mathrm{vec}(\mathbf{x}^\delta) = 0$ for all $|\delta| = t+1$. Fix $\delta \in \mathbb{N}^n_{t+1}$. Say, $\delta_k \geq 1$ and $\mathbf{x}^{\delta-e_k} - u \in \mathrm{Ker}\, M_t$, where $u \in \mathrm{Span}_{\mathbb{R}}(\mathcal{B})$. Then, $D\mathrm{vec}(\mathbf{x}^\delta) = M_t\mathrm{vec}(\mathbf{x}_k u)$ by construction. Using the above, this implies $\mathrm{vec}(\mathbf{x}_i r - \mathbf{x}_j s)^T D\mathrm{vec}(\mathbf{x}^\delta) = \mathrm{vec}(\mathbf{x}_i r - \mathbf{x}_j s)^T M_t\mathrm{vec}(\mathbf{x}_k u) = 0$. □

We now verify that the matrix N is a moment matrix, i.e., that N satisfies the conditions (i), (ii) from Lemma 5.15.

LEMMA 5.17.
(i) *$N_{\gamma,\delta} = N_{\gamma+e_i,\delta-e_i}$ for $\gamma, \delta \in \mathbb{N}^n_{t+1}$ with $\delta_i \geq 1$ and $|\gamma| \leq t$.*
(ii) *$N_{\gamma,\delta} = N_{\gamma-e_j+e_i,\delta+e_j-e_i}$ for $\gamma, \delta \in \mathbb{N}^n_{t+1}$ with $\gamma_j \geq 1$, $\delta_i \geq 1$, $|\gamma| = |\delta| = t+1$.*

Proof. (i) Assume $\mathbf{x}^{\delta-e_i}-r, \mathbf{x}^\gamma - s \in \operatorname{Ker} M_t$ for some $r, s \in \operatorname{Span}_{\mathbb{R}}(\mathcal{B})$; then $\mathbf{x}^\delta - \mathbf{x}_i r$, $\mathbf{x}_i\mathbf{x}^\gamma - \mathbf{x}_i s \in \operatorname{Ker} N$, by construction. We have

$$\begin{aligned}\operatorname{vec}(\mathbf{x}^\gamma)^T N \operatorname{vec}(\mathbf{x}^\delta) &= \operatorname{vec}(\mathbf{x}^\gamma)^T N \operatorname{vec}(\mathbf{x}_i r) = \operatorname{vec}(\mathbf{x}^\gamma)^T M_t \operatorname{vec}(\mathbf{x}_i r)\\ &= s^T M_t \operatorname{vec}(\mathbf{x}_i r) = \operatorname{vec}(\mathbf{x}_i s)^T M_t r = \operatorname{vec}(\mathbf{x}_i s)^T M_t \operatorname{vec}(\mathbf{x}^{\delta-e_i})\\ &= \operatorname{vec}(\mathbf{x}_i s)^T N \operatorname{vec}(\mathbf{x}^{\delta-e_i}) = \operatorname{vec}(\mathbf{x}_i\mathbf{x}^\gamma)^T N \operatorname{vec}(\mathbf{x}^{\delta-e_i}).\end{aligned}$$

This shows $N_{\gamma,\delta} = N_{\gamma+e_i,\delta-e_i}$.
(ii) Let $r, s \in \operatorname{Span}_{\mathbb{R}}(\mathcal{B})$ for which $\mathbf{x}^{\delta-e_i} - r, \mathbf{x}^{\gamma-e_j} - s \in \operatorname{Ker} M_t$. Then, $\mathbf{x}^\delta - \mathbf{x}_i r$, $\mathbf{x}_j\mathbf{x}^{\delta-e_i} - \mathbf{x}_j r$, $\mathbf{x}^\gamma - \mathbf{x}_j s$, $\mathbf{x}_i\mathbf{x}^{\gamma-e_j} - \mathbf{x}_i s \in \operatorname{Ker} N$ by construction. We have

$$\begin{aligned}\operatorname{vec}(\mathbf{x}^\gamma)^T N \operatorname{vec}(\mathbf{x}^\delta) &= \operatorname{vec}(\mathbf{x}_j s)^T N \operatorname{vec}(\mathbf{x}_i r) = \operatorname{vec}(\mathbf{x}_j s)^T M_t \operatorname{vec}(\mathbf{x}_i r)\\ &= \operatorname{vec}(\mathbf{x}_i s)^T M_t \operatorname{vec}(\mathbf{x}_j r) = \operatorname{vec}(\mathbf{x}_i s)^T N \operatorname{vec}(\mathbf{x}_j r)\\ &= \operatorname{vec}(\mathbf{x}^{\gamma-e_j+e_i})^T N \operatorname{vec}(\mathbf{x}^{\delta-e_i+e_j}),\end{aligned}$$

which shows $N_{\gamma,\delta} = N_{\gamma-e_j+e_i,\delta+e_j-e_i}$. □

This concludes the first proof of Theorem 5.14.

Second proof. The following proof of Theorem 5.14 is from Schweighofer [134]; it is less technical than the proof just presented, but uses some basic knowledge about Gröbner bases. (Cf. e.g. [21] for the undefined notions used in the proof below.)

LEMMA 5.18. *Suppose $y \in \mathbb{N}^n_{2t}$ and $M_t(y)$ is a flat extension of $M_{t-1}(y)$. Then*

$$U := \{f \in \mathbb{R}[\mathbf{x}]_{2t} \mid y^T(fg) = 0 \text{ for all } g \in \mathbb{R}[\mathbf{x}] \text{ with } fg \in \mathbb{R}[\mathbf{x}]_{2t}\}$$

is a linear subspace of $\mathbb{R}[\mathbf{x}]_{2t}$ with

$$U \cap \mathbb{R}[\mathbf{x}]_t = \operatorname{Ker} M_t(y) \quad \text{and} \tag{5.3}$$

$$fg \in U \text{ for all } f \in U \text{ and } g \in \mathbb{R}[\mathbf{x}] \text{ with } fg \in \mathbb{R}[\mathbf{x}]_{2t}. \tag{5.4}$$

For every fixed total degree monomial ordering, there exists a Gröbner basis G of the ideal $\mathcal{I} := (\operatorname{Ker} M_t(y)) \subseteq \mathbb{R}[\mathbf{x}]$ such that $G \subseteq \operatorname{Ker} M_t(y)$. In particular,

$$\operatorname{Ker} M_t(y) = \mathcal{I} \cap \mathbb{R}[\mathbf{x}]_t \subseteq \mathcal{I} \cap \mathbb{R}[\mathbf{x}]_{2t} \subseteq U. \tag{5.5}$$

Proof. To prove (5.3), we fix $f \in \mathbb{R}[\mathbf{x}]_t$. Suppose first that $f \in U$. By Lemma 4.1 and the definition of U, we have $g^T M_t(y) f = y^T \operatorname{vec}(fg) = 0$ for all $g \in \mathbb{R}[\mathbf{x}]_t$. Hence $f \in \operatorname{Ker} M_t(y)$. Conversely, suppose now $M_t(y)f = 0$ and $f \neq 0$. For every $\alpha \in \mathbb{N}^n$ with $|\alpha| + \deg(f) \leq 2t$, we can write $\mathbf{x}^\alpha = \mathbf{x}^\beta\mathbf{x}^\gamma$ with $\mathbf{x}^\beta, \mathbf{x}^\gamma f \in \mathbb{R}[\mathbf{x}]_t$. By Lemma 5.7(ii), we get $\mathbf{x}^\gamma f \in \operatorname{Ker} M_t(y)$

and therefore $y^T\text{vec}(f\mathbf{x}^\alpha) = \text{vec}(\mathbf{x}^\beta)^T M_t(y)\text{vec}(\mathbf{x}^\gamma f) = 0$ as desired. (5.4) is clear from the definition of U.

Take a finite set F of polynomials that generates Ker $M_t(y)$ as a vector space and contains for each $\alpha \in \mathbb{N}_t^n$ a polynomial of the form $\mathbf{x}^\alpha - p$ with $p \in \mathbb{R}[\mathbf{x}]_{t-1}$. Using the Buchberger algorithm, one can complete F to a Gröbner basis G of the ideal $\mathcal{I}$. We claim that all polynomials in G still lie in Ker $M_t(y)$, provided one uses a total degree monomial ordering. Indeed, every S-polynomial of two polynomials in Ker $M_t(y)$ lies in U by (5.3) and (5.4). In the Buchberger algorithm, such an S-polynomial will be reduced by F to a polynomial of degree at most t. Since $F \subseteq U$, this reduced S-polynomial will again lie in U by (5.4). Hence all polynomials added to F by the Buchberger algorithm lie in $U \cap \mathbb{R}[\mathbf{x}]_t = \text{Ker } M_t(y)$ by (5.3). This shows that we find $G \subseteq \text{Ker } M_t(y)$.

It remains only to show that $\mathcal{I} \cap \mathbb{R}[\mathbf{x}]_{2t} \subseteq U$, since this will imply $\mathcal{I} \cap \mathbb{R}[\mathbf{x}]_t \subseteq \text{Ker } M_t(y)$ by (5.3). We use the Gröbner basis G to show this. Let $f \in \mathcal{I} \cap \mathbb{R}[\mathbf{x}]_{2t}$ be given, $f \neq 0$. The division algorithm described on page 172 yields $f = \sum_{g \in G} u_g g$ where $u_g \in \mathbb{R}[\mathbf{x}]$ and $\deg(u_g g) \leq \deg(f) \leq 2t$ for all $g \in G$. By (5.4), we have $u_g g \in U$ for all $g \in G$. Hence $f \in U$. □

Now we can conclude the second proof of Theorem 5.14. In fact, we will extend the given vector $y \in \mathbb{N}_{2t}^n$ all at once to an infinite vector $\tilde{y} \in \mathbb{N}^n$ such that the infinite moment matrix $M(\tilde{y})$ is a flat extension of $M_t(y)$. For $\alpha \in \mathbb{N}^n$, we define $\tilde{y}_\alpha := y^T p^{(\alpha)}$, where $p^{(\alpha)} \in \mathbb{R}[\mathbf{x}]_t$ is chosen such that $\mathbf{x}^\alpha - p^{(\alpha)} \in \mathcal{I} = (\text{Ker } M_t(y))$. This is well defined since such $p^{(\alpha)}$ exists and, for $p, q \in \mathbb{R}[\mathbf{x}]_t$ with $\mathbf{x}^\alpha - p, \mathbf{x}^\alpha - q \in \mathcal{I}$, we have $p - q \in \mathcal{I} \cap \mathbb{R}[\mathbf{x}]_t \subseteq U$, giving $y^T p = y^T q$. Observe first that

$$\tilde{y}_\alpha = y_\alpha \ \forall \alpha \in \mathbb{N}_{2t}^n. \tag{5.6}$$

Indeed, for $|\alpha| \leq 2t$, $\tilde{y}_\alpha - y_\alpha = y^T\text{vec}(p^{(\alpha)} - \mathbf{x}^\alpha) = 0$ since $p^{(\alpha)} - \mathbf{x}^\alpha \in \mathcal{I} \cap \mathbb{R}[\mathbf{x}]_{2t} \subseteq U$ (by Lemma 5.18). Next observe that

$$\tilde{y}^T q = 0 \ \forall q \in \mathcal{I}. \tag{5.7}$$

For this, let $q = \sum_\alpha q_\alpha \mathbf{x}^\alpha \in \mathcal{I}$. Then, $\tilde{y}^T q = \sum_\alpha q_\alpha \tilde{y}_\alpha = \sum_\alpha q_\alpha y^T p^{(\alpha)} = y^T(\sum_\alpha q_\alpha p^{(\alpha)})$. As the polynomial $\sum_\alpha q_\alpha p^{(\alpha)} = \sum_\alpha q_\alpha (p^{(\alpha)} - \mathbf{x}^\alpha) + q$ lies in $\mathbb{R}[\mathbf{x}]_t \cap \mathcal{I} \subseteq U$, we find $\tilde{y}^T q = 0$, thus showing (5.7). From (5.7) we derive that $\mathcal{I} \subseteq \text{Ker } M(\tilde{y})$. We now verify that $M(\tilde{y})$ is a flat extension of $M_t(y)$. Indeed, for $\alpha \in \mathbb{N}^n$, we have $M(\tilde{y})\text{vec}(x^\alpha) = M(\tilde{y})\text{vec}(x^\alpha - p^{(\alpha)} + p^{(\alpha)}) = M(\tilde{y})\text{vec}(p^{(\alpha)})$, since $x^\alpha - p^{(\alpha)} \in \mathcal{I}$. This shows that all columns of $M(\tilde{y})$ are linear combinations of the columns indexed by $\mathbb{N}_t^n$, i.e. $M(\tilde{y})$ is a flat extension of $M_t(\tilde{y})$ and thus of $M_t(y)$ (by (5.6)). This concludes the second proof of Theorem 5.14.

5.4. Flat extensions and representing measures. We group here several results about the truncated moment problem. The first result from Theorem 5.19 essentially follows from the flat extension theorem (Theorem

5.14) combined with Theorem 5.1 about finite rank (infinite) moment matrices. This result is in fact the main ingredient that will be used for the extraction procedure of global minimizers in the polynomial optimization problem (see Section 6.7).

THEOREM 5.19. *Let* $y \in \mathbb{R}^{\mathbb{N}^n_{2t}}$ *for which* $M_t(y) \succeq 0$ *and* $\operatorname{rank} M_t(y) = \operatorname{rank} M_{t-1}(y)$*. Then* y *can be extended to a (unique) vector* $\tilde{y} \in \mathbb{R}^{\mathbb{N}^n}$ *satisfying* $M(\tilde{y}) \succeq 0$, $\operatorname{rank} M(\tilde{y}) = \operatorname{rank} M_t(y)$, *and* $\operatorname{Ker} M(\tilde{y}) = (\operatorname{Ker} M_t(y))$, *the ideal generated by* $\operatorname{Ker} M_t(y)$*. Moreover, any set* $\mathcal{B} \subseteq \mathbb{T}^n_{t-1}$ *indexing a maximum nonsingular principal submatrix of* $M_{t-1}(y)$ *is a basis of* $\mathbb{R}[\mathbf{x}]/(\operatorname{Ker} M_t(y))$*. Finally,* $\tilde{y}$*, and thus* y*, has a (unique) representing measure* μ*, which is* r*-atomic with* $\operatorname{supp}(\mu) = V_{\mathbb{C}}(\operatorname{Ker} M_t(y))$*.*

Proof. Applying iteratively Theorem 5.14 we find an extension $\tilde{y} \in \mathbb{R}^{\mathbb{N}^n}$ of y for which $M(\tilde{y})$ is a flat extension of $M_t(y)$; thus $\operatorname{rank} M(\tilde{y}) = \operatorname{rank} M_t(y) =: r$ and $M(\tilde{y}) \succeq 0$. By Theorem 5.1, $\tilde{y}$ has a (unique) representing measure μ, which is r-atomic and satisfies $\operatorname{supp}(\mu) = V_{\mathbb{C}}(\operatorname{Ker} M(\tilde{y}))$. To conclude the proof, it suffices to verify that $(\operatorname{Ker} M_t(y)) = \operatorname{Ker} M(\tilde{y})$, as this implies directly $\operatorname{supp}(\mu) = V_{\mathbb{C}}(\operatorname{Ker} M(\tilde{y})) = V_{\mathbb{C}}(\operatorname{Ker} M_t(y))$. Obviously, $\operatorname{Ker} M_t(y) \subseteq \operatorname{Ker} M(\tilde{y})$, implying $(\operatorname{Ker} M_t(y)) \subseteq \operatorname{Ker} M(\tilde{y})$. We now show the reverse inclusion. Let $\mathcal{B} \subseteq \mathbb{T}^n_{t-1}$ index a maximum nonsingular principal submatrix of $M_{t-1}(y)$. Thus $|\mathcal{B}| = r$ and $\mathcal{B}$ also indexes a maximum nonsingular principal submatrix of $M(\tilde{y})$. Hence, by Lemma 5.3, $\mathcal{B}$ is a basis of $\mathbb{R}[\mathbf{x}]/\operatorname{Ker} M(\tilde{y})$. We show that $\mathcal{B}$ is a generating set in $\mathbb{R}[\mathbf{x}]/(\operatorname{Ker} M_t(y))$; that is, for all $\beta \in \mathbb{N}^n$,

$$\mathbf{x}^\beta \in \operatorname{Span}_{\mathbb{R}}(\mathcal{B}) + (\operatorname{Ker} M_t(y)). \tag{5.8}$$

We prove (5.8) using induction on $|\beta|$. If $|\beta| \leq t$, (5.8) holds since $\mathcal{B}$ indexes a basis of the column space of $M_t(y)$. Assume $|\beta| \geq t+1$. Write $\mathbf{x}^\beta = \mathbf{x}_i \mathbf{x}^\gamma$ where $|\gamma| = |\beta| - 1$. By the induction assumption, $\mathbf{x}^\gamma = \sum_{\mathbf{x}^\alpha \in \mathcal{B}} \lambda_\alpha \mathbf{x}^\alpha + q$, where $\lambda_\alpha \in \mathbb{R}$ and $q \in (\operatorname{Ker} M_t(y))$. Then, $\mathbf{x}^\beta = \mathbf{x}_i \mathbf{x}^\gamma = \sum_{\mathbf{x}^\alpha \in \mathcal{B}} \lambda_\alpha \mathbf{x}_i \mathbf{x}^\alpha + \mathbf{x}_i q$. Obviously, $\mathbf{x}_i q \in (\operatorname{Ker} M_t(y))$. For $\mathbf{x}^\alpha \in \mathcal{B}$, $\deg(\mathbf{x}_i \mathbf{x}^\alpha) \leq t$ and, therefore, $\mathbf{x}_i \mathbf{x}^\alpha \in \operatorname{Span}_{\mathbb{R}}(\mathcal{B}) + (\operatorname{Ker} M_t(y))$. From this follows that $\mathbf{x}^\beta \in \operatorname{Span}_{\mathbb{R}}(\mathcal{B}) + (\operatorname{Ker} M_t(y))$. Thus (5.8) holds for all $\beta \in \mathbb{N}^n$. Take $p \in \operatorname{Ker} M(\tilde{y})$. In view of (5.8), we can write $p = p_0 + q$, where $p_0 \in \operatorname{Span}_{\mathbb{R}}(\mathcal{B})$ and $q \in (\operatorname{Ker} M_t(y))$. Hence, $p - q \in \operatorname{Ker} M(\tilde{y}) \cap \operatorname{Span}_{\mathbb{R}}(\mathcal{B})$, which implies $p - q = 0$, since $\mathcal{B}$ is a basis of $\mathbb{R}[\mathbf{x}]/\operatorname{Ker} M(\tilde{y})$. Therefore, $p = q \in (\operatorname{Ker} M_t(y))$, which concludes the proof for equality $\operatorname{Ker} M(\tilde{y}) = (\operatorname{Ker} M_t(y))$. □

We now give several results characterizing existence of a finite atomic measure for truncated sequences. By Lemma 4.2 (i), a necessary condition for the existence of a finite atomic reprenting measure μ for a sequence $y \in \mathbb{R}^{\mathbb{N}^n_{2t}}$ is that its moment matrix $M_t(y)$ has rank at most $|\operatorname{supp}(\mu)|$. Theorem 5.20 below gives a characterization for the existence of a *minimum* atomic measure, i.e., satisfying $|\operatorname{supp}(\mu)| = \operatorname{rank} M_t(y)$. Then Theorem 5.21 deals with the general case of existence of a finite atomic representing measure and Theorems 5.23 and 5.24 give the analogous results for existence of a measure supported by a prescribed semialgebraic set. In these results, the notion of *flat extension* studied in the preceding section plays a central role.

THEOREM 5.20. [23] *The following assertions are equivalent for $y \in \mathbb{R}^{\mathbb{N}^n_{2t}}$.*

(i) *y has a $(\text{rank}\, M_t(y))$-atomic representing measure.*

(ii) *$M_t(y) \succeq 0$ and one can extend y to a vector $\tilde{y} \in \mathbb{R}^{\mathbb{N}^n_{2t+2}}$ in such a way that $M_{t+1}(\tilde{y})$ is a flat extension of $M_t(y)$.*

Proof. Directly from Theorems 5.1 and 5.14. □

THEOREM 5.21. [24, 38] *Let $y \in \mathbb{R}^{\mathbb{N}^n_{2t}}$, $r := \text{rank}\, M_t(y)$ and $v := |V_{\mathbb{R}}(\text{Ker}\, M_t(y))| \leq \infty$; thus $r \leq v$ (by Lemma 4.2 (i)). Consider the following assertions:*

(i) *y has a representing measure.*

(ii) *y has a $\binom{n+2t}{2t}$-atomic representing measure.*

(iii) *$M_t(y) \succeq 0$ and there exists an integer $k \geq 0$ for which y can be extended to a vector $\tilde{y} \in \mathbb{R}^{\mathbb{N}^n_{2(t+k+1)}}$ in such a way that $M_{t+k}(\tilde{y}) \succeq 0$ and $M_{t+k+1}(\tilde{y})$ is a flat extension of $M_{t+k}(\tilde{y})$.*

(iv) *When $v < \infty$, y can be extended to a vector $\tilde{y} \in \mathbb{R}^{\mathbb{N}^n_{2(t+v-r+1)}}$ in such a way that $M_{t+v-r+1}(\tilde{y}) \succeq 0$ and $\text{rank}\, M_{t+v-r+1}(\tilde{y}) \leq |V_{\mathbb{R}}(\text{Ker}\, M_{t+v-r+1}(\tilde{y}))|$.*

Then, (i) $\Longleftrightarrow$ (ii) $\Longleftrightarrow$ (iii) and, when $v < \infty$, (i) $\Longleftrightarrow$ (iv). Moreover one can assume in (iii) that $k \leq \binom{n+2t}{2t} - r$ and, when $v < \infty$, that $k \leq v - r$.

Proof. The equivalence of (i) and (ii) follows from Theorem 5.8 and the implication (iii) $\Longrightarrow$ (i) follows from Theorem 5.19. Assume now that (ii) holds; that is, y has a finite atomic representing measure μ with $|\text{supp}(\mu)| \leq \binom{n+2t}{2t}$. Thus y can be extended to the sequence $\tilde{y} \in \mathbb{R}^{\mathbb{N}^n}$ consisting of all the moments of the measure μ. By Theorem 5.1 (ii), $M(\tilde{y}) \succeq 0$ and rank $M(\tilde{y}) = |\text{supp}(\mu)|$. Moreover, for any integer $k \geq 0$, $\text{rank}\, M_{t+k}(\tilde{y}) \leq |\text{supp}(\mu)| \leq |V_{\mathbb{R}}(\text{Ker}\, M_{t+k}(\tilde{y}))|$ (by Lemma 4.2 (i)). When $v < \infty$, we find (iv) by letting $k := v - r + 1$. Let $k \geq 0$ be the smallest integer for which rank $M_{t+k+1}(\tilde{y}) = \text{rank}\, M_{t+k}(\tilde{y})$ (whose existence follows from the fact that $r \leq \text{rank}\, M_{t+k}(\tilde{y}) \leq \binom{n+2t}{2t}$ for all $k \geq 0$). Then, $M_{t+k+1}(\tilde{y})$ is a flat extension of $M_{t+k}(\tilde{y})$, which shows (iii). Moreover, $\text{rank}\, M_{t+k+1}(\tilde{y}) \geq \text{rank}\, M_t(y) + k = r + k$ which, together with $\text{rank}\, M_{t+k+1}(\tilde{y}) \leq \binom{n+2t}{2t}$, gives the claimed bound $k \leq \binom{n+2t}{2t} - r$. As $V_{\mathbb{R}}(\text{Ker}\, M_{t+k+1}(\tilde{y})) \subseteq V_{\mathbb{R}}(\text{Ker}\, M_t(y))$ since $\text{Ker}\, M_t(y) \subseteq \text{Ker}\, M_{t+k+1}(\tilde{y})$, we find $r + k \leq \text{rank}\, M_{t+k+1}(\tilde{y}) \leq |V_{\mathbb{R}}(\text{Ker}\, M_{t+k+1}(\tilde{y}))| \leq |V_{\mathbb{R}}(\text{Ker}\, M_t(y))| = v$ and thus $k \leq v - r$ in the case when $v < \infty$.

Finally assume $v < \infty$ and (iv) holds. Using again the fact that $V_{\mathbb{R}}(\text{Ker}\, M_{t+v-r+1}(\tilde{y})) \subseteq V_{\mathbb{R}}(\text{Ker}\, M_t(y))$, we find $\text{rank}\, M_{t+v-r+1}(\tilde{y}) \leq |V_{\mathbb{R}}(\text{Ker}\, M_{t+v-r+1}(\tilde{y}))| \leq |V_{\mathbb{R}}(\text{Ker}\, M_t(y))| = v$. Therefore, there exists $k \in \{0, \ldots, v-r\}$ for which $\text{rank}\, M_{t+k+1}(\tilde{y}) = \text{rank}\, M_{t+k}(\tilde{y})$ for, if not, we would have $\operatorname{rank} M_{t+v-r+1}(\tilde{y}) \geq \operatorname{rank} M_t(y) + v - r + 1 = v + 1$, contradicting $\operatorname{rank} M_{t+v-r+1}(\tilde{y}) \leq v$. This shows that (iii) holds (and again that we can choose $k \leq v - r$ in (iii)). □

REMARK 5.22. Theorem 5.21 provides conditions characterizing the existence of a representing measure for a truncated sequence. It is however not clear how to check these conditions and the smallest integer k for which (iii) holds as well as the gap $v - r$ may be large. We refer to Fialkow [38] for a detailed treatment of such issues.

Let us observe here that in some instances the bound $v - r$ is better than the bound $\binom{n+2t}{2t} - r$. For instance, as observed in [38], in the 2-dimensional case ($n = 2$), $v \leq t^2$ by Bezout theorem, implying $\binom{2t+2}{2t} - v \geq \binom{2t+2}{2t} - t^2 = t^2 + 3t + 1$. Moreover, Fialkow [38] constructs an instance with large gap $v - r \geq \binom{t-1}{2}$. For this choose two polynomials $p, q \in \mathbb{R}[x_1, x_2]_t$ having t^2 common zeros in $\mathbb{R}^2$, i.e., $|V_{\mathbb{R}}(p,q)| = t^2$. Let μ be a measure on $\mathbb{R}^2$ with support $V_{\mathbb{R}}(p,q)$ and let y be its sequence of moments. Then, $V_{\mathbb{R}}(\operatorname{Ker} M_t(y)) = V_{\mathbb{R}}(p,q)$ and thus $v = t^2$. Indeed, $t^2 = |\operatorname{supp}(\mu)| \leq |V_{\mathbb{R}}(\operatorname{Ker} M_t(y))|$ and $V_{\mathbb{R}}(\operatorname{Ker} M_t(y)) \subseteq V_{\mathbb{R}}(p,q)$ since $p, q \in \operatorname{Ker} M_t(y)$. Moreover, $r = \operatorname{rank} M_t(y) \leq |\mathbb{N}^2_t| - 2 = \binom{t+2}{2} - 2$ which implies $v - r \geq t^2 - \binom{t+2}{2} + 2 = \binom{t-1}{2}$.

The next two theorems (from Curto and Fialkow [25]) extend the results from Theorems 5.20 and 5.21 to truncated sequences having a finite atomic representing measure whose support is contained in a prescribed semialgebraic set K. As indicated in [80], they can be derived easily from Theorems 5.20 and 5.21 using Lemma 5.6. In what follows K is as in (1.2) and d_K as in (1.10). One may assume w.l.o.g. that the polynomials g_j defining K are not constant; thus $d_{g_j} \geq 1$. For convenience we set $d_K := 1$ if $m = 0$, i.e., if there are no constraints defining the set K, in which case $K = \mathbb{R}^n$.

THEOREM 5.23. [25] *Let K be the set from (1.2) and $d_K = \max_{j=1,\ldots,m} d_{g_j}$. The following assertions are equivalent for $y \in \mathbb{R}^{\mathbb{N}^n_{2t}}$.*

(i) *y has a (rank $M_t(y)$)-atomic representing measure μ whose support is contained in K.*
(ii) *$M_t(y) \succeq 0$ and y can be extended to a vector $\tilde{y} \in \mathbb{R}^{\mathbb{N}^n_{2(t+d_K)}}$ in such a way that $M_{t+d_K}(\tilde{y})$ is a flat extension of $M_t(y)$ and $M_t(g_j\tilde{y}) \succeq 0$ for $j = 1, \ldots, m$.*

Then, setting $r_j := \operatorname{rank} M_t(g_j\tilde{y})$, exactly $r - r_j$ of the atoms in the support of μ belong to the set of roots of the polynomial $g_j(x)$. Moreover μ is a representing measure for $\tilde{y}$.

Proof. The implication (i) $\Longrightarrow$ (ii) follows from Theorem 5.20 ((i) $\Longrightarrow$ (ii)) together with Lemma 4.2 (ii). Conversely, assume that (ii) holds and set $r := \operatorname{rank} M_t(y)$. By Theorem 5.20 ((ii) $\Longrightarrow$ (i)), y has a r-atomic representing measure μ; say, $\mu = \sum_{v \in S} \lambda_v \delta_v$ where $\lambda_v > 0$, $|S| = r$. We prove that $S \subseteq K$; that is, $g_j(v) \geq 0$ for all $v \in S$. By Lemma 5.6, there exist interpolation polynomials p_v ($v \in S$) having degree at most t. Then, $p_v^T M_t(g_j y) p_v = \sum_{u \in V} (p_v(u))^2 g_j(u) \lambda_u = g_j(v) \lambda_v \geq 0$, since $M_t(g_j y) \succeq 0$. This implies that $g_j(v) \geq 0$ for all $j = 1, \ldots, m$ and $v \in S$, and thus $S \subseteq K$. That is, the measure μ is supported by the set K.

We now verify that $r - r_j$ of the points of S are zeros of the polynomial g_j. Denote by $\tilde{y} \in \mathbb{R}^{\mathbb{N}^n}$ the (infinite) sequence of moments of the measure μ; then $g_j \tilde{y}$ is the (infinite) sequence of moments of the measure $\mu_j := \sum_{v \in S} \lambda_v g_j(v) \delta_v$. Thus, $\tilde{y}$ (resp., $g_j \tilde{y}$) is an extension of y (resp., $g_j y$). Moreover, rank $M(g_j \tilde{y}) = |\{v \in S \mid g_j(v) > 0\}|$. We now verify that $M(g_j \tilde{y})$ is a flat extension of $M_t(g_j \tilde{y})$, which implies that $r_j = |\{v \in S \mid g_j(v) > 0\}|$, giving the desired result. For this we note that $\operatorname{Ker} M(\tilde{y}) \subseteq \operatorname{Ker} M(g_j \tilde{y})$. Indeed, if $p \in \operatorname{Ker} M(\tilde{y})$ then, using Lemma 4.1, $p^T M(g_j \tilde{y}) p = \operatorname{vec}(p g_j)^T M(\tilde{y}) p = 0$. Now, as $M(\tilde{y})$ is a flat extension of $M_t(y)$, it follows that $M(g_j \tilde{y})$ is a flat extension of $M_t(g_j \tilde{y})$ too. □

THEOREM 5.24. [25] *Let K be the set from (1.2) and $d_K = \max_{j=1,\ldots,m} d_{g_j}$. The following assertions are equivalent for $y \in \mathbb{R}^{\mathbb{N}^n_{2t}}$.*

(i) *y has a (finite atomic) representing measure whose support is contained in K.*

(ii) *$M_t(y) \succeq 0$ and there exists an integer $k \geq 0$ for which y can be extended to a vector $\tilde{y} \in \mathbb{R}^{\mathbb{N}^n_{2(t+k+d_K)}}$ in such a way that $M_{t+k+d_K}(\tilde{y}) \succeq 0$, $M_{t+k+d_K}(\tilde{y})$ is a flat extension of $M_{t+k}(\tilde{y})$, and $M_{t+k}(g_j \tilde{y}) \succeq 0$ for $j = 1, \ldots, m$.*

Proof. Analogously using Theorems 5.21 and 5.23. □

Part 2: Application to Optimization

6. Back to the polynomial optimization problem.

6.1. Hierarchies of relaxations. We consider again the optimization problem (1.1). Following Lasserre [65] and as explained earlier, hierarchies of semidefinite programming relaxations can be constructed for (1.1); namely, the SOS relaxations (3.8) (introduced in Section 3.4), that are based on relaxing polynomial positivity by sums of squares representations, and the moment relaxations (4.7) (introduced in Section 4.2), that are based on relaxing existence of a representing measure by positive semidefiniteness of moment matrices. For convenience we repeat the formulation of the bounds p_t^{sos} from (3.8) and p_t^{mom} from (4.7). Recall

$$d_p = \lceil \deg(p)/2 \rceil,\ d_{g_j} = \lceil \deg(g_j)/2 \rceil,\ d_K = \begin{cases} \max(d_{g_1}, \ldots, d_{g_m}) \\ 1 \text{ if } m = 0 \end{cases} \tag{6.1}$$

Then for any integer $t \geq \max(d_p, d_K)$,

$$\begin{aligned} p_t^{\text{sos}} &= \sup \rho \text{ s.t. } \ p - \rho \in \mathbf{M}_{2t}(g_1, \ldots, g_m) \\ &= \sup \rho \text{ s.t. } \ p - \rho = s_0 + \textstyle\sum_{j=1}^m s_j g_j \ \text{ for some } s_0, s_j \in \Sigma \\ &\qquad\qquad \text{with } \deg(s_0), \deg(s_j g_j) \leq 2t. \end{aligned} \tag{6.2}$$

$$\begin{aligned} p_t^{\text{mom}} &= \inf_{L \in (\mathbb{R}[\mathbf{x}]_{2t})^*} L(p) \text{ s.t. } \ L(1) = 1, \\ &\qquad\qquad L(f) \geq 0 \ \forall f \in \mathbf{M}_{2t}(g_1, \ldots, g_m) \\ &= \inf_{y \in \mathbb{R}^{\mathbb{N}^n_{2t}}} p^T y \text{ s.t. } \ y_0 = 1,\ M_t(y) \succeq 0, \\ &\qquad\qquad M_{t-d_{g_j}}(g_j y) \succeq 0 \ (j = 1, \ldots, m). \end{aligned} \tag{6.3}$$

We refer to program (6.2) as the *SOS relaxation* of order t, and to program (6.3) as the *moment relaxation* of order t. The programs (6.2) and (6.3) are semidefinite programs involving matrices of size $\binom{n+t}{t} = O(n^t)$ and $O(n^{2t})$ variables. Hence, for any *fixed* t, p_t^{mom} and p_t^{sos} can be computed in polynomial time (to any precision). In the remaining of Section 6 we study in detail some properties of these bounds. In particular,

(i) ***Duality:*** $p_t^{\text{sos}} \leq p_t^{\text{mom}}$ and, under some condition on the set K, the two bounds p_t^{mom} and p_t^{sos} coincide.

(ii) ***Convergence:*** Under certain conditions on the set K, there is **asymptotic** (sometimes even **finite**) **convergence** of the bounds p_t^{mom} and p_t^{sos} to $p^{\min}$.

(iii) ***Optimality certificate:*** Under some conditions, the relaxations are exact, i.e. $p_t^{\text{sos}} = p_t^{\text{mom}} = p^{\min}$ (or at least $p_t^{\text{mom}} = p^{\min}$).

(iv) ***Finding global minimizers:*** Under some conditions, one is able to extract some global minimizers for the original problem (1.1) from an optimum solution to the moment relaxation (6.3).

6.2. Duality. One can verify that the two programs (6.3) and (6.2) are dual semidefinite programs (cf. [65]), which implies $p_t^{\text{sos}} \leq p_t^{\text{mom}}$ by weak duality; this inequality also follows directly as noted earlier in (4.8). We now give a condition ensuring that strong duality holds, i.e. there is no duality gap between (6.3) and (6.2).

THEOREM 6.1. *[65, 132] If K has a nonempty interior (i.e. there exists a full dimensional ball contained in K), then $p_t^{mom} = p_t^{sos}$ for all $t \geq \max(d_p, d_K)$. Moreover, if (6.2) is feasible then it attains its supremum.*

Proof. We give two arguments. The first argument comes from [132] and relies on Theorem 3.33. Let $\rho > p_t^{\text{sos}}$, i.e. $p - \rho \notin \mathbf{M}_{2t}(g_1, \ldots, g_m)$. As $\mathbf{M}_{2t}(g_1, \ldots, g_m)$ is a closed convex cone (by Theorem 3.33), there exists a hyperplane strictly separating $p - \rho$ from $\mathbf{M}_{2t}(g_1, \ldots, g_m)$; that is, there exists $y \in \mathbb{R}^{\mathbb{N}^n_{2t}}$ with

$$y^T \text{vec}(p - \rho) < 0 \text{ and } y^T \text{vec}(f) \geq 0 \, \forall f \in \mathbf{M}_{2t}(g_1, \ldots, g_m). \tag{6.4}$$

If $y_0 > 0$ then we may assume $y_0 = 1$ by rescaling. Then y is feasible for (6.3), which implies $p_t^{\text{mom}} \leq y^T \text{vec}(p) < \rho$. As this is true for all $\rho > p_t^{\text{sos}}$, we deduce that $p_t^{\text{mom}} \leq p_t^{\text{sos}}$ and thus equality holds. Assume now $y_0 = 0$. Pick $x \in K$ and set $z := y + \epsilon \zeta_{2t,x}$ where $\zeta_{2t,x} = (x^\alpha)_{|\alpha| \leq 2t}$. Then, $z^T \text{vec}(p - \rho) < 0$ if we choose $\epsilon > 0$ small enough and $z^T \text{vec}(f) \geq 0$ for all $f \in \mathbf{M}_{2t}(g_1, \ldots, g_m)$, that is, z satisfies (6.4). As $z_0 = \epsilon > 0$, the previous argument (applied to z in place of y) yields again $p_t^{\text{mom}} = p_t^{\text{sos}}$. Finally, if (6.2) is feasible then it attains its supremum since $\mathbf{M}_{2t}(g_1, \ldots, g_m)$ is closed and one can bound the variable ρ.

The second argument, taken from [65], works under the assumption that (6.2) is feasible and uses the strong duality theorem for semidefinite programming. Indeed, by Lemma 4.4, the program (6.3) is strictly feasible and thus, by Theorem 1.3, there is no duality gap and (6.2) attains its supremum. □

PROPOSITION 6.2.

(i) If $\mathbf{M}(g_1, \ldots, g_m)$ is Archimedean, then the SOS relaxation (6.2) is feasible for t large enough.

(ii) If the ball constraint $R^2 - \sum_{i=1}^n \mathbf{x}_i^2 \geq 0$ is present in the description of K, then the feasible region of the moment relaxation (6.3) is bounded and the infimum is attained in (6.3).

Proof. (i) Using (3.17), $p + N \in \mathbf{M}(g_1, \ldots, g_m)$ for some N and thus $-N$ is feasible for (6.2) for t large enough.
(ii) Let y be feasible for (6.3). With $g := R^2 - \sum_{i=1} \mathbf{x}_i^2$, $(gy)_{2\beta} = R^2 y_{2\beta} - \sum_{i=1}^n y_{2\beta+2e_i}$. Thus the constraint $M_{t-1}(gy) \succeq 0$ implies $y_{2\beta+2e_i} \leq R^2 y_{2\beta}$ for all $|\beta| \leq t-1$ and $i = 1, \ldots, n$. One can easily derive (using induction on $|\beta|$) that $y_{2\beta} \leq R^{2|\beta|}$ for $|\beta| \leq t$. This in turn implies $|y_\gamma| \leq R^{|\gamma|}$ for $|\gamma| \leq 2t$. Indeed, write $\gamma = \alpha + \beta$ with $|\alpha|, |\beta| \leq t$; then as $M_t(y) \succeq 0$, $y_{\alpha+\beta}^2 \leq y_{2\alpha} y_{2\beta} \leq R^{2|\alpha|} R^{2|\beta|}$, giving $|y_\gamma| \leq R^{|\gamma|}$. This shows that the

feasible region to (6.3) is bounded and thus compact (as it is closed). Thus (6.3) attains its infimum. □

The next example (taken from [132]) shows that the infimum may not be attained in (6.3) even when K has a nonempty interior.

EXAMPLE 6.3. Consider the problem $p^{\min} := \inf_{x\in K} x_1^2$, where $K \subseteq \mathbb{R}^2$ is defined by the polynomial $g_1 = \mathbf{x}_1\mathbf{x}_2 - 1 \geq 0$. Then $p^{\min} = p_t^{\mathrm{mom}} = 0$ for any $t \geq 1$, but these optimum values are not attained. Indeed, for small $\epsilon > 0$, the point $x := (\epsilon, 1/\epsilon)$ lies in K, which implies $p^{\min} \leq \epsilon^2$. As $p_t^{\mathrm{mom}} \geq 0$ (since $y_{20} \geq 0$ for any y feasible for (6.3)), this gives $p_t^{\mathrm{mom}} = p^{\min} = 0$. On the other hand $y_{20} > 0$ for any feasible y for (6.3); indeed $M_0(g_1y) \succeq 0$ implies $y_{11} \geq 1$, and $y_{20} = 0$ would imply $y_{11} = 0$ since $M_1(y) \succeq 0$. Thus the infimum is not attained in (6.3) in this example. Note that the above still holds if we add the constraints $-2 \leq \mathbf{x}_1 \leq 2$ and $2 \leq \mathbf{x}_2 \leq 2$ to the description of K to make it compact.

On the other hand, when K has an empty interior, the duality gap may be infinite. We now give such an instance (taken from [132]) where $-\infty = p_t^{\mathrm{sos}} < p_t^{\mathrm{mom}} = p^{\min}$.

EXAMPLE 6.4. Consider the problem $p^{\min} := \min_{x\in K} x_1x_2$, where $K := \{x \in \mathbb{R}^2 \mid g_1(x), g_2(x), g_3(x) \geq 0\}$ with $g_1 := -\mathbf{x}_2^2$, $g_2 := 1 + \mathbf{x}_1$, $g_3 := 1 - \mathbf{x}_1$. Thus $K = [-1,1] \times \{0\}$. Obviously, $p^{\min} = 0$. We verify that $p_1^{\mathrm{mom}} = 0$, $p_1^{\mathrm{sos}} = -\infty$. For this let y be feasible for the program (6.3) for order $t = 1$; we show that $y_{e_1+e_2} = 0$. Indeed, $(M_1(y))_{e_2,e_2} = y_{2e_2} \geq 0$ and $(M_0(g_1y))_{0,0} = -y_{2e_2} \geq 0$ imply $y_{2e_2} = 0$. Thus the e_2th column of $M_1(y)$ is zero, which gives $y_{e_1+e_2} = (M_1(y))_{e_1,e_2} = 0$. Assume now that ρ is feasible for the program (6.2) at order $t = 1$. That is, $\mathbf{x}_1\mathbf{x}_2 - \rho = \sum_i (a_i + b_i\mathbf{x}_1 + c_i\mathbf{x}_2)^2 - e_1\mathbf{x}_2^2 + e_2(1+\mathbf{x}_1) + e_3(1-\mathbf{x}_1)$ for some $a_i, b_i, c_i \in \mathbb{R}$ and $e_1, e_2, e_3 \in \mathbb{R}_+$. Looking at the coefficient of $\mathbf{x}_1^2$ we find $0 = \sum_i b_i^2$ and thus $b_i = 0$ for all i. Looking at the coefficient of $\mathbf{x}_1\mathbf{x}_2$ we find $1 = 0$, a contradiction. Therefore there is no feasible solution, i.e., $p_1^{\mathrm{sos}} = -\infty$. On the other hand, $p_2^{\mathrm{sos}} = 0$ since, for all $\epsilon > 0$, $p_2^{\mathrm{sos}} \geq -\epsilon$ as $\mathbf{x}_1\mathbf{x}_2 + \epsilon = \frac{(\mathbf{x}_2+2\epsilon)^2}{8\epsilon}(\mathbf{x}_1+1) + \frac{(\mathbf{x}_2-2\epsilon)^2}{8\epsilon}(-\mathbf{x}_1+1) - \frac{1}{4\epsilon}\mathbf{x}_2^2$.

What if K has an empty interior? When K has a nonempty interior the moment/SOS relaxations behave nicely; indeed there is no duality gap (Theorem 6.1) and the optimum value is attained under some conditions (cf. Proposition 6.2). Marshall [88] has studied in detail the case when K has an empty interior. He proposes to exploit the presence of equations to sharpen the SOS/moment bounds, in such a way that there is no duality gap between the sharpened bounds. Consider an ideal $\mathcal{J} \subseteq \mathcal{I}(K)$, where $\mathcal{I}(K) = \{f \in \mathbb{R}[\mathbf{x}] \mid f(x) = 0\ \forall x \in K\}$ is the vanishing ideal of K; thus $\mathcal{I}(K) = \{0\}$ if K has a nonempty interior. Marshall makes the following assumption:

$$\mathcal{J} \subseteq \mathbf{M}(g_1, \ldots, g_m). \tag{6.5}$$

If this assumption does not hold and $\{h_1, \ldots, h_{m_0}\}$ is a system of generators of the ideal $\mathcal{J}$, it suffices to add the polynomials $\pm h_1, \ldots, \pm h_{m_0}$ in order to obtain a representation of K that fulfills (6.5). Now one may work with polynomials modulo the ideal $\mathcal{J}$. Let

$$\mathbf{M}'_{2t}(g_1, \ldots, g_m) := \{p' \mid p \in \mathbf{M}_{2t}(g_1, \ldots, g_m)\} \subseteq \mathbb{R}[\mathbf{x}]_{2t}/\mathcal{J}$$

be the image of $\mathbf{M}_{2t}(g_1, \ldots, g_m)$ under the map $p \mapsto p' := p \mod \mathcal{J}$ from $\mathbb{R}[\mathbf{x}]$ to $\mathbb{R}[\mathbf{x}]/\mathcal{J}$. (This set was introduced in (3.22) for the ideal $\mathcal{J} = \mathcal{I}(K)$.) Consider the following refinement of the SOS relaxation (6.2)

$$\begin{aligned} p_t^{\text{sos,eq}} &:= \sup \rho \ \text{ s.t. } (p-\rho)' \in \mathbf{M}'_{2t}(g_1, \ldots, g_m) \\ &= \sup \rho \ \text{ s.t. } p - \rho \in \mathbf{M}_{2t}(g_1, \ldots, g_m) + \mathcal{J}. \end{aligned} \tag{6.6}$$

For the analogue of (6.3), we now consider linear functionals on $\mathbb{R}[\mathbf{x}]_{2t}/\mathcal{J}$, i.e. linear functionals on $\mathbb{R}[\mathbf{x}]_{2t}$ vanishing on $\mathcal{J} \cap \mathbb{R}[\mathbf{x}]_{2t}$, and define

$$\begin{aligned} p_t^{\text{mom,eq}} := \inf_{L \in (\mathbb{R}[\mathbf{x}]_{2t}/\mathcal{J})^*} & \ L(f) \\ \text{s.t. } & L(1) = 1, L(f) \geq 0 \ \forall f \in \mathbf{M}'_{2t}(g_1, \ldots, g_m). \end{aligned} \tag{6.7}$$

Then, $p_t^{\text{sos}} \leq p_t^{\text{sos,eq}} \leq p^{\text{sos}}$, where the last inequality follows using (6.5); $p_t^{\text{sos,eq}} \leq p_t^{\text{mom,eq}}$, $p_t^{\text{mom}} \leq p_t^{\text{mom,eq}} \leq p^{\min}$. Moreover, $p_t^{\text{mom}} = p_t^{\text{mom,eq}}$, $p_t^{\text{sos}} = p_t^{\text{sos,eq}}$ if K has a nonempty interior since then $\mathcal{J} = \mathcal{I}(K) = \{0\}$. Marshall [88] shows the following extension of Theorem 6.1, which relies on Theorem 3.35 showing that $\mathbf{M}'_{2t}(g_1, \ldots, g_m)$ is closed when $\mathcal{J} = \mathcal{I}(K)$. We omit the details of the proof which are similar to those for Theorem 6.1.

THEOREM 6.5. *[88] When $\mathcal{J} = \mathcal{I}(K)$ satisfies (6.5), $p_t^{sos,eq} = p_t^{mom,eq}$ for all $t \geq \max(d_p, d_K)$.*

As a consequence,

$$\sup_t p_t^{\text{mom}} = p^{\text{sos}} \ \text{ if } \ \mathcal{J} = \mathcal{I}(K) \subseteq \mathbf{M}(g_1, \ldots, g_m).$$

Indeed, $p_t^{\text{mom}} \leq p_t^{\text{mom,eq}} = p_t^{\text{sos,eq}} \leq p^{\text{sos}}$, yielding $p^{\text{mom}} \leq p^{\text{sos}}$ and thus equality holds.

If we know a basis $\{h_1, \ldots, h_{m_0}\}$ of $\mathcal{J}$ then we can add the equations $h_j = 0$ ($j \leq m_0$), leading to an enriched representation for the set K of the form (2.5). Assuming $\mathcal{J} = \mathcal{I}(K)$, the SOS/moment bounds with respect to the description (2.5) of K are related to the bounds (6.6), (6.7) by

$$p_t^{\text{sos}} \leq p_t^{\text{mom}} \leq p_t^{\text{mom,eq}} = p_t^{\text{sos,eq}}. \tag{6.8}$$

LEMMA 6.6. *Assume that $\mathcal{J} = \mathcal{I}(K)$, $\{h_1, \ldots, h_{m_0}\}$ is a Gröbner basis of $\mathcal{J}$ for a total degree ordering, and $\deg(h_j)$ is even $\forall j \leq m_0$. Then equality holds throughout in (6.8).*

Proof. Let ρ be feasible for (6.6); we show that ρ is feasible for (6.2), implying $p_t^{\text{sos,eq}} = p_t^{\text{sos}}$. We have $p - \rho = \sum_{j=0}^m s_j g_j + q$ where $s_j \in \Sigma$, $\deg(s_j g_j) \leq 2t$ and $q \in \mathcal{J}$. Then $q = \sum_{j=1}^{m_0} u_j h_j$ with $\deg(u_j h_j) \leq 2t$ (since the h_j's form a Gröbner basis for a total degree ordering) and thus $\deg(u_j) \leq 2(t - d_{h_j})$ (since $\deg(h_j) = 2d_{h_j}$ is even), i.e. ρ is feasible for (6.2). □

REMARK 6.7. As each equation $h_j = 0$ is treated like two inequalities $\pm h_j \geq 0$, we have $f \in \mathbf{M}_{2t}(g_1, \ldots, g_m, \pm h_1, \ldots, \pm h_{m_0})$ if and only if $f = \sum_{j=0}^m s_j g_j + \sum_{j=1}^{m_0}(u_j' - u_j'')h_j$ for some $s_j, u_j', u_j'' \in \Sigma$ with $\deg(s_j g_j), \deg(u_j' h_j), \deg(u_j'' h_j) \leq 2t$. As $\deg(u_j' h_j), \deg(u_j'' h_j) \leq 2t$ is equivalent to $\deg(u_j'), \deg(u_j'') \leq 2(t - d_{h_j})$, one may equivalently write $\sum_{j=1}^{m_0}(u_j' - u_j'')h_j = \sum_{j=1}^{m_0} u_j h_j$ where $u_j \in \mathbb{R}[\mathbf{x}]_{2(t-d_{h_j})}$. Note that $\deg(u_j) \leq 2(t - d_{h_j})$ implies $\deg(u_j h_j) \leq 2t$, but the reverse does not hold, except if at least one of $\deg(u_j)$, $\deg(h_j)$ is even. This is why we assume in Lemma 6.6 that $\deg(h_j)$ is even. As an illustration, consider again Example 6.4, where $\mathcal{I}(K) = (\mathbf{x}_2)$. If we add the equation $\mathbf{x}_2 = 0$ to the description of K, we still get $p_1^{\text{sos}} = -\infty$ (since the multiplier of $\mathbf{x}_2$ in a decomposition of $\mathbf{x}_1\mathbf{x}_2 - \rho \in \mathbf{M}_2(\mathbf{x}_1 + 1, 1 - \mathbf{x}_1, \pm\mathbf{x}_2)$ should be a scalar), while $p_1^{\text{sos,eq}} = 0$ (since $\mathbf{x}_1$ is now allowed as multiplier of $\mathbf{x}_2$).

6.3. Asymptotic convergence. The asymptotic convergence of the SOS/moment bounds to $p^{\min}$ follows directly from Putinar's theorem (Theorem 3.20); recall Definition 3.18 for an Archimedean quadratic module.

THEOREM 6.8. [65] *If $\mathbf{M}(g_1, \ldots, g_m)$ is Archimedean, then $p^{sos} = p^{mom} = p^{min}$, i.e. $\lim_{t\to\infty} p_t^{sos} = \lim_{t\to\infty} p_t^{mom} = p^{min}$.*

Proof. Given $\epsilon > 0$, the polynomial $p - p^{\min} + \epsilon$ is positive on K. By Theorem 3.20, it belongs to $\mathbf{M}(g_1, \ldots, g_m)$ and thus the scalar $p^{\min} - \epsilon$ is feasible for the program (6.2) for some t. Therefore, there exists t for which $p_t^{\text{sos}} \geq p^{\min} - \epsilon$. Letting ϵ go to 0, we find that $p^{\text{sos}} = \lim_{t\to\infty} p_t^{\text{sos}} \geq p^{\min}$, implying $p^{\text{sos}} = p^{\text{mom}} = p^{\min}$. □

Note that if we would have a representation result valid for *nonnegative* (instead of *positive*) polynomials, this would immediately imply the *finite convergence* of the bounds $p_t^{\text{sos}}, p_t^{\text{mom}}$ to $p^{\min}$. For instance, Theorem 2.4 in Section 2.1 gives such a reprentation result in the case when the description of K involves a set of polynomial equations generating a zero-dimensional radical ideal. Thus we have the following result.

COROLLARY 6.9. *Assume K is as in (2.5) and $h_1, \ldots, h_{m_0}$ generate a zero-dimensional radical ideal. Then, $p_t^{sos} = p_t^{mom} = p^{min}$ for t large enough.*

Proof. Directly from Theorem 2.4, as in the proof of Theorem 6.8. □

In the non-compact case, convergence to $p^{\min}$ may fail. For instance, Marshall [88] shows that when K contains a full dimensional cone then, for all $t \geq \max(d_p, d_K)$, $p_t^{\text{sos}} = p_t^{\text{mom}}$, which can be strictly smaller than $p^{\min}$. This applies in particular to the case $K = \mathbb{R}^n$.

6.4. Approximating the unique global minimizer via the moment relaxations. Here we prove that when (1.1) has a *unique* global minimizer, then this minimizer can be approximated from the optimum solutions to the moment relaxations (6.3). We show in fact a stronger result (Theorem 6.11); this result is taken from Schweighofer [132] (who however formulates it in a slightly more general form in [132]). Recall the definition of the quadratic module $\mathbf{M}(g_1, \ldots, g_m)$ from (3.14) and of its truncation $\mathbf{M}_t(g_1, \ldots, g_m)$ from (3.20). Define the set of global minimizers of (1.1)

$$K_p^{\min} := \{x \in K \mid p(x) = p^{\min}\}. \tag{6.9}$$

DEFINITION 6.10. *Given* $y^{(t)} \in \mathbb{R}^{\mathbb{N}^n_{2t}}$, $y^{(t)}$ *is* nearly optimal *for (6.3) if* $y^{(t)}$ *is feasible for (6.3) and* $\lim_{t\to\infty} p^T y^{(t)} = \lim p_t^{mom}$.

THEOREM 6.11. *[132] Assume* $\mathbf{M}(g_1, \ldots, g_m)$ *is Archimedian,* $K_p^{\min} \neq \emptyset$, *and let* $y^{(t)}$ *be nearly optimal solutions to (6.3). Then,* $\forall \epsilon > 0\ \ \exists t_0 \geq \max(d_p, d_K)\ \ \forall t \geq t_0\ \ \exists \mu$ *probability measure on* $K_p^{\min}$ *such that* $\max_{i=1,\ldots,n} |y_{e_i}^{(t)} - \int x_i \mu(dx)| \leq \epsilon$.

Proof. As $\mathbf{M}(g_1, \ldots, g_m)$ is Archimedian, we deduce from (3.17) that

$$\forall k \in \mathbb{N}\ \ \exists N_k \in \mathbb{N}\ \ \forall \alpha \in \mathbb{N}_k^n\ \ N_k \pm \mathbf{x}^\alpha \in M_{N_k}(g_1, \ldots, g_m). \tag{6.10}$$

Define the sets $Z := \prod_{\alpha \in \mathbb{N}^n} [-N_{|\alpha|}, N_{|\alpha|}]$, $C_0 := \{z \in Z \mid z_0 = 1\}$, $C_f := \{z \in Z \mid z^T f \geq 0\}$ for $f \in \mathbf{M}(g_1, \ldots, g_m)$, $C_\delta := \{z \in Z \mid |z^T p - p^{\min}| \leq \delta\}$ for $\delta > 0$, and

$$C := \{z \in Z \mid \ \max_{i=1,\ldots,n} |z_{e_i} - \textstyle\int x_i \mu(dx)| > \epsilon \ \ \forall \mu \text{ probability measure on } K_p^{\min}\}.$$

CLAIM 6.12. $\bigcap_{f \in \mathbf{M}(g_1, \ldots, g_m)} C_f \cap \bigcap_{\delta > 0} C_\delta \bigcap C_0 \bigcap C = \emptyset$.

Proof. Assume z lies in the intersection. As $z \in C_0 \cap \bigcap_{f \in \mathbf{M}(g_1, \ldots, g_m)} C_f$, we deduce using (4.6) that $z \in \mathcal{M}_{\succeq}^{put}(g_1, \ldots, g_m)$ (recall (4.16)). Hence, by Theorem 4.11, $z \in \mathcal{M}_K$, i.e. z has a representing measure μ which is a probability measure on the set K. As $z \in \cap_{\delta > 0} C_\delta$, we have $p^T z = p^{\min}$, i.e. $\int (p(x) - p^{\min}) \mu(dx) = 0$, which implies that the support of μ is contained in the set $K_p^{\min}$, thus contradicting the fact that $z \in C$. □

As Z is a compact set (by Tychonoff's theorem) and all the sets C_f, C_δ, C_0, C are closed subsets of Z, there exists a finite collection of those sets having an empty intersection. That is, there exist $f_1, \ldots, f_s \in \mathbf{M}(g_1, \ldots, g_m)$, $\delta > 0$ such that

$$C_{f_1} \cap \ldots \cap C_{f_s} \cap C_\delta \cap C_0 \cap C = \emptyset. \tag{6.11}$$

Choose an integer $t_1 \geq \max(d_p, d_K)$ such that $f_1, \ldots, f_s \in M_{2t_1}(g_1, \ldots, g_m)$. Then choose an integer t_0 such that $t_0 \geq t_1$, $2t_0 \geq \max(N_k \mid k \leq 2t_1)$ (recall (6.10)) and $|p^T y^{(t)} - p^{\min}| \leq \delta$ for all $t \geq t_0$. We now verify that this t_0 satisfies the conclusion of the theorem. For this fix $t \geq t_0$. Consider the vector $z \in \mathbb{R}^{\mathbb{N}^n}$ defined by $z_\alpha := y_\alpha^{(t)}$ if $|\alpha| \leq 2t_1$, and $z_\alpha := 0$ otherwise.

CLAIM 6.13. $z \in Z$.

Proof. Let $\alpha \in \mathbb{N}^n$ with $|\alpha| =: k \leq 2t_1$. Then $N_k \pm \mathbf{x}^\alpha \in M_{N_k}(g_1, \ldots, g_m) \subseteq M_{2t_0}(g_1, \ldots, g_m) \subseteq \mathbf{M}_{2t}(g_1, \ldots, g_m)$. As $y^{(t)}$ is feasible for (6.3) we deduce that $(y^{(t)})^T \mathrm{vec}(N_k \pm \mathbf{x}^\alpha) \geq 0$, implying $|y_\alpha^{(t)}| \leq N_k = N_{|\alpha|}$. □

Obviously $z \in C_0$. Next $z \in C_\delta$ since $|z^T p - p^{\min}| = |(y^{(t)})^T p - p^{\min}| \leq \delta$ as $2t_1 \geq \deg(p)$. Finally, for any $r = 1, \ldots, s$, $z \in C_{f_r}$ since $z^T f_r = (y^{(t)})^T f_r \geq 0$ as $f_r \in M_{2t_1}(g_1, \ldots, g_m) \subseteq \mathbf{M}_{2t}(g_1, \ldots, g_m)$. As the set in (6.11) is empty, we deduce that $z \notin C$. Therefore, there exists a probability measure μ on $K_p^{\min}$ for which $\max_i |y_{e_i}^{(t)} - \int x_i \mu(dx)| = \max_i |z_{e_i} - \int x_i \mu(dx)| \leq \epsilon$. This concludes the proof of Theorem 6.11. □

COROLLARY 6.14. *Assume $\mathbf{M}(g_1, \ldots, g_m)$ is Archimedian and problem (1.1) has a unique minimizer x^*. Let $y^{(t)}$ be nearly optimal solutions to (6.3). Then $\lim_{t\to\infty} y_{e_i}^{(t)} = x_i^*$ for each $i = 1, \ldots, n$.*

Proof. Directly from Theorem 6.11 since the Dirac measure δ_{x^*} at x^* is the unique probability measure on $K_p^{\min}$. □

6.5. Finite convergence. Here we show finite convergence for the moment/SOS relaxations, when the description of the semialgebraic set K contains a set of polynomial equations $h_1 = 0, \ldots, h_{m_0} = 0$ generating a zero-dimensional ideal. (Recall Corollary 6.9 for the *radical* zero-dimensional case.) Theorem 6.15 below extends a result of Laurent [81] and uses ideas from Lasserre et al. [75].

THEOREM 6.15. *Consider the problem (1.1) of minimizing $p \in \mathbb{R}[\mathbf{x}]$ over the set $K = \{x \in \mathbb{R}^n \mid h_j(x) = 0 \ (j = 1, \ldots, m_0),\ g_j(x) \geq 0 \ (j = 1, \ldots, m)\}$ (as in (2.5)). Set $\mathcal{J} := (h_1, \ldots, h_{m_0})$.*
(i) If $|V_\mathbb{C}(\mathcal{J})| < \infty$, then $p^{min} = p_t^{mom} = p_t^{sos}$ for t large enough.
(ii) If $|V_\mathbb{R}(\mathcal{J})| < \infty$, then $p^{min} = p_t^{mom}$ for t large enough.

Proof. Fix $\epsilon > 0$. The polynomial $p - p^{\min} + \epsilon$ is positive on K. For the polynomial $u := -\sum_{j=1}^{m_0} h_j^2$, the set $\{x \in \mathbb{R}^n \mid u(x) \geq 0\} = V_\mathbb{R}(\mathcal{J})$ is compact (in fact, finite under (i) or (ii)) and u belongs to the quadratic module generated by the polynomials $\pm h_1, \ldots, \pm h_{m_0}$. Hence we can apply Theorem 3.20 and, therefore, there is a decomposition

$$p - p^{\min} + \epsilon = s_0 + \sum_{j=1}^{m} s_j g_j + q, \tag{6.12}$$

where s_0, s_j are sums of squares and $q \in \mathcal{J}$. To finish the proof we distinguish the two cases (i), (ii).

Consider first the case (i) when $|V_{\mathbb{C}}(\mathcal{J})| < \infty$. Let $\{f_1, \ldots, f_L\}$ be a Gröbner basis of $\mathcal{J}$ for a total degree monomial ordering, let $\mathcal{B}$ be a basis of $\mathbb{R}[\mathbf{x}]/\mathcal{J}$, and set $d_{\mathcal{B}} := \max_{b \in \mathcal{B}} \deg(b)$ (which is well defined as $|\mathcal{B}| < \infty$ since $\mathcal{J}$ is zero-dimensional). Consider the decomposition (6.12). Say, $s_j = \sum_i s_{i,j}^2$ and write $s_{i,j} = r_{i,j} + q_{i,j}$, where $r_{i,j}$ is a linear combination of members of $\mathcal{B}$ and $q_{i,j} \in \mathcal{J}$; thus $\deg(r_{i,j}) \leq d_{\mathcal{B}}$. In this way we obtain another decomposition:

$$p - p^{\min} + \epsilon = s_0' + \sum_{j=1}^{m} s_j' g_j + q', \tag{6.13}$$

where s_0', s_j' are sums of squares, $q' \in \mathcal{J}$ and $\deg(s_0'), \deg(s_j') \leq 2d_{\mathcal{B}}$. Set

$$T_0 := \max(2d_p, 2d_{\mathcal{B}}, 2d_{\mathcal{B}} + 2d_{g_1}, \ldots, 2d_{\mathcal{B}} + 2d_{g_m}). \tag{6.14}$$

Then, $\deg(s_0'), \deg(s_j' g_j), \deg(p - p^{\min} + \epsilon) \leq T_0$ and thus $\deg(q') \leq T_0$. Therefore, q' has a decomposition $q' = \sum_{l=1}^{L} u_l f_l$ with $\deg(u_l f_l) \leq \deg(q') \leq T_0$ (because we use a total degree monomial ordering). We need to find a decomposition of q' with bounded degrees in the original basis $\{h_1, \ldots, h_{m_0}\}$ of $\mathcal{J}$. For this, write $f_l = \sum_{j=1}^{m_0} a_{l,j} h_j$ where $a_{l,j} \in \mathbb{R}[\mathbf{x}]$. Then, $q' = \sum_{l=1}^{L} u_l (\sum_{j=1}^{m_0} a_{l,j} h_j) = \sum_{j=1}^{m_0} (\sum_{l=1}^{L} a_{l,j} u_l) h_j =: \sum_{j=1}^{m_0} b_j h_j$, setting $b_j := \sum_{l=1}^{L} a_{l,j} u_l$. As $\deg(u_l) \leq T_0$, we have $\deg(b_j h_j) \leq 2d_{h_j} + T_0 + \max_{l=1}^{L} \deg(a_{l,j})$. Thus, $\deg(b_j h_j) \leq T_g$ after setting $T_g := T_0 + \max_{l,j}(\deg(a_{l,j}) + 2d_{h_j})$, which is a constant not depending on ϵ. Therefore we can conclude that $p^{\min} - \epsilon$ is feasible for the program (6.2) for all $t \geq T_1 := \lceil T_g/2 \rceil$. This implies $p_t^{\mathrm{sos}} \geq p^{\min} - \epsilon$ for all $t \geq T_1$. Letting ϵ go to zero, we find $p_t^{\mathrm{sos}} \geq p^{\min}$ and thus $p_t^{\mathrm{sos}} = p^{\min}$ for $t \geq T_1$, which concludes the proof in case (i).

Consider now the case (ii) when $|V_{\mathbb{R}}(\mathcal{J})| < \infty$. Let y be feasible for the program (6.3); that is, $y \in \mathbb{R}^{\mathbb{N}^n_{2t}}$ satisfies $y_0 = 1$, $M_t(y) \succeq 0$, $M_{t-d_{h_j}}(h_j y) = 0$ $(j = 1, \ldots, m_0)$, $M_{t-d_{g_j}}(g_j y) \succeq 0$ $(j = 1, \ldots, m)$. We show $p^T y \geq p^{\min}$ for t large enough. We need the following observations about the kernel of $M_t(y)$. First, for $j = 1, \ldots, m_0$, $h_j \in \operatorname{Ker} M_t(y)$ for $t \geq 2d_{h_j}$ (directly, from the fact that $M_{t-d_{h_j}}(h_j y) = 0$). Moreover, for t large enough, $\operatorname{Ker} M_t(y)$ contains any given finite set of polynomials of $\mathcal{I}(V_{\mathbb{R}}(\mathcal{J}))$.

Claim 6.16. *Let $f_1, \ldots, f_L \in \mathcal{I}(V_{\mathbb{R}}(\mathcal{J}))$. There exists $t_1 \in \mathbb{N}$ such that $f_1, \ldots, f_L \in \operatorname{Ker} M_t(y)$ for all $t \geq t_1$.*

Proof. Fix $l = 1, \ldots, L$. As $f_l \in \mathcal{I}(V_{\mathbb{R}}(\mathcal{J}))$, by the Real Nullstellensatz (Theorem 2.1), $f_l^{2m_l} + \sum_i p_{l,i}^2 = \sum_{j=1}^{m} u_{l,j} h_j$ for some $p_{l,i}, u_{l,j} \in \mathbb{R}[\mathbf{x}]$ and $m_l \in \mathbb{N} \setminus \{0\}$. Set $t_1 := \max(\max_{j=1}^{m_0} 2d_{h_j}, 1 + \max_{l \leq L, j \leq m_0} \deg(u_{l,j} h_j))$ and let $t \geq t_1$. Then, each $u_{l,j}h_j$ lies in $\operatorname{Ker} M_t(y)$ by Lemma 5.7. Therefore, $f_l^{2m_l} + \sum_i p_{l,i}^2 \in \operatorname{Ker} M_t(y)$, which implies $f_l^{m_l}, p_{l,i} \in \operatorname{Ker} M_t(y)$. An easy induction permits to conclude that $f_l \in \operatorname{Ker} M_t(y)$. □

Let $\{f_1, \ldots, f_L\}$ be a Gröbner basis of $\mathcal{I}(V_{\mathbb{R}}(\mathcal{J}))$ for a total degree monomial ordering, let $\mathcal{B}$ be a basis of $\mathbb{R}[\mathbf{x}]/\mathcal{I}(V_{\mathbb{R}}(\mathcal{J}))$, and set $d_{\mathcal{B}} := \max_{b \in \mathcal{B}} \deg(b)$ (which is well defined since $|\mathcal{B}| < \infty$ as $\mathcal{I}(V_{\mathbb{R}}(\mathcal{J}))$ is zero-dimensional). Given $\epsilon > 0$, consider the decomposition (6.12) where s_0, s_j are sums of squares and $q \in \mathcal{J}$. As in case (i), we can derive another decomposition (6.13) where s_0', s_j' are s.o.s., $q' \in \mathcal{I}(V_{\mathbb{R}}(\mathcal{J}))$, and $\deg(s_0'), \deg(s_j') \leq 2d_{\mathcal{B}}$. Then, $\deg(s_0'), \deg(s_j'g_j), \deg q' \leq T_0$ with T_0 being defined as in (6.14) and we can write $q' = \sum_{l=1}^{L} u_l f_l$ with $\deg(u_l f_l) \leq T_0$. Fix $t \geq \max(T_0+1, t_1)$. Then, $u_l f_l \in \operatorname{Ker} M_t(y)$ (by Lemma 5.7 and Claim 6.16) and thus $q' \in \operatorname{Ker} M_t(y)$. Moreover, $\operatorname{vec}(1)^T M_t(y) \operatorname{vec}(s_j' g_j) \geq 0$; to see it, write $s_j' = \sum_i a_{i,j}^2$ and note that $\operatorname{vec}(1)^T M_t(y) \operatorname{vec}(s_j' g_j) = \sum_i a_{i,j}^T M_{t-d_{g_j}}(g_j y) a_{i,j} \geq 0$ since $M_{t-d_{g_j}}(g_j y) \succeq 0$. Therefore, we deduce from (6.13) that $\operatorname{vec}(1)^T M_t(y) \operatorname{vec}(p - p^{\min} + \epsilon) \geq 0$, which gives $p^T y = 1^T M_t(y) p \geq p^{\min} - \epsilon$ and thus $p_t^{\text{mom}} \geq p^{\min} - \epsilon$. Letting ϵ go to 0, we obtain $p_t^{\text{mom}} \geq p^{\min}$ and thus $p_t^{\text{mom}} = p^{\min}$. □

QUESTION 6.17. *Does there exist an example with $|V_{\mathbb{C}}(\mathcal{J})| = \infty$, $|V_{\mathbb{R}}(\mathcal{J})| < \infty$ and where $p_t^{sos} < p^{min}$ for all t ?*

The finite convergence result from Theorem 6.15 applies, in particular, to the case when K is contained in a discrete grid $K_1 \times \ldots \times K_n$ with $K_i \subseteq \mathbb{R}$ finite, considered by Lasserre [67], and by Lasserre [66] in the special case $K \subseteq \{0,1\}^n$. We will come back to the topic of exploiting equations in Section 8.2.

6.6. Optimality certificate. We now formulate some stopping criterion for the moment hierarchy (6.3), i.e. some condition permitting to claim that the moment relaxation (6.3) is in fact exact, i.e. $p_t^{\text{mom}} = p^{\min}$, and to extract some global minimizer for (1.1).

A first easy such condition is as follows. Let y be an optimum solution to (6.3) and $x^* := (y_{10\ldots0}, \ldots, y_{0\ldots01})$ the point in $\mathbb{R}^n$ consisting of the coordinates of y indexed by $\alpha \in \mathbb{N}^n$ with $|\alpha| = 1$. Then

$$\begin{aligned} x^* \in K \text{ and } p_t^{\text{mom}} = p(x^*) \Longrightarrow p_t^{\text{mom}} = p^{\min} \text{ and} \\ x^* \text{ is a global minimizer of } p \text{ over } K. \end{aligned} \tag{6.15}$$

Indeed $p^{\min} \leq p(x^*)$ as $x \in K$, which together with $p(x^*) = p_t^{\text{mom}} \leq p^{\min}$ implies equality $p_t^{\text{mom}} = p^{\min}$ and x^* is a minimizer of p over K. Note that $p_t^{\text{mom}} = p(x^*)$ automatically holds if p is linear. According to Theorem 6.11 this condition has a good chance to be satisfied (approximatively) when problem (1.1) has a unique minimizer. See Examples 6.24, 6.25 for instances where the criterion (6.15) is satisfied.

We now see another stopping criterion, which may work when problem (1.1) has a *finite* number of global minimizers. This stopping criterion, which has been formulated by Henrion and Lasserre [54], deals with the rank of the moment matrix of an optimal solution to (6.3) and is based on the result of Curto and Fialkow from Theorem 5.23. As in (6.9), $K_p^{\min}$ denotes the set of global minimizers of p over the set K. Thus $K_p^{\min} \neq \emptyset$, e.g., when K is compact. The next result is based on [54] combined with ideas from [75].

THEOREM 6.18. *Let $t \geq \max(d_p, d_K)$ and let $y \in \mathbb{R}^{\mathbb{N}^n_{2t}}$ be an optimal solution to the program (6.3). Assume that the following rank condition holds:*

$$\exists s \text{ s.t. } \max(d_p, d_K) \leq s \leq t \text{ and } \operatorname{rank} M_s(y) = \operatorname{rank} M_{s-d_K}(y). \quad (6.16)$$

Then $p_t^{mom} = p^{min}$ and $V_{\mathbb{C}}(\operatorname{Ker} M_s(y)) \subseteq K_p^{\min}$. Moreover, equality $V_{\mathbb{C}}(\operatorname{Ker} M_s(y)) = K_p^{\min}$ holds if $\operatorname{rank} M_t(y)$ is maximum among all optimal solutions to (6.3).

Proof. By assumption, $p_t^{\text{mom}} = p^T y$, $M_t(y) \succeq 0$, rank $M_s(y) =$ rank $M_{s-d_K}(y) =: r$ and $M_{s-d_K}(g_j y) \succeq 0$ for $j = 1, \ldots, m$, where $\max(d_p, d_K) \leq s \leq t$. As $s \geq d_K$, we can apply Theorem 5.23 and conclude that the sequence $(y_\alpha)_{\alpha \in \mathbb{N}^n_{2s}}$ has a r-atomic representing measure $\mu = \sum_{i=1}^r \lambda_i \delta_{v_i}$, where $v_i \in K$, $\lambda_i > 0$ and $\sum_{i=1}^r \lambda_i = 1$ (since $y_0 = 1$). As $s \geq d_p$, $p_t^{\text{mom}} = p^T y = \sum_{|\alpha| \leq 2s} p_\alpha y_\alpha = \sum_{i=1}^r \lambda_i p(v_i) \geq p^{\min}$, since $p(v_i) \geq p^{\min}$ for all i. On the other hand, $p^{\min} \geq p_t^{\text{mom}}$. This implies that $p^{\min} = p_t^{\text{mom}}$ and that each v_i is a minimizer of p over the set K, i.e., $\operatorname{supp}(\mu) = \{v_1, \ldots, v_r\} \subseteq K_p^{\min}$. As $\operatorname{supp}(\mu) = V_{\mathbb{C}}(\operatorname{Ker} M_s(y))$ by Theorem 5.19, we obtain $V_{\mathbb{C}}(\operatorname{Ker} M_s(y)) \subseteq K_p^{\min}$.

Assume now that rank $M_t(y)$ is maximum among all optimal solutions to (6.3). By Lemma 1.4, $\operatorname{Ker} M_t(y) \subseteq \operatorname{Ker} M_t(y')$ for any other optimal solution y' to (6.3). For any $v \in K_p^{\min}$, $y' := \zeta_{2t,v}$ is feasible for (6.3) with objective value $p^T y' = p(v) = p^{\min}$; thus y' is an optimal solution and thus $\operatorname{Ker} M_t(y) \subseteq \operatorname{Ker} M_t(\zeta_{2t,v})$. This implies $\operatorname{Ker} M_t(y) \subseteq \bigcap_{v \in K_p^{\min}} \operatorname{Ker} M_t(\zeta_{2t,v}) \subseteq \mathcal{I}(K_p^{\min})$. Therefore, $\operatorname{Ker} M_s(y) \subseteq \operatorname{Ker} M_t(y) \subseteq \mathcal{I}(K_p^{\min})$, which implies $K_p^{\min} \subseteq V_{\mathbb{C}}(\operatorname{Ker} M_s(y))$ and thus equality $V_{\mathbb{C}}(\operatorname{Ker} M_s(y)) = K_p^{\min}$ holds. □

Hence, if at some order $t \geq \max(d_p, d_K)$ one can find a maximum rank optimal solution to the moment relaxation (6.3) which satisfies the rank condition (6.16), then one can find *all* the global minimizers of p over the set K, by computing the common zeros to the polynomials in $\operatorname{Ker} M_s(y)$. In view of Theorem 5.19 and Lemma 5.2, the ideal $(\operatorname{Ker} M_s(y))$ is (real) radical and zero-dimensional. Hence its variety $V_{\mathbb{C}}(\operatorname{Ker} M_s(y))$ is finite. Moreover one can determine this variety with the eigenvalue method, described in Section 2.4. This extraction procedure is presented in Henrion and Lasserre [54] and is implemented in their optimization software GloptiPoly.

The second part of Theorem 6.18, asserting that all global minimizers are found when having a maximum rank solution, relies on ideas from [75]. When p is the constant polynomial 1 and K is defined by the equations $h_1 = 0, \ldots, h_{m_0} = 0$, then $K_p^{\min}$ is the set of all common real roots of the h_j's. The paper [75] explains in detail how the moment methodology applies to finding real roots, and [76] extends this to complex roots.

As we just proved, if (6.16) holds for a maximum rank optimal solution y to (6.3), then $K_p^{\min} = V_{\mathbb{C}}(\operatorname{Ker} M_s(y))$ is finite. Hence the conditions of Theorem 6.18 can apply *only* when p has finitely many global minimizers over the set K. We will give in Example 6.24 an instance with infinitely many global minimizers and thus, as predicted, the rank condition (6.16) is not satisfied. We now see an example where the set $K_p^{\min}$ of global minimizers is finite but yet the conditions of Theorem 6.18 are never met.

EXAMPLE 6.19. We give here an example where $|K_p^{\min}| < \infty$ and $p_t^{\text{mom}} = p_t^{\text{sos}} < p^{\min}$; hence condition (6.16) does not hold. Namely consider the problem

$$p^{\min} = \min_{x \in K} p(x) \quad \text{where } K := \{x \in \mathbb{R}^n \mid g_1(x) := 1 - \sum_{i=1}^{n} x_i^2 \geq 0\}.$$

Assume that p is a homogeneous polynomial which is positive (i.e., $p(x) > 0$ for all $x \in \mathbb{R}^n \setminus \{0\}$), but not a sum of squares. Then, $p^{\min} = 0$ and the origin is the unique minimizer, i.e., $K_p^{\min} = \{0\}$. Consider the moment relaxation (6.3) and the dual SOS relaxation (6.2) for $t \geq d_p$. As $\mathbf{M}(g_1)$ is Archimedean, the SOS relaxation (6.2) is feasible for t large enough. Moreover, as K has a nonempty interior, there is no duality gap, i.e. $p_t^{\text{mom}} = p_t^{\text{sos}}$, and the supremum is attained in (6.2) (apply Theorem 6.1). We now verify that $p_t^{\text{sos}} = p_t^{\text{mom}} < p^{\min} = 0$. Indeed, if $p_t^{\text{sos}} = 0$, then $p = s_0 + s_1(1 - \sum_{i=1}^n \mathbf{x}_i^2)$ where $s_0, s_1 \in \mathbb{R}[\mathbf{x}]$ are sums of squares. It is not difficult to verify that this implies that p must be a sum of squares (see [30, Prop. 4]), yielding a contradiction. Therefore, on this example, $p_t^{\text{mom}} = p_t^{\text{sos}} < p^{\min}$ and thus the rank condition (6.16) cannot be satisfied. This situation is illustrated in Example 6.25. There we choose $p = M + \epsilon(\mathbf{x}_1^6 + \mathbf{x}_2^6 + \mathbf{x}_3^6)$ where M is the Motzkin form (introduced in Example 3.7). Thus p is a homogeneous positive polynomial and there exists $\epsilon > 0$ for which p_ϵ is not SOS (for if not $M = \lim_{\epsilon \to 0} p_\epsilon$ would be SOS since the cone $\Sigma_{3,6}$ is closed).

On the other hand, we now show that the rank condition (6.16) in Theorem 6.18 holds for t large enough when the description of the set K comprises a system of equations $h_1 = 0, \ldots, h_{m_0} = 0$ having finitely many real zeros. Note that the next result also provides an alternative proof for Theorem 6.15 (ii), which does not use Putinar's theorem but results about moment matrices instead.

THEOREM 6.20. *[75, Prop. 4.6] Let K be as in (2.5), let $\mathcal{J}$ be the ideal generated by $h_1, \ldots, h_{m_0}$ and assume that $|V_{\mathbb{R}}(\mathcal{J})| < \infty$. For t large enough,* *there exists an integer* s, $\max(d_K, d_p) \leq s \leq t$, *such that* $\operatorname{rank} M_s(y) = \operatorname{rank} M_{s-d_K}(y)$ *for any feasible solution* y *to (6.3).*

Proof. As in the proof of Theorem 6.15 (ii), let $\{f_1, \ldots, f_L\}$ be a Gröbner basis of $\mathcal{I}(V_{\mathbb{R}}(\mathcal{J}))$ for a total degree monomial ordering. By Claim 6.16, there exists $t_1 \in \mathbb{N}$ such that $f_1, \ldots, f_L \in \operatorname{Ker} M_t(y)$ for all $t \geq t_1$. Let $\mathcal{B}$ be a basis of $\mathbb{R}[\mathbf{x}]/\mathcal{I}(V_{\mathbb{R}}(\mathcal{J}))$ and $d_{\mathcal{B}} := \max_{b \in \mathcal{B}} \deg(b)$. Write any monomial as $x^\alpha = r^{(\alpha)} + \sum_{l=1}^{L} p_l^{(\alpha)} f_l$, where $r^{(\alpha)} \in \operatorname{Span}_{\mathbb{R}}(\mathcal{B})$, $p_l^{(\alpha)} \in \mathbb{R}[\mathbf{x}]$. Set $t_2 := \max(t_1, d_{\mathcal{B}} + d_K, d_p)$ and $t_3 := 1 + \max(t_2, \deg(p_l^{(\alpha)} f_l)$ for $l \leq L, |\alpha| \leq t_2)$. Fix $t \geq t_3$ and let y be feasible for (6.3). We claim that $\operatorname{rank} M_{t_2}(y) = \operatorname{rank} M_{t_2 - d_K}(y)$. Indeed, consider $\alpha \in \mathbb{N}^n_{t_2}$. As $\deg(p_l^{(\alpha)} f_l) \leq t - 1$ and $f_l \in \operatorname{Ker} M_t(y)$, we deduce (using Lemma 5.7) that $p_l^{(\alpha)} f_l \in \operatorname{Ker} M_t(y)$ and thus $\mathbf{x}^\alpha - r^{(\alpha)} \in \operatorname{Ker} M_t(y)$. As $\deg(r^{(\alpha)}) \leq d_{\mathcal{B}} \leq t_2 - d_K$, this shows that the αth column of $M_t(y)$ can be written as a linear combination of columns of $M_{t_2 - d}(y)$; that is, $\operatorname{rank} M_{t_2}(y) = \operatorname{rank} M_{t_2 - d_K}(y)$. □

Let us conclude this section with a brief discussion about the assumptions made in Theorem 6.18. A first basic assumption we made there is that the moment relaxation (6.3) attains its minimum. This is the case, e.g., when the feasible region of (6.3) is bounded (which happens e.g. when a ball constraint is present in the description of K, cf. Proposition 6.2), or when program (6.2) is strictly feasible (recall Theorem 1.3). A second basic question is to find conditions ensuring that there is no duality gap, i.e. $p_t^{\text{mom}} = p_t^{\text{sos}}$, since this is needed if one wants to solve the semidefinite programs using a primal-dual interior point algorithm. This is the case, e.g. when K has a nonempty interior (by Theorem 6.1) or when any of the programs (6.3) or (6.2) is strictly feasible (recall Theorem 1.3).

Another question raised in Theorem 6.18 is to find an optimum solution to a semidefinite program with maximum rank. It is in fact a property of most interior-point algorithms that they return a maximum rank optimal solution. This is indeed the case for the SDP solver SeDuMi used within GloptiPoly. More precisely, when both primal and dual problems (6.3) and (6.2) are strictly feasible, then the interior-point algorithm SeDuMi constructs a sequence of points on the so-called central path, which has the property of converging to an optimal solution of maximum rank. SeDuMi also finds a maximum rank optimal solution under the weaker assumption that (6.3) is feasible and attains its minimum, (6.2) is feasible, and $p_t^{\text{mom}} = p_t^{\text{sos}} < \infty$. Indeed SeDuMi applies the so-called extended self-dual embedding technique, which consists of embedding the given program into a new program satisfying the required strict feasibility property; a maximum rank optimal solution for the original problem can then be derived from a maximum rank optimal solution to the embedded problem. See [28, Ch. 4], [156, Ch. 5] for details. (This issue is also discussed in [75] in the context of solving systems of polynomial equations.)

There are many further numerical issues arising for the practical implementation of the SOS/moment method. Just to name a few, the numerical instability of linear algebra dealing with matrices with a Hankel type structure, or the numerically sensitive issue of computing ranks, etc. To address the first issue, Löfberg and Parrilo [84] suggest to use sampling to represent polynomials and other non-monomial bases of the polynomial ring $\mathbb{R}[\mathbf{x}]$; see also [29] where promising numerical results are reported for the univariate case, and [122].

6.7. Extracting global minimizers. We explain here how to extract global minimizers to the problem (1.1) assuming we are in the situation of Theorem 6.18. That is, $y \in \mathbb{R}^{\mathbb{N}^n_{2t}}$ is an optimal solution to the program (6.3) satisfying the rank condition (6.16). Then, as claimed in Theorem 6.18 (and its proof), $p_t^{\text{mom}} = p^{\min}$, y has a r-atomic representing measure $\mu = \sum_{i=1}^r \lambda_i \delta_{v_i}$, where $\lambda_i > 0$, $\sum_{i=1}^r \lambda_i = 1$, $r = \operatorname{rank} M_s(y)$, and $V_{\mathbb{C}}(\operatorname{Ker} M_s(y)) = \{v_1, \ldots, v_r\} \subseteq K_p^{\min}$, the set of optimal solutions to (1.1). The question we now address is how to find the v_i's from the moment matrix $M_s(y)$. We present the method proposed by Henrion and Lasserre [54], although our description differs in some steps and follows the implementation proposed by Jibetean and Laurent [60] and presented in detail in Lasserre et al. [75].

Denote by $\tilde{y}$ the (infinite) sequence of moments of the measure μ. Then, $M(\tilde{y})$ is a flat extension of $M_s(y)$. Hence, by Theorem 5.19, $\mathcal{I} := \operatorname{Ker} M(\tilde{y}) = (\operatorname{Ker} M_s(y))$ and any set $\mathcal{B} \subseteq \mathbb{T}^n_{s-1}$ indexing a maximum nonsingular principal submatrix of $M_{s-1}(y)$ is a basis of $\mathbb{R}[\mathbf{x}]/\mathcal{I}$. One can now determine $V_{\mathbb{C}}(\operatorname{Ker} M_s(y)) = V_{\mathbb{C}}(\mathcal{I})$ with the eigenvalue method presented in Section 2.4.

In a first step we determine a subset $\mathcal{B} \subseteq \mathbb{T}^n_{s-d_K}$ indexing a maximum nonsingular principal submatrix of $M_s(y)$. We can find such set $\mathcal{B}$ in a 'greedy manner', by computing the successive ranks of the north-east corner principal submatrices of $M_{s-d_K}(y)$. Starting from the constant monomial 1, we insert in $\mathcal{B}$ as many low degree monomials as possible.

In a second step, for each $i = 1, \ldots, n$, we construct the multiplication matrix $M_{\mathbf{x}_i}$ of the 'multiplication by $\mathbf{x}_i$' operator $m_{\mathbf{x}_i}$ (recall (2.6)) with respect to the basis $\mathcal{B}$ of $\mathbb{R}[\mathbf{x}]/\mathcal{I}$. By definition, for $\mathbf{x}^\beta \in \mathcal{B}$, the $\mathbf{x}^\beta$th column of $M_{\mathbf{x}_i}$ contains the residue of the monomial $\mathbf{x}_i\mathbf{x}^\beta$ modulo $\mathcal{I}$ w.r.t. the basis $\mathcal{B}$. That is, setting $M_{\mathbf{x}_i} := (a^{(i)}_{\alpha,\beta})_{\mathbf{x}^\alpha,\mathbf{x}^\beta \in \mathcal{B}}$, the polynomial $\mathbf{x}_i\mathbf{x}^\beta - \sum_{\mathbf{x}^\alpha \in \mathcal{B}} a^{(i)}_{\alpha,\beta}\mathbf{x}^\alpha$ belongs to $\mathcal{I}$ and thus to $\operatorname{Ker} M_s(y)$. From this we immediately derive the following explicit characterization for $M_{\mathbf{x}_i}$ from the moment matrix $M_s(y)$.

LEMMA 6.21. *Let M_0 denote the principal submatrix of $M_s(y)$ indexed by $\mathcal{B}$ and let U_i denote the submatrix of $M_s(y)$ whose rows are indexed by $\mathcal{B}$ and whose columns are indexed by the set $\mathbf{x}_i\mathcal{B} := \{\mathbf{x}_i\mathbf{x}^\alpha \mid \mathbf{x}^\alpha \in \mathcal{B}\}$. Then, $M_{\mathbf{x}_i} = M_0^{-1}U_i$.*

Given a polynomial $h \in \mathbb{R}[\mathbf{x}]$, the multiplication matrix of the 'multiplication by h' operator w.r.t. the basis $\mathcal{B}$ is then given by $M_h = h(M_{\mathbf{x}_1}, \ldots, M_{\mathbf{x}_n})$. In view of Theorem 2.9, the eigenvectors of M_h^T are the vectors $\zeta_{\mathcal{B},v} = (v^\alpha)_{\mathbf{x}^\alpha \in \mathcal{B}}$ with respective eigenvalues $h(v)$ for $v \in V_{\mathbb{C}}(\mathcal{I})$. A nice feature of the ideal $\mathcal{I} = \operatorname{Ker} M(\tilde{y}) = (\operatorname{Ker} M_s(y))$ is that it is (real) radical. Hence, if the values $h(v)$ for $v \in V_{\mathbb{C}}(\mathcal{I})$ are all distinct, then the matrix M_h^T is non-derogatory, i.e., its eigenspaces are 1-dimensional and spanned by the vectors $\zeta_{\mathcal{B},v}$ (for $v \in V_{\mathbb{C}}(\mathcal{I})$) (recall Lemma 2.12). In that case, one can recover the vectors $\zeta_{\mathcal{B},v}$ directly from the right eigenvectors of M_h. Then it is easy - in fact, immediate if $\mathcal{B}$ contains the monomials $\mathbf{x}_1, \ldots, \mathbf{x}_n$ - to recover the components of v from the vector $\zeta_{\mathcal{B},v}$. According to [20], if we choose h as a random linear combination of the monomials $\mathbf{x}_1, \ldots, \mathbf{x}_n$ then, with high probability, the values $h(v)$ at the distinct points of $V_{\mathbb{C}}(\mathcal{I})$ are all distinct.

6.8. Software and examples. Several software packages have been developed for computing sums of squares of polynomials and optimizing polynomials over semialgebraic sets.

- **GloptiPoly**, developed by Henrion and Lasserre [53], implements the moment/SOS hierarchies (6.3), (6.2), and the techniques described in this section for testing optimality and extracting global optimizers. See `http://www.laas.fr/~henrion/software/gloptipoly/` The software has been recently updated to treat more general moment problems; cf. [55].
- **SOSTOOLS**, developed by Prajna, Papachristodoulou, Seiler and Parrilo [113], is dedicated to formulate and compute sums of squares optimization programs. See `http://www.mit.edu/~parrilo/sostools/`
- **SparsePOP**, developed by Waki, Kim, Kojima and Muramatsu [153], implements sparse moment/SOS relaxations for polynomial optimization problems having some sparsity pattern (see Section 8.1). See `http://www.is.titech.ac.jp/~kojima/SparsePOP/`
- **Yalmip**, developed by Löfberg, is a MATLAB toolbox for rapid prototyping of optimization problems, which implements in particular the sum-of-squares and moment based approaches. See `http://control.ee.ethz.ch/~joloef/yalmip.php`

We conclude with some small examples. See e.g. [65, 54] for more examples.

EXAMPLE 6.22. Consider the problem:

$$\begin{aligned}
\min \quad & p(x) = -25(x_1-2)^2 - (x_2-2)^2 - (x_3-1)^2 \\
& \qquad -(x_4-4)^2 - (x_5-1)^2 - (x_6-4)^2 \\
\text{s.t.} \quad & (x_3-3)^2 + x_4 \geq 4, \quad (x_5-3)^2 + x_6 \geq 4 \\
& x_1 - 3x_2 \leq 2, \quad -x_1 + x_2 \leq 2, \; x_1 + x_2 \leq 6, \\
& x_1 + x_2 \geq 2, \; 1 \leq x_3 \leq 5, \; 0 \leq x_4 \leq 6, \\
& 1 \leq x_5 \leq 5, \; 0 \leq x_6 \leq 10, \; x_1, x_2 \geq 0.
\end{aligned}$$

As shown in Table 1, GloptiPoly finds the optimum value -310 and a global minimizer $(5,1,5,0,5,10)$ at the relaxation of order $t = 2$. This involves then the computation of a SDP with 209 variables, one semidefinite constraint involving a matrix of size 28 (namley, $M_2(y) \succeq 0$) and 16 semidefinite constraints involving matrices size 7 (namely, $M_1(g_j y) \succeq 0$, corresponding to the 16 constraints $g_j \geq 0$ of degree 1 or 2).

TABLE 1
Moment relaxations for Example 6.22.

order t	rank sequence	bound p_t^{mom}	solution extracted
1	(1,7)	unbounded	none
2	**(1,1, 21)**	-310	(5,1,5,0,5,10)

EXAMPLE 6.23. Consider the problem

$$\begin{aligned} \min \quad & p(x) = -x_1 - x_2 \\ \text{s.t.} \quad & x_2 \leq 2x_1^4 - 8x_1^3 + 8x_1^2 + 2 \\ & x_2 \leq 4x_1^4 - 32x_1^3 + 88x_1^2 - 96x_1 + 36 \\ & 0 \leq x_1 \leq 3,\ 0 \leq x_2 \leq 4. \end{aligned}$$

As shown in Table 2, GloptiPoly solves the problem at optimality at the relaxation of order $t = 4$.

TABLE 2
Moment relaxations for Example 6.23.

order t	rank sequence	bound p_t^{mom}	solution extracted
2	(1,1,4)	-7	none
3	(1,2,2,4)	-6.6667	none
4	**(1,1,1,6)**	-5.5080	(2.3295,3.1785)

EXAMPLE 6.24. Consider the problem:

$$\begin{aligned} \min \quad & p(x) = x_1^2 x_2^2 (x_1^2 + x_2^2 - 3x_3^2) + x_3^6 \\ \text{s.t.} \quad & x_1^2 + x_2^2 + x_3^2 \leq 1, \end{aligned}$$

of minimizing the Motzkin form over the unit ball. As we see in Table 3, the moment bounds p_t^{mom} converge to $p^{\min} = 0$, but optimality cannot be detected via the rank condition (6.16) since it is never satisfied. This is to be expected since p has infinitely many global minimizers over the unit ball. However the criterion (6.15) applies here; indeed GloptiPoly returns that the relaxed vector $x^* := (y_{e_i})_{i=1}^3$ (where y is the optimum solution to the moment relaxation) is feasible (i.e. lies in the unit ball) and reaches

TABLE 3
Moment relaxations for Example 6.24.

order t	rank sequence	bound p_t^{mom}	value reached by moment vector
3	(1,4, 9, 13)	−0.0045964	7 10^{-26}
4	(1, 4, 10,20, 29)	−0.00020329	3 10^{-30}
5	(1, 4, 10, 20, 34, 44)	−2.8976 10^{-5}	3 10^{-36}
6	(1, 4, 10, 20, 34, 56, 84)	−6.8376 10^{-6}	6 10^{-42}
7	(1, 4, 10, 20, 35, 56, 84, 120)	−2.1569 10^{-6}	4 10^{-43}

the objective value which is mentioned in the last column of Table 3; here $x^* \sim 0$.

EXAMPLE 6.25. Consider the problem

$$\begin{aligned} \min \quad & p(x) = x_1^2x_2^2(x_1^2+x_2^2-3x_3^2)+x_3^6+\epsilon(x_1^6+x_2^6+x_3^6) \\ \text{s.t.} \quad & x_1^2+x_2^2+x_3^2 \le 1, \end{aligned}$$

of minimizing the perturbed Motzkin form over the unit ball. For any $\epsilon > 0$, $p^{\min} = 0$ and p is positive, i.e. the origin is the unique global minimizer. Moreover, p_ϵ is SOS if and only if $\epsilon \geq \epsilon^* \sim 0.01006$ [152]. Hence, as explained in Example 6.19, it is to be expected that for $\epsilon < \epsilon^*$, the rank condition (6.16) does not hold. This is confirmed in Table 4 which gives results for $\epsilon = 0.01$. Again the criterion (6.15) applies, i.e. the moment vector y yields the global minimizer $x^* = (y_{e_i})_{i=1}^3$, $x^* \sim 0$, and the last column gives the value of p_ϵ evaluated at x^*.

TABLE 4
Moment relxations for Example 6.25.

t	rank sequence	bound p_t^{mom}	value reached by moment vector
3	(1, 4, 9, 13)	−2.11 10^{-5}	1.67 10^{-44}
4	(1, 4, 10, 20, 35)	−1.92 10^{-9}	4.47 10^{-60}
5	(1, 4, 10,20, 35, 56)	2.94 10^{-12}	1.26 10^{-44}
6	(1, 4, 10, 20, 35, 56, 84)	3.54 10^{-12}	1.5 10^{-44}
7	(1, 4, 10, 20, 35, 56, 84, 120)	4.09 10^{-12}	2.83 10^{-43}
8	(1, 4, 10, 20, 35, 56, 84, 120, 165)	4.75 10^{-12}	5.24 10^{-44}

7. Application to optimization - Some further selected topics.

7.1. Approximating positive polynomials by sums of squares. We now come back to the comparison between nonnegative polynomials and sums of squares of polynomials. As we saw earlier, the parameters (n, d) for which every nonnegative polynomial of degree d in n variables is a sum of squares have been characterized by D. Hilbert; namely, they are ($n = 1, d$ even), ($n \geq 1, d = 2$), and ($n = 2, d = 4$). Thus for any other pair (n, d) (d even) there exist nonnegative polynomials that cannot be written as a sum of squares. A natural question is whether there are many or few such polynomials. Several answers may be given depending whether the degree and the number of variables are fixed or not. First, on the negative side, Blekherman [13] has shown that when the degree d is fixed but the number n of variables grows, then there are significantly more positive polynomials than sums of squares. More precisely, for $d \in \mathbb{N}$ even, consider the cone $\mathbf{H}_d$ (resp., $\mathbf{\Sigma}_d$) of homogeneous polynomials $p \in \mathbb{R}[\mathbf{x}]$ of degree d that are nonnegative on $\mathbb{R}^n$ (resp., a sum of squares). In order to compare the two cones, Blekherman considers their sections $\widehat{\mathbf{H}}_d := \mathbf{H}_d \cap H$ and $\widehat{\mathbf{\Sigma}}_d := \mathbf{\Sigma}_d \cap H$ by the hyperplane $H := \{p \mid \int_{S^{n-1}} p(x)\mu(dx) = 1\}$, where μ is the rotation invariant probability measure on the unit sphere S^{n-1}.

THEOREM 7.1. *[13] There exist constants $C_1, C_2 > 0$ depending on d only such that for any n large enough,*

$$C_1 n^{(d/2-1)/2} \leq \left(\frac{vol\ \widehat{\mathbf{H}}_d}{vol\ \widehat{\mathbf{\Sigma}}_d} \right)^{1/D} \leq C_2 n^{(d/2-1)/2},$$

where $D := \binom{n+d-1}{d} - 1$.

However, on the positive side, Berg [8] has shown that, when the number of variables is fixed but the degree is variable, then the cone of sums of squares is dense in the cone of polynomials nonnegative on $[-1, 1]^n$. While Berg's result is existential, Lasserre and Netzer [77] have provided an explicit and very simple sum of squares approximation, which we present in Theorem 7.2 below. Previously, Lasserre [71] had given an analogous result for polynomials nonnegative on the whole space $\mathbb{R}^n$, presented in Theorem 7.3 below. To state the results we need the following polynomials for any $t \in \mathbb{N}$

$$\theta_t := \sum_{k=0}^{t} \sum_{i=1}^{n} \frac{\mathbf{x}_i^{2k}}{k!}, \quad \Theta_t := 1 + \sum_{i=1}^{n} \mathbf{x}_i^{2t}, \tag{7.1}$$

THEOREM 7.2. *[77] Let $f \in \mathbb{R}[\mathbf{x}]$ be a polynomial nonnegative on $[-1, 1]^n$ and let Θ_t be as in (7.1). For any $\epsilon > 0$, there exists $t_0 \in \mathbb{N}$ such that the polynomial $f + \epsilon\Theta_t$ is a sum of squares for all $t \geq t_0$.*

THEOREM 7.3. *[71] Let $f \in \mathbb{R}[\mathbf{x}]$ be a polynomial nonnegative on $\mathbb{R}^n$ and let θ_t be as in (7.1). For any $\epsilon > 0$, there exists $t_0 \in \mathbb{N}$ such that $f + \epsilon\theta_t$ is a sum of squares for all $t \geq t_0$.*

In both cases the proof relies on a result about existence of a representing measure, combined with some elementary bounds on the entries of positive semidefinite moment matrices. For Theorem 7.2 we need only the (quite elementary) result from Theorem 4.7 about existence of a representing measure for bounded sequences. On the other hand, for Theorem 7.3, we need the following (non-elementary) result of Carleman (for the case $n = 1$) and Nussbaum (for $n \geq 1$). Recall that $e_1, \ldots, e_n$ denote the standard unit vectors in $\mathbb{R}^n$. Thus, for $y \in \mathbb{R}^{\mathbb{N}^n}$, y_{2ke_i} is its entry indexed by $2ke_i$, i.e. $y_{2ke_i} = y_{(0,\ldots,0,2k,0,\ldots,0)}$ where $2k$ is at the ith position.

THEOREM 7.4. *[102] Given $y \in \mathbb{R}^{\mathbb{N}^n}$, if $M(y) \succeq 0$ and*

$$\sum_{k=0}^{\infty} y_{2ke_i}^{-1/2k} = \infty \quad (i = 1, \ldots, n) \tag{7.2}$$

then y has a (unique) representing measure.

In what follows we first give the proof of Theorem 7.2, which is simpler, and then we prove Theorem 7.3. We begin with some elementary bounds from [71, 77] on the entries of $M_t(y)$. As we now see, when $M_t(y) \succeq 0$, all entries y_α can be bounded in terms of y_0 and $y_{(2t,0,\ldots,0)}, \ldots, y_{(0,\ldots,0,2t)}$, the entries indexed by the constant monomial 1 and the highest order monomials $\mathbf{x}_1^{2t}, \ldots, \mathbf{x}_n^{2t}$. For $0 \leq k \leq t$, set

$$\tau_k := \max(y_{(2k,0,\ldots,0)}, \ldots, y_{(0,\ldots,0,2k)}) = \max_{i=1,\ldots,n} y_{2ke_i};$$

thus $\tau_0 = y_0$. We will use the inequality $y_{\alpha+\beta}^2 \leq y_{2\alpha}y_{2\beta}$ (for $\alpha, \beta \in \mathbb{N}_t^n$), which follows from the fact that the submatrix of $M_t(y)$ indexed by $\{\alpha, \beta\}$ is positive semidefinite.

LEMMA 7.5. *Assume $M_t(y) \succeq 0$ and $n = 1$. Then $y_{2k} \leq \max(\tau_0, \tau_t)$ for $0 \leq k \leq t$.*

Proof. The proof is by induction on $t \geq 0$. If $t = 0, 1$, the result is obvious. Assume $t \geq 1$ and the result holds for $t - 1$, i.e. $y_0, \ldots, y_{2t-2} \leq \max(y_0, y_{2t-2})$; we show that $y_0, \ldots, y_{2t} \leq \max(y_0, y_{2t})$. This is obvious if $y_0 \geq y_{2t-2}$. Assume now $y_0 \leq y_{2t-2}$. As $y_{2t-2}^2 \leq y_{2t-4}y_{2t} \leq y_{2t-2}y_{2t}$, we deduce $y_{2t-2} \leq y_{2t}$ and thus $y_0, \ldots, y_{2t} \leq y_{2t} = \max(y_0, y_{2t})$. □

LEMMA 7.6. *Assume $M_t(y) \succeq 0$. Then $y_{2\alpha} \leq \tau_k$ for all $|\alpha| = k \leq t$.*

Proof. The case $n = 1$ being obvious, we first consider the case $n = 2$. Say $s := \max_{|\alpha|=k} y_{2\alpha}$ is attained at $y_{2\alpha^*}$. As $2\alpha_1^* \geq k \iff 2\alpha_2^* \leq k$, we may assume w.l.o.g. $2\alpha_1^* \geq k$. Write $2\alpha^* = (k, 0) + (2\alpha_1^* - k, 2\alpha_2^*) = (k, 0) + (k - 2\alpha_2^*, 2\alpha_2^*)$. Then $y_{2\alpha^*}^2 \leq y_{(2k,0)}y_{(2k-4\alpha_2^*,4\alpha_2^*)}$. Now $y_{2\alpha^*}^2 = s^2$,

$y_{(2k-4\alpha_2^*,4\alpha_2^*)} \leq s$, $y_{(2k,0)} \leq \tau_k$, which implies $s \leq \tau_k$. This shows the result in the case $n = 2$.

Assume now $n \geq 3$ and the result holds for $n-1$. Thus $y_{2\alpha} \leq \tau_k$ if $|\alpha| = k$ and $\alpha_i = 0$ for some i. Assume now $1 \leq \alpha_1 \leq \ldots \leq \alpha_n$. Consider the sequences $\gamma := (2\alpha_1, 0, \alpha_3 + \alpha_2 - \alpha_1, \alpha_4, \ldots, \alpha_n)$ and $\gamma' := (0, 2\alpha_2, \alpha_3 + \alpha_1 - \alpha_2, \alpha_4, \ldots, \alpha_n)$. Thus $|\gamma| = |\gamma'| = |\alpha| = k$, $\gamma + \gamma' = 2\alpha$. As $\gamma_2 = \gamma'_1 = 0$, we have $y_{2\gamma}, y_{2\gamma'} \leq \tau_k$. Hence $y_{2\alpha}^2 = y_{\gamma+\gamma'}^2 \leq y_{2\gamma}y_{2\gamma'} \leq \tau_k^2$, implying $y_{2\alpha} \leq \tau_k$. □

COROLLARY 7.7. *Assume $M_t(y) \succeq 0$. Then $|y_\alpha| \leq \max_{0 \leq k \leq t} \tau_k = \max(\tau_0, \tau_t)$.*

Proof. Using Lemma 7.5, we see that $y_{(2k,0,\ldots,0)} \leq \max(y_0, y_{2t,0,\ldots,0)}) \leq \max(\tau_0, \tau_t)$, implying $\tau_k \leq \max(\tau_0, \tau_t)$ and thus $\max_{0 \leq k \leq t} \tau_k = \max(\tau_0, \tau_t) =: \tau$. By Lemma 7.6 we deduce $y_{2\alpha} \leq \tau$ for $|\alpha| \leq t$. Consider now $|\gamma| \leq 2t$. Write $\gamma = \alpha + \beta$ with $|\alpha|, |\beta| \leq t$. Then $y_\gamma^2 \leq y_{2\alpha}y_{2\beta} \leq \tau^2$, giving $|y_\gamma| \leq \tau$. □

We mention for completeness another result for bounding entries of a positive semidefinite moment matrix. This result is used in [73] for giving an explicit set of conditions ensuring that a polynomial p is a sum of squares, the conditions depending only on the coefficients of p.

LEMMA 7.8. *[73] If $M_t(y) \succeq 0$ and $y_0 = 1$, then $|y_\alpha|^{1/|\alpha|} \leq \tau_t^{1/2t}$ for all $|\alpha| \leq 2t$.*

Proof. Use induction on $t \geq 1$. The result holds obviously for $t = 1$. Assume the result holds for $t \geq 1$ and let $M_{t+1}(y) \succeq 0$, $y_0 = 1$. By the induction assumption, $|y_\alpha|^{1/|\alpha|} \leq \tau_t^{1/2t}$ for $|\alpha| \leq 2t$. By Lemma 7.6, $|y_\alpha| \leq \tau_{t+1}$ for $|\alpha| = 2t+2$. We first show $\tau_t^{1/t} \leq \tau_{t+1}^{1/(t+1)}$. For this, say $\tau_t = y_{2te_1}$; then $\tau_t^2 = y_{2te_1}^2 \leq y_{2(t+1)e_1}y_{2(t-1)e_1} \leq \tau_{t+1}\tau_t^{(2t-2)/2t}$, which gives $\tau_t^{1/t} \leq \tau_{t+1}^{1/(t+1)}$. Remains only to show that $|y_\alpha|^{1/|\alpha|} \leq \tau_{t+1}^{1/t+1}$ for $|\alpha| = 2t+1$ (as the case $|\alpha| \leq 2t$ follows using the induction assumption, and $|\alpha| = 2t+2$ has been settled above). Say, $|\alpha| = 2t+1$ and $\alpha = \beta + \gamma$ with $|\beta| = t$, $|\gamma| = t+1$. Then $y_\alpha^2 \leq y_{2\beta}y_{2\gamma} \leq \tau_t\tau_{t+1} \leq \tau_{t+1}^{t/(t+1)}\tau_{t+1} = \tau_{t+1}^{(2t+1)/(t+1)}$, giving the desired result. □

The following result is crucial for the proof of Theorem 7.2.

PROPOSITION 7.9. *Given a polynomial $f \in \mathbb{R}[\mathrm{x}]$ consider the program*

$$\epsilon_t^* := \inf f^T y \text{ s.t. } M_t(y) \succeq 0,\ y^T \Theta_t \leq 1 \tag{7.3}$$

for any integer $t \geq d_f = \lceil \deg(f)/2 \rceil$. Recall the polynomial Θ_t from (7.1). Then,

(i) *$-\infty < \epsilon_t^* \leq 0$ and the infimum is attained in (7.3).*

(ii) *For $\epsilon \geq 0$, the polynomial $f + \epsilon\Theta_t$ is a sum of squares if and only if $\epsilon \geq -\epsilon_t^*$. In particular, f is a sum of squares if and only if $\epsilon_t^* = 0$.*

(iii) *If the polynomial* $f \in \mathbb{R}[\mathbf{x}]$ *is nonnegative on* $[-1,1]^n$, *then* $\lim_{t\to\infty} \epsilon_t^* = 0$.

Proof. Let y be feasible for the program (7.3). Then $0 \leq y_0, y_{(2t,0,\ldots,0)}, \ldots, y_{(0,\ldots,0,2t)} \leq 1$ (from the linear constraint $y^T\Theta_t \leq 1$) which, using Corollary 7.7, implies $|y_\alpha| \leq 1$ for all α. Hence the feasible region of (7.3) is bounded and nonempty (as $y = 0$ is feasible). Therefore the infimum is attained in program (7.3) and $-\infty < \epsilon_t^* \leq 0$, showing (i). One can verify that the dual semidefinite program of (7.3) reads

$$d_t^* := \sup_{\lambda \geq 0} -\lambda \ \text{ s.t. } \ f + \lambda\Theta_t \ \text{ is a sum of squares.}$$

As the program (7.3) is strictly feasible (choose for y the sequence of moments of a measure with positive density on $\mathbb{R}^n$, with finite moments up to order $2t$, rescaled so as to satisfy $y^T\Theta_t \leq 1$), its dual semidefinite program attains it supremum and there is no duality gap, i.e. $\epsilon_t^* = d_t^*$. Thus $f + \epsilon\Theta_t$ is a sum of squares if and only if $-\epsilon \leq d_t^* = \epsilon_t^*$, i.e. $\epsilon \geq -\epsilon_t^*$, showing (ii).

We now show (iii). Say $\epsilon_t^* = f^T y^{(t)}$, where $y^{(t)}$ is an optimum solution to (7.3) with, as we saw above, $y^{(t)} \in [-1,1]^{\mathbb{N}^n_{2t}}$. Complete $y^{(t)}$ to a sequence $\tilde{y}^{(t)} = (y^{(t)}, 0, \ldots, 0) \in [-1,1]^{\mathbb{N}^n}$. As $[-1,1]^{\mathbb{N}^n}$ is compact, there exists a converging subsequence $(y^{(t_l)})_{l\geq 0}$, converging to $y^* \in [-1,1]^{\mathbb{N}^n}$ in the product topology. In particular, there is coordinate-wise convergence, i.e. $(y_\alpha^{(t_l)})_{l\geq 0}$ converges to y_α^*, for all α. Therefore $M(y^*) \succeq 0$. As $y^* \in [-1,1]^{\mathbb{N}^n}$, we deduce using Theorem 4.7 that y^* has a representing measure μ on $[-1,1]^n$. In particular, $\epsilon_{t_l}^* = f^T y^{(t_l)}$ converges to $f^T y^* = \int_{[-1,1]^n} f(x)\mu(dx)$. By assumption, $f \geq 0$ on $[-1,1]^n$ and thus $f^T y^* \geq 0$. On the other hand, $f^T y^* \leq 0$ since $\epsilon_t^* \leq 0$ for all t. Thus $f^T y^* = 0$. This shows that the only accumulation point of the sequence ϵ_t is 0 and thus ϵ_t converges to 0. □

We can now conclude the proof of Theorem 7.2. Let $\epsilon > 0$. By Proposition 7.9 (iii), $\lim_{t\to\infty} \epsilon_t^* = 0$. Hence there exists $t_0 \in \mathbb{N}$ such that $\epsilon_t^* \geq -\epsilon$ for all $t \geq t_0$. Applying Proposition 7.9 (ii), we deduce that $f + \epsilon\Theta_t$ is a sum of squares.

As an example, consider the univariate polynomial $f = 1 - \mathbf{x}^2$, obviously nonnegative on $[-1,1]$. Then, for $\epsilon \geq (t-1)^{t-1}/t^t$, the polynomial $f + \epsilon\mathbf{x}^{2t}$ is nonnegative on $\mathbb{R}$ and thus a sum of squares (see [77] for details).

We now turn to the proof of Theorem 7.3, whose details are a bit more technical. Given an integer $M > 0$, consider the program

$$\mu_M^* := \inf_\mu \int f(x)\mu(dx) \ \text{ s.t. } \int \sum_{i=1}^n e^{x_i^2}\mu(dx) \leq n e^{M^2}, \qquad (7.4)$$

where the infimum is taken over all probability measures μ on $\mathbb{R}^n$.

LEMMA 7.10. *Let $f \in \mathbb{R}[\mathbf{x}]$ and assume $f^{min} := \inf_{x\in\mathbb{R}^n} f(x) > -\infty$. Then $\mu_M^* \downarrow f^{min}$ as $M \to \infty$.*

Proof. Obviously, the sequence $(\mu_M^*)_M$ is monotonically non-increasing and $\mu_M^* \geq f^{\min}$. Next observe that $\mu_M^* \leq \inf_{\|x\|_\infty \leq M} f(x)$ since the Dirac measure $\mu = \delta_x$ at any point x with $\|x\|_\infty \leq M$ is feasible for (7.4) with objective value $f(x)$. Now $\inf_{\|x\|_\infty \leq M} f(x)$ converges to $f^{\min}$ as $M \to \infty$, which implies that $\mu_M^* \downarrow f^{\min}$ as $M \to \infty$. □

The idea is now to approach the optimum value of (7.4) via a sequence of moment relaxations. Namely, for any integer $t \geq d_f = \lceil \deg(f)/2 \rceil$, consider the semidefinite program

$$\epsilon_{t,M}^* := \inf f^T y \text{ s.t. } M_t(y) \succeq 0,\ y_0 = 1,\ y^T\theta_t \leq ne^{M^2} \tag{7.5}$$

whose dual reads

$$d_{t,M}^* := \sup_{\gamma,\lambda} \gamma - \lambda ne^{M^2} \text{ s.t. } \lambda \geq 0,\ \gamma + \lambda\theta_r \text{ is a sum of squares.} \tag{7.6}$$

The next result is crucial for the proof of Theorem 7.3.

PROPOSITION 7.11. *Fix $M > 0$, $t \geq d_f$, and assume $f^{min} > -\infty$. The following holds for the programs (7.5) and (7.6).*

(i) *The optimum is attained in both programs (7.5) and (7.6) and there is no duality gap, i.e. $\epsilon_{t,M}^* = d_{t,M}^*$.*

(ii) *$\epsilon_{t,M}^* \uparrow \mu_M^*$ as $t \to \infty$.*

Proof. (i) As (7.5) is strictly feasible, its dual (7.6) attains its optimum and there is no duality gap. The infimum is attained in program (7.5) since the feasible region is bounded (directly using the constraint $y^T\theta_t \leq ne^{M^2}$ together with Corollary 7.7) and nonempty (as $y = (1, 0, \ldots, 0)$ is feasible for (7.5)).

(ii) We begin with observing that $(\epsilon_{t,M}^*)_t$ is monotonically non-decreasing; hence $\lim_{t\to\infty} \epsilon_{t,M}^* = \sup_t \epsilon_{t,M}^*$. Let μ be feasible for (7.4) and let y be its sequence of moments. Then, for any integer $t \geq d_f$, $\int f(x)\mu(dx) = f^T y$, $M_t(y) \succeq 0$, $y_0 = 1$ and, as $\sum_{k=0}^\infty x_i^{2k}/k! = e^{x_i^2}$, the constraint $\int \sum_{i=1}^n e^{x_i^2} \mu(dx) \leq ne^{M^2}$ implies $y^T\theta_t \leq ne^{M^2}$. That is, y is feasible for (7.5) and thus $\mu_M^* \geq \epsilon_{t,M}^*$. This shows $\mu_M^* \geq \lim_{t\to\infty} \epsilon_{t,M}^*$.

We now show the reverse inequality. For this we first note that if y is feasible for (7.5), then $\max_{i\leq n, k\leq t} y_{2ke_i} \leq t!ne^{M^2} =: \sigma_t$ and thus $\max_{|\alpha|\leq 2t} |y_\alpha| \leq \sigma_t$ (by Corollary 7.7). Moreover, for any $s \leq t$, $|y_\alpha| \leq \sigma_s$ for $|\alpha| \leq 2s$ (since the restriction of y to $\mathbb{R}^{\mathbb{N}^n_{2s}}$ is feasible for the program (7.5) with s in place of t).

Say $\epsilon_{t,M}^* = f^T y^{(t)}$, where $y^{(t)}$ is an optimum solution to (7.5) (which is attained by (i)). Define $\tilde{y}^{(t)} = (y^{(t)}, 0 \ldots 0) \in \mathbb{R}^{\mathbb{N}^n}$ and $\hat{y}^{(t)} \in \mathbb{R}^{\mathbb{N}^n}$ by $\hat{y}_\alpha^{(t)} := \tilde{y}_\alpha^{(t)}/\sigma_s$ if $2s - 1 \leq |\alpha| \leq 2s$, $s \geq 0$. Thus each $\hat{y}^{(t)}$ lies in the compact set $[-1, 1]^{\mathbb{N}^n}$. Hence there is a converging subsequence $(\hat{y}^{(t_l)})_{l\geq 0}$, converging say to $\hat{y} \in [-1,1]^{\mathbb{N}^n}$. In particular, $\lim_{l\to\infty} \hat{y}_\alpha^{(t_l)} = \hat{y}_\alpha$ for all α. Define $y^* \in \mathbb{R}^{\mathbb{N}^n}$ by $y^*_\alpha := \sigma_s \hat{y}_\alpha$ for $2s-1 \leq |\alpha| \leq 2s$, $s \geq 0$. Then $\lim_{l\to\infty} \tilde{y}_\alpha^{(t_l)} = y^*_\alpha$ for all α and $\lim_{l\to\infty} y_\alpha^{(t_l)} = y^*_\alpha$ for all $|\alpha| \leq 2t_l$. From this follows that $M(y^*) \succeq 0$, $y^*_0 = 1$, and $(y^*)^T\theta_r \leq ne^{M^2}$ for any $r \geq 0$. In particular, $\sum_{k=0}^\infty \sum_{i=1}^n \frac{y^*_{2ke_i}}{k!} \leq ne^{M^2}$, which implies[2] $\sum_{k=0}^\infty (y_{2ke_i})^{-1/2k} = \infty$ for all i. That is, condition (7.2) holds and thus, by Theorem 7.4, y^* has a representing measure μ. As μ is feasible for (7.4), this implies $\mu^*_M \leq \int f(x)\mu(dx) = f^Ty^* = \lim_{l\to\infty} f^Ty^{(t_l)} = \lim_{l\to\infty} \epsilon^*_{t_l,M}$. Hence we find $\mu^*_M \leq \lim_{l\to\infty} \epsilon^*_{t_l,M} \leq \lim_{t\to\infty} \epsilon^*_{t,M} \leq \mu^*_M$ and thus equality holds throughout, which shows (ii). □

We can now conclude the proof of Theorem 7.3. We begin with two easy observations. First it suffices to show the existence of $t_0 \in \mathbb{N}$ for which $f + \epsilon\theta_{t_0}$ is a sum of squares (since this obviously implies that $f + \epsilon\theta_t$ is a sum of squares for all $t \geq t_0$). Second we note that it suffices to show the result for the case $f^{\min} > 0$. Indeed, if $f^{\min} = 0$, consider the polynomial $g := f + n\epsilon/2$ with $g^{\min} = n\epsilon/2 > 0$. Hence, for some $t_0 \in \mathbb{N}$, $g + (\epsilon/2)\theta_{t_0}$ is a sum of squares. As $(\epsilon/2)(\theta_{t_0} - n)$ is a sum of squares, we find that $f + \epsilon\theta_{t_0} = g + (\epsilon/2)\theta_{t_0} + (\epsilon/2)(\theta_{t_0} - n)$ is a sum of squares.

So assume $f^{\min} > 0$ and $f^{\min} > 1/M$, where M is a positive integer. By Proposition 7.11 (ii), $\epsilon^*_{t_M,M} \geq \mu^*_M - 1/M \geq f^{\min} - 1/M > 0$ for some integer t_M. By Proposition 7.11 (i), we have $\epsilon^*_{t_M,M} = \gamma_M - \lambda_M ne^{M^2}$, where $\lambda_M \geq 0$ and $f - \gamma_M + \lambda_M\theta_{t_M} =: q$ is a sum of squares. Hence $f + \lambda_M\theta_{t_M} = q + \gamma_M$ is a sum of squares, since $\gamma_M = n\lambda_M e^{M^2} + \epsilon^*_{t_M,M} \geq 0$. Moreover, evaluating at the point 0, we find $f(0) - \gamma_M + \lambda_M n = q(0) \geq 0$, i.e. $f(0) - f^{\min} + f^{\min} - \epsilon^*_{t_M,M} - \lambda_M ne^{M^2} + \lambda_M n \geq 0$. As $f^{\min} - \epsilon^*_{t_M,M} \leq 1/M$, this implies $\lambda_M \leq \frac{1/M + f(0) - f^{\min}}{n(e^{M^2}-1)}$. Therefore, $\lim_{M\to\infty} \lambda_M = 0$. We can now conclude the proof: Given $\epsilon > 0$, choose M in such a way that $f^{\min} > 1/M$ and $\lambda_M < \epsilon$. Then $f + \epsilon\theta_{t_M}$ is a sum of squares.

We refer to [69], [77] for further approximation results by sums of squares for polynomials nonnegative on a basic closed semialgebraic set.

7.2. Unconstrained polynomial optimization. In this section we come back to the unconstrained minimization problem (1.3) which, given a polynomial $p \in \mathbb{R}[\mathbf{x}]$, asks for its infimum $p^{\min} = \inf_{x\in\mathbb{R}^n} p(x)$. There is quite a large literature on this problem; we sketch here only some of the methods that are most relevant to the topic of this survey. We first make some general easy observations. To begin with, we may assume that $\deg(p) =: 2d$ is even, since otherwise $p^{\min} = -\infty$. Probably the most natural idea is to search for global minimizers of p within the critical points of p. One should be careful however. Indeed p may have infinitely many

[2]Indeed if $a_k > 0$, $C \geq 1$ satisfy $a_k \leq Ck!$ for all $k \geq 1$, then $a_k \leq Ck^k$, implying $a_k^{-1/2k} \geq C^{-1/2k}/\sqrt{k}$ and thus $\sum_{k\geq 1} a_k^{-1/2k} = \infty$.

global minimizers, or p may have none! The latter happens, for instance, for the polynomial $p = \mathbf{x}_1^2 + (\mathbf{x}_1\mathbf{x}_2 - 1)^2$; then for $\epsilon > 0$, $p(\epsilon, 1/\epsilon) = \epsilon^2$ converges to 0 as $\epsilon \to 0$, showing $p^{\min} = 0$ but the infimum is *not* attained. Next, how can one recognize whether p has a global minimizer? As observed by Marshall [88], the highest degree homogeneous component of p plays an important role.

LEMMA 7.12. *[88] For a polynomial $p \in \mathbb{R}[\mathbf{x}]$, let $\tilde{p}$ be its highest degree component, consisting of the sum of the terms of p with maximum degree, and let $\tilde{p}_S^{\min}$ denote the minimum of $\tilde{p}$ over the unit sphere.*

(i) *If $\tilde{p}_S^{\min} < 0$ then $p^{min} = -\infty$.*

(ii) *If $\tilde{p}_S^{\min} > 0$ then p has a global minimizer. Moreover any global minimizer x satisfies $\|x\| \le \max\left(1, \frac{1}{\tilde{p}_S^{\min}} \sum_{1\le|\alpha|<\deg(p)} |p_\alpha|\right)$.*

Proof. (i) is obvious. (ii) Set $\deg(p) =: d$, $p = \tilde{p} + g$, where all terms of g have degree $\le d-1$. If $p^{\min} = p(0)$, 0 is a global minimizer and we are done. Otherwise let $x \in \mathbb{R}^n$ with $p(x) \le p(0)$. Then, $\tilde{p}(x) \le g(x) - p(0) \le \sum_{1\le|\alpha|\le d-1} |p_\alpha||x^\alpha|$. Combined with $\tilde{p}(x) = \tilde{p}(x/\|x\|)\|x\|^d \ge \tilde{p}_S^{\min}\|x\|^d$, and $|x^\alpha| \le \|x\|^{|\alpha|}$ if $\|x\| \ge 1$, this gives $\|x\| \le \max\left(\frac{1}{\tilde{p}_S^{\min}}, \sum_{|\alpha|<\deg(p)} |p_\alpha|, 1\right)$. □

No conclusion can be drawn when $\tilde{p}_S^{\min} = 0$; indeed p may have a minimum (e.g. for $p = \mathbf{x}_1^2\mathbf{x}_2^2$), or a finite infimum (e.g. for $p = \mathbf{x}_1^2 + (\mathbf{x}_1\mathbf{x}_2 - 1)^2$), or an infinite infimum (e.g. for $p = \mathbf{x}_1^2 + \mathbf{x}_2$).

We now see how we can apply the general relaxation scheme from Section 6 to the problem (1.3). As there are no constraints, we find just one lower bound for $p^{\min}$:

$$p_t^{\rm sos} = p_t^{\rm mom} = p_d^{\rm sos} = p_d^{\rm mom} \le p^{\min} \text{ for all } t \ge d,$$

with equality $p_d^{\rm sos} = p^{\min}$ if and only if $p - p^{\min}$ is a sum of squares. Indeed, $p_t^{\rm sos} = p_d^{\rm sos}$ since the degree of a sum of squares decomposition of $p - \rho$ ($\rho \in \mathbb{R}$) is bounded by $2d$. Moreover, as (6.3) is strictly feasible, the supremum is attained in (6.2), there is no duality gap, i.e. $p_t^{\rm sos} = p_t^{\rm mom}$, and $p_d^{\rm sos} = p^{\min}$ if and only if $p - p^{\min}$ is a sum of squares. Therefore, if $p - p^{\min}$ is a sum of squares, the infimum $p^{\min}$ of p can be found via the semidefinite program (6.3) at order $t = d$. Otherwise, we just find one lower bound for $p^{\min}$. One may wonder when is this lower bound nontrivial, i.e., when is $p_d^{\rm sos} \ne -\infty$, or in other words when does there exist a scalar ρ for which $p - \rho$ is a sum of squares. Marshall [89] gives an answer which involves again the highest degree component of p.

PROPOSITION 7.13. *[89] Let $p \in \mathbb{R}[\mathbf{x}]_{2d}$, $\tilde{p}$ its highest degree component, and $\Sigma_{n,2d}$ the cone of homogeneous polynomials of degree $2d$ that are sums of squares.*

(i) *If* $p_d^{sos} \neq -\infty$ *then* $\tilde{p}$ *is a sum of squares, i.e.* $\tilde{p} \in \Sigma_{n,2d}$.

(ii) *If* $\tilde{p}$ *is an interior point of* $\Sigma_{n,2d}$ *then* $p_d^{sos} \neq -\infty$.

For instance, $\sum_{i=1}^n \mathbf{x}_i^{2d}$, $(\sum_{i=1}^d \mathbf{x}_i^2)^d$ are interior points of $\Sigma_{n,2d}$. See [89] for details.

EXAMPLE 7.14. *Here are some examples taken from [89]. For the Motzkin polynomial* $p = p_M := \mathbf{x}^4\mathbf{y}^2 + \mathbf{x}^2\mathbf{y}^4 - 3\mathbf{x}^2\mathbf{y}^2 + 1$, $p^{min} = 0$, $\tilde{p} = \mathbf{x}^4\mathbf{y}^2 + \mathbf{x}^2\mathbf{y}^4$ *is a sum of squares, and* $p_3^{sos} = -\infty$. *Thus the necessary condition from Proposition 7.13 is not sufficient.*

For $p = (\mathbf{x} - \mathbf{y})^2$, $p^{min} = p_1^{sos} = 0$, *and* $\tilde{p} = p$ *lies on the boundary of* $\Sigma_{2,2}$. *Thus the sufficient condition of Proposition 7.13 is not necessary.*

For $p = p_M + \epsilon(\mathbf{x}^6 + \mathbf{y}^6)$, *where* p_M *is the Motzkin polynomial,* $p^{min} = \epsilon/(1+\epsilon)$, $\tilde{p}_\epsilon = \mathbf{x}^4\mathbf{y}^2 + \mathbf{x}^2\mathbf{y}^4 + \epsilon(\mathbf{x}^6 + \mathbf{y}^6)$ *is an interior point of* $\Sigma_{3,6}$. *Thus* $p_{\epsilon,3}^{sos} \neq -\infty$. *Yet* $\lim_{\epsilon \to 0} p_{\epsilon,3}^{sos} = -\infty$ *for otherwise* $p_M + \rho$ *would be a sum of squares for some* ρ *(which is not possible, as observed in Example 3.7).*

Thus arises naturally the question of designing alternative relaxation schemes to get better approximations for $p^{\min}$. A natural idea is to try to transform the unconstrained problem (1.3) into a *constrained* problem. We now start with the most favourable situation when p has a minimum and moreover some information is known about the position of a global minimizer.

Assume p attains its minimum and one can locate a global minimizer. If p attains its minimum and if some bound R is known on the Euclidian norm of a global minimizer, then (1.3) can be reformulated as the constrained minimization problem over the ball

$$p^{\min} = p^{\text{ball}} := \min\ p(x) \text{ s.t. } \sum_{i=1}^n x_i^2 \leq R^2. \tag{7.7}$$

We can now apply the relaxation scheme from Section 6 to the semialgebraic set $K = \{x \mid \sum_{i=1}^n x_i^2 \leq R^2\}$ which obviously satisfies Putinar's assumption (3.15); thus the moment/SOS bounds converge to $p^{\text{ball}} = p^{\min}$. This approach seems to work well if the radius R is not too large.

What if no information is known about the norm of a global minimizer? Nie, Demmel and Sturmfels [99] propose an alternative way of transforming (1.3) into a constrained problem when p attains its infimum. Define the *gradient ideal* of p

$$\mathcal{I}_p^{\text{grad}} := \left(\frac{\partial p}{\partial x_i}\ (i = 1, \ldots, n) \right), \tag{7.8}$$

as the ideal generated by the partial derivatives of p. Since all global minimizers of p are critical points, i.e. they lie in $V_{\mathbb{R}}(\mathcal{I}_p^{\text{grad}})$, the (real) gradient variety of p, the unconstrained minimization problem (1.3) can be reformulated as the constrained minimization problem over the gradient variety

$$p^{\min} = p^{\text{grad}} := \min_{x \in V_{\mathbb{R}}(\mathcal{I}_p^{\text{grad}})} p(x). \tag{7.9}$$

Note that the equality $p^{\min} = p^{\text{grad}}$ may not hold if p has no minimum. E.g. for $p = \mathbf{x}_1^2 + (1-\mathbf{x}_1\mathbf{x}_2)^2$, $p^{\min} = 0$ while $p^{\text{grad}} = 1$ as $V_{\mathbb{C}}(\mathcal{I}_p^{\text{grad}}) = \{0\}$.

We can compute the moments/SOS bounds obtained by applying the relaxation scheme from Section 6 to the semialgebraic set $V_{\mathbb{R}}(\mathcal{I}_p^{\text{grad}})$. However in general this set may not satisfy the assumption (3.15), hence we cannot apply Theorem 6.8 (which relies on Theorem 3.20) to show the asymptotic convergence of the moment/SOS bounds to p^{grad}. Yet asymptotic convergence *does hold* and sometimes even *finite* convergence. Nie et al. [99] show the representation results from Theorems 7.15-7.16 below, for positive (nonnegative) polynomials on their gradient variety as sums of squares modulo their gradient ideal. As an immediate application, there is asymptotic convergence (moreover, finite convergence when $\mathcal{I}_p^{\text{grad}}$ is radical) of the moment/SOS bounds from the programs (6.3), (6.2) (applied to the polynomial constraints $\partial p/\partial x_i = 0$ ($i = 1, \ldots, n$)) to the parameter p^{grad} (hence to $p^{\min}$ when p is assumed to attain its minimum).

THEOREM 7.15. *[99] If $p(x) > 0$ for all $x \in V_{\mathbb{R}}(\mathcal{I}_p^{\text{grad}})$, then p is a sum of squares modulo its gradient ideal $\mathcal{I}_p^{\text{grad}}$, i.e., $p = s_0 + \sum_{i=1}^n s_i \partial p/\partial x_i$, where $s_i \in \mathbb{R}[\mathbf{x}]$ and s_0 is a sum of squares.*

THEOREM 7.16. *[99] Assume $\mathcal{I}_p^{\text{grad}}$ is a radical ideal and $p(x) \geq 0$ for all $x \in V_{\mathbb{R}}(\mathcal{I}_p^{\text{grad}})$. Then p is a sum of squares modulo its gradient ideal $\mathcal{I}_p^{\text{grad}}$, i.e., $p = s_0 + \sum_{i=1}^n s_i \partial p/\partial x_i$, where $s_i \in \mathbb{R}[\mathbf{x}]$ and s_0 is a sum of squares.*

We postpone the proofs of these two results, which need some algebraic tools, till Section 7.3. The following example of C. Scheiderer shows that the assumption that $\mathcal{I}_p^{\text{grad}}$ is radical cannot be removed in Theorem 7.16.

EXAMPLE 7.17. Consider the polynomial $p = \mathbf{x}^8 + \mathbf{y}^8 + \mathbf{z}^8 + M$, where $M = \mathbf{x}^4\mathbf{y}^2 + \mathbf{x}^2\mathbf{y}^4 + \mathbf{z}^6 - 3\mathbf{x}^2\mathbf{y}^2\mathbf{z}^2$ is the Motzkin form. As observed earlier, M is nonnegative on $\mathbb{R}^3$ but not a sum of squares. The polynomial p is nonnegative over $\mathbb{R}^3$, thus over $V_{\mathbb{R}}(\mathcal{I}_p^{\text{grad}})$, but it is not a sum of squares modulo $\mathcal{I}_p^{\text{grad}}$. Indeed one can verify that $p - M/4 \in \mathcal{I}_p^{\text{grad}}$ and that M is not a sum of squares modulo $\mathcal{I}_p^{\text{grad}}$ (see [99] for details); thus $\mathcal{I}_p^{\text{grad}}$ is not radical.

Let us mention (without proof) a related result of Marshall [89] which shows a representation result related to that of Theorem 7.16 but under a different assumption.

THEOREM 7.18. *[89] Assume p attains its minimum and the matrix $\left(\frac{\partial^2 p}{\partial x_i \partial x_j}(x)\right)_{i,j=1}^n$ is positive definite at every global minimizer x of p. Then*

$p - p^{min}$ is a sum of squares modulo $\mathcal{I}_p^{\rm grad}$.

Summarizing, the above results of Nie et al. [99] show that the parameter $p^{\rm grad}$ can be approximated via converging moment/SOS bounds; when p has a minimum, then $p^{\min} = p^{\rm grad}$ and thus $p^{\min}$ too can be approximated.

How to deal with polynomials that do not have a minimum? A first strategy is to perturb the polynomial in such a way that the perturbed polynomial has a minimum. For instance, Hanzon and Jibetean [50], Jibetean [59] propose the following perturbation

$$p_\epsilon := p + \epsilon\Big(\sum_{i=1}^n \mathbf{x}_i^{2d+2}\Big)$$

if p has degree $2d$, where $\epsilon > 0$. Then the perturbed polynomial p_ϵ has a minimum (e.g. because the minimum of $\sum_i \mathbf{x}_i^{2d+2}$ over the unit sphere is equal to $1/n^d > 0$; recall Lemma 7.12) and $\lim_{\epsilon\to 0} p_\epsilon^{\min} = p^{\min}$.

For fixed $\epsilon > 0$, $p_\epsilon^{\min} = p_\epsilon^{\rm grad}$ can be obtained by minimizing p_ϵ over its gradient variety and the asymptotic convergence of the moment/SOS bounds to $p_\epsilon^{\rm grad}$ follows from the above results of Nie et al. [99]. Alternatively we may observe that the gradient variety of p_ϵ is finite. Indeed, $\partial p_\epsilon/\partial x_i = (2d+2)\mathbf{x}_i^{2d+1} + \partial p/\partial x_i$, where $\deg(\partial p/\partial \mathbf{x}_i) < 2d$. Hence, $|V_\mathbb{C}(\mathcal{I}_{p_\epsilon}^{\rm grad})| \leq \dim \mathbb{R}[\mathbf{x}]/\mathcal{I}_{p_\epsilon}^{\rm grad} \leq (2d+1)^n$. By Theorem 6.15, we can conclude to the *finite* convergence of the moment/SOS bounds to $p_\epsilon^{\rm grad} = p_\epsilon^{\min}$. Jibetean and Laurent [60] have investigated this approach and present numerical results. Moreover they propose to exploit the equations defining the gradient variety to reduce the number of variables in the moment relaxations.

Hanzon and Jibetean [50] and Jibetean [59] propose in fact an exact algorithm for computing $p^{\min}$. Roughly speaking they exploit the fact (recall Theorem 2.9) that the points of the gradient variety of p_ϵ can be obtained as eigenvalues of the multiplication matrices in the quotient space $\mathbb{R}[\mathbf{x}]/\mathcal{I}_{p_\epsilon}^{\rm grad}$ and they study the behaviour of the limits as $\epsilon \to 0$. In particular they show that when p has a minimum, the limit set as $\epsilon \to 0$ of the set of global minimizers of p_ϵ is contained in the set of global minimizers of p, and each connected component of the set of global minimizers of p contains a point which is the limit of a branch of minimizers of p_ϵ. Their method however has a high computational cost and is thus not practical.

Schweighofer [133] proposes a different strategy for dealing with the case when p has no minimum. Namely he proposes to minimize p over the following semialgebraic set

$$K_{\nabla p} := \Big\{x \in \mathbb{R}^n \,\Big|\, \Big(\sum_{i=1}^n \Big(\frac{\partial p}{\partial x_i}(x)\Big)^2\Big)\Big(\sum_{i=1}^n x_i^2\Big) \leq 1\Big\},$$

which contains the gradient variety. Schweighofer [133] shows that, if $p^{\min} > -\infty$, then $p^{\min} = \inf_{x \in K_{\nabla p}} p(x)$. Moreover, he shows the following representation theorem, thus leading to a hierarchy of SOS/moment approximations for $p^{\min}$, also in the case when the infimum is not attained; the result holds under some technical condition, whose precise definition can be found in [133].

THEOREM 7.19. *[133] Assume $p^{min} > -\infty$. Furthermore assume that, either p has only isolated singularities at infinity (which is always true if $n = 2$), or $K_{\nabla p}$ is compact. Then the following assertions are equivalent.*

(i) $p \geq 0$ on $\mathbb{R}^n$;

(ii) $p \geq 0$ on $K_{\nabla p}$;

(iii) $\forall \epsilon > 0\ \exists s_0, s_1 \in \Sigma\ \ p + \epsilon = s_0 + s_1 \Big(1 - (\sum_{i=1}^n (\partial p/\partial x_i)^2)(\sum_{i=1}^n \mathbf{x}_i^2)\Big)$.

7.3. Sums of squares over the gradient ideal. We give here the proofs for Theorems 7.15 and 7.16 about sums of squares representations modulo the gradient ideal, following Nie et al. [99] (although our proof slightly differs at some places). We begin with the following lemma which can be seen as an extension of Lemma 2.3 about existence of interpolation polynomials. Recall that a set $V \subseteq \mathbb{C}^n$ is a variety if $V = V_{\mathbb{C}}(\{p_1, \ldots, p_s\})$ for some polynomials $p_i \in \mathbb{C}[\mathbf{x}]$. When all p_i's are real polynomials, i.e. $p_i \in \mathbb{R}[\mathbf{x}]$, then $V = \overline{V} := \{\overline{v} \mid v \in V\}$, i.e. $v \in V \Leftrightarrow \overline{v} \in V$.

LEMMA 7.20. *Let $V_1, \ldots, V_r$ be pairwise disjoint varieties in $\mathbb{C}^n$ such that $V_i = \overline{V}_i := \{\overline{v} \mid v \in V_i\}$ for all i. There exist polynomials $p_1, \ldots, p_r \in \mathbb{R}[\mathbf{x}]$ such that $p_i(V_j) = \delta_{i,j}$ for $i, j = 1, \ldots r$; that is, $p_i(v) = 1$ if $v \in V_i$ and $p_i(v) = 0$ if $v \in V_j$ $(j \neq i)$.*

Proof. The ideal $\mathcal{I}_i := \mathcal{I}(V_i) \subseteq \mathbb{C}[\mathbf{x}]$ is radical with $V_{\mathbb{C}}(\mathcal{I}_i) = V_i$. We have $V_{\mathbb{C}}(\mathcal{I}_i + \bigcap_{j \neq i} \mathcal{I}_j) = V_{\mathbb{C}}(\mathcal{I}_i) \cap V_{\mathbb{C}}(\bigcap_{j \neq i} \mathcal{I}_j) = V_{\mathbb{C}}(\mathcal{I}_i) \cap (\bigcup_{j \neq i} V_{\mathbb{C}}(\mathcal{I}_j)) = V_i \cap (\bigcup_{j \neq i} V_j) = \emptyset$. Hence, by Hilbert's Nullstellensatz (Theorem 2.1 (i)), $1 \in \mathcal{I}_i + \bigcap_{j \neq i} \mathcal{I}_j$; say $1 = q_i + p_i$, where $q_i \in \mathcal{I}_i$ and $p_i \in \bigcap_{j \neq i} \mathcal{I}_j$. Hence $p_i(V_j) = \delta_{i,j}$ (since q_i vanishes on V_i and p_i vanishes on V_j for $j \neq i$). As $V_i = \overline{V}_i$ for all i, we can replace p_i by its real part to obtain polynomials satisfying the properties of the lemma. □

A variety $V \subseteq \mathbb{C}^n$ is irreducible if any decomposition $V = V_1 \cup V_2$, where V_1, V_2 are varieties, satisfies $V_1 = V$ or $V_2 = V$. It is a known fact that any variety can be written (in a unique way) as a finite union of irreducible varieties (known as its irreducible components) (see e.g. [21, Chap. 4]). Let $V_{\mathbb{C}}(\mathcal{I}_p^{\text{grad}}) = \bigcup_{l=1}^L V_l$ be the decomposition of the gradient variety into irreducible varieties. The following fact is crucial for the proof.

LEMMA 7.21. *The polynomial p is constant on each irreducible component of its gradient variety $V_{\mathbb{C}}(\mathcal{I}_p^{\text{grad}})$.*

Proof. Fix an irreducible component V_l. We use the fact[3] that V_l is connected by fintely many differentiable paths. Given $x, y \in V_l$, assume that there exists a continuous differentiable function $\varphi : [0,1] \to V_l$ with $\varphi(0) = x$ and $\varphi(1) = y$; we show that $p(x) = p(y)$, which will imply that p is constant on V_l. Applying the mean value theorem to the function $t \mapsto g(t) := p(\varphi(t))$, we find that $g(1) - g(0) = g'(t^*)$ for some $t^* \in (0,1)$. Now $g(t) = \sum_\alpha p_\alpha \varphi(t)^\alpha$, $g'(t) = \sum_\alpha p_\alpha (\sum_{i=1}^n \alpha_i \varphi_i'(t) \frac{\varphi(t)}{\varphi_i(t)}) = \sum_{i=1}^n \frac{\partial p}{\partial x_i}(\varphi(t))\varphi_i'(t)$, which implies $g'(t^*) = 0$ as $\varphi(t^*) \in V_l \subseteq V_{\mathbb{C}}(\mathcal{I}_p^{\text{grad}})$. Therefore, $0 = g(1) - g(0) = p(y) - p(x)$. □

We now group the irreducible components of $V_{\mathbb{C}}(\mathcal{I}_p^{\text{grad}})$ in the following way:

$$V_{\mathbb{C}}(\mathcal{I}_p^{\text{grad}}) = W_0 \cup W_1 \cup \ldots \cup W_r,$$

where $W_0 := \bigcup_{l | p(V_l) \in \mathbb{C}\setminus\mathbb{R}} V_l$ (thus $W_0 \cap \mathbb{R}^n = \emptyset$), p takes a constant value a_i on each W_i ($i = 1, \ldots, r$), and $a_1, \ldots, a_r$ are all distinct. Then, $W_0, W_1, \ldots, W_r$ are pairwise disjoint, $a_1, \ldots, a_r \in \mathbb{R}$, and $\overline{W_i} = W_i$ for $0 \leq i \leq r$. Hence we can apply Lemma 7.20 and deduce the existence of polynomials $p_0, p_1, \ldots, p_r \in \mathbb{R}[\mathbf{x}]$ satisfying $p_i(W_j) = \delta_{i,j}$ for $i, j = 0, \ldots, r$.

LEMMA 7.22. *$p = s_0$ modulo $\mathcal{I}(W_0)$, where s_0 is a sum of squares.*

Proof. We apply the Real Nullstellensatz (Theorem 2.1 (ii)) to the ideal $\mathcal{I} := \mathcal{I}(W_0) \subseteq \mathbb{R}[\mathbf{x}]$. As $V_{\mathbb{R}}(\mathcal{I}) = W_0 \cap \mathbb{R}^n = \emptyset$, we have $\sqrt[\mathbb{R}]{\mathcal{I}} = \mathcal{I}(V_{\mathbb{R}}(\mathcal{I})) = \mathbb{R}[\mathbf{x}]$. Hence, $-1 \in \sqrt[\mathbb{R}]{\mathcal{I}}$; that is, $-1 = s + q$, where s is a sum of squares and $q \in \mathcal{I}$. Writing $p = p_1 - p_2$ with p_1, p_2 sums of squares, we find $p = p_1 + sp_2 + p_2 q$, where $s_0 := p_1 + sp_2$ is a sum of squares and $p_2 q \in \mathcal{I} = \mathcal{I}(W_0)$. □

We can now conclude the proof of Theorem 7.16. By assumption, p is nonnegative on $V_{\mathbb{R}}(\mathcal{I}_p^{\text{grad}})$. Hence, the values $a_1, \ldots, a_r$ taken by p on $W_1, \ldots, W_r$ are nonnegative numbers. Consider the polynomial $q := s_0 p_0^2 + \sum_{i=1}^r a_i p_i^2$, where $p_0, p_1, \ldots, p_r$ are derived from Lemma 7.20 as indicated above and s_0 is as in Lemma 7.22. By construction, q is a sum of squares. Moreover, $p - q$ vanishes on $V_{\mathbb{C}}(\mathcal{I}_p^{\text{grad}}) = W_0 \cup W_1 \cup \ldots \cup W_r$, since $q(x) = s_0(x) = p(x)$ for $x \in W_0$ (by Lemma 7.22) and $q(x) = a_i = p(x)$ for $x \in W_i$ ($i = 1, \ldots, r$). As $\mathcal{I}_p^{\text{grad}}$ is radical, we deduce that $p - q \in \mathcal{I}(V_{\mathbb{C}}(\mathcal{I}_p^{\text{grad}})) = \mathcal{I}_p^{\text{grad}}$, which shows that p is a sum of squares modulo $\mathcal{I}_p^{\text{grad}}$ and thus concludes the proof of Theorem 7.16.

We now turn to the proof of Theorem 7.15. Our assumption now is that p is positive on $V_{\mathbb{R}}(\mathcal{I}_p^{\text{grad}})$; that is, $a_1, \ldots, a_r > 0$. Consider a

[3]This is a nontrivial result of algebraic geometry; we thank M. Schweighofer for communicating us the following sketch of proof. Let V be an irreducible variety in $\mathbb{C}^n$. Then V is connected with respect to the usual norm topology of $\mathbb{C}^n$ (see e.g. [136]). Viewing V as a connected semialgebraic set in $\mathbb{R}^{2n}$, it follows that V is connected by a semialgebraic continuous path (see e.g. [14]). Finally, use the fact that a semialgebraic continuous path is piecewise differentiable (see [151, Chap. 7, 2, Prop. 2.5.]).

primary decomposition of the ideal $\mathcal{I}_p^{\text{grad}}$ (see [21, Chap. 4]) as $\mathcal{I}_p^{\text{grad}} = \bigcap_{h=1}^k \mathcal{I}_h$. Then each variety $V_{\mathbb{C}}(\mathcal{I}_h)$ is irreducible and thus contained in W_i for some $i = 0, \ldots, r$. For $i = 0, \ldots, r$, set $\mathcal{J}_i := \bigcap_{h | V_{\mathbb{C}}(\mathcal{I}_h) \subseteq W_i} \mathcal{I}_h$. Then, $\mathcal{I}_p^{\text{grad}} = \mathcal{J}_0 \cap \mathcal{J}_1 \cap \ldots \cap \mathcal{J}_r$, with $V_{\mathbb{C}}(\mathcal{J}_i) = W_i$ for $0 \leq i \leq r$. As $V_{\mathbb{C}}(\mathcal{J}_i + \mathcal{J}_j) = V_{\mathbb{C}}(\mathcal{J}_i) \cap V_{\mathbb{C}}(\mathcal{J}_j) = W_i \cap W_j = \emptyset$, we have $\mathcal{J}_i + \mathcal{J}_j = \mathbb{R}[\mathbf{x}]$ for $i \neq j$. The next result follows from the Chinese reminder theorem, but we give the proof for completeness.

LEMMA 7.23. *Given $s_0, \ldots, s_r \in \mathbb{R}[\mathbf{x}]$, there exists $s \in \mathbb{R}[\mathbf{x}]$ satisfying $s - s_i \in \mathcal{J}_i$ ($i = 0, \ldots, r$). Moreover, if each s_i is a sum of squares then s too can be chosen to be a sum of squares.*

Proof. The proof is by induction on $r \geq 1$. Assume first $r = 1$. As $\mathcal{J}_0 + \mathcal{J}_1 = \mathbb{R}[\mathbf{x}]$, $1 = u_0 + u_1$ for some $u_0 \in \mathcal{J}_0$, $u_1 \in \mathcal{J}_1$. Set $s := u_0^2 s_1 + u_1^2 s_0$; thus s is a sum of squares if s_0, s_1 are sums of squares. Moreover, $s - s_0 = u_0^2 s_1 + s_0(u_1^2 - 1) = u_0^2 s_1 - u_0(u_1 + 1)s_0 \in \mathcal{J}_0$. Analogously, $s - s_1 \in \mathcal{J}_1$.

Let s be the polynomial just constructed, satisfying $s - s_0 \in \mathcal{J}_0$ and $s - s_1 \in \mathcal{J}_1$. Consider now the ideals $\mathcal{J}_0 \cap \mathcal{J}_1, \mathcal{J}_2, \ldots, \mathcal{J}_r$. As $(\mathcal{J}_0 \cap \mathcal{J}_1) + \mathcal{J}_i = \mathbb{R}[\mathbf{x}]$ ($i \geq 2$), we can apply the induction assumption and deduce the existence of $t \in \mathbb{R}[\mathbf{x}]$ for which $t - s \in \mathcal{J}_0 \cap \mathcal{J}_1$, $t - s_i \in \mathcal{J}_i$ ($i \geq 2$). Moreover, t is a sum of squares if $s, s_2, \ldots, s_r$ are sums of squares, which concludes the proof. $\square$

The above lemma shows that the mapping

$$\begin{array}{ccc} \mathbb{R}[\mathbf{x}]/\mathcal{I}_p^{\text{grad}} = \mathbb{R}[\mathbf{x}]/\cap_{i=0}^r \mathcal{J}_i & \to & \prod_{i=0}^r \mathbb{R}[\mathbf{x}]/\mathcal{J}_i \\ s \mod \mathcal{I}_p^{\text{grad}} & \mapsto & (s_i \mod \mathcal{J}_i | i = 0, \ldots, r) \end{array}$$

is a bijection. Moreover if, for all $i = 0, \ldots, r$, $p - s_i \in \mathcal{J}_i$ with s_i sum of squares, then there exists a sum of squares s for which $p - s \in \mathcal{I}_p^{\text{grad}}$. Therefore, to conclude the proof of Theorem 7.15, it suffices to show that p is a sum of squares modulo each ideal $\mathcal{J}_i$. For $i = 0$, as $V_{\mathbb{R}}(\mathcal{J}_0) = \emptyset$, this follows from the Real Nullstellensatz (same argument as for Lemma 7.22). The next lemma settles the case $i \geq 1$ and thus the proof of Theorem 7.15.

LEMMA 7.24. *p is a sum of squares modulo $\mathcal{J}_i$, for $i = 1, \ldots, r$.*

Proof. By assumption, $p(x) = a_i > 0$ for all $x \in V_{\mathbb{C}}(\mathcal{J}_i) = W_i$. Hence the polynomial $u := p/a_i - 1$ vanishes on $V_{\mathbb{C}}(\mathcal{J}_i)$ and thus $u \in \mathcal{I}(V_{\mathbb{C}}(\mathcal{J}_i)) = \sqrt{\mathcal{J}_i}$; that is, using Hilbert's Nullstellensatz (Theorem 2.1 (i)), $u^m \in \mathcal{J}_i$ for some integer $m \geq 1$. The identity

$$1 + u = \left(\sum_{k=0}^{m-1} \binom{1/2}{k} u^k \right)^2 + q u^m \tag{7.10}$$

(where $q \in \text{Span}_{\mathbb{R}}(u^i \mid i \geq 0)$) gives directly that $p/a_i = 1 + u$ is a sum of squares modulo $\mathcal{J}_i$. To show (7.10), write $\left(\sum_{k=0}^{m-1} \binom{1/2}{k} u^k \right)^2 =$

$\sum_{j=0}^{2m-2} c_j u^j$, where $c_j := \sum_k \binom{1/2}{k}\binom{1/2}{j-k}$ with the summation over k satisfying $\max(0, j-m+1) \leq k \leq \min(j, m-1)$. We now verify that $c_j = 1$ for $j = 0, 1$ and $c_j = 0$ for $j = 2, \ldots, m-1$, which implies (7.10). For this fix $0 \leq j \leq m-1$ and consider the univariate polynomial $g_j := \sum_{h=0}^{j} \binom{\mathtt{t}}{h}\binom{\mathtt{t}}{j-h} - \binom{2\mathtt{t}}{j} \in \mathbb{R}[\mathtt{t}]$; as g_j vanishes at all $t \in \mathbb{N}$, g_j is identically zero and thus $g_j(1/2) = 0$, which gives $c_j = \binom{1}{j}$ for $j \leq m-1$, i.e. $c_0 = c_1 = 1$ and $c_j = 0$ for $2 \leq j \leq m-1$. □

8. Exploiting algebraic structure to reduce the problem size.

In the previous sections we have seen how to construct moment/SOS approximations for the infimum of a polynomial over a semialgebraic set. The simplest instance is the unconstrained minimization problem (1.3) of computing $p^{\min}$ $(= \inf_{x\in\mathbb{R}^n} p(x))$ where p is a polynomial of degree $2d$, its moment relaxation p_d^{mom} $(= \inf p^T y \text{ s.t. } M_d(y) \succeq 0,\ y_0 = 1)$, and its SOS relaxation p_d^{sos} $(= \sup \rho \text{ s.t. } p - \rho$ is a sum of squares). Recall that $p_d^{\text{mom}} = p_d^{\text{sos}}$. To compute $p_d^{\text{mom}} = p_d^{\text{sos}}$ one needs to solve a semidefinite program involving a matrix indexed by $\mathbb{N}_d^n$, thus of size $\binom{n+d}{d}$. This size becomes prohibitively large as soon as n or d is too large. It is thus of crucial importance to have methods permitting to reduce the size of this semidefinite program. For this one can exploit the specific structure of the problem at hand. For instance, the problem may have some symmetry, or may have some sparsity pattern, or may contain equations, all features which can be used to reduce the number of variables and sometimes the size of the matrices involved. See e.g. Parrilo [106] for an overview about exploiting algebraic structure in SOS programs. Much research has been done in the recent years about such issues, which we cannot cover in detail in this survey. We will only treat certain chosen topics.

8.1. Exploiting sparsity.

Using the Newton polynomial. Probably one of the first results about exploiting sparsity is a result of Reznick [118] about Newton polytopes of polynomials. For a polynomial $p = \sum_{|\alpha|\leq d} p_\alpha \mathbf{x}^\alpha$, its *Newton polytope* is defined as

$$N(p) := \text{conv}(\alpha \in \mathbb{N}_d^n \mid p_\alpha \neq 0).$$

Reznick [118] shows the following properties for Newton polytopes.

THEOREM 8.1. *[118] Given $p, q, f_1, \ldots, f_m \in \mathbb{R}[\mathbf{x}]$.*
(i) *$N(pq) = N(p) + N(q)$ and, if p, q are nonnegative on $\mathbb{R}^n$ then $N(p) \subseteq N(p+q)$.*
(ii) *If $p = \sum_{j=1}^m f_j^2$, then $N(f_j) \subseteq \frac{1}{2}N(p)$ for all j.*
(iii) *$N(p) \subseteq \text{conv}(2\alpha \mid p_{2\alpha} > 0)$.*

We illustrate the result on the following example taken from [106].

EXAMPLE 8.2. Consider the polynomial $p = (\mathbf{x}_1^4+1)(\mathbf{x}_2^4+1)(\mathbf{x}_3^4+1)(\mathbf{x}_4^4+1) + 2\mathbf{x}_1 + 3\mathbf{x}_2 + 4\mathbf{x}_3 + 5\mathbf{x}_4$ of degree $2d = 16$ in $n = 4$ variables.

Suppose we wish to find a sum of squares decomposition $p = \sum_j f_j^2$. A priori, each f_j has degree at most 8 and thus may involve the $495 = \binom{4+8}{4}$ monomials $\mathbf{x}^\alpha$ with $|\alpha| \in \mathbb{N}_8^4$. The polynomial p is however very sparse; it has only 20 terms, thus much less than the total number $4845 = \binom{4+16}{16}$ of possible terms. As a matter of fact, using the above result of Reznick, one can restrict the support of f_j to the 81 monomials $\mathbf{x}^\alpha$ with $\alpha \in \{0,1,2\}^4$. Indeed the Newton polytope of p is the cube $[0,4]^4$, thus $\frac{1}{2}N(p) = [0,2]^4$ and $\mathbb{N}^4 \cap \frac{1}{2}N(p) = \{0,1,2\}^4$.

Kojima, Kim and Waki [61] further investigate effective methods for reducing the support of polynomials entering the sum of square decomposition of a sparse polynomial, which are based on Theorem 8.1 and further refinements.

Structured sparsity on the constraint and objective polynomials. We now consider the polynomial optimization problem (1.1) where some sparsity structure is assumed on the polynomials $p, g_1, \ldots, g_m$. Roughly speaking we assume that each g_j uses only a small set of variables and that p can be separated into polynomials using only these small specified sets of variables. Then under some assumption on these specified sets, when searching for a decomposition $p = s_0 + \sum_{j=1}^m s_j g_j$ with all s_j sums of squares, we may restrict our search to polynomials s_j using again the specified sets of variables. We now give the precise definitions.

For a set $I \subseteq \{1, \ldots, n\}$, let $\mathbf{x}_I$ denote the set of variables $\{\mathbf{x}_i \mid i \in I\}$ and $\mathbb{R}[\mathbf{x}_I]$ the polynomial ring in those variables. Assume $\{1, \ldots, n\} = I_1 \cup \ldots \cup I_k$ where the I_h's satisfy the property

$$\forall h \in \{1, \ldots, k-1\} \ \exists r \in \{1, \ldots, h\} \ \ I_{h+1} \cap (I_1 \cup \ldots \cup I_h) \subseteq I_r. \tag{8.1}$$

Note that (8.1) holds automatically for $k \leq 2$. We make the following assumptions on the polynomials $p, g_1, \ldots, g_m$:

$$p = \sum_{h=1}^k p_h \ \text{ where } p_h \in \mathbb{R}[\mathbf{x}_{I_h}] \tag{8.2}$$

$$\{1, \ldots, m\} = J_1 \cup \ldots \cup J_k \ \text{ and } g_j \in \mathbb{R}[\mathbf{x}_{I_h}] \text{ for } j \in J_h, \ 1 \leq h \leq k. \tag{8.3}$$

REMARK 8.3. If $I_1, \ldots, I_k$ are the maximal cliques of a chordal graph, then $k \leq n$ and (8.1) is satisfied (after possibly reordering the I_h's) and is known as the *running intersection property.* Cf. e.g. [12] for details about chordal graphs. The following strategy is proposed in [153] for identifying a sparsity structure like (8.2)-(8.3). Define the (correlative sparsity) graph $G = (V, E)$ where $V := \{1, \ldots, n\}$ and there is an edge $ij \in E$ if some term of p uses both variables $\mathbf{x}_i, \mathbf{x}_j$, or if both variables $\mathbf{x}_i, \mathbf{x}_j$ are used by some g_l $(l = 1, \ldots, m)$. Then find a chordal extension G' of G and choose the maximal cliques of G' as $I_1, \ldots, I_k$.

EXAMPLE 8.4. For instance, the polynomials $p = \mathbf{x}_1^2\mathbf{x}_2\mathbf{x}_3 + \mathbf{x}_3\mathbf{x}_4^2 + \mathbf{x}_3\mathbf{x}_5 + \mathbf{x}_6$, $g_1 = \mathbf{x}_1\mathbf{x}_2 - 1$, $g_2 = \mathbf{x}_1^2 + \mathbf{x}_2\mathbf{x}_3 - 1$, $g_3 = \mathbf{x}_2 + \mathbf{x}_3^2\mathbf{x}_4$, $g_4 = \mathbf{x}_3 + \mathbf{x}_5$, $g_5 = \mathbf{x}_3\mathbf{x}_6$, $g_6 = \mathbf{x}_2\mathbf{x}_3$ satisfy conditions (8.2), (8.3) after setting $I_1 = \{1,2,3\}$, $I_2 = \{2,3,4\}$, $I_3 = \{3,5\}$, $I_4 = \{3,6\}$.

EXAMPLE 8.5. The so-called chained singular function: $p = \sum_{i=1}^{n-3}(\mathbf{x}_i + 10\mathbf{x}_{i+1})^2 + 5(\mathbf{x}_{i+2} - \mathbf{x}_{i+3})^2 + (\mathbf{x}_{i+1} - 2\mathbf{x}_{i+2})^4 + 10(\mathbf{x}_i - 10\mathbf{x}_{i+3})^4$ satisfies (8.2) with $I_h = \{h, h+1, h+2, h+3\}$ ($h = 1, \ldots, n-3$). Cf. [153] for computational results.

Let us now formulate the sparse moment and SOS relaxations for problem (1.1) for any order $t \geq \max(d_p, d_{g_1}, \ldots, d_{g_m})$. For $\alpha \in \mathbb{N}^n$, set $\text{supp}(\alpha) = \{i \in \{1, \ldots, n\} \mid \alpha_i \geq 1\}$. For $t \in \mathbb{N}$ and a subset $I \subseteq \{1, \ldots, n\}$ set $\Lambda_t^I := \{\alpha \in \mathbb{N}_t^n \mid \text{supp}(\alpha) \subseteq I\}$. Finally set $\Lambda_t := \cup_{h=1}^k \Lambda_t^{I_h}$. The sparse moment relaxation of order t involves a variable $y \in \mathbb{R}^{\Lambda_{2t}}$, thus having entries y_α only for $\alpha \in \mathbb{N}_{2t}^n$ with supp(α) contained in some I_h; moreover, it involves the matrices $M_t(y, I_h)$, where $M_t(y, I_h)$ is the submatrix of $M_t(y)$ indexed by $\Lambda_t^{I_h}$. The sparse moment relaxation of order t reads as follows

$$\begin{aligned} \widehat{p_t^{\text{mom}}} := \inf\ p^T y \text{ s.t. } \quad & y_0 = 1,\ M_t(y, I_h) \succeq 0\ (h = 1, \ldots, k) \\ & M_{t-d_{g_j}}(g_j y, I_h) \succeq 0\ (j \in J_h, h = 1, \ldots, k) \end{aligned} \tag{8.4}$$

where the variable y lies in $\mathbb{R}^{\Lambda_{2t}}$. The corresponding sparse SOS relaxation of order t reads

$$\begin{aligned} \widehat{p_t^{\text{sos}}} := \sup \rho \text{ s.t. } \quad & p - \rho = \textstyle\sum_{h=1}^k \left(u_h + \sum_{j \in J_h} u_{jh} g_j\right) \\ & u_h, u_{jh}\ (j \in J_h) \text{ sums of squares in } \mathbb{R}[\mathbf{x}_{I_h}] \\ & \deg(u_h), \deg(u_{jh} g_j) \leq 2t\ (h = 1, \ldots, k). \end{aligned} \tag{8.5}$$

Obviously,

$$\widehat{p_t^{\text{sos}}} \leq \widehat{p_t^{\text{mom}}} \leq p^{\min},\ \widehat{p_t^{\text{mom}}} \leq p_t^{\text{mom}},\ \widehat{p_t^{\text{sos}}} \leq p_t^{\text{sos}}.$$

The sparse relaxation is in general weaker than the dense relaxation. However when all polynomials p_h, g_j are quadratic then the sparse and dense relaxations are equivalent (cf. [153, §4.5], also [98, Th. 3.6]). We sketch the details below.

LEMMA 8.6. *Assume $p = \sum_{h=1}^k p_h$ where $p_h \in \mathbb{R}[\mathbf{x}_{I_h}]$ and the sets I_h satisfy (8.1). If $\deg(p_h) \leq 2$ for all h and p is a sum of squares, then p has a sparse sum of squares decomposition, i.e. of the form*

$$p = \sum_{h=1}^k s_h \quad \textit{where } s_h \in \mathbb{R}[\mathbf{x}_{I_h}] \textit{ and } s_h \textit{ is SOS.} \tag{8.6}$$

Proof. Consider the dense/sparse SOS/moment relaxations of order 1 of the problem $\min_{x \in \mathbb{R}^n} p(x)$, with optimum values p_1^{sos}, p_1^{mom}, $\widehat{p_1^{\text{sos}}}$, $\widehat{p_1^{\text{mom}}}$.

The strict feasibility of the moment relaxations implies that $p_1^{\text{sos}} = p_1^{\text{mom}}$, $\widehat{p_1^{\text{sos}}} = \widehat{p_1^{\text{mom}}}$, the optimum is attained in the dense/sparse SOS relaxations, p SOS $\Longleftrightarrow p_1^{\text{sos}} \geq 0$, and p has a sparse SOS decomposition (8.6) $\Longleftrightarrow$ $\widehat{p_1^{\text{sos}}} \geq 0$. Thus it suffices to show that $p_1^{\text{mom}} \leq \widehat{p_1^{\text{mom}}}$. For this let y be feasible for the program defining $\widehat{p_1^{\text{mom}}}$, i.e. $y_0 = 1$, $M_1(y, I_h) \succeq 0$ for all $h = 1, \ldots, k$. Using a result of Grone et al. [44] (which claims that any partial positive semidefinite matrix whose specified entries form a chordal graph can be completed to a fully specified positive semidefinite matrix), we can complete y to a vector $\tilde{y} \in \mathbb{R}^{\mathbb{N}_2^n}$ satisfying $M_1(\tilde{y}) \succeq 0$. Thus $\tilde{y}$ is feasible for the program defining p_1^{mom}, which shows $p_1^{\text{mom}} \leq \widehat{p_1^{\text{mom}}}$. □

COROLLARY 8.7. *Consider the problem (1.1) and assume that (8.1), (8.2), (8.3) hold. If all p_h, g_j are quadratic, then $p_1^{mom} = \widehat{p_1^{mom}}$ and $p_1^{sos} = \widehat{p_1^{sos}}$.*

Proof. Assume y is feasible for the program defining $\widehat{p_1^{\text{mom}}}$; that is, $y_0 = 1$, $M_1(y, I_h) \succeq 0$ $(h = 1, \ldots, k)$ and $(g_j y)_0 (= \sum_\alpha (g_j)_\alpha y_\alpha) \geq 0$ $(j = 1, \ldots, m)$. Using the same argument as in the proof of Lemma 8.6 we can complete y to $\tilde{y} \in \mathbb{R}^{\mathbb{N}_2^n}$ such that $M_1(\tilde{y}) \succeq 0$ and thus $\tilde{y}$ is feasible for the program defining p_1^{mom}, which shows $p_1^{\text{mom}} \leq \widehat{p_1^{\text{mom}}}$. Assume now $\rho \in \mathbb{R}$ is feasible for the program defining p_1^{sos}; that is, $p - \rho = s_0 + \sum_{j=1}^m s_j g_j$ where s is a sum of squares in $\mathbb{R}[\mathbf{x}]$ and $s_j \in \mathbb{R}_+$. Now the polynomial $p - \rho - \sum_{j=1}^m s_j g_j$ is separable (i.e. can be written as a sum of polynomials in $\mathbb{R}[\mathbf{x}_{I_h}]$); hence, by Lemma 8.6, it has a sparse sum of squares decomposition, of the form $\sum_{h=1}^k s_h$ with $s_h \in \mathbb{R}[\mathbf{x}_{I_h}]$ SOS. This shows that ρ is feasible for the program defining $\widehat{p_1^{\text{sos}}}$, giving the desired inequality $p_1^{\text{sos}} \leq \widehat{p_1^{\text{sos}}}$. □

EXAMPLE 8.8. We give an example (mentioned in [98, Ex. 3.5]) showing that the result of Lemma 8.6 does not hold for polynomials of degree more than 2. Consider the polynomial $p = p_1 + p_2$, where $p_1 = \mathbf{x}_1^4 + (\mathbf{x}_1\mathbf{x}_2 - 1)^2$ and $p_2 = \mathbf{x}_2^2\mathbf{x}_3^2 + (\mathbf{x}_3^2 - 1)^2$. Waki [152] verified that $0 = \widehat{p_2^{\text{sos}}} < p_2^{\text{sos}} = p^{\min} \sim 0.84986$.

Waki et al. [153] have implemented the above sparse SDP relaxations. Their numerical results show that they can be solved much faster than the dense relaxations and yet they give very good approximations of $p^{\min}$. Lasserre [72, 74] proved the theoretical convergence, i.e. $\lim_{t\to\infty} \widehat{p_t^{\text{sos}}} = \lim_{t\to\infty} \widehat{p_t^{\text{mom}}} = p^{\min}$, under the assumption that K has a nonempty interior and that a ball constraint $R_h^2 - \sum_{i \in J_h} \mathbf{x}_i \geq 0$ is present in the description of K for each $h = 1, \ldots, k$. Kojima and Muramatsu [62] proved the result for compact K with possibly empty interior. Grimm, Netzer and Schweighofer [43] give a simpler proof, which does not need the presence of ball constraints in the description of K but instead assumes that each set of polynomials g_j $(j \in J_h)$ generates an Archimedean module.

THEOREM 8.9. *[43] Assume that, for each $h = 1, \ldots, k$, the quadratic module $\mathbf{M}_h := \mathbf{M}(g_j \mid j \in J_h)$ generated by g_j $(j \in J_h)$ is Archimedean.*

Assume that (8.1) holds and that $p, g_1, \ldots, g_m$ satisfy (8.2), (8.3). If p is positive on the set $K = \{x \in \mathbb{R}^n \mid g_j(x) \geq 0 \ (j = 1, \ldots, m)\}$, then $p \in \mathbf{M}_1 + \ldots + \mathbf{M}_k$; that is, $p = \sum_{h=1}^{k} \left(u_h + \sum_{j \in J_h} u_{jh} g_j\right)$, where u_h, u_{jh} are sums of squares in $\mathbb{R}[\mathbf{x}_{I_h}]$.

Before proving the theorem we state the application to asymptotic convergence.

COROLLARY 8.10. *Under the assumptions of Theorem 8.9, we have $\lim_{t \to \infty} \widehat{p_t^{sos}} = \lim_{t \to \infty} \widehat{p_t^{mom}} = p^{min}$.*

Proof. Fix $\epsilon > 0$. As $p - p^{\min} + \epsilon$ is positive on K and satisfies (8.2), we deduce from Theorem 8.9 that $p - p^{\min} + \epsilon \in \sum_{h=1}^{k} \mathbf{M}_h$. Thus $p^{\min} - \epsilon$ is feasible for (8.5) for some t. Hence, for every $\epsilon > 0$, there exists $t \in \mathbb{N}$ with $p^{\min} - \epsilon \leq \widehat{p_t^{\rm sos}} \leq p^{\min}$. This shows that $\lim_{t \to \infty} \widehat{p_t^{\rm sos}} = p^{\rm mom}$. □

Proof of Theorem 8.9. We give the proof of [43] which is elementary except it uses the following special case of Schmüdgen's theorem (Theorem 3.16): For $p \in \mathbb{R}[\mathbf{x}]$,

$$p>0 \text{ on } \{x \mid R^2 - \sum_{i=1}^{n} x_i^2 \geq 0\} \Longrightarrow \exists\, s_0, s_1 \in \Sigma \ \ p = s_0 + s_1 (R^2 - \sum_{i=1}^{n} \mathbf{x}_i^2). \quad (8.7)$$

We start with some preliminary results.

LEMMA 8.11. *Let $C \subseteq \mathbb{R}$ be compact. Assume $p = p_1 + \ldots + p_k$ where $p_h \in \mathbb{R}[\mathbf{x}_{I_h}]$ $(h = 1, \ldots, k)$ and $p > 0$ on C^n. Then $p = f_1 + \ldots + f_k$ where $f_h \in \mathbb{R}[\mathbf{x}_{I_h}]$ and $f_h > 0$ on C^{I_h} $(h = 1, \ldots, k)$.*

Proof. We use induction on $k \geq 2$. Assume first $k = 2$. Let $\epsilon > 0$ such that $p = p_1 + p_2 \geq \epsilon$ on C^n. Define the function F on $\mathbb{R}^{I_1 \cap I_2}$ by

$$F(y) := \min_{x \in C^{I_1 \setminus I_2}} p_1(x, y) - \frac{\epsilon}{2} \quad \text{for } y \in \mathbb{R}^{I_1 \cap I_2}.$$

The function F is continuous on $C^{I_1 \cap I_2}$. Indeed for $y, y' \in C^{I_1 \cap I_2}$ and $x, x' \in C^{I_1 \setminus I_2}$ minimizing respectively $p_1(x, y)$ and $p_1(x', y')$, we have

$$|F(y) - F(y')| \leq \max(|p_1(x, y) - p_1(x, y')|, |p_1(x', y) - p_1(x', y')|),$$

implying the uniform continuity of F on $C^{I_1 \cap I_2}$ since p_1 is uniform continuous on C^{I_1}. Next we claim that

$$p_1(x, y) - F(y) \geq \frac{\epsilon}{2}, \ p_2(y, z) + F(y) \geq \frac{\epsilon}{2} \ \forall x \in \mathbb{R}^{I_1 \setminus I_2}, y \in \mathbb{R}^{I_1 \cap I_2}, z \in \mathbb{R}^{I_2 \setminus I_1}.$$

The first follows from the definition of F. For the second note that $p_2(y, z) + F(y) = p_2(y, z) + p_1(x, y) - \frac{\epsilon}{2}$ (for some $x \in C^{I_1 \setminus I_2}$), which in turn is equal to $p(x, y, z) - \frac{\epsilon}{2} \geq \epsilon - \frac{\epsilon}{2} = \frac{\epsilon}{2}$. By the Stone-Weierstrass theorem, F can be uniformly approximated by a polynomial $f \in \mathbb{R}[\mathbf{x}_{I_1 \cap I_2}]$ satisfying

$|F(y) - f(y)| \leq \frac{\epsilon}{4}$ for all $y \in C^{I_1 \cap I_2}$. Set $f_1 := p_1 - f$ and $f_2 := p_2 + f$. Thus $p = f_1 + f_2$; $f_1 > 0$ on C^{I_1} since $f_1(x,y) = p_1(x,y) - f(y) = p_1(x,y) - F(y) + F(y) - f(y) \geq \frac{\epsilon}{2} - \frac{\epsilon}{4} = \frac{\epsilon}{4}$; $f_2 > 0$ on C^{I_2} since $f_2(y,z) = p_2(y,z) + f(y) = p_2(y,z) + F(y) + f(y) - F(y) \geq \frac{\epsilon}{2} - \frac{\epsilon}{4} = \frac{\epsilon}{4}$. Thus the lemma holds in the case $k = 2$.

Assume now $k \geq 3$. Write $\tilde{I} := \cup_{h=1}^{k-1} I_h$, $\tilde{p} := p_1 + \ldots + p_{k-1} \in \mathbb{R}[\mathbf{x}_{\tilde{I}}]$, so that $p = \tilde{p} + f_k$. By the above proof, there exists $f \in \mathbb{R}[\mathbf{x}_{\tilde{I} \cap I_k}]$ such that $\tilde{p} - f > 0$ on $C^{\tilde{I}}$ and $p_k + f > 0$ on C^{I_k}. Using (8.1), it follows that $\tilde{I} \cap I_k \subseteq I_{h_0}$ for some $h_0 \leq k-1$. Hence $f \in \mathbb{R}[\mathbf{x}_{I_{h_0}}] \cap \mathbb{R}[\mathbf{x}_{I_k}]$ and $\tilde{p} - f$ is a sum of polynomials in $\mathbb{R}[\mathbf{x}_{I_h}]$ $(h = 1, \ldots, k-1)$. Using the induction assumption for the case $k-1$, we deduce that $\tilde{p} - f = f_1 + \ldots + f_{k-1}$ where $f_h \in \mathbb{R}[\mathbf{x}_{I_h}]$ and $f_h > 0$ on C^{I_h} for each $h \leq k-1$. This gives $p = \tilde{p} + p_k = \tilde{p} - f + f + p_k = f_1 + \ldots + f_{k-1} + f + p_k$ which is the desired conclusion since $f + p_k \in \mathbb{R}[\mathbf{x}_{I_k}]$ and $f + p_k > 0$ on C^{I_k}. □

LEMMA 8.12. *Assume $p = p_1 + \ldots + p_k$ where $p_h \in \mathbb{R}[\mathbf{x}_{I_h}]$ and $p > 0$ on the set K. Let B be a bounded set in $\mathbb{R}^n$. There exist $t \in \mathbb{N}$, $\lambda \in \mathbb{R}$ with $0 < \lambda \leq 1$, and polynomials $f_h \in \mathbb{R}[\mathbf{x}_{I_h}]$ such that $f_h > 0$ on B and*

$$p = \sum_{j=1}^{m} (1 - \lambda g_j)^{2t} g_j + f_1 + \ldots + f_k. \tag{8.8}$$

Proof. Choose a compact set $C \subseteq \mathbb{R}$ such that $B \subseteq C^n$ and choose $\lambda \in \mathbb{R}$ such that $0 < \lambda \leq 1$ and $\lambda g_j(x) \leq 1$ for all $x \in C^n$ and $j = 1, \ldots, m$. For $t \in \mathbb{N}$ set

$$F_t := p - \sum_{j=1}^{m} (1 - \lambda g_j)^{2t} g_j.$$

Obviously $F_t \leq F_{t+1}$ on C^n. First we claim

$$\forall x \in C^n \ \exists t \in \mathbb{N}^n \ F_t(x) > 0. \tag{8.9}$$

We use the fact that $(1 - \lambda g_j(x))^{2t} g_j(x)$ goes to 0 as t goes to ∞ if $g_j(x) \geq 0$, and to ∞ otherwise. If $x \in K$ then $\lim_{t \to \infty} F_t(x) = p(x)$ and thus $F_t(x) > 0$ for t large enough. If $x \in C^n \setminus K$ then $\lim_{t \to \infty} F_t(x) = \infty$ and thus $F_t(x) > 0$ again for t large enough. This shows (8.9). Next we claim

$$\exists t \in \mathbb{N} \ \forall x \in C^n \ F_t(x) > 0. \tag{8.10}$$

By (8.9), for each $x \in C^n$ there exists an open ball B_x containing x and $t_x \in \mathbb{N}$ such that $F_{t_x} > 0$ on B_x. Thus $C^n \subseteq \cup_{x \in C^n} B_x$. As C^n is compact, we must have $C^n \subseteq B_{x_1} \cup \ldots \cup B_{x_N}$ for finitely many x_i. As $F_t > 0$ on B_{x_i} for all $t \geq t_{x_i}$, we deduce that $F_t > 0$ on C^n for all $t \geq \max_{i=1,\ldots,N} t_{x_i}$, which shows (8.10). Hence we have found the decomposition $p = \sum_{j=1}^{m} (1 - \lambda g_j)^{2t} g_j + F_t$ where $F_t > 0$ on C^n. As F_t is a sum of polynomials in

$\mathbb{R}[\mathbf{x}_{I_h}]$ and $F_t > 0$ on C^n, we can apply Lemma 8.11 and deduce that $F_t = f_1 + \ldots + f_k$ where $f_h \in \mathbb{R}[\mathbf{x}_{I_h}]$ and $f_h > 0$ on C^{I_h} and thus on B. Thus (8.8) holds. □

We can now conclude the proof of Theorem 8.9. As each module $\mathbf{M}_h$ is Archimedean, we can find $R > 0$ for which $R^2 - \sum_{i \in I_h} \mathbf{x}_i^2 \in \mathbf{M}_h$ for each $h = 1, \ldots, k$. By assumption, $p > 0$ on K. We apply Lemma 8.12 to the closed ball B in $\mathbb{R}^n$ of radius R. Thus we find a decomposition as in (8.8). As $f_h > 0$ on B we deduce that $f_h \in \mathbf{M}_h$ using (8.7). Finally observe that $\sum_{j=1}^m (1 - \lambda g_j)^{2t} g_j = \sum_{h=1}^k u_h$ where $u_h := \sum_{j \in J_h} (1 - \lambda g_j)^{2t} g_j \in \mathbf{M}_h$. This concludes the proof of Theorem 8.9.

Extracting global minimizers. In some cases one is also able to extract global minimizers for the original problem (1.1) from the sparse SDP relaxation (8.4). Namely assume y is an optimum solution to the sparse moment ralaxation (8.4) and that the following rank conditions hold:

$$\operatorname{rank} M_s(y, I_h) = \operatorname{rank} M_{s-a_h}(y, I_h) \ \forall h = 1, \ldots, k, \quad (8.11)$$

$$\operatorname{rank} M_s(y, I_h \cap I_{h'}) = 1 \ \forall h \neq h' = 1, \ldots, k \ \text{ with } I_h \cap I_{h'} \neq \emptyset, \quad (8.12)$$

setting $a_h := \max_{j \in J_h} d_{g_j}$. Then we can apply the results from Sections 5.2, 6.6 to extract solutions. Namely for each $h \leq k$, by (8.11), the restriction of y to $\mathbb{R}^{\Lambda_{2t}^{I_h}}$ has a unique representing measure with support $\Delta^h \subseteq \mathbb{R}^{I_h}$. Moreover, by (8.12), if $I_h \cap I_{h'} \neq \emptyset$, then the restriction of y to $\mathbb{R}^{\Lambda_{2t}^{I_h \cap I_{h'}}}$ has a unique representing measure which is a Dirac measure at a point $x^{(hh')} \in \mathbb{R}^{I_h \cap I_{h'}}$. Therefore, any $x^{(h)} \in \Delta^h$, $x^{(h')} \in \Delta^{h'}$ coincide on $I_h \cap I_{h'}$, i.e. $x_i^{(h)} = x_i^{(h')} = x_i^{(hh')}$ for $i \in I_h \cap I_{h'}$. Therefore any point $x^* \in \mathbb{R}^n$ obtained by setting $x_i^* := x_i^{(h)}$ $(i \in I_h)$ for some $x^{(h)} \in \Delta^h$, is an optimum solution to the original problem (1.1). The rank conditions (8.11)-(8.12) are however quite restrictive.

Here is another situation when one can extract a global minimizer; namely when (1.1) has a unique global minimizer. Assume that for all t large enough we have a near optimal solution $y^{(t)}$ to the sparse moment relaxation of order t; that is, $y^{(t)}$ is feasible for (8.4) and $p^T y^{(t)} \leq \widehat{p^{\text{mom}}}_t + 1/t$. Lasserre [72] shows that, if problem (1.1) has a unique global minimizer x^*, then the vectors $(y_{e_i}^{(t)})_{i=1}^n$ converge to the global minimizer x^* as t goes to ∞.

SparsePOP software. Waki, Kim, Kojima, Muramatsu, and Sugimoto have developed the software SparsePOP, which implements the sparse moment and SOS relaxations (8.4)-(8.5) proposed in [153] for the problem (1.1). The software can be downloaded from the website `http://www.is.titech.ac.jp/~kojima/SparsePOP/`.

We also refer to [153] where another technique is proposed, based on perturbing the objective function in (1.1) which, under some conditions, permits the extraction of an approximate global minimizer.

For a detailed presentation of several examples together with computational numerical results, see in particular [153]; see also [98], and [97] for instances arising from sensor network localization (which is an instance of the distance realization problem described in Section 1).

8.2. Exploiting equations. Here we come back to the case when the semialgebraic set K is as in (2.5), i.e. there are explicit polynomial equations $h_1 = 0, \ldots, h_{m_0} = 0$ present in its decription. Let $\mathcal{J} := (h_1, \ldots, h_{m_0})$ be the ideal generated by these polynomials. As noted in Section 6.2 one can formulate SOS/moment bounds by working in the quotient ring $\mathbb{R}[\mathbf{x}]/\mathcal{J}$, which leads to a saving in the number of variables and thus in the complexity of the SDP's to be solved. Indeed suppose we know a (linear) basis $\mathcal{B}$ of $\mathbb{R}[\mathbf{x}]/\mathcal{J}$, so that $\mathbb{R}[\mathbf{x}] = \text{Span}_{\mathbb{R}}(\mathcal{B}) \oplus \mathcal{J}$. Then, for $p \in \mathbb{R}[\mathbf{x}]$,

$$p \text{ SOS} \mod \mathcal{J} \Longleftrightarrow p = \textstyle\sum_l u_l^2 + q \text{ with } u_l \in \text{Span}_{\mathbb{R}}(\mathcal{B}), q \in \mathcal{J}. \quad (8.13)$$

(This is obvious: If $p = \sum_l f_l^2 + g$ with $f_l \in \mathbb{R}[\mathbf{x}]$, $g \in \mathcal{J}$, write $f_l = u_l + v_l$ with $u_l \in \text{Span}_{\mathbb{R}}(\mathcal{B})$ and $v_l \in \mathcal{J}$, so that $p = \sum_l u_l^2 + q$ after setting $q := g + \sum_l v_l^2 + 2u_l v_l \in \mathcal{J}$.) Hence to check the existence of a SOS decomposition modulo $\mathcal{J}$, we can apply the Gram-matrix method from Section 3.3 working with matrices indexed by $\mathcal{B}$ (or a subset of it) instead of the full set of monomials. Moreover, when formulating the moment relaxations, one can use the equations $h_j = 0$ to eliminate some variables within $y = (y_\alpha)_\alpha$. Let us illustrate this on an example (taken from [106]).

EXAMPLE 8.13. Suppose we want to minimize the polynomial $p = 10 - \mathbf{x}_1^2 - \mathbf{x}_2$ over $\{(x, y) \in \mathbb{R}^2 \mid g_1 := x_1^2 + x_2^2 - 1 = 0\}$ (the unit circle). To get a lower bound on $p^{\min}$, one can compute the largest ρ for which $p - \rho$ is SOS modulo the ideal $\mathcal{J} = (\mathbf{x}_1^2 + \mathbf{x}_2^2 - 1)$. As $\mathcal{B} := \{\mathbf{x}_1^i, \mathbf{x}_2\mathbf{x}_1^i \mid i \geq 0\}$ is a basis of $\mathbb{R}[\mathbf{x}]/\mathcal{J}$ (it is the set of standard monomials w.r.t. a graded lex monomial ordering), one can first try to find a decomposition as in (8.13) using only monomials in the subset $\{1, \mathbf{x}_1, \mathbf{x}_2\} \subseteq \mathcal{B}$. Namely, find the largest scalar ρ for which

$$\begin{aligned} 10 - \mathbf{x}_1^2 - \mathbf{x}_2 - \rho &= \begin{pmatrix} 1 \\ \mathbf{x}_1 \\ \mathbf{x}_2 \end{pmatrix}^T \underbrace{\begin{pmatrix} a & b & c \\ b & d & e \\ c & e & f \end{pmatrix}}_{X \succeq 0} \begin{pmatrix} 1 \\ \mathbf{x}_1 \\ \mathbf{x}_2 \end{pmatrix} \mod \mathcal{J} \\ &= a + f + (d - f)\mathbf{x}_1^2 + 2b\mathbf{x}_1 + 2c\mathbf{x}_2 + 2e\mathbf{x}_1\mathbf{x}_2 \mod \mathcal{J} \end{aligned}$$

giving $10 - \rho - \mathbf{x}_1^2 - \mathbf{x}_2 = a + f + (d - f)\mathbf{x}_1^2 + 2b\mathbf{x}_1 + 2c\mathbf{x}_2 + 2e\mathbf{x}_1\mathbf{x}_2$. Equating coefficients in both sides, we find

$$X = \begin{pmatrix} 10 - f - \rho & 0 & -1/2 \\ 0 & f - 1 & 0 \\ -1/2 & 0 & f \end{pmatrix}.$$

One can easily verify that the largest ρ for which $X \succeq 0$ is $\rho = 35/4$, obtained for $f = 1$, in which case $X = L^T L$ with $L = \begin{pmatrix} -1/2 & 0 & 1 \end{pmatrix}$, giving $p - 35/4 = (\mathbf{x}_2 - 1/2)^2 \mod \mathcal{J}$. This shows $p^{\min} \geq 35/4$. Equality holds since $p(x_1, x_2) = 35/4$ for $(x_1, x_2) = (\pm\sqrt{7}/2, -1/2)$.

On the moment side, the following program

$$\inf\ 10 - y_{20} - y_{01} \text{ s.t. } \begin{pmatrix} 1 & y_{10} & y_{01} \\ y_{10} & y_{20} & y_{11} \\ y_{01} & y_{11} & 1 - y_{20} \end{pmatrix} \succeq 0$$

gives a lower bound for $p^{\min}$. Here we have used the condition $0 = (g_1 y)_{00} = y_{20} + y_{02} - y_{00}$ stemming from the equation $\mathbf{x}_1^2 + \mathbf{x}_2^2 - 1 = 0$, which thus permits to eliminate the variable y_{02}. One can easily check that the optimum of this program is again 35/4, obtained for $y_{10} = y_{11} = 0$, $y_{01} = 1/2$, $y_{20} = 3/4$.

The zero-dimensional case. When $\mathcal{J}$ is zero-dimensional, $\mathcal{B}$ is a finite set; say $\mathcal{B} = \{b_1, \ldots, b_N\}$ where $N := \dim \mathbb{R}[\mathbf{x}]/\mathcal{J} \geq |V_{\mathbb{C}}(\mathcal{J})|$. For convenience assume $\mathcal{B}$ contains the constant monomial 1, say $b_1 = 1$. By Theorem 6.15, there is finite convergence of the SOS/moment hierarchies and thus problem (1.1) can be reformulated as the semidefinite program (6.2) or (6.3) for t large enough. Moreover the SOS bound

$$\begin{aligned} p^{\text{sos}} &= \sup \rho \text{ s.t. } p - \rho \in \mathbf{M}(g_1, \ldots, g_m, \pm h_1, \ldots, \pm h_{m_0}) \\ &= \sup \rho \text{ s.t. } p - \rho = \textstyle\sum_{j=0}^m s_j g_j \mod \mathcal{J} \quad \text{for some } s_j \in \Sigma \end{aligned}$$

can be computed via a semidefinite program involving $N \times N$ matrices in view of (the argument for) (8.13), and $p^{\text{sos}} = p^{\min}$ by Theorem 6.8, since the quadratic module $\mathbf{M}(g_1, \ldots, g_m, \pm h_1, \ldots, \pm h_{m_0})$ is Archimedean as $\mathcal{J}$ is zero-dimensional. Therefore, $p^{\text{sos}} = p^{\text{mom}} = p^{\min}$.

We now give a direct argument for equality $p^{\text{mom}} = p^{\min}$, relying on Theorem 5.1 (about finite rank moment matrices, instead of Putinar's theorem) and giving an explicit moment SDP formulation for (1.1) using $N \times N$ matrices; see (8.15). Following [81], we use a so-called combinatorial moment matrix which is simply a moment matrix in which some variables are eliminated using the equations $h_j = 0$. For $f \in \mathbb{R}[\mathbf{x}]$, $\text{res}_{\mathcal{B}}(f)$ denotes the unique polynomial in $\text{Span}_{\mathbb{R}}(\mathcal{B})$ such that $f - \text{res}_{\mathcal{B}}(f) \in \mathcal{J}$. Given $y \in \mathbb{R}^N$, define the linear operator L_y on $\text{Span}_{\mathbb{R}}(\mathcal{B})$ ($\simeq \mathbb{R}[\mathbf{x}]/\mathcal{J}$) by $L_y(\sum_{i=1}^N \lambda_i b_i) := \sum_{i=1}^N \lambda_i y_i$ ($\lambda \in \mathbb{R}^N$) and extend L_y to a linear operator on $\mathbb{R}[\mathbf{x}]$ by setting $L_y(f) := L_y(\text{res}_{\mathcal{B}}(f))$ ($f \in \mathbb{R}[\mathbf{x}]$). Then define the $N \times N$ matrix $M_{\mathcal{B}}(y)$ (the *combinatorial moment matrix* of y) whose (i,j)th entry is $L_y(b_i b_j)$. Consider first for simplicity the problem of minimizing $p \in \mathbb{R}[\mathbf{x}]$ over $V_{\mathbb{R}}(\mathcal{J})$, obviously equivalent to minimizing $\text{res}_{\mathcal{B}}(p)$ over $V_{\mathbb{R}}(\mathcal{J})$. With $\text{res}_{\mathcal{B}}(p) := \sum_{i=1}^N c_i b_i$ where $c \in \mathbb{R}^N$, we have $p(v) = [\text{res}_{\mathcal{B}}(p)](v) = c^T \zeta_{\mathcal{B},v}$ $\forall v \in V_{\mathbb{R}}(\mathcal{J})$, after setting $\zeta_{\mathcal{B},v} := (b_i(v))_{i=1}^N$. Hence

$$p^{\min} = \min_{x \in V_{\mathbb{R}}(\mathcal{J})} p(x) = \min c^T y \text{ s.t. } y \in \text{conv}(\zeta_{\mathcal{B},v} \mid v \in V_{\mathbb{R}}(\mathcal{J})). \quad (8.14)$$

The next result implies a semidefinite programming formulation for (8.14) and its proof implies $p^{\text{mom}} = p^{\text{min}}$.

PROPOSITION 8.14. *[81, Th. 14] A vector $y \in \mathbb{R}^N$ lies in the polytope conv$(\zeta_{\mathcal{B},v} \mid v \in V_{\mathbb{R}}(\mathcal{J}))$ if and only if $M_{\mathcal{B}}(y) \succeq 0$ and $y_1 = 1$.*

Proof. Let U denote the $N \times |\mathbb{N}^n|$ matrix whose αth column is the vector containing the coordinates of $\text{res}_{\mathcal{B}}(x^\alpha)$ in the basis $\mathcal{B}$. Define $\tilde{y} := U^T y \in \mathbb{R}^{\mathbb{N}^n}$ with $\tilde{y}_\alpha = L_y(\mathbf{x}^\alpha)$ $\forall \alpha \in \mathbb{N}^n$. One can verify that $M(\tilde{y}) = U^T M_{\mathcal{B}}(y) U$, $\mathcal{J} \subseteq \text{Ker}\, M(\tilde{y})$, and $\tilde{y}^T \text{vec}(p) = y^T c$ with $\text{res}_{\mathcal{B}}(p) = \sum_{i=1}^N c_i b_i$. Consider the following assertions (i)-(iv): (i) $y \in \mathbb{R}_+(\zeta_{\mathcal{B},v} \mid v \in V_{\mathbb{R}}(\mathcal{J}))$; (ii) $M_{\mathcal{B}}(y) \succeq 0$; (iii) $M(\tilde{y}) \succeq 0$; and (iv) $\tilde{y} \in \mathbb{R}_+(\zeta_v \mid v \in V_{\mathbb{R}}(\mathcal{J}))$. Then, (i) $\Longrightarrow$ (ii) [since $M_{\mathcal{B}}(\zeta_{\mathcal{B},v}) = \zeta_{\mathcal{B},v}\zeta_{\mathcal{B},v}^T \succeq 0$]; (ii) $\Longrightarrow$ (iii) [since $M(\tilde{y}) = U^T M_{\mathcal{B}}(y) U$]; (iii) $\Longrightarrow$ (iv) [by Theorem 5.1, since $\text{rank}\, M(\tilde{y}) < \infty$ as $\mathcal{J} \subseteq \text{Ker}\, M(\tilde{y})$]; and (iv) $\Longrightarrow$ (i), because $\tilde{y} = \sum_{v \in V_{\mathbb{R}}(\mathcal{J})} a_v \zeta_v \Longrightarrow y = \sum_{v \in V_{\mathbb{R}}(\mathcal{J})} a_v \zeta_{\mathcal{B},v}$ [since $\sum_v a_v b_i(v) = \sum_v a_v \text{vec}(b_i)^T \zeta_v = \text{vec}(b_i)^T \tilde{y} = \sum_\alpha (b_i)_\alpha L_y(\mathbf{x}^\alpha) = L_y(b_i) = y_i$]. Finally, as $b_1 = 1$, $y_1 = 1$ means $\sum_v a_v = 1$, corresponding to having a convex combination when $a_v \geq 0$. □

Inequalities $g_j \geq 0$ are treated in the usual way; simply add the conditions $M_{\mathcal{B}}(g_j y) \succeq 0$ to the system $M_{\mathcal{B}}(y) \succeq 0$, $y_1 = 1$, after setting $g_j y := M_{\mathcal{B}}(y) c^{(j)}$ where $\text{res}_{\mathcal{B}}(g_j) = \sum_{i=1}^N c_i^{(j)} b_i$, $c^{(j)} = (c_i^{(j)})_{i=1}^N$. Summarizing we have shown

$$p^{\min} = \min\ c^T y \ \text{ s.t. } y_1 = 1, M_{\mathcal{B}}(y) \succeq 0, M_{\mathcal{B}}(g_j y) \succeq 0\ (\ \forall j \leq m). \quad (8.15)$$

This idea of using equations to reduce the number of variables has been applied e.g. by Jibetean and Laurent [60] in relation with unconstrained minimization. Recall (from Section 7.2, page 241) that for $p \in \mathbb{R}[\mathbf{x}]_{2d}$, $p^{\min} = \inf_{x \in \mathbb{R}^n} p(x)$ can be approximated by computing the minimum of p over the variety $V_{\mathbb{R}}(\mathcal{J})$ with $\mathcal{J} := ((2d{+}2)x_i^{2d+1} {+} \partial p/\partial x_i\ (i = 1, \ldots, n))$ for small $\epsilon > 0$. Then $\mathcal{J}$ is zero-dimensional, $\mathcal{B} = \{\mathbf{x}^\alpha \mid 0 \leq \alpha_i \leq 2d\ \forall i \leq n\}$ is a basis of $\mathbb{R}[\mathbf{x}]/\mathcal{J}$, and the equations in $\mathcal{J}$ give a direct algorithm for computing residues modulo $\mathcal{J}$ and thus the combinatorial moment matrix $M_{\mathcal{B}}(y)$. Such computation can however be demanding for large n, d. We now consider the 0/1 case where the residue computation is trivial.

The 0/1 case. A special case, which is particularly relevant to applications in combinatorial optimization, concerns the minimization of a polynomial p over the 0/1 points in a semialgebraic set K. In other words, the equations $\mathbf{x}_i^2 - \mathbf{x}_i = 0$ $(i = 1, \ldots, n)$ are present in the description of K; thus $\mathcal{J} = (\mathbf{x}_1^2 - \mathbf{x}_1, \ldots, \mathbf{x}_n^2 - \mathbf{x}_n)$ with $V_{\mathbb{C}}(\mathcal{J}) = \{0,1\}^n$. Using the equations $\mathbf{x}_i^2 = \mathbf{x}_i$, we can reformulate all variables y_α $(\alpha \in \mathbb{N}^n)$ in terms of the 2^n variables y_β $(\beta \in \{0,1\}^n)$ via $y_\alpha = y_\beta$ with $\beta_i := \min(\alpha_i, 1)$ $\forall i$.

With $\mathcal{P}(V)$ denoting the collection of all subsets of $V := \{1, \ldots, n\}$, the set $\mathcal{B} := \{\mathbf{x}_I := \prod_{i \in I} \mathbf{x}_i \mid I \in \mathcal{P}(V)\}$ is a basis of $\mathbb{R}[\mathbf{x}]/\mathcal{J}$ and $\dim \mathbb{R}[\mathbf{x}]/\mathcal{J} =$

$|\mathcal{P}(V)| = 2^n$. It is convenient to index a combinatorial moment matrix $M_{\mathcal{B}}(y)$ and its argument y by the set $\mathcal{P}(V)$. The matrix $M_{\mathcal{B}}(y)$ has a particularly simple form, since its (I,J)th entry is $y_{I\cup J}$ $\forall I, J \in \mathcal{P}(V)$. Set

$$\Delta_V := \operatorname{conv}(\zeta_{\mathcal{B},v} \mid v \in \{0,1\}^n) \subseteq \mathbb{R}^{\mathcal{P}(V)}. \tag{8.16}$$

We now give a different, elementary, proof[4] for Proposition 8.14.

LEMMA 8.15. $\Delta_V = \{y \in \mathbb{R}^{\mathcal{P}(V)} \mid y_\emptyset = 1, M_{\mathcal{B}}(y) \succeq 0\} = \{y \in \mathbb{R}^{\mathcal{P}(V)} \mid y_\emptyset = 1, \sum_{J\subseteq V \mid I\subseteq J}(-1)^{|J\setminus I|} y_J \geq 0 \; \forall I \subseteq V\}$.

Proof. Let $Z_{\mathcal{B}}$ be the $2^n \times 2^n$ matrix[5] with columns the vectors $\zeta_{\mathcal{B},v} = (\prod_{i\in I} v_i)_{I\in\mathcal{P}(V)}$ $(v \in \{0,1\}^n)$. Given $y \in \mathbb{R}^{\mathcal{P}(V)}$, let D denote the diagonal matrix whose diagonal entries are the coordinates of the vector $Z_{\mathcal{B}}^{-1}y$. As $M_{\mathcal{B}}(y) = Z_{\mathcal{B}} D Z_{\mathcal{B}}^T$ (direct verification, using the fact that $\mathcal{J}$ is radical), $M_{\mathcal{B}}(y) \succeq 0 \iff D \succeq 0 \iff Z_{\mathcal{B}}^{-1} y \geq 0 \iff y = Z_{\mathcal{B}}(Z_{\mathcal{B}}^{-1}y)$ is a conic combination of the vectors $\zeta_{\mathcal{B},v}$ $(v \in \{0,1\}^n)$. Finally use the form of $Z_{\mathcal{B}}^{-1}$ mentioned in the footnote. □

EXAMPLE 8.16. Consider the stable set problem. Using the formulation (1.5) for $\alpha(G)$, we derive using Lemma 8.15 that $\alpha(G)$ is given by the program

$$\max_{y\in\mathbb{R}^{\mathcal{P}(V)}} \sum_{i\in V} y_{\{i\}} \text{ s.t. } y_\emptyset = 1,\; M_{\mathcal{B}}(y) \succeq 0,\; y_{\{i,j\}} = 0 \; (ij \in E). \tag{8.17}$$

Thus $\alpha(G)$ can be computed via a semidefinite program with a matrix of size 2^n, or via an LP with 2^n linear inequalities and variables. As this is too large for practical purpose, one can instead consider *truncated* combinatorial moment matrices $M_{\mathcal{B}_t}(y)$, indexed by $\mathcal{B}_t := \{\mathbf{x}_I \mid I \in \mathcal{P}(V), |I| \leq t\} \subseteq \mathcal{B}$, leading to the following upper bound on $\alpha(G)$

$$\max \sum_{i\in V} y_{\{i\}} \text{ s.t. } y_\emptyset = 1,\; M_{\mathcal{B}_t}(y) \succeq 0,\; y_{\{i,j\}} = 0 \; (ij \in E). \tag{8.18}$$

For $t = 1$ this upper bound is the well known theta number $\vartheta(G)$ introduced by Lovász [85]. See [78, 83] and references therein for more details.

EXAMPLE 8.17. Consider the *max-cut* problem, introduced in (1.8). We are now dealing with the ideal $\mathcal{J} = (\mathbf{x}_1^2 - 1, \ldots, \mathbf{x}_n^2 - 1)$ with $V_{\mathbb{C}}(\mathcal{J}) = \{\pm 1\}^n$. The above treatment for the 0/1 case extends in the obvious way

[4] This proof applies more general to any zero-dimensional radical ideal $\mathcal{J}$ (cf. [81]).

[5] This matrix is also known as the *Zeta matrix* of the lattice $\mathcal{P}(V)$ of subsets of $V = \{1,\ldots,n\}$ and its inverse $Z_{\mathcal{B}}^{-1}$ as the Möbius matrix; cf. [86]. This fact motivates the name *Zeta vector* chosen in [81] for the vectors $\zeta_{\mathcal{B},v}$ and by extension for the vectors ζ_v. We may identify each $v \in \{0,1\}^n$ with its support $J := \{i \in \{1,\ldots,n\} \mid v_i = 1\}$; the (I,J)th entry of $Z_{\mathcal{B}}$ (resp., of $Z_{\mathcal{B}}^{-1}$) is 1 (resp., is $(-1)^{|J\setminus I|}$) if $I \subseteq J$ and 0 otherwise.

to the ± 1 case after defining $M_{\mathcal{B}}(y) := (y_{I\Delta J})_{I,J\in\mathcal{P}(V)}$ ($I\Delta J$ denotes the symmetric difference of I, J). For any integer t,

$$\max \sum_{ij\in E} (w_{ij}/2)(1-y_{\{i,j\}}) \text{ s.t. } y_\emptyset = 1,\ M_{\mathcal{B}_t}(y) = (y_{I\Delta J})_{|I|,|J|\le t} \succeq 0$$

gives an upper bound for $\mathrm{mc}(G,w)$, equal to it when $t = n$; moreover, $\mathrm{mc}(G,w)$ can reformulated[6] as

$$\max \sum_{ij\in E} (w_{ij}/2)(1-y_{\{i,j\}}) \text{ s.t. } y_\emptyset = 1,\ \sum_{J\subseteq V} (-1)^{|I\cap J|} y_J \ge 0\ \forall I \subseteq V.$$

For $t = 1$, the above moment relaxation is the celebrated SDP relaxation for max-cut used by Goemans and Williamson [42] for deriving the first nontrivial approximation algorithm for max-cut (still with the best performance guarantee as of today). Cf. e.g. [78, 79, 83] and references therein for more details.

Several other combinatorial methods have been proposed in the literature for constructing hierarchies of (LP or SDP) bounds for $p^{\min}$ in the 0/1 case; in particular, by Sherali and Adams [137] and by Lovász and Schrijver [86]. It turns out that the hierarchy of SOS/moment bounds described here refines these other hierarchies; see [78, 83] for a detailed comparison.

Exploiting sparsity in the 0/1 case. Here we revisit exploiting sparsity in the 0/1 case. Namely, consider problem (1.1) where the equations $\mathbf{x}_i^2 = \mathbf{x}_i$ ($i \le n$) are present in the description of K and there is a sparsity structure, i.e. (8.1), (8.2), (8.3) hold. By Corollary 8.10 there is asymptotic convergence to $p^{\min}$ of the sparse SOS/moment bounds. We now give an elementary argument showing *finite* convergence, as well as a sparse semidefinite programming (and linear programming) formulation for (1.1).

Given $v \in \{0,1\}^n$ with support $J = \{i \in V \mid v_i = 1\}$, it is convenient to rename $\zeta_{\mathcal{B},v}$ as $\zeta_J^V \in \{0,1\}^{\mathcal{P}(V)}$ (thus with Ith entry 1 if $I \subseteq J$ and 0 otherwise, for $I \in \mathcal{P}(V)$). Extend the notation (8.16) to any $U \subseteq V$, setting $\Delta_U := \mathrm{conv}(\zeta_J^U \mid J \subseteq U) \subseteq \mathbb{R}^{\mathcal{P}(U)}$. The next lemma[7] shows that two vectors in Δ_{I_1} and in Δ_{I_2} can be merged to a new vector in $\Delta_{I_1\cup I_2}$ when certain obvious compatibility conditions hold.

LEMMA 8.18. *Assume $V = I_1 \cup \ldots \cup I_k$ where the I_h's satisfy (8.1) and, for $1 \le h \le k$, let $y^{(h)} \in \Delta_{I_h}$ satisfying $y_I^{(h)} = y_I^{(h')}$ for all $I \subseteq I_h \cap I_{h'}, 1 \le h, h' \le k$. Then there exists $y \in \Delta_V$ which is a common extension of the $y^{(h)}$'s, i.e. $y_I = y_I^{(h)}$ for all $I \subseteq I_h$, $1 \le h \le k$.*

---

[6]Use here the analogue of Lemma 8.15 for the ± 1 case which claims $M_{\mathcal{B}}(y) = (y_{I\Delta J})_{I,J\subseteq V} \succeq 0 \Longleftrightarrow \sum_{J\in\mathcal{P}(V)} (-1)^{|I\cap J|} y_J \ge 0$ for all $I \in \mathcal{P}(V)$ (cf. [79]).

[7]Lasserre [72] uses the analogue of this result for non-atomic measures, which is a nontrivial result, while the proof in the 0/1 case is elementary.

Proof. Consider first the case $k=2$. Set $I_0 := I_1 \cap I_2$ and, for $h=1,2$, write $y^{(h)} = \sum_{I\subseteq I_h} \lambda_I^h \zeta_I^{I_h} = \sum_{H\subseteq I_0} \sum_{I\subseteq I_h | I\cap I_0=H} \lambda_I^h \zeta_I^{I_h}$ for some $\lambda_I^h \geq 0$ with $\sum_{I\subseteq I_h} \lambda_I^h = 1$. Taking the projection on $\mathbb{R}^{\mathcal{P}(I_0)}$, we obtain

$$\sum_{H\subseteq I_0} \Big(\sum_{I\subseteq I_1 | I\cap I_0=H} \lambda_I^1 \Big) \zeta_H^{I_0} = \sum_{H\subseteq I_0} \Big(\sum_{J\subseteq I_2 | J\cap I_0=H} \lambda_J^2 \Big) \zeta_H^{I_0},$$

which implies $\sum_{I\subseteq I_1 | I\cap I_0=H} \lambda_I^1 = \sum_{J\subseteq I_2 | J\cap I_0=H} \lambda_J^2 =: \lambda_H$ $\forall H \subseteq I_0$, since the vectors $\zeta_H^{I_0}$ $(H\subseteq I_0)$ are linearly independent. One can verify that

$$y := \sum_{H\subseteq I_0 | \lambda_H>0} \frac{1}{\lambda_H} \sum_{I\subseteq I_1, J\subseteq I_2 | I\cap I_0 = J\cap I_0 = H} \lambda_I^1 \lambda_J^2 \zeta_{I\cup J}^{I_1\cup I_2} \in \mathbb{R}^{\mathcal{P}(I_1\cup I_2)}$$

lies in $\Delta_{I_1\cup I_2}$ and that y extends each $y^{(h)}$, $h=1,2$.

In the general case $k\geq 2$ we show, using induction on j, $1\leq j\leq k$, that there exists $z^{(j)} \in \Delta_{I_1\cup\ldots\cup I_j}$ which is a common extension of $y^{(1)},\ldots,y^{(j)}$. Assuming $z^{(j)}$ has been found, we derive from the above case $k=2$ applied to $z^{(j)}$ and $y^{(j+1)}$ the existence of $z^{(j+1)}$. □

COROLLARY 8.19. *Assume $V = I_1\cup\ldots\cup I_k$ where (8.1) holds, let $\mathcal{P}_0 := \cup_{h=1}^k \mathcal{P}(I_h)$ and $y\in\mathbb{R}^{\mathcal{P}_0}$ with $y_\emptyset = 1$. Then, y has an extension $\tilde{y}\in\Delta_V \iff M_{\mathcal{B}}(y,I_h) := (y_{I\cup J})_{I,J\in\mathcal{P}(I_h)} \succeq 0$ for all $h=1,\ldots,k$.*

Proof. Directly from Lemma 8.18 combined with Lemma 8.15. □

As an application one can derive an explicit sparse LP formulation for several graph optimization problems for partial κ-trees; we illustrate this on the stable set and max-cut problems. Let $G=(V,E)$ be a graph satisfying

$$V = I_1\cup\ldots\cup I_k \text{ and (8.1) holds,} \tag{8.19}$$

$$\forall ij\in E\ \exists h\in\{1,\ldots,k\} \text{ s.t. } i,j\in I_h. \tag{8.20}$$

First consider the formulation (1.5) for the stability number $\alpha(G)$; as (8.20) holds, this formulation satisfies the sparsity assumptions (8.2) and (8.3). Hence, using Lemma 8.15 combined with Corollary 8.19, we deduce that $\alpha(G)$ can be obtained by maximizing the linear objective function $\sum_{i\in V} y_{\{i\}}$ over the set of $y\in\mathbb{R}^{\mathcal{P}_0}$ satisfying $y_\emptyset = 1$, $y_{\{i,j\}} = 0$ for $ij\in E$, and any one of the following equivalent conditions (8.21) or (8.22)

$$M_{\mathcal{B}}(y,I_h)\succeq 0 \text{ for all } 1\leq h\leq k, \tag{8.21}$$

$$\sum_{J\in\mathcal{P}(I_h)|I\subseteq J} (-1)^{|J\setminus I|} y_J \geq 0 \text{ for all } I\in\mathcal{P}(I_h), 1\leq h\leq k. \tag{8.22}$$

More generally, given weights c_i $(i\in V)$ attached to the nodes of G, one can find $\alpha(G,c)$, the maximum weight $\sum_{i\in S} c_i$ of a stable set S, by maximizing the linear objective function $\sum_{i\in V} c_i y_{\{i\}}$ over the above LP. Analogously, the objective function in the formulation (1.8) of the max-cut problem satisfies (8.2) and thus the max-cut value $\mathrm{mc}(G,w)$ can be obtained by maximizing the linear objective function $\sum_{ij\in E}(w_{ij}/2)(1-y_{\{i,j\}})$ over the set of $y \in \mathbb{R}^{\mathcal{P}_0}$ satisfying $y_\emptyset = 1$ and

$$\sum_{J\in\mathcal{P}(I_h)} (-1)^{|I\cap J|} y_J \geq 0 \quad \text{for all } I \in \mathcal{P}(I_h),\ 1 \leq h \leq k. \tag{8.23}$$

With $\max_{h=1}^k |I_h| \leq \kappa$, we find for both the stable set and max-cut problems an LP formulation involving $O(k2^\kappa)$ linear inequalities and variables. This applies in particular when G is a partial κ-tree (i.e. G is a subgraph of a chordal graph with maximum clique size κ). Indeed, then (8.19)-(8.20) hold with $\max_h |I_h| \leq \kappa$ and $k \leq n$, and thus $\alpha(G,c)$, $\mathrm{mc}(G,w)$ can be computed via an LP with $O(n2^\kappa)$ inequalities and variables. As an application, for fixed κ, $\alpha(G,c)$ and $\mathrm{mc}(G,w)$ can be computed in polynomial time[8] for the class of partial κ-trees. This is a well known result; cf. eg. [15, 146, 155].

8.3. Exploiting symmetry. Another useful property that can be exploited to reduce the size of the SOS/moment relaxations is to use the presence of structural symmetries in the polynomials $p, g_1, \ldots, g_m$. This relies on combining ideas from group representation and invariance theory, as explained in particular in the work of Gaterman and Parrilo [41] (see also Vallentin [149]). We will only sketch some ideas illustrated on some examples as a detailed treatment of this topic is out of the scope of this paper.

Group action. Let $\mathcal{G}$ be a finite group acting on $\mathbb{R}^N$ ($N \geq 1$) via an action $\rho_0 : \mathcal{G} \to \mathrm{GL}(\mathbb{R}^N)$. This induces an action $\rho : \mathcal{G} \to \mathrm{Aut}(\mathrm{Sym}_N)$ on Sym_N, the space of $N \times N$ symmetric matrices, defined by $\rho(g)(X) := \rho_0(g)^T X \rho_0(g)$ for $g \in \mathcal{G}$, $X \in \mathrm{Sym}_N$. This also induces an action on PSD_N, the set of $N \times N$ positive semidefinite matrices. We assume here that each $\rho_0(g)$ is an orthogonal matrix. Then, a matrix $X \in \mathbb{R}^{N\times N}$ is invariant under action of $\mathcal{G}$, i.e. $\rho(g)(X) = X\ \forall g \in \mathcal{G}$, if and only if X belongs to the commutant algebra

$$\mathcal{A}^G := \{X \in \mathbb{R}^{N\times N} \mid \rho_0(g)X = X\rho_0(g)\ \forall g \in G\}. \tag{8.24}$$

Note that the commutant algebra also depends on the specific action ρ_0.

Invariant semidefinite program. Consider a semidefinite program

$$\max\ \langle C, X\rangle \ \text{ s.t. } \langle A_r, X\rangle = b_r\ (r = 1, \ldots, m), X \in \mathrm{PSD}_N, \tag{8.25}$$

in the variable $X \in \mathrm{Sym}_N$, where $C, A_r \in \mathrm{Sym}_N$ and $b_r \in \mathbb{R}$. Assume that this semidefinite program is invariant under action of $\mathcal{G}$; that is, C is

[8] in fact, in strongly polynomial time, since all coefficients in (8.22), (8.23) are $0, \pm 1$; see [128, §15.2].

invariant, i.e. $C \in \mathcal{A}^{\mathcal{G}}$, and the feasible region is globally invariant, i.e. X feasible for (8.25) $\Longrightarrow \rho(g)(X)$ feasible for (8.25) $\forall g \in \mathcal{G}$. Let X be feasible for (8.25). An important consequence of the convexity of the feasible region is that the new matrix $X_0 := \frac{1}{|G|}\sum_{g\in G}\rho(g)(X)$ is again feasible; moreover X_0 is invariant under action of $\mathcal{G}$ and it has the same objective value as X. Therefore, we can w.l.o.g. require that X is invariant in (8.25), i.e. we can add the constraint $X \in \mathcal{A}^{\mathcal{G}}$ (which is linear in X) to (8.25) and get an equivalent program.

Action induced by permutations. An important special type of action is when $\mathcal{G}$ is a subgroup of $\mathcal{S}_N$, the group of permutations on $\{1,\ldots,N\}$. Then each $g \in \mathcal{S}_N$ acts naturally on $\mathbb{R}^N$ by $\rho_0(g)(x) := (x_{g(i)})_{i=1}^N$ for $x = (x_i)_{i=1}^N \in \mathbb{R}^N$, and on $\mathbb{R}^{N\times N}$ by $\rho(g)(X) := (X_{g(i),g(j)})_{i,j=1}^N$ for $X = (X_{i,j})_{i,j=1}^N$; that is, $\rho(g)(X) = M_g X M_g^T$ after defining M_g as the $N \times N$ matrix with $(M_g)_{i,j} = 1$ if $j = g(i)$ and 0 otherwise.

For $(i,j) \in \{1,\ldots,N\}^2$, its orbit under action of $\mathcal{G}$ is the set $\{(g(i),g(j)) \mid g \in \mathcal{G}\}$. Let ω denote the number of orbits of $\{1,\ldots,N\}^2$ and, for $l = 1,\ldots,\omega$, define the $N \times N$ matrix $\tilde{D}_l$ by $(\tilde{D}_l)_{i,j} := 1$ if the pair (i,j) belongs to the lth orbit, and 0 otherwise. Following de Klerk, Pasechnik and Schrijver [33], define $D_l := \frac{\tilde{D}_l}{\sqrt{\langle \tilde{D}_l, \tilde{D}_l\rangle}}$ for $l = 1,\ldots,\omega$, the multiplication parameters $\gamma_{i,j}^l$ by

$$D_i D_j = \sum_{l=1}^{\omega} \gamma_{i,j}^l D_l \quad \text{for } i,j = 1,\ldots,\omega,$$

and the $\omega \times \omega$ matrices $L_1,\ldots,L_\omega$ by $(L_l)_{i,j} := \gamma_{l,j}^i$ for $i,j,k = 1,\ldots,\omega$. Then the commutant algebra from (8.24) is

$$\mathcal{A}^G = \Big\{\sum_{l=1}^{\omega} x_l D_l \mid x_l \in \mathbb{R}\Big\}$$

and thus $\dim \mathcal{A}^G = \omega$.

THEOREM 8.20. *[33] The mapping $D_l \mapsto L_l$ is a $*$-isomorphism, known as the regular $*$-representation of $\mathcal{A}^G$. In particular, given $x_1,\ldots,x_\omega \in \mathbb{R}$,*

$$\sum_{l=1}^{\omega} x_l D_l \succeq 0 \Longleftrightarrow \sum_{l=1}^{\omega} x_l L_l \succeq 0. \tag{8.26}$$

An important application of this theorem is that it provides an *explicit* equivalent formulation for an invariant SDP, using only ω variables and a matrix of order ω. Indeed, assume (8.25) is invariant under action of $\mathcal{G}$. Set $c := (\langle C, D_l\rangle)_{l=1}^{\omega}$ so that $C = \sum_{l=1}^{\omega} c_l D_l$, and $a_r := (\langle A_r, D_l\rangle)_{l=1}^{\omega}$. As observed above the matrix variable X can be assumed to lie in $\mathcal{A}^G$ and thus to be of the form $X = \sum_{l=1}^{\omega} x_l D_l$ for some scalars $x_l \in \mathbb{R}$. Therefore, using (8.26), (8.25) can be equivalently reformulated as

$$\max \sum_{l=1}^{\omega} c_l x_l \text{ s.t. } a_r^T x = b_r \ (r = 1, \ldots, m), \sum_{l=1}^{\omega} x_l L_l \succeq 0. \tag{8.27}$$

The new program (8.27) involves a $\omega \times \omega$ matrix and ω variables and can thus be much more compact than (8.25). Theorem 8.20 is used in [33] to compute the best known bounds for the crossing number of complete bipartite graphs. It is also applied in [82] to the stable set problem for the class of Hamming graphs as sketched below.

EXAMPLE 8.21. Given $\mathcal{D} \subseteq \{1, \ldots, n\}$, let $G(n, \mathcal{D})$ be the graph with node set $\mathcal{P}(V)$ (the collection of all subsets of $V = \{1, \ldots, n\}$) and with an edge (I, J) when $|I \Delta J| \in \mathcal{D}$. (Computing the stability number of $G(n, \mathcal{D})$ is related to finding large error correcting codes in coding theory; cf. e.g. [82, 129]). Consider the moment relaxation of order t for $\alpha(G(n, \mathcal{D}))$ as defined in (8.18); note that it involves a matrix of size $O(\binom{|\mathcal{P}(V)|}{t}) = O((2^n)^t)$, which is exponentially large in n. However, as shown in [82], this semidefinite program is invariant under action of the symmetric group $\mathcal{S}_n$, and there are $O(n^{2^{2t-1}-1})$ orbits. Hence, by Theorem 8.20, there is an equivalent SDP whose size is $O(n^{2^{2t-1}-1})$, thus polynomial in n for any fixed t, which implies that the moment upper bound on $\alpha(G(n, \mathcal{D}))$ can be computed in polynomial time for any fixed t.

Block-diagonalization. Theorem 8.20 gives a first, explicit, symmetry reduction for matrices in $\mathcal{A}^G$. Further reduction is possible. Indeed, using Schur's lemma from representation theory (cf. e.g. Serre [135]), it can be shown that all matrices in $\mathcal{A}^G$ can be put in block-diagonal form by a linear change of coordinates. Namely, there exists a unitary complex matrix T and positive integers $h, n_1, \ldots, n_h, m_1, \ldots, m_h$ such that the set $T^* \mathcal{A}^G T := \{T^* X T \mid X \in \mathcal{A}^G\}$ coincides with the set of the block-diagonal matrices

$$\begin{pmatrix} C_1 & 0 & \ldots & 0 \\ 0 & C_2 & \ldots & 0 \\ \vdots & \vdots & \ddots & \vdots \\ 0 & 0 & \ldots & C_h \end{pmatrix},$$

where each C_i $(i = 1, \ldots, h)$ is a block-diagonal matrix with m_i identical blocks on its diagonal, all equal to some $B_i \in \mathbb{R}^{n_i \times n_i}$. The above parameters have the following interpretation: $N = \sum_{i=1}^{h} m_i n_i$, $\dim \mathcal{A}^G = \sum_{i=1}^{h} n_i^2$, there are h nonequivalent irreducible representations $\theta_1, \ldots, \theta_h$ for the group $\mathcal{G}$, with respective representation dimensions $n_1, \ldots, n_h$ so that $\rho = m_1 \theta_1 \oplus \ldots \oplus m_h \theta_h$, where $m_1, \ldots, m_h$ are the multiplicities. We refer to Gaterman and Parrilo [41], Vallentin [149] for details and further references therein. To be able to apply this for practical computation one needs to know the explicit block-diagonalization. Several examples are treated in detail in [41]. Here is a small (trivial) example as illustration.

EXAMPLE 8.22. Consider the semidefinite program

$$\min\ d+f \text{ s.t. } X := \begin{pmatrix} a & b & c \\ b & d & e \\ c & e & f \end{pmatrix} \succeq 0,\ d+f+2e-b-c=0 \tag{8.28}$$

It is invariant under action of the group $\{1,\sigma\} \sim \mathcal{S}_2$, where σ permutes simultaneously the last two rows and columns of X. Thus we may assume in (8.28) that X is invariant under this action, i.e. $d=f$ and $b=c$. This reduces the number of variables from 6 to 4. Next we give the explicit block-diagonalization. Namely, consider the orthogonal matrix $T := \begin{pmatrix} 1 & 0 & 0 \\ 0 & u & u \\ 0 & u & -u \end{pmatrix}$ where $u := 1/\sqrt{2}$, and observe that

$$T^*XT = \begin{pmatrix} a & \sqrt{2}b & 0 \\ \sqrt{2}b & d+e & 0 \\ 0 & 0 & d-e \end{pmatrix}.$$

We now mention the following example due to Schrijver [129], dealing with the block-diagonalization of the Terwilliger algebra.

EXAMPLE 8.23. Consider the permutation group $\mathcal{S}_n$ acting on $V = \{1,\dots,n\}$. Then each $g \in \mathcal{S}_n$ acts in the obvious way on $\mathcal{P}(V)$ (by $g(I) := \{g(i) \mid i \in I\}$ for $I \subseteq V$) and thus on matrices indexed by $\mathcal{P}(V)$. The orbit of $(I,J) \in \mathcal{P}(V)\times\mathcal{P}(V)$ depends on the triple $(|I|,|J|,|I\cap J|)$. Therefore, the commutant algebra, consisting of the matrices $X \in \mathbb{R}^{\mathcal{P}(V)\times\mathcal{P}(V)}$ that are invariant under action of $\mathcal{S}_n$, is

$$\Big\{ \sum_{i,j,t\in\mathbb{N}} \lambda_{i,j}^t M_{i,j}^t \mid \lambda_{i,j}^t \in \mathbb{R} \Big\},$$

known as the Terwilliger algebra. Here $M_{i,j}^t$ denotes the matrix indexed by $\mathcal{P}(V)$ with (I,J)th entry 1 if $|I|=i$, $|J|=j$ and $|I\cap J|=t$, and 0 otherwise. Schrijver [129] has computed the explicit block-diagonalization for the Terwilliger algebra and used it for computing sharp SDP bounds for the stability number $\alpha(G(n,\mathcal{D}))$, also considered in Example 8.21. As explained in [81] this new bound lies between the moment bound of order 1 and the moment bound of order 2. See also [149] for an exposition of symmetry reduction with application to the Terwilliger algebra.

Symmetry in polynomial optimization. When the polynomial optimization problem (1.1) is invariant under action of some finite group $\mathcal{G}$, it is natural to search for relaxation schemes that inherit the symmetry pattern of the polynomials $p, g_1, \ldots, g_m$. For instance, if p is a symmetric polynomial which is a SOS, one may wonder about the existence of a sum of symmetric squares. One has to be careful however. For instance, as noted in [41], the univariate polynomial $p = \mathbf{x}^2 + (\mathbf{x} - \mathbf{x}^3)^2 = \mathbf{x}^6 - 2\mathbf{x}^4 + 2\mathbf{x}^2$ is invariant under the action $x \mapsto -x$, but there is no sum of square decomposition $p = \sum_l u_l^2$ where each u_l is invariant under this action as well (for otherwise, u_l should be a polynomial of degree 3 in $\mathbf{x}^2$, an obvious contradiction). Yet symmetry of p does imply some special symmetry structure for the squares; we refer to Gaterman and Parrilo [41] for a detailed account.

Jansson et al. [58] study how symmetry carries over to the moment relaxations of problem (1.1). Say, the polynomials $p, g_1, \ldots, g_m$ are invariant under action of a group $\mathcal{G} \subseteq \mathcal{S}_n$; i.e. $p(x) = p(\rho_0(g)(x))\ \forall g \in \mathcal{G}$, where $\rho_0(g)(x) = (x_{g(i)})_{i=1}^n$, and analogously for the g_j's. For instance the following problem, studied in [58],

$$\min \sum_{i=1}^n x_i^q \text{ s.t. } \sum_{i=1}^n x_i^j = b_j \ (j = 1, \ldots, m) \tag{8.29}$$

with $q \in \mathbb{N}$, $b_j \in \mathbb{R}$, falls in this setting with $\mathcal{G} = \mathcal{S}_n$. Then some symmetry carries over to the moment relaxations (6.3). Indeed, if x is a global minimizer of p over K, then each $\rho_0(g)(x)$ (for $g \in \mathcal{G}$) too is a global minimizer. Thus the sequence y of moments of the measure $\mu := \frac{1}{|\mathcal{G}|}\sum_{g\in\mathcal{G}} \delta_{\rho_0(g)(x)}$ is feasible for any moment relaxation, with optimum value $p^{\min}$. In other words, we can add the invariance condition

$$y_\alpha = y_{\rho_0(g)(\alpha)}, \text{ i.e. } y_{(\alpha_1,\ldots,\alpha_n)} = y_{(\alpha_{g(1)},\ldots,\alpha_{g(n)})}\ \forall g \in \mathcal{G}$$

on the entries of variable y to the formulation of the moment relaxation (6.3) of any order t. For instance, when $\mathcal{G} = \mathcal{S}_n$, one can require that $y_{e_1} = \ldots = y_{e_n}$, i.e. all y_α take a common value for any $|\alpha| = 1$, that all y_α take a common value for any $|\alpha| = 2$, etc. Thus the moment matrix of order 1 is of the form

$$M_1(y) = \begin{pmatrix} a & b & b & b & \ldots & b \\ b & c & d & d & \ldots & d \\ b & d & c & d & \ldots & d \\ \vdots & \vdots & & \ddots & \ddots & \vdots \\ b & d & \ldots & d & c & d \\ b & d & \ldots & d & d & c \end{pmatrix}.$$

It is explained in [58] how to find the explicit block-diagonalization for such symmetric $M_t(y)$ ($t = 1, 2$, etc). This is not difficult in the case $t = 1$; using a Schur complement with respect to the upper left corner, one deduces easily that $M_1(y) \succeq 0 \iff c + (n-1)d - nb^2/a \geq 0$ and $c - d \geq 0$.

The details for $t = 2$ are already more complicated and need information about the irreducible representations of the symmetric group $\mathcal{S}_n$.

In conclusion, exploiting symmetry within polynomial optimization and, more generally, semidefinite programming, has spurred recently lots of interesting research activity, with many exciting new developments in various areas. Let us just mention pointers to a few papers dealing with symmetry reduction in various contexts; the list is not exclusive. In particular, Bachoc and Vallentin [2–4] study the currently best known bounds for spherical codes and the kissing number; Bai et al. [5] deal with truss topology optimization; de Klerk and Sotirov [34] study lower bounds for quadratic assignment; Gvozdenović and Laurent [47, 48] compute approximations for the chromatic number of graphs; Vallentin [148] considers the minimum distortion of embeddings of highly regular graphs in the Euclidean space.

Acknowledgements. I am very grateful to Markus Schweighofer for discussion and his help with some proofs in the survey, and to Hayato Waki for his careful reading and for his useful questions and suggestions. I also thank Claus Scheiderer, Lars Schewe, Konrad Schmüdgen, Achill Schürmann, Rekha Thomas, Frank Vallentin for their comments.

REFERENCES

[1] N.I. AKHIEZER. *The classical moment problem,* Hafner, New York, 1965.

[2] C. BACHOC AND F. VALLENTIN, *New upper bounds for kissing numbers from semidefinite programming,* Journal of the American Mathematical Society, to appear.

[3] ——, *Semidefinite programming, multivariate orthogonal polynomials, and codes in spherical caps,* arXiv:math.MG/0610856, 2006.

[4] ——, *Optimality and uniqueness of the* $(4,10,1/6)$*-spherical code,* arXiv:math.MG/0708.3947, 2007.

[5] Y. BAI, E. DE KLERK, D.V. PASECHNIK, R. SOTIROV, *Exploiting group symmetry in truss topology optimization,* CentER Discussion paper 2007-17, Tilburg University, The Netherlands, 2007.

[6] S. BASU, R. POLLACK, AND M.-F. ROY, Algorithms in Real Algebraic Geometry, Springer, 2003.

[7] C. BAYER AND J. TEICHMANN, *The proof of Tchakaloff's theorem,* Proceedings of the American Mathematical Society **134**:3035–3040, 2006.

[8] C. BERG. *The multidimensional moment problem and semi-groups,* Proc. Symp. Appl. Math. **37**:110–124, 1987.

[9] C. BERG, J.P.R. CHRISTENSEN, AND C.U. JENSEN, *A remark on the multidimensional moment problem,* Mathematische Annalen **243**:163–169, 1979.

[10] C. BERG, J.P.R. CHRISTENSEN, AND P. RESSEL, *Positive definite functions on Abelian semigroups,* Mathematische Annalen **223**:253–272, 1976.

[11] C. BERG AND P.H. MASERICK, *Exponentially bounded positive definite functions,* Illinois Journal of Mathematics **28**:162–179, 1984.

[12] J.R.S. BLAIR AND B. PEYTON, *An introduction to chordal graphs and clique trees,* In *Graph Theory and Sparse Matrix Completion,* A. George, J.R. Gilbert, and J.W.H. Liu, eds, Springer-Verlag, New York, pp 1–29, 1993.

[13] G. BLEKHERMAN, *There are significantly more nonnegative polynomials than sums of squares, Isreal Journal of Mathematics,* **153**:355–380, 2006.

[14] J. Bochnak, M. Coste, and M.-F. Roy, *Géométrie Algébrique Réelle*, Springer, Berlin, 1987. (*Real algebraic geometry*, second edition in english, Springer, Berlin, 1998.)

[15] H.L. Bodlaender and K. Jansen, *On the complexity of the maximum cut problem*, Nordic Journal of Computing **7(1)**:14-31, 2000.

[16] H. Bosse, *Symmetric positive definite polynomials which are not sums of squares*, preprint, 2007.

[17] M.-D. Choi and T.-Y. Lam, *Extremal positive semidefinite forms*, Math. Ann. **231**:1–18, 1977.

[18] M.-D. Choi, T.-Y. Lam, and B. Reznick, *Sums of squares of real polynomials*, Proceedings of Symposia in Pure mathematics **58(2)**:103–126, 1995.

[19] A.M. Cohen, H. Cuypers, and H. Sterk (eds.), Some Tapas of Computer Algebra, Springer, Berlin, 1999.

[20] R.M. Corless, P.M. Gianni, and B.M. Trager, *A reordered Schur factorization method for zero-dimensional polynomial systems with multiple roots*, Proceedings ACM International Symposium Symbolic and Algebraic Computations, Maui, Hawaii, 133–140, 1997.

[21] D.A. Cox, J.B. Little, and D. O'Shea, *Ideals, Varieties, and Algorithms: An Introduction to Computational Algebraic Geometry and Commutative Algebra*, Springer, 1997.

[22] ———, *Using Algebraic Geometry*, Graduate Texts in Mathematics **185**, Springer, New York, 1998.

[23] R.E. Curto and L.A. Fialkow, *Solution of the truncated complex moment problem for flat data*, Memoirs of the American Mathematical Society **119**(568), 1996.

[24] ———, *Flat extensions of positive moment matrices: recursively generated relations*, Memoirs of the American Mathematical Society, **136**(648), 1998.

[25] ———, *The truncated complex K-moment problem* Transactions of the American Mathematical Society **352**:2825–2855, 2000.

[26] ———, *An analogue of the Riesz-Haviland theorem for the truncated moment problem*, preprint, 2007.

[27] R.E. Curto, L.A. Fialkow, and H.M. Möller, *The extremal truncated moment problem*, Integral Equations and Operator Theory, to appear.

[28] E. de Klerk, *Aspects of Semidefinite Programming - Interior Point Algorithms and Selected Applications*, Kluwer, 2002.

[29] E. de Klerk, D. den Hertog, and G. Elabwabi, *Optimization of univariate functions on bounded intervals by interpolation and semidefinite programming*, CentER Discussion paper 2006-26, Tilburg University, The Netherlands, April 2006.

[30] E. de Klerk, M. Laurent, and P. Parrilo, *On the equivalence of algebraic approaches to the minimization of forms on the simplex*, In *Positive Polynomials in Control*, D. Henrion and A. Garulli (eds.), Lecture Notes on Control and Information Sciences **312**:121–133, Springer, Berlin, 2005.

[31] ———, *A PTAS for the minimization of polynomials of fixed degree over the simplex*, Theoretical Computer Science **361(2-3)**:210–225, 2006.

[32] E. de Klerk and D.V. Pasechnik, *Approximating the stability number of a graph via copositive programming*, SIAM Journal on Optimization **12**:875–892, 2002.

[33] E. de Klerk, D.V. Pasechnik, and A. Schrijver, *Reduction of symmetric semidefinite programs using the regular *-representation*, Mathematical Programming B **109**: 613-624, 2007.

[34] E. de Klerk and R. Sotirov, *Exploiting group symmetry in semidefinite programming relaxations of the quadratic assignment problem*, Optimization Online, 2007.

[35] J.L. Krivine, *Anneaux préordonnés*, J. Anal. Math. **12**:307–326, 1964.

[36] ———, *Quelques propriétés des préordres dans les anneaux commutatifs unitaires*, C.R. Académie des Sciences de Paris **258**:3417–3418, 1964.

[37] A. Dickenstein and I. Z. Emiris (eds.). *Solving Polynomial Equations: Foundations, Algorithms, and Applications*, Algorithms and Computation in Mathematics 14, Springer-Verlag, 2005.

[38] L.A. Fialkow, *Truncated multivariate moment problems with finite variety*, *Journal of Operator Theory*, to appear.

[39] B. Fuglede, *The multidimensional moment problem*, Expositiones Mathematicae **1**:47–65, 1983.

[40] M.R. Garey and D.S. Johnson, *Computers and Intractability: A Guide to the Theory of NP-Completeness*, San Francisco, W.H. Freeman & Company, Publishers, 1979.

[41] K. Gaterman and P. Parrilo, *Symmetry groups, semidefinite programs and sums of squares*, Journal of Pure and Applied Algebra **192**:95–128, 2004.

[42] M.X. Goemans and D. Williamson, *Improved approximation algorithms for maximum cuts and satisfiability problems using semidefinite programming*, Journal of the ACM **42**:1115–1145, 1995.

[43] D. Grimm, T. Netzer, and M. Schweighofer, *A note on the representation of positive polynomials with structured sparsity*, Archiv der Mathematik, to appear.

[44] B. Grone, C.R. Johnson, E. Marques de Sa, and H. Wolkowicz, *Positive definite completions of partial Hermitian matrices*, Linear Algebra and its Applications **58**:109–124, 1984.

[45] M. Grötschel, L. Lovász, and A. Schrijver, *Geometric Algorithms and Combinatorial Optimization*, Springer-Verlag, Berlin, 1988.

[46] N. Gvozdenović and M. Laurent, *Semidefinite bounds for the stability number of a graph via sums of squares of polynomials*, Mathematical Programming **110(1)**:145–173, 2007.

[47] ———, *The operator Ψ for the chromatic number of a graph*, SIAM Journal on Optimization, to appear.

[48] ———, *Computing semidefinite programming lower bounds for the (fractional) chromatic number via block-diagonalization*, SIAM Journal on Optimization, to appear.

[49] D. Handelman, *Representing polynomials by positive linear functions on compact convex polyhedra*, Pacific Journal of Mathematics **132(1)**:35–62, 1988.

[50] B. Hanzon and D. Jibetean, *Global minimization of a multivariate polynomial using matrix methods*, Journal of Global Optimization **27**:1–23, 2003.

[51] E.K. Haviland, *On the momentum problem for distributions in more than one dimension*, American Journal of Mathematics **57**:562–568, 1935.

[52] J.W. Helton and M. Putinar, *Positive polynomials in scalar and matrix variables, the spectral theorem and optimization*. In *Operator Theory, Structured Matrices, and Dilations: Tiberiu Constantinescu Memorial Volume*, M. Bakonyi, A. Gheondea, M. Putinar, and J. Rovnyak (eds.), Theta, Bucharest, pp. 229–306, 2007.

[53] D. Henrion and J.-B. Lasserre, *GloptiPoly: Global optimization over polynomials with Matlab and SeDuMi*, ACM Transactions Math. Soft. **29**:165–194, 2003.

[54] ———, *Detecting global optimality and extracting solutions in GloptiPoly*, In *Positive Polynomials in Control*, D. Henrion and A. Garulli (eds.), Lecture Notes on Control and Information Sciences **312**:293–310, Springer, Berlin, 2005.

[55] D. Henrion, J.-B. Lasserre, and J. Löfberg, *GloptiPoly 3: moments, optimization and semidefinite programming*, arXiv:0709.2559, 2007.

[56] D. Hilbert, *Über die Darstellung definiter Formen als Summe von Formenquadraten*, Mathematische Annalen **32**:342–350, 1888. See Ges. Abh. **2**:154–161, Springer, Berlin, reprinted by Chelsea, New York, 1981.

[57] T. JACOBI AND A. PRESTEL, *Distinguished representations of strictly positive polynomials*, Journal für die Reine und Angewandte Mathematik **532**:223–235, 2001.

[58] L. JANSSON, J.B. LASSERRE, C. RIENER, AND T. THEOBALD, *Exploiting symmetries in SDP-relaxations for polynomial optimization*, Optimization Online, 2006.

[59] D. JIBETEAN, *Algebraic Optimization with Applications to System Theory*, Ph.D thesis, Vrije Universiteit, Amsterdam, The Netherlands, 2003.

[60] D. JIBETEAN AND M. LAURENT, *Semidefinite approximations for global unconstrained polynomial optimization*, SIAM Journal on Optimization **16**:490–514, 2005.

[61] M. KOJIMA, S. KIM, AND H. WAKI, *Sparsity in sums of squares of polynomials*, Mathematical Programming **103**:45–62, 2005.

[62] M. KOJIMA AND M. MURAMATSU, *A note on sparse SOS and SDP relaxations for polynomial optimization problems over symmetric cones*, Research Report B-421, Department of Mathematical and Computing Sciences, Tokyo Institute of Technology, 2006.

[63] S. KUHLMAN AND M. MARSHALL, *Positivity, sums of squares and the multi-dimensional moment problem*, Transactions of the American Mathematical Society **354**:4285–4301, 2002.

[64] H. LANDAU (ed.), *Moments in Mathematics*, Proceedings of Symposia in Applied Mathematics **37**, 1–15, AMS, Providence, 1987.

[65] J.B. LASSERRE, *Global optimization with polynomials and the problem of moments*, SIAM Journal on Optimization **11**:796–817, 2001.

[66] ———, *An explicit exact SDP relaxation for nonlinear $0-1$ programs* In K. AARDAL AND A.M.H. GERARDS (EDS.), Lecture Notes in Computer Science **2081**:293–303, 2001.

[67] ———, *Polynomials nonnegative on a grid and discrete representations*, Transactions of the American Mathematical Society **354**:631–649, 2001.

[68] ———, *Semidefinite programming vs LP relaxations for polynomial programming*, Mathematics of Operations Research **27**:347–360, 2002.

[69] ———, *SOS approximations of polynomials nonnegative on a real algebraic set*, SIAM Journal on Optimization **16**:610–628, 2005.

[70] ———, *Polynomial programming: LP-relaxations also converge*, SIAM Journal on Optimization **15**:383-393, 2005.

[71] ———, *A sum of squares approximation of nonnegative polynomials*, SIAM Journal on Optimization **16**:751–765, 2006.

[72] ———, *Convergent semidefinite relaxations in polynomial optimization with sparsity*, SIAM Journal on Optimization **17**:822–843, 2006.

[73] ———, *Sufficient conditions for a real polynomial to be a sum of squares*, Archiv der Mathematik, to appear.

[74] ———, *A Positivstellensatz which preserves the coupling pattern of variables*, Preprint.

[75] J.B. LASSERRE, M. LAURENT, AND P. ROSTALSKI, *Semidefinite characterization and computation of real radical ideals*, Foundations of Computational Mathematics, to appear.

[76] ———, *A unified approach for real and complex zeros of zero-dimensional ideals*, this volume, IMA Volumes in Mathematics and its Applications, *Emerging Applications of Algebraic Geometry*, M. Putinar and S. Sullivant, eds., Springer Science+Business Media, New York, 2008.

[77] J.B. LASSERRE AND T. NETZER, *SOS approximations of nonnegative polynomials via simple high degree perturbations*, Mathematische Zeitschrift **256**:99–112, 2006.

[78] M. LAURENT, *A comparison of the Sherali-Adams, Lovász-Schrijver and Lasserre relaxations for 0-1 programming*, Mathematics of Operations Research **28**(3):470–496, 2003.

[79] ———, *Semidefinite relaxations for Max-Cut*, In The Sharpest Cut: The Impact of Manfred Padberg and His Work. M. Grötschel, ed., pp. 257–290, MPS-SIAM Series in Optimization 4, 2004.

[80] ———, *Revisiting two theorems of Curto and Fialkow on moment matrices*, Proceedings of the American Mathematical Society **133**(10):2965–2976, 2005.

[81] ———, Semidefinite representations for finite varieties. *Mathematical Programming*, 109:1–26, 2007.

[82] ———, *Strengthened semidefinite programming bounds for codes*, Mathematical Programming B 109:239–261, 2007.

[83] M. LAURENT AND F. RENDL, *Semidefinite Programming and Integer Programming*, In Handbook on Discrete Optimization, K. Aardal, G. Nemhauser, R. Weismantel (eds.), pp. 393–514, Elsevier B.V., 2005.

[84] J. LÖFBERG AND P. PARRILO, *From coefficients to samples: a new approach to SOS optimization*, 43rd IEEE Conference on Decision and Control, Vol. **3**, pp. 3154–3159, 2004.

[85] L. LOVÁSZ, *On the Shannon capacity of a graph*. IEEE Transactions on Information Theory **IT-25**:1–7, 1979.

[86] L. LOVÁSZ AND A. SCHRIJVER, *Cones of matrices and set-functions and* $0-1$ *optimization*, SIAM Journal on Optimization **1**:166–190, 1991.

[87] M. MARSHALL, *Positive polynomials and sums of squares*, Dottorato de Ricerca in Matematica, Dipartimento di Matematica dell'Università di Pisa, 2000.

[88] ———, *Optimization of polynomial functions*, Canadian Math. Bull. **46**:575–587, 2003.

[89] ———, *Representation of non-negative polynomials, degree bounds and applications to optimization*, Canad. J. Math., to appear.

[90] ———, *Positive Polynomials and Sums of Squares*, AMS Surveys and Monographs, forthcoming book.

[91] H.M. MÖLLER, *An inverse problem for cubature formulae*, Vychislitel'nye Tekhnologii (Computational Technologies) **9**:13–20, 2004.

[92] H.M. MÖLLER AND H.J. STETTER, *Multivariate polynomial equations with multiple zeros solved by matrix eigenproblems*, Numerische Mathematik **70**:311–329, 1995.

[93] T.S. MOTZKIN AND E.G. STRAUS, *Maxima for graphs and a new proof of a theorem of Túran*, Canadian Journal of Mathematics **17**:533–540, 1965.

[94] K.G. MURTY AND S.N. KABADI, *Some NP-complete problems in quadratic and nonlinear programming*, Mathematical Programming **39**:117–129, 1987.

[95] Y.E. NESTEROV, *Squared functional systems and optimization problems*, In J.B.G. FRENK, C. ROOS, T. TERLAKY, AND S. ZHANG (EDS.), *High Performance Optimization*, 405–440, Kluwer Academic Publishers, 2000.

[96] Y.E. NESTEROV AND A. NEMIROVSKI, *Interior Point Polynomial Methods in Convex Programming*, Studies in Applied Mathematics, vol. 13, SIAM, Philadelphia, PA, 1994.

[97] J. NIE, *Sum of squares method for sensor network localization*, Optimization Online, 2006,

[98] J. NIE AND J. DEMMEL, *Sparse SOS Relaxations for Minimizing Functions that are Summation of Small Polynomials*, ArXiv math.OC/0606476, 2006.

[99] J. NIE, J. DEMMEL, AND B. STURMFELS, *Minimizing polynomials via sums of squares over the gradient ideal*, Mathematical Programming Series A **106**:587-606, 2006.

[100] J. NIE AND M. SCHWEIGHOFER, *On the complexity of Putinar's Positivstellensatz*, Journal of Complexity **23(1)**:135–150, 2007.

[101] J. NOCEDAL AND S.J. WRIGHT, *Numerical Optimization*, Springer Verlag, 2000.

[102] A.E. NUSSBAUM, *Quasi-analytic vectors*, Archiv. Mat. **6**:179–191, 1966.

[103] P.A. PARRILO, *Structured semidefinite programs and semialgebraic geometry methods in robustness and optimization*, PhD thesis, California Institute of Technology, 2000.

[104] ——, *Semidefinite programming relaxations for semialgebraic problems*, Mathematical Programming B **96**:293–320, 2003.

[105] ——, *An explicit construction of distinguished representations of polynomials nonnegative over finite sets*, IfA Technical Report AUT02-02, ETH Zürich, 2002.

[106] ——, *Exploiting algebraic structure in sum of squares programs*, In *Positive Polynomials in Control*, D. Henrion and A. Garulli, eds., LNCIS **312**:181–194, 2005.

[107] P.A. Parrilo and B. Sturmfels, *Minimizing polynomial functions*, In *Algorithmic and Quantitative Real Algebraic geometry*, S. Basu and L. Gonzáles-Vega, eds., DIMACS Series in Discrete Mathematics and Theoretical Computer Science, vol. 60, pp. 83–99, 2003.

[108] G. Pólya, *Über positive Darstellung von Polynomen*, Vierteljahresschrift der Naturforschenden Gesellschaft in Zürich **73**:141–145, 1928. Reprinted in Collected Papers, vol. 2, MIT Press, Cambridge, MA, pp. 309–313, 1974.

[109] V. Powers and B. Reznick, *A new bound for Pólya's theorem with applications to polynomials positive on polyhedra*, Journal of Pure and Applied Algebra **164**(1-2):221–229, 2001.

[110] V. Powers, B. Reznick, C. Scheiderer, and F. Sottile, *A new approach to Hilbert's theorem on ternary quartics*, C. R. Acad. Sci. Paris, Ser. I **339**:617–620, 2004.

[111] V. Powers and C. Scheiderer, *The moment problem for non-compact semialgebraic sets*, Adv. Geom. **1**:71–88, 2001.

[112] V. Powers and T. Wörmann, *An algorithm for sums of squares of real polynomials*, Journal of Pure and Applied Algebra **127**:99–104, 1998.

[113] S. Prajna, A. Papachristodoulou, P. Seiler, and P.A. Parrilo, *SOSTOOLS (Sums of squares optimization toolbox for MATLAB) User's guide*, `http://www.cds.caltech.edu/sostools/`

[114] A. Prestel and C.N. Delzell, *Positive Polynomials – From Hilbert's 17th Problem to Real Algebra*, Springer, Berlin, 2001.

[115] M. Putinar, *Positive polynomials on compact semi-algebraic sets*, Indiana University Mathematics Journal **42**:969–984, 1993.

[116] ——, *A note on Tchakaloff's theorem*, *Proceedings of the American Mathematical Society* **125**:2409–2414, 1997.

[117] J. Renegar, *A Mathematical View of Interior-Point Methods in Convex Optimization*, MPS-SIAM Series in Optimization, 2001.

[118] B. Reznick, *Extremal PSD forms with few terms*, Duke Mathematical Journal **45**(2):363–374, 1978.

[119] ——, *Uniform denominators in Hilbert's Seventeenth Problem*, Mathematische Zeitschrift **220**: 75-98, 1995.

[120] ——, *Some concrete aspects of Hilbert's 17th problem*, In *Real Algebraic Geometry and Ordered Structures*, C.N. Delzell and J.J. Maddén (eds.), Contemporary Mathematics **253**:251–272, 2000.

[121] ——, On Hilbert's construction of positive polynomials, 2007, `http://front.math.ucdavis.edu/0707.2156`.

[122] T. Roh and L. Vandenberghe, *Discrete transforms, semidefinite programming and sum-of-squares representations of nonnegative polynomials*, SIAM Journal on Optimization **16**:939–964, 2006.

[123] J. Saxe, *Embeddability of weighted graphs in k-space is strongly NP-hard*, In *Proc. 17th Allerton Conf. in Communications, Control, and Computing*, Monticello, IL, pp. 480–489, 1979.

[124] C. Scheiderer, Personal communication, 2004.

[125] ——*Positivity and sums of squares: A guide to recent results*, this volume, IMA Volumes in Mathematics and its Applications, *Emerging Applications of Algebraic Geometry*, M. Putinar and S. Sullivant, eds., Springer Science+Business Media, New York, 2008.

[126] K. SCHMÜDGEN, *The K-moment problem for compact semi-algebraic sets*, Mathematische Annalen **289**:203–206, 1991.
[127] ——*Noncommutative real algebraic geometry - Some basic concepts and first ideas*, this volume, IMA Volumes in Mathematics and its Applications, *Emerging Applications of Algebraic Geometry*, M. Putinar and S. Sullivant, eds., Springer Science+Business Media, New York, 2008.
[128] A. SCHRIJVER, *Theory of Linear and Integer Programming*, Wiley, 1979.
[129] ——, *New code upper bounds from the Terwilliger algebra and semidefinite programming*, IEEE Trans. Inform. Theory **51**:2859–2866, 2005.
[130] M. SCHWEIGHOFER, *An algorithmic approach to Schmüdgen's Positivstellensatz*, Journal of Pure and Applied Algebra, **166**:307–319, 2002.
[131] ——, *On the complexity of Schmüdgen's Positivstellensatz*, Journal of Complexity **20**:529–543, 2004.
[132] ——, *Optimization of polynomials on compact semialgebraic sets*, SIAM Journal on Optimization **15**(3):805–825, 2005.
[133] ——, *Global optimization of polynomials using gradient tentacles and sums of squares*, SIAM Journal on Optimization **17**(3):920–942, 2006.
[134] ——, *A Gröbner basis proof of the flat extension theorem for moment matrices*, Preprint, 2008.
[135] J.-P. SERRE, *Linear Representation of Finite Groups*, Graduate Texts in Mathematics, Vol. **42**, Springer Verlag, New York, 1977.
[136] I.R. SHAFAREVICH, *Basic Algebraic Geometry*, Springer, Berlin, 1994.
[137] H.D. SHERALI AND W.P. ADAMS, *A hierarchy of relaxations between the continuous and convex hull representations for zero-one programming problems*, SIAM Journal on Discrete Mathematics, **3**:411–430, 1990.
[138] N.Z. SHOR, *An approach to obtaining global extremums in polynomial mathematical programming problems*, Kibernetika, **5**:102–106, 1987.
[139] ——, *Class of global minimum bounds of polynomial functions*, Cybernetics **23(6)**:731–734, 1987. (Russian orig.: Kibernetika **6**:9–11, 1987.)
[140] ——, *Quadratic optimization problems*, Soviet J. Comput. Systems Sci. **25**:1–11, 1987.
[141] ——, *Nondifferentiable Optimization and Polynomial Problems*, Kluwer, Dordrecht, 1998.
[142] G. STENGLE, *A Nullstellensatz and a Positivstellensatz in semialgebraic geometry*, Math. Ann. **207**:87–97, 1974.
[143] J. STOCHEL, *Solving the truncated moment problem solves the moment problem*, Glasgow Journal of Mathematics, **43**:335–341, 2001.
[144] B. STURMFELS, *Solving Systems of Polynomial Equations.* CBMS, Regional Conference Series in Mathematics, Number 97, AMS, Providence, 2002.
[145] V. TCHAKALOFF, Formules de cubatures mécaniques à coefficients non négatifs. *Bulletin des Sciences Mathématiques*, **81**:123–134, 1957.
[146] J.A. TELLE AND A. PROSKUROWSKI, *Algorithms for vertex partitioning problems on partial k-trees*, SIAM Journal on Discrete Mathematics **10(4)**:529–550, 1997.
[147] M. TODD, *Semidefinite optimization*, Acata Numer. **10**:515–560, 2001.
[148] F. VALLENTIN, *Optimal Embeddings of Distance Regular Graphs into Euclidean Spaces*, Journal of Combinatorial Theory, Series B **98**:95–104, 2008.
[149] ——, *Symmetry in semidefinite programs*, arXiv:0706.4233, 2007.
[150] L. VANDENBERGHE AND S. BOYD, *Semidefinite programming*, SIAM Review **38**:49–95, 1996.
[151] L. VAN DEN DRIES, *Tame topology and o-minimal structures*, Cambridge University Press, Cambridge, 1998.
[152] H. WAKI, Personal communication, 2007.
[153] H. WAKI, S. KIM, M. KOJIMA, AND M. MURAMATSU, *Sums of squares and semidefinite programming relaxations for polynomial optimization problems with structured sparsity*, SIAM Journal on Optimization **17**(1):218–242, 2006.

[154] H. WHITNEY, *Elementary structure of real algebraic varieties*, Annals of Mathematics **66**(3):545–556, 1957.
[155] T. WIMER, *Linear Time Algorithms on k-Terminal Graphs*, Ph.D. thesis, Clemson University, Clemson, SC, 1988.
[156] H. WOLKOWICZ, R. SAIGAL, AND L. VANDEBERGHE (eds.), *Handbook of Semidefinite Programming*, Boston, Kluwer Academic, 2000.

POSITIVITY AND SUMS OF SQUARES: A GUIDE TO RECENT RESULTS*

CLAUS SCHEIDERER†

Abstract. This paper gives a survey, with detailed references to the literature, on recent developments in real algebra and geometry concerning the polarity between positivity and sums of squares. After a review of foundational material, the topics covered are Positiv- and Nichtnegativstellensätze, local rings, Pythagoras numbers, and applications to moment problems.

Key words. Positive polynomials, sums of squares, Positivstellensätze, real algebraic geometry, local rings, moment problems.

AMS(MOS) 2000 Subject Classifications. 14P05, 11E25, 44A60, 14P10.

In this paper I will try to give an overview, with detailed references to the literature, of recent developments, results and research directions in the field of real algebra and geometry, as far as they are directly related to the concepts mentioned in the title. Almost everything discussed here (except for the first section) is less than 15 years old, and much of it less than 10 years. This illustrates the rapid development of the field in recent years, a fact which may help to excuse that this article does not do justice to all facets of the subject (see more on this below).

Naturally, new results and techniques build upon established ones. Therefore, even though this is a report on recent progress, there will often be need to refer to less recent work. Sect. 1 of this paper is meant to facilitate such references, and also to give the reader a coherent overview of the more classical parts of the field. Generally, this section collects fundamental concepts and results which date back to 1990 and before. (A few more pre-1990 results will be discussed in later sections as well.) It also serves the purpose of introducing and unifying matters of notation and definition.

The polarity between positive polynomials and sums of squares of polynomials is what this survey is all about. After Sect. 1, I decided to divide the main body of the material into two parts: ***Positivstellensätze*** (Sect. 2) and *Nichtnegativstellensätze* (Sect. 3), which refer to strict, resp. non-strict, positivity. There is not always a well-defined borderline between the two, but nevertheless this point of view seems to be useful for purposes of exposition.

---

*On the occasion of the Algebraic Geometry tutorial, I spent a few inspiring days at the IMA at Minneapolis in April 2007. I would like to thank the institute for the kind invitation.

†Fachbereich Mathematik und Statistik, Universität Konstanz, 78457 Konstanz, Germany (claus.scheiderer@uni-konstanz.de).

There are two further sections. Sect. 4 is concerned with positivity in the context of local rings. These results are without doubt interesting enough by themselves. But an even better justification for including them here is that they have immediate significance for global questions, in particular via various local-global principles, as is demonstrated in this paper.

The final part (Sect. 5) deals with applications to moment problems. Since Schmüdgen's groundbreaking contribution in 1991, the interplay between algebraic and analytic methods in this field has proved most fruitful. My account here does by far not exhaust all important aspects of current work on moment problems, and I refer to the article [Lt] by Laurent (in this volume) for complementary information.

Speaking about what is not in this text, a major omission is the application of sums of squares methods to polynomial optimization. Again, I strongly recommend to consult Laurent's article [Lt], together with the literature mentioned there. Other topics which I have not touched are linear matrix inequality (LMI) representations of semi-algebraic sets, or positivity and sums of squares in non-commuting variables. For the first one may consult the recent survey [HMPV]. Summaries of current work on the non-commutative side are contained in the surveys [HP] and [Sm3]. Both are recommended as excellent complementary reading to this article.

Throughout, I am not aiming at the greatest possible generality. Original references should be consulted for stronger or for more complete versions. Also, the survey character of this text makes it mostly impossible to include proofs.

A first version of this article was written in 2003 for the web pages of the European network RAAG. For the present version, many updates have been incorporated which take into account what has happened in the meantime, but the overall structure of the original survey has been left unchanged.

Contents:

1. Preliminaries and 'classical' results. In this section we introduce basic concepts and review some fundamental 'classical' results (classical roughly meaning from before 1990).

1.1. Hilbert's seventeenth Problem. Even though it is generally known so well, this seems a good point of departure. Hilbert's occupation with sums of squares representations of positive polynomials has, in many ways, formed the breeding ground for what we consider today as modern real algebra, even though significant elements of real algebra had been in the air well before Hilbert.

A polynomial $f \in \mathbb{R}[x_1, \ldots, x_n]$ is said to be *positive semidefinite* (*psd* for short) if it has non-negative values on all of $\mathbb{R}^n$. Of course, if f is a sum of squares of polynomials, then f is psd. If $n = 1$, then conversely every psd polynomial is a sum of squares of polynomials, by an elementary argument. For every $n \geq 2$, Hilbert [H1] showed in 1888 that there exist psd polynomials in n variables which cannot be written as a sum of squares of polynomials. After further reflecting upon the question, he was able to prove in the case $n = 2$ that every psd polynomial is a sum of squares of rational functions [H2] (1893). For more than two variables, however, he found himself unable to prove this, and included the question as the seventeenth on his famous list of twenty-three mathematical problems (1900). The question was later decided in the positive by Artin [A]:

THEOREM 1.1.1 (Artin 1927). *Let R be a real closed field, and let f be a psd polynomial in $R[x_1, \ldots, x_n]$. Then there exists an identity*

$$fh^2 = h_1^2 + \cdots + h_r^2$$

where $h, h_1, \ldots, h_r \in R[x_1, \ldots, x_n]$ and $h \neq 0$.

REMARKS 1.1.2.

1. Motzkin (1967) was the first to publish an example of a psd polynomial f which is not a sum of squares of polynomials, namely

$$f(x_1, x_2) = 1 + x_1^2 x_2^2 \cdot (x_1^2 + x_2^2 - 3).$$

Although we know many constructions today which produce such examples, there is still an interest in them. Generally it is considered non-trivial to produce explicit examples of psd polynomials which fail to be sums of squares. We refer the reader to one of the available surveys on Hilbert's 17th problem and its consequences. In particular, we recommend the account written by Reznick [Re2] and the references given there; see also [Re4].

2. The previous remark might suggest that 'most' psd polynomials are sums of squares (sos). But there is more than one answer to the question for the quantitative relation between psd and sos polynomials. On the one hand, if one fixes the degree, there are results by Blekherman [Bl] saying that, in a precise quantitative sense, there exist significantly more psd polynomials than sums of squares. He also gives asymptotic bounds for the sizes of these sets, showing that the discrepancy grows with the number of variables. On the other hand, if the degree is kept variable, there are results showing that sums of squares are ubiquitous among all psd polynomials. Berg et al. ([BChR] Thm. 9.1) proved that sums of squares are dense among the polynomials which are non-negative on the unit cube $[-1, 1]^n$, with

respect to the l_1-norm of coefficients. A simple explicit version of this result is given by Lasserre and Netzer in [LN]. Explicit coefficient-wise approximations of globally non-negative polynomials are given by Lasserre in [La1] and [La2]. Of course, the degrees of the approximating sums of squares go to infinity in all these results.

1.2. 'Classical' Stellensätze. Let R always denote a real closed field. The Stellensätze, to be recalled here, date back to the 1960s and 70s. They can be considered to be generalizations of Artin's theorem 1.1.1 while, at the same time, they are refinements of this theorem. A common reference is [BCR] ch. 4. (Other accounts of real algebra proceed directly to the 'abstract' versions of Sect. 1.3 below, using the real spectrum.)

1.2.1. Given a set F of polynomials in $R[x_1,\dots,x_n]$, we denote the set of common zeros of the elements of F in R^n by

$$\mathscr{Z}(F) := \{x \in R^n\colon f(x) = 0 \text{ for every } f \in F\}.$$

This is a real algebraic (Zariski closed) subset of R^n. Conversely, given a subset S of R^n, write

$$\mathscr{I}(S) := \{f \in R[x_1,\dots,x_n]\colon f(x) = 0 \text{ for every } x \in S\}$$

for the vanishing ideal of S. So $\mathscr{Z}(\mathscr{I}(S))$ is the Zariski closure of S in R^n. On the other hand, if F is a subset of $R[x_1,\dots,x_n]$, the ideal $\mathscr{I}(\mathscr{Z}(F))$ (of polynomials which vanish on the real zero set of F) is described by (the 'geometric', or 'strong', version of) the real Nullstellensatz:[1]

PROPOSITION 1.2.2 (Real Nullstellensatz, geometric version). *Let I be an ideal of $R[x_1,\dots,x_n]$ and let $f \in R[x_1,\dots,x_n]$. Then $f \in \mathscr{I}(\mathscr{Z}(I))$ if and only if there is an identity*

$$f^{2N} + g_1^2 + \cdots + g_r^2 \in I$$

in which N, $r \geq 0$ and $g_1,\dots,g_r \in R[x_1,\dots,x_n]$.

This was first proved by Krivine (1964), and later found again independently by Dubois (1969) and Risler (1970). The ideal $\mathscr{I}(\mathscr{Z}(I))$ is the real radical of I, see 1.3.5 below. Note the analogy to the Hilbert Nullstellensatz in classical (complex) algebraic geometry.

1.2.3. In real algebraic geometry it is not enough to study sets defined by polynomial equations $f = 0$. Rather, the solution sets of inequalities $f \geq 0$ or $f > 0$ cannot be avoided. Therefore, given a subset F of $R[x_1,\dots,x_n]$, we write

$$\mathscr{S}(F) := \{x \in R^n\colon f(x) \geq 0 \text{ for every } f \in F\}.$$

[1] Bearing in mind that $\mathscr{Z}(F) = \mathscr{Z}(I)$ where $I := (F)$ is the ideal generated by F.

This is a closed subset of R^n (in the topology defined by the ordering of R). Here we are interested in the case where F is finite, so that $\mathscr{S}(F)$ is a *basic closed* semi-algebraic set.[2] In order to characterize the polynomials which are strictly (resp. non-strictly) positive on $\mathscr{S}(F)$, one needs to introduce the preordering generated by F. For the general notion of preorderings see 1.3.6 below. Here, if $F = \{f_1, \ldots, f_r\}$, we define the *preordering* generated by F to be the subset

$$PO(f_1, \ldots, f_r) := \left\{ \sum_{e \in \{0,1\}^r} s_e f_1^{e_1} \cdots f_r^{e_r} : \text{the } s_e \text{ are sums of squares in } R[\mathrm{x}] \right\}$$

of $R[\mathrm{x}] := R[x_1, \ldots, x_n]$.

As the name indicates, the Positivstellensatz (resp. Nichtnegativstellensatz) describes the polynomials which are strictly (resp. non-strictly) positive on the set $\mathscr{S}(f_1, \ldots, f_r)$:

PROPOSITION 1.2.4. *Let $f_1, \ldots, f_r \in R[x_1, \ldots, x_n]$. Put $K = \mathscr{S}(f_1, \ldots, f_r)$, and let $T = PO(f_1, \ldots, f_r)$ be the preordering generated by the f_i. Let $f \in R[x_1, \ldots, x_n]$.*

(a) (Positivstellensatz, geometric version) *$f > 0$ on K iff there is an identity $sf = 1 + t$ with $s, t \in T$.*

(b) (Nichtnegativstellensatz, geometric version) *$f \geq 0$ on K iff there is an identity $sf = f^{2N} + t$ with $N \geq 0$ and $s, t \in T$.*

REMARKS 1.2.5.

1. In 1964, Krivine [Kr] proved essentially the real spectrum version of (a) for arbitrary rings (see 1.3.9 below), and could have deduced the geometric formulations 1.2.4 above. These were first proved by Stengle in 1974 [St1], who was unaware of Krivine's work. In each of the three Stellensätze 1.2.2 and 1.2.4, the 'if' part of the statement is trivial, whereas the 'only if' part requires work. The upshot of each of the Stellensätze is, therefore, that for any sign condition satisfied by $f|_K$, there exists an explicit certificate (in the form of an identity) which makes this sign condition obvious.

2. In terminology introduced further below (1.3.11), the Nichtnegativstellensatz describes the saturated preordering generated by F.

3. In 1.2.4(b), note that the identity $sf = f^{2N} + t$ implies $\mathscr{Z}(s) \cap K \subset \mathscr{Z}(f)$.

4. The particular case $r = 1$, $f_1 = 1$ of 1.2.4(b) gives again the solution of Hilbert's 17th problem (Theorem 1.1.1), after multiplying the identity $sf = f^{2N} + t$ with s. Using the previous remark, one gets actually a strengthening of Artin's theorem, to the effect that $\mathscr{Z}(h) \subset \mathscr{Z}(f)$ can be achieved in 1.1.1.

[2] If F is an arbitrary infinite set, it is generally more reasonable to look at $\mathscr{X}_F$, the real spectrum counterpart of $\mathscr{S}(F)$ (1.3.7).

5. The three Stellensätze 1.2.4 and 1.2.2 can be combined into a single one, the general (geometric) real Stellensatz. See [BCR] Thm. 4.4.2, [S] p. 94, or 1.3.10 below.

1.2.6. The proofs of the real Nullstellensatz 1.2.2 and of the Stellensätze 1.2.4 do not give a clue on complexity or effectiveness questions. These issues are not yet well understood. Suppose, for example, that $f_1, \dots, f_r$ are polynomials in $R[x_1, \dots, x_n]$ with $\mathscr{S}(f_1, \dots, f_r) = \varnothing$. By 1.2.4(b) there are sums of squares of polynomials s_e ($e \in \{0,1\}^r$) such that $1 + \sum_e s_e f_1^{e_1} \cdots f_r^{e_r} = 0$. Can one give a bound d such that there exist necessarily such s_e with $\deg(s_e) \le d$?

It is not hard to see (fixing n, r and the $\deg(f_i)$) that such a bound d must exist. But to find one explicitly is much more difficult. Lombardi and Roy have announced around 1993 that there is a bound which is five-fold exponential in n and in the degrees of the f_i. It seems that this has never been published. Schmid ([Sd], 1998) has proved related results on the complexity of Hilbert's 17th problem (1.1.1) and the real Nullstellensatz. For 1.1.1, he has established a bound for the $\deg(h_i)$ which is n-fold exponential in $\deg(f)$.

Both the high complexity and the uncertainty about its precise magnitude are in sharp contrast with the very precise results on the complexity of the classical Hilbert Nullstellensatz (see [Ko]).

1.3. Orderings and preorderings of rings. To proceed further, we have to follow a more abstract approach now. All rings will be assumed to be commutative and to have a unit. The Zariski spectrum of A, denoted $\operatorname{Spec} A$, is the set of all prime ideals of A, equipped with the Zariski topology. Unless mentioned otherwise, A is an arbitrary ring now.

1.3.1. We briefly recall a few basic notions of real algebra (see any of [BCR], [KS] or [PD] for more details). Let A be a ring. The *real spectrum* of A, denoted $\operatorname{Sper} A$, is the set consisting of all pairs $\alpha = (\mathfrak{p}, \omega)$ where $\mathfrak{p} \in \operatorname{Spec} A$ and ω is an ordering of the residue field of $\mathfrak{p}$. The prime ideal $\mathfrak{p}$ is called the *support* of α, written $\mathfrak{p} = \operatorname{supp}(\alpha)$. A prime ideal of A is called *real* if it supports an element of $\operatorname{Sper} A$, i.e., if its residue field can be ordered.

For $f \in A$ and $\alpha = (\mathfrak{p}, \omega) \in \operatorname{Sper} A$, the notation '$f(\alpha) \ge 0$' (resp., '$f(\alpha) > 0$') indicates that the residue class f mod $\mathfrak{p}$ is non-negative (resp., positive) with respect to ω. The *Harrison topology* on $\operatorname{Sper} A$ is defined to have the collection of sets

$$U(f) := \{\alpha \in \operatorname{Sper} A \colon f(\alpha) > 0\}$$

($f \in A$) as a subbasis of open sets. The support map supp: $\operatorname{Sper} A \to \operatorname{Spec} A$ is continuous. A subset of $\operatorname{Sper} A$ is called *constructible* if it is a finite boolean combination of sets $U(f)$, $f \in A$, that is, if it can be described by imposing sign conditions on finitely many elements of A.

1.3.2. A convenient alternative description of the real spectrum is by orderings. By a *subsemiring* of A we mean a subset $P \subset A$ containing 0, 1 and satisfying $P+P \subset P$ and $PP \subset P$. An *ordering*[3] of A is a subsemiring P of A which satisfies $P \cup (-P) = A$ and $a^2 \in P$ for every $a \in A$, such that in addition $\mathrm{supp}(P) := P \cap (-P)$ is a prime ideal of A. The elements of $\mathrm{Sper}\, A$ are in bijective correspondence with the set of all orderings P of A, the ordering corresponding to $\alpha \in \mathrm{Sper}\, A$ being $P_\alpha := \{f \in A\colon f(\alpha) \geq 0\}$. Therefore, the real spectrum of A is often defined through orderings.

1.3.3. If $R[\mathrm{x}] = R[x_1, \ldots, x_n]$ is the polynomial ring over a real closed field, one can naturally identify R^n with a subset of $\mathrm{Sper}\, R[\mathrm{x}]$, by making a point $a \in R^n$ correspond to the ordering $P_a := \{f \in R[\mathrm{x}]\colon f(a) \geq 0\}$ of $R[\mathrm{x}]$. The map $a \mapsto P_a$ is in fact a topological embedding $R^n \hookrightarrow \mathrm{Sper}\, R[\mathrm{x}]$. As a consequence of the celebrated Artin–Lang theorem (see [BCR] 4.1 or [KS] II.11, for example), every non-empty constructible set in $\mathrm{Sper}\, R[\mathrm{x}]$ contains a point of R^n. Therefore,

$$K \mapsto K \cap R^n$$

is a bijection between the constructible subsets K of $\mathrm{Sper}\, R[\mathrm{x}]$ and the semi-algebraic subsets S of R^n. The inverse bijection is traditionally denoted by the 'operator tilda', $S \mapsto \widetilde{S}$. Thus, $\widetilde{S}$ is the unique constructible subset of $\mathrm{Sper}\, R[\mathrm{x}]$ with $\widetilde{S} \cap R^n = S$.

1.3.4. Back to arbitrary rings A. Given an ideal I of A, the *real radical* $\sqrt[\mathrm{re}]{I}$ of I is defined to be the intersection of all real prime ideals of A which contain I. The real radical is described by the weak real Nullstellensatz, due to Stengle (1974):

PROPOSITION 1.3.5 (Weak (or abstract) real Nullstellensatz). *Let I be an ideal of A, let ΣA^2 be the set of all sums of squares of elements of A. Then*

$$\sqrt[\mathrm{re}]{I} = \{f \in A\colon \exists\, N \geq 0\ \exists\, s \in \Sigma A^2\ f^{2N} + s \in I\}.$$

Thus, $f \in A$ lies in the real radical of I if and only if $-f^{2N}$ is a sum of squares modulo I, for some $N \geq 0$. For a proof see [St1], [BCR] 4.1.7, [KS] p. 105, or [PD] 4.2.5.

1.3.6. A subsemiring T of A is called a *preordering* of A if $a^2 \in T$ for every $a \in A$. The preordering T is called *proper* if $-1 \notin T$.[4] Every preordering of A contains ΣA^2, the set of all sums of squares of A, and ΣA^2 is the smallest preordering of A. Any intersection of preorderings is again a preordering. Therefore it is clear what is meant by the preordering

[3] Also called a *prime (positive) cone*.

[4] In some texts, preorderings are proper by definition. If A contains $\frac{1}{2}$, then $T = A$ is the only improper preordering in A, according to the identity $x = (\frac{x+1}{2})^2 - (\frac{x-1}{2})^2$.

generated by a subset F of A. It is denoted by $PO(F)$ or $PO_A(F)$, and consists of all finite sums of products

$$a^2 f_1 \cdots f_m$$

where $a \in A$, $m \geq 0$ and $f_1, \dots, f_m \in F$. If $F = \{f_1, \dots, f_r\}$ is finite, one also writes $PO(f_1, \dots, f_r) := PO(F)$. A preordering is called *finitely generated* if it can be generated by finitely many elements.

1.3.7. Let F be any subset of A. With F one associates the closed subset

$$\begin{aligned} \mathscr{X}_F = \mathscr{X}_F(A) &:= \{\alpha \in \operatorname{Sper} A \colon f(\alpha) \geq 0 \text{ for every } f \in F\} \\ &= \{P \in \operatorname{Sper} A \colon F \subset P\} \end{aligned}$$

of $\operatorname{Sper} A$.[5] If $T = PO(F)$ is the preordering generated by F then clearly $\mathscr{X}_F = \mathscr{X}_T$. Note that if $A = R[x_1, \dots, x_n]$ and the set $F = \{f_1, \dots, f_r\}$ is finite, then $\mathscr{X}_F = \widetilde{\mathscr{S}(F)}$, the constructible subset of $\operatorname{Sper} R[x_1, \dots, x_n]$ associated with the semi-algebraic set $\mathscr{S}(F) = \{x \colon f_1(x) \geq 0, \dots, f_r(x) \geq 0\}$ in R^n (see 1.3.3).

The next result is easy to prove, but of central importance:

PROPOSITION 1.3.8. *If T is a preordering of A and $\mathscr{X}_T = \varnothing$, then $-1 \in T$.*

By a Zorn's lemma argument, proving Proposition 1.3.8 means to show that every maximal proper preordering is an ordering. The argument for this is elementary (see, e.g., [KS] p. 141). Note however that the proof of 1.3.8 is a pure existence proof. It does not give any hint how to find a concrete representation of -1 as an element of T (c.f. also 1.2.6).

From Proposition 1.3.8 one can immediately derive the following 'abstract' versions of the various Stellensätze 1.2.2 and 1.2.4:

COROLLARY 1.3.9. *Let T be a preordering of A, and let $f \in A$.*

(a) (Positivstellensatz) *$f > 0$ on $\mathscr{X}_T \Leftrightarrow \exists\, s, t \in T \;\; sf = 1 + t$.*

(b) (Nichtnegativstellensatz) *$f \geq 0$ on $\mathscr{X}_T \Leftrightarrow \exists\, s, t \in T \; \exists N \geq 0 \;\; sf = f^{2N} + t$.*

(c) (Nullstellensatz) *$f \equiv 0$ on $\mathscr{X}_T \Leftrightarrow \exists N \geq 0 \;\; -f^{2N} \in T$.*

See, e.g., [KS] III §9 or [PD] §4.2. The implications '$\Leftarrow$' are obvious. To prove '$\Rightarrow$', apply 1.3.8 to the preordering $T - fT$ in the case of (a). For (b), work with the preordering T_f generated by T in the localized ring A_f, and apply (a) using $f > 0$ on $\mathscr{X}_{T_f}$. To get (c), apply (b) to $-f^2$.

The Positivstellensatz 1.3.9(a) was essentially proved by Krivine [Kr] in 1964.

[5] Here, of course, the second description refers to the description of $\operatorname{Sper} A$ as the set of orderings of A.

REMARKS 1.3.10.

1. We deduced the statements of 1.3.9 from 1.3.8. Conversely, setting $f = -1$ exhibits 1.3.8 as a particular case of each of the three statements in 1.3.9.

2. The Nullstellensatz 1.3.9(c) also generalizes the weak real Nullstellensatz 1.3.5, as one sees by applying 1.3.9(c) to $T = I + \Sigma A^2$, where $I \subset A$ is an ideal.

3. The geometric versions 1.2.2 and 1.2.4 result immediately from the corresponding abstract versions 1.3.9 via the Artin–Lang density property (1.3.3).

4. The three Stellensätze 1.3.9 can be combined into a single one: Given subsets F, G, H of A, the subset

$$\bigcap_{f\in F}\{\alpha\colon f(\alpha)=0\}\cap\bigcap_{g\in G}\{\alpha\colon g(\alpha)\geq 0\}\cap\bigcap_{h\in H}\{\alpha\colon h(\alpha)>0\}$$

of Sper A is empty if and only if there exists an identity

$$a+b+c=0$$

in which $a \in \sum_{f\in F} Af$ (the ideal generated by F), $b \in PO(G \cup H)$, and c lies in the multiplicative monoid (with unit) generated by H. (Compare [BCR] 4.4.1.)

1.3.11. If X is any subset of Sper A, we can associate with X the preordering

$$\mathscr{P}(X) := \{f \in A\colon f(\alpha) \geq 0 \text{ for every } \alpha \in X\} = \bigcap_{P\in X} P$$

of A. The two operators $\mathscr{X}$ and $\mathscr{P}$ interact as follows.

A subset of Sper A is called *pro-basic closed* if it has the form $\mathscr{X}_F$ for some subset F of A, i.e., if it can be described by a (possibly infinite) conjunction of non-strict inequalities. Given a subset X of Sper A, the set $\mathscr{X}_{\mathscr{P}(X)}$ is the smallest pro-basic closed subset of Sper A which contains X. On the other hand, a preordering T is called *saturated* if it is an intersection of orderings, or equivalently, if it has the form $T = \mathscr{P}(Z)$ for some subset Z of Sper A. It is also equivalent that $T = \mathscr{P}(\mathscr{X}_T)$, i.e., that T contains every element which is non-negative on $\mathscr{X}_T$. Given a subset F of A, the preordering $\mathscr{P}(\mathscr{X}_F)$ is the smallest saturated preordering of A which contains F, and is called the *saturation* of F, denoted $\mathrm{Sat}(F) := \mathscr{P}(\mathscr{X}_F)$. If $F = T$ is itself a preordering, then the Nichtnegativstellensatz 1.3.9(b) tells us that

$$\mathrm{Sat}(T) = \{f \in A\colon \exists\, s,\, t \in T\ \ \exists\, N \geq 0\ \ fs = f^{2N} + t\}.$$

1.3.12. If A is a field, every preordering T is saturated. For other types of rings, this is usually far from true. The study of the gap between T and $\mathrm{Sat}(T)$ often leads to interesting and difficult questions (see, e. g., Sections 3 and 5). As a rule, even if T is finitely generated, its saturation $\mathrm{Sat}(T)$ won't usually be. For a basic example take $T_0 = \Sigma A^2$, the preordering of all sums of squares in A. The saturation of T_0 is the preordering

$$\mathrm{Sat}(T_0) = \mathscr{P}(\mathrm{Sper}\, A) = \bigcap_{P \in \mathrm{Sper}\, A} P =: A_+$$

consisting of all *positive semidefinite* (or *psd*) *elements* of A. To study the gap between T_0 and $\mathrm{Sat}(T_0)$ means to ask which psd elements of A are sums of squares. We will say that 'psd = sos *holds in A*' if T_0 is saturated, i. e., if every psd element in A is a sum of squares. As remarked in the beginning, Hilbert proved that this property fails for all polynomial rings $\mathbb{R}[x_1, \dots, x_n]$ in $n \geq 2$ variables.[6] However, there are non-trivial and interesting classes of examples where psd = sos holds, see Section 3.

1.3.13. The relation between the operators $\mathscr{X}$ and $\mathscr{P}$ described in 1.3.11 can be summarized by saying that these operators set up a 'Galois adjunction' between the subsets of A and the subsets of $\mathrm{Sper}\, A$.[7] The closed objects of this adjunction are the saturated preorderings of A on the one side and the pro-basic closed subsets of $\mathrm{Sper}\, A$ on the other. Hence $\mathscr{X}$ and $\mathscr{P}$ restrict to mutually inverse bijections between these two classes of objects.

1.3.14. A preordering T of A is *generating* if $T - T = A$. This property always holds if $\frac{1}{2} \in A$. The *support* of T is defined as $\mathrm{supp}(T) := T \cap (-T)$. If T is generating, this is an ideal of A (namely the largest ideal contained in T), and one has

$$\sqrt{\mathrm{supp}(T)} = \sqrt[\mathrm{re}]{\mathrm{supp}(T)} = \bigcap_{\alpha \in \mathscr{X}_T} \mathrm{supp}(\alpha).$$

(The first inclusion '$\supset$' follows from the weak real Nullstellensatz 1.3.5, the second '$\supset$' from the more general abstract Nullstellensatz 1.3.9(c). The inclusions '$\subset$' are obvious.)

1.4. Modules and semiorderings in rings. 1.4.1. The concept of preorderings has important generalizations in two directions: Modules and preprimes. We first discuss the latter.

---

[6]As a matter of fact, the saturation $\mathrm{Sat}(T_0)$ fails to be finitely generated in these polynomial rings ([Sch2] Thm. 6.4).

[7]For $F \subset A$ and $X \subset \mathrm{Sper}\, A$ one has $F \subset \mathscr{P}(X) \Leftrightarrow X \subset \mathscr{X}(F)$, and both are equivalent to $F|_X \geq 0$. We have $\mathscr{P} \circ \mathscr{X} \circ \mathscr{P} = \mathscr{P}$ and $\mathscr{X} \circ \mathscr{P} \circ \mathscr{X} = \mathscr{X}$. The operators $F \mapsto \mathscr{P} \circ \mathscr{X}(F)$ and $X \mapsto \mathscr{X} \circ \mathscr{P}(X)$ are closure operators: The first sends F to the saturated preordering generated by F, the second sends X to the pro-basic closed subset generated by X.

Let k be a (base) ring (usually $k = \mathbb{Z}$, or $k = R$, a real closed field), and let A be a k-algebra.[8] A subsemiring P of A is called a *k-preprime*[9] of A if $a^2 \in P$ for every $a \in k$. The preprime P is said to be *generating* if $P - P = A$.

By definition, the preorderings of A are the A-preprimes of A. If $k = \mathbb{Z}$, the $\mathbb{Z}$-preprimes of A are often just called *preprimes.* These are just the subsemirings of A.

Any intersection of k-preprimes is again a k-preprime. The k-preprime generated by a subset F of A is denoted $PP_k(F)$. The smallest k-preprime in A is the image of Σk^2 in A.

1.4.2. Let P be a preprime of A. A subset M of A is called a *P-module* if $1 \in M$, $M + M \subset M$ and $PM \subset M$ hold. If $-1 \notin M$ then M is called *proper.* The *support* of M is the additive subgroup $\operatorname{supp}(M) := M \cap (-M)$ of A; this is an ideal of A if P is generating.

Particularly important is the case $P = \Sigma A^2$. The ΣA^2-modules of A are called the *quadratic modules* of A. Given a subset F of A, we denote by $QM(F)$ the quadratic module generated by F in A. Thus $QM(f_1, \dots, f_r) = \Sigma A^2 + f_1 \Sigma A^2 + \cdots + f_r \Sigma A^2$.

Let us assume that $\frac{1}{2} \in A$ and M is a quadratic module. Then $\operatorname{supp}(M)$ is an ideal of A, and

$$\sqrt{\operatorname{supp}(M)} = \sqrt[\text{re}]{\operatorname{supp}(M)} \subset \bigcap_{\alpha \in \mathscr{X}_M} \operatorname{supp}(\alpha).$$

This should be compared to 1.3.14. Other than for preorderings, the second inclusion can be strict. In particular, it can happen that $-1 \notin M$ but $\mathscr{X}_M = \varnothing$. An example is the quadratic module $M = QM(x - 1, y - 1, -xy)$ in $\mathbb{R}[x, y]$.[10] However, equality can be recovered if $\mathscr{X}_M$ (the set of all orderings which contain M) is replaced by the larger set $\mathscr{Y}_M$ of all semiorderings which contain M, as we shall explain now.

1.4.3. Semiorderings are objects that relate to quadratic modules in the same way as orderings relate to preorderings. A *semiordering* of a ring A is a quadratic module S of A with $S \cup (-S) = A$, for which the ideal $\operatorname{supp}(S)$ is prime (and necessarily real).

Every ordering is a semiordering. With respect to a fixed semiordering S, every $f \in A$ has a unique sign in $\{-1, 0, 1\}$, as for orderings. However, this sign fails to be multiplicative with respect to f, unless S is an ordering.

Given any subset F of A, we write

$$\mathscr{Y}_F := \{S \colon S \text{ is a semiordering of } A \text{ with } F \subset S\}$$

[8] So A is a ring together with a fixed ring homomorphismus $k \to A$. Usually one can think of k as being a subring of A.

[9] The term preprime goes back to Harrison.

[10] One uses valuation theory to show $-1 \notin M$. See also [PD] exerc. 5.5.7.

for the semiorderings analogue of $\mathscr{X}_F$. The following is the analogue of Proposition 1.3.8:

PROPOSITION 1.4.4. *If M is a quadratic module in A and $\mathscr{Y}_M = \varnothing$, then $-1 \in M$.*

An equivalent formulation is that every maximal proper quadratic module is a semiordering. See, e.g., [PD] p. 114 for the proof. One can derive abstract Stellensätze from 1.4.4 in exactly the same way as 1.3.9 was obtained from 1.3.8. They apply to quadratic modules and refer to semiorderings, instead of orderings:

COROLLARY 1.4.5. *Let M be a quadratic module of A, and let $f \in A$.*

(a) (Positivstellensatz) $f > 0$ *on* $\mathscr{Y}_M \;\Leftrightarrow\; \exists\, s \in \Sigma A^2\ \exists\, m \in M\ \ fs = 1 + m$.

(b) (Nichtnegativstellensatz) $f \geq 0$ *on* $\mathscr{Y}_M \;\Leftrightarrow\; \exists\, s \in \Sigma A^2\ \exists\, m \in M\ \exists\, N \geq 0\ \ fs = f^{2N} + m$.

(c) (Nullstellensatz) $f \equiv 0$ *on* $\mathscr{Y}_M \;\Leftrightarrow\; \exists\, N \geq 0\ \ -f^{2N} \in M$.

(Reference for (a): [PD] 5.1.10.) The question arises how to decide in a concrete situation whether $f|_{\mathscr{Y}_M} > 0$ holds. We will give an answer below (1.4.11).

REMARKS 1.4.6.

1. Nullstellensatz 1.4.5(c) says $\sqrt{\operatorname{supp}(M)} = \bigcap_{\beta \in \mathscr{Y}_M} \operatorname{supp}(\beta)$.

2. If the quadratic module M is archimedean (see 1.5.2 below), then every maximal element in $\mathscr{Y}_M$ is an ordering, and not just a semiordering ([PD] 5.3.5). For archimedean M, therefore, one can replace $\mathscr{Y}_M$ by $\mathscr{X}_M$ in 1.4.4 and in 1.4.5(a).

3. Given a quadratic module M and $g \in A$, let $M(g) = M + g \cdot \Sigma A^2$, the quadratic module generated by M and g. The Positivstellensatz 1.4.5(a) can be rephrased as

$$f > 0 \text{ on } \mathscr{Y}_M \quad \Leftrightarrow \quad -1 \in M(-f).$$

The right hand condition says that the quadratic module $M(-f)$ is improper.

1.4.7. It is an important fact that (im-) properness of quadratic modules is a condition that can be 'localized'. This uses quadratic form theory. If k is a field (with $\operatorname{char}(k) \neq 2$), then by a quadratic form over k we always mean a nonsingular quadratic form in finitely many variables. If $a_1, \ldots, a_n \in k^*$, then $\langle a_1, \ldots, a_n \rangle$ denotes the 'diagonal' quadratic form $a_1 x_1^2 + \cdots + a_n x_n^2$ (in n variables) over k. A quadratic form $q = q(x_1, \ldots, x_n)$ over k is said to be *isotropic* if there is $0 \neq w \in k^n$ with $q(w) = 0$, that is, if q represents zero non-trivially. The form q is called *weakly isotropic* if there are finitely many non-zero vectors $w_1, \ldots, w_N \in k^n$ with $q(w_1) + \cdots + q(w_N) = 0$.

If q is weakly isotropic, then clearly q is indefinite with respect to every ordering of k. The converse is not true in general, but it becomes true if orderings are replaced by semiorderings. In other words, q is weakly isotropic iff q is indefinite with respect to every semiordering of k ([PD] Lemma 6.1.1).

We are now going to explain how the properness condition for quadratic modules of a ring A can be localized. To begin with, one has the following reduction to the residue fields:

LEMMA 1.4.8. *Let $M = QM(g_1, \ldots, g_m)$, a finitely generated quadratic module of A. Then $-1 \in M$ if and only if for every real prime ideal $\mathfrak{p}$ of A, the quadratic form $<1, g_1, \ldots, g_m>^*$ over the residue field $\kappa(\mathfrak{p})$ of $\mathfrak{p}$ is weakly isotropic.*

Here we keep writing g_i (instead of $\overline{g}_i$) for the residue class of g_i in $\kappa(\mathfrak{p})$. The notation $<1, g_1, \ldots, g_m>^*$ means that those entries g_i which are zero in $\kappa(\mathfrak{p})$ (i. e., lie in $\mathfrak{p}$) should be left away. The proof of the non-trivial implication in 1.4.8 is easy using 1.4.4.

The condition in Lemma 1.4.8 can be localized even further. This is the content of an important local-global principle for weak isotropy of quadratic forms over fields, due to Bröcker and Prestel. The 'local objects' for this principle are the henselizations of the field with respect to certain (Krull) valuations. For this exposition we prefer a technically simpler formulation which avoids the notion of henselization:

THEOREM 1.4.9 (Bröcker, Prestel, 1974). *Let q be a quadratic form over a field k with* $\mathrm{char}(k) \neq 2$. *Then q is weakly isotropic if and only if the following two conditions hold:*

(1) q is indefinite with respect to every ordering of k;

(2) for every valuation v of k with real residue field κ for which q has at least two residue forms with respect to v, at least one of these residue forms is weakly isotropic (over κ).

1.4.10. For the proof see [Br] and [Pr], or [Scha] 3.7.12. We briefly explain the notion of residue forms that was used in the statement of 1.4.9 (see [Scha] for more details). Let $v\colon k^* \to \Gamma$ be a (Krull) valuation of k, where Γ is an ordered abelian group, written additively. Given a quadratic form q over k, one can diagonalize q in the form

$$q \cong \bigoplus_{i=1}^{r} c_i <u_{i1}, \ldots, u_{in_i}>$$

with r, $n_i \geq 1$ and c_i, $u_{ij} \in k^*$, such that $v(u_{ij}) = 0$ for all i and j (that is, the u_{ij} are v-units), and such that $v(c_i) \not\equiv v(c_j)$ (mod 2Γ) for $i \neq j$. The r quadratic forms

$$\overline{q}_i := <\overline{u}_{i1}, \ldots, \overline{u}_{in_i}>$$

$(i = 1, \ldots, r)$ over the residue field κ of v are called the *residue forms* of q (with respect to v).

Although these residue forms may depend on the chosen diagonalization, the question whether or not one of them is weakly isotropic does not.

1.4.11. Let $M = QM(f_1, \ldots, f_r)$ be a finitely generated quadratic module. Given $f \in A$, the local-global principle for weak isotropy can be used to decide whether $f > 0$ on $\mathcal{Y}_M$. Indeed, this holds if and only if the form $<1, -f, f_1, \ldots, f_r>^*$ is weakly isotropic in $\kappa(\mathfrak{p})$ for every real prime ideal $\mathfrak{p}$ of A (by 1.4.6.3 and 1.4.8). And the local-global principle 1.4.9 allows to reformulate the last condition. (Compare [PD] Th. 6.2.1.)

Applications of these ideas will be given in 2.3 below.

1.4.12. The relations between the various concepts discussed so far are symbolically displayed in the following picture:

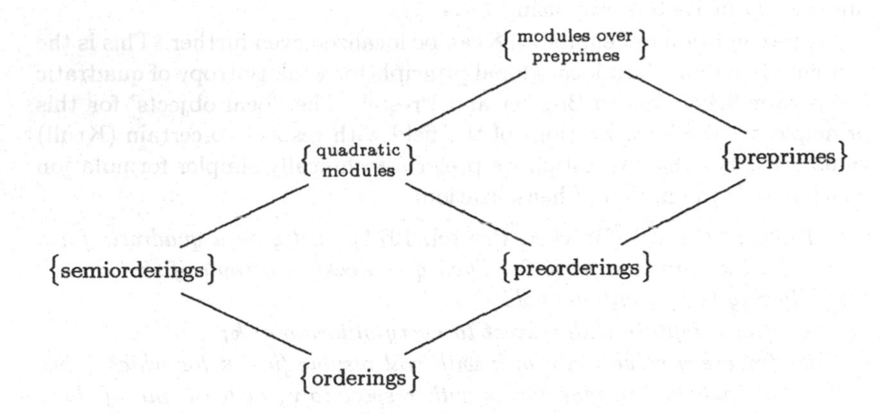

1.5. The representation theorem. The Axiom of Archimedes says that for any two positive real numbers a and b there exists a natural number n such that $na > b$. This is a fundamental property which turns out to be of great importance in more abstract and general settings.

1.5.1. Given an ordered field $(K, \leq)$ and a subring k of K, recall that K is said to be *(relatively) archimedean over* k (with respect to $\leq$) if for every $x \in K$ there exists $a \in k$ with $\pm x \leq a$. The ordered field $(K, \leq)$ is called (absolutely) *archimedean* if it is relatively archimedean over $\mathbb{Z}$. A classical (elementary) result of Hölder says that every archimedean ordered field has a unique order-compatible embedding into $\mathbb{R}$, the field of real numbers.

Generalizing this, let now k be any ring and let A be a k-algebra. A subset X of $\operatorname{Sper} A$ is *bounded over* k if, for every $a \in A$, there is $b \in k$ with $|a| \leq b$ on X. Assuming that X is closed in $\operatorname{Sper} A$, one proves easily that X is bounded over k iff the ordered residue field of every closed point of X is relatively archimedean over (the image of) k ([KS] III.11).

1.5.2. Let k be a ring, A a k-algebra and M a module over some preprime of A.

(a) M is called *archimedean over* k if for every $a \in A$ there exists $b \in k$ with $b \pm a \in M$.

(b) M is called *weakly archimedean over* k if for every $a \in A$ there exists $b \in k$ with $b \pm a \geq 0$ on $\mathscr{X}_M$.

The most important case is $k = \mathbb{Z}$. Then one simply says *archimedean*, instead of archimedean over $\mathbb{Z}$, and similarly for weakly archimedean.

Obviously, archimedean implies weakly archimedean. By definition, the module M is weakly archimedean over k iff the pro-basic closed set $\mathscr{X}_M$ is bounded over k, iff the saturated preordering $\mathrm{Sat}(M)$ is archimedean over k. So the weak archimedean property of M depends only on the set $\mathscr{X}_M$. In contrast, the archimedean property of M is stronger and much more subtle. It depends not only on the set $\mathscr{X}_M$, but also on the 'inequalities' used for its definition, that is, on the module M.

If $A = \mathbb{R}[x_1, \dots, x_n]$ and M is a module in A for which the subset $\mathscr{X}_M$ of $\widetilde{\mathbb{R}}^n = \mathrm{Sper}\, A$ is constructible (for example, M could be a finitely generated quadratic module), say $\mathscr{X}_M = \widetilde{K}$ where K is a semi-algebraic subset of $\mathbb{R}^n$, then M is weakly archimedean (over $\mathbb{Z}$) if and only if the set K is compact. In general, the weak archimedean property should be thought of as an abstract kind of compactness property of $\mathscr{X}_M$.

1.5.3. Assume that A is generated by $x_1, \dots, x_n$ as a k-algebra. Then it is obvious that a module M in A is weakly archimedean over k iff there is $a \in k$ with $\sum_{i=1}^n x_i^2 \leq a$ on $\mathscr{X}_M$. Criteria for the archimedean property are somewhat more subtle. For simplicity assume $\frac{1}{2} \in k$. If P is a k-preprime in A, then P is archimedean over k iff there is $a_i \in k$ with $a_i \pm x_i \in P$, for $i = 1, \dots, n$. If M is a quadratic module, then M is archimedean over k iff $a - \sum_i x_i^2 \in M$ for some $a \in k$. (Compare [BW1], Lemmas 1 and 2.)

1.5.4. By $(\mathrm{Sper}\, A)^{\max}$ we denote the set of closed points of $\mathrm{Sper}\, A$. Considered as a topological subspace of $\mathrm{Sper}\, A$, $(\mathrm{Sper}\, A)^{\max}$ is a compact (Hausdorff) space ([KS] p. 126). More generally, if X is any closed subset of $\mathrm{Sper}\, A$, we denote by $X^{\max} = X \cap (\mathrm{Sper}\, A)^{\max}$ the space of closed points of X.

On the other hand, let us regard $\mathrm{Hom}(A, \mathbb{R})$ (the set of ring homomorphisms $A \to \mathbb{R}$) as a closed subset of the direct product space $\mathbb{R}^A = \prod_A \mathbb{R}$. Then naturally $\mathrm{Hom}(A, \mathbb{R}) \subset (\mathrm{Sper}\, A)^{\max}$. This inclusion respects the topologies, and it identifies $\mathrm{Hom}(A, \mathbb{R})$ with the set of orderings of A whose ordered residue field is archimedean. For any closed subset X of $\mathrm{Sper}\, A$, it follows that X is bounded over $\mathbb{Z}$ if and only if $X \cap \mathrm{Hom}(A, \mathbb{R})$ is compact (as a subset of $\mathrm{Hom}(A, \mathbb{R})$), if and only if $X \cap \mathrm{Hom}(A, \mathbb{R}) = X^{\max}$.

1.5.5. Let now $X \subset \operatorname{Sper} A$ be a closed subset that is bounded over $\mathbb{Z}$. As just remarked, to every $\alpha \in X^{\max} = X \cap \operatorname{Hom}(A, \mathbb{R})$ corresponds a unique ring homomorphism $\rho_\alpha \colon A \to \mathbb{R}$ which induces the ordering α on A. In this way, every $f \in A$ defines a function

$$\Phi_X(f) \colon X^{\max} \to \mathbb{R}, \quad \alpha \mapsto \rho_\alpha(f),$$

and $\Phi_X(f)$ is clearly continuous. We thus have a representation

$$\Phi_X : A \to C(X^{\max}, \mathbb{R}) \tag{1}$$

of A by continuous functions on the compact topological space $X^{\max}$. Note that $\Phi_X(f) \geq 0$ iff $(1 + nf)|_X \geq 0$ for every $n \in \mathbb{N}$. From the Stone–Weierstraß approximation theorem one concludes:

PROPOSITION 1.5.6. *Let A be a ring containing $\frac{1}{q}$ for some $q \in \mathbb{N}$, $q > 1$, and let X be a closed subset of $\operatorname{Sper} A$ that is bounded over $\mathbb{Z}$. Then every continuous function $X^{\max} \to \mathbb{R}$ can be uniformly approximated by elements of A (via the representation Φ_X).* □

1.5.7. The celebrated Representation Theorem, to be discussed next, has been discovered and improved by many mathematicians over the years. The history of its genesis is a complicated one. An ur-version is due to M. Stone (1940). Later it was generalized by Kadison (1951) and Dubois (1967). Independently, Krivine found essentially the full version discussed here in 1964. A purely algebraic proof was given by Becker and Schwartz in 1983 [BS]. There, as in many other places of the literature, the result goes under the name 'Kadison–Dubois theorem'. We refer to [PD] 5.6 for a more detailed historical account.

Let M be a module over some preprime of A. We want to study representation (1) for the (pro-basic) closed set $\mathscr{X}_M$. So we have to assume that M is weakly archimedean (over $\mathbb{Z}$). As explained in 1.5.4, this means that the set

$$X_M := \{\alpha \in \operatorname{Hom}(A, \mathbb{R}) \colon \alpha|_M \geq 0\}$$

is compact and equal to $\mathscr{X}_M^{\max}$.

The Representation Theorem characterizes the elements $f \in A$ for which $\Phi_{\mathscr{X}_M}(f)$ is non-negative, or strictly positive, via identities involving elements of M. So it can be considered to be a Nichtnegativ- or a Positivstellensatz. However, this theorem requires the *archimedean*, and not just the weak archimedean property:

THEOREM 1.5.8 (Representation Theorem, 1st version). *Let A be a ring, and let M be a module over an archimedean preprime P of A. For every $f \in A$ we have:*

$$f \geq 0 \text{ on } X_M = \mathscr{X}_M^{\max} \quad \Leftrightarrow \quad \forall n \in \mathbb{N}\ \exists m \in \mathbb{N}\ \ m(1 + nf) \in M.$$

If P contains $\frac{1}{q}$ for some integer $q > 1$, it is also equivalent that $1+nf \in M$ *for every* $n \in \mathbb{N}$.

Note that it is the preprime P that is required to be archimedean. The archimedean property for the module M alone is not enough. However, there is an important case in which the archimedean property of M suffices for the conclusion, namely when M is a quadratic module. This is a more recent result and will be discussed in 2.3 below.

Note also that the set $\{f \in A\colon f \geq 0 \text{ on } X_M\}$ described by 1.5.8 can be viewed as kind of a bi-dual of the cone M.

While '$\Leftarrow$' in 1.5.8 is trivial, the converse is not obvious at all, even though the proof is not too hard. This converse is essentially equivalent to the '$\Rightarrow$' implications in

THEOREM 1.5.9 (Representation Theorem, 2nd version). *Let A be a ring, let M be a module over an archimedean preprime P of A, and let $f \in A$.*

(a) $f > 0$ *on* $\mathscr{X}_M \;\Leftrightarrow\; \exists n \in \mathbb{N}\;\; nf \in 1 + M$.

(b) If P contains $\frac{1}{q}$ for some $1 < q \in \mathbb{N}$, then

$$f > 0 \text{ on } \mathscr{X}_M \quad \Leftrightarrow \quad \exists r \in \mathbb{N}\;\; f \in q^{-r} + M.$$

In case (b), in particular, every $f \in A$ with $f > 0$ on $\mathscr{X}_M$ is contained in M.

For both versions, note that $\frac{1}{q} \in P$ is automatic if P is a k-preprime and $\frac{1}{q} \in k$. We refer to [BS] for the proof of 1.5.8 and 1.5.9. See also [PD] for the case $M = P$ (5.2.6 for preorderings and 5.4.4 for preprimes).

Version 1.5.9 exhibits the Representation Theorem as a Positivstellensatz. One should compare it to the general Positivstellensatz 1.3.9(a) for preorderings. The latter gives, for every $f \in A$ with $f > 0$ on $\mathscr{X}_M$, a representation of f (in terms of M) *with denominator*. Of course, this result does not require any archimedean property. In contrast, the Representation Theorem (at least in its version 1.5.9(b)) gives a *denominator-free* representation, and it requires the archimedean hypothesis.

As a corollary one gets the following characterization of archimedean preprimes:

COROLLARY 1.5.10. *Let A be a ring, and let P be a preprime of A with $\frac{1}{q} \in P$ for some $q > 1$. The following conditions are equivalent:*

(i) P is archimedean;

(ii) P is weakly archimedean and contains every $f \in A$ with $f > 0$ on $\mathscr{X}_P$.

Of course, in most applications of geometric origin, one will have $\frac{1}{q} \in P$ for all $q \in \mathbb{N}$. It may nevertheless be worthwile to record the more general case as well, with applications to arithmetic situations in mind.

2. Positivstellensätze. We now start reviewing results which are more recent. The common feature of much of what is assembled in this section is that these results give *denominator-free* expressions for *strictly positive* functions, usually in terms of weighted sums of squares. A good part can be regarded as applications of the Representation Theorem (Section 1.5). For the sake of concreteness, we will often (but not always) state results in geometric form (polynomials, semi-algebraic sets etc. over the reals), even when more abstract versions are available.

2.0. We will have to speak of affine algebraic varieties (over the reals). For this, the language used in some of the textbooks on real algebraic geometry is not always suitable. Instead, we prefer to use what has long become standard in algebraic geometry, the Grothendieck language of schemes. It will be needed only in its most basic form: Given a field k, by an *affine k-variety* we mean an affine k-scheme V of finite type. This means $V = \operatorname{Spec} A$ where A is a finitely generated k-algebra (not necessarily a domain, not even necessarily reduced). One writes $A =: k[V]$, and calls the elements of $k[V]$ the regular functions on V. Thus, in this text, *an affine k-variety is usually neither irreducible nor reduced.*

Given any field extension E/k, one writes $V(E) := \operatorname{Hom}_k(k[V], E)$. This is the set of E-rational points of V. If $k[V]$ is generated by $a_1, \dots, a_n$ as a k-algebra, and if the kernel of the k-homomorphism

$$k[x_1, \dots, x_n] \to k[V], \quad x_i \mapsto a_i \quad (i = 1, \dots, n)$$

is generated by $f_1, \dots, f_m$ as an ideal, we may identify $V(E)$ with the ('k-algebraic') subset $\{x \in E^n : f_1(x) = \cdots = f_m(x) = 0\}$ of E^n. In particular, if $E = R$ is a real closed field, we have a natural (order) topology on $V(R)$, and have a natural notion of semi-algebraic subsets of $V(R)$. Neither of them depends on the choice of the generators $a_1, \dots, a_n$ of $k[V]$.

Recall that a subset K of $V(\mathbb{R})$ is called *basic closed* if there are $f_1, \dots, f_r \in \mathbb{R}[V]$ with $K = \mathscr{S}(f_1, \dots, f_r) = \{x \in V(\mathbb{R}) : f_i(x) \geq 0, i = 1, \dots, r\}$.

2.1. Schmüdgen's Positivstellensatz. Since its appearance, this result has triggered much activity and stimulated new directions of research, some of which we shall try to record here. For all of what follows it matters that $\mathbb{R}$ is the field of classical real numbers (or a real closed subfield thereof). Results become false in general over real closed fields that are non-archimedean.

In 1991, Schmüdgen proved

THEOREM 2.1.1 ([Sm1]). *Let $f_1, \dots, f_r \in \mathbb{R}[x_1, \dots, x_n]$, and assume that the semi-algebraic set $K = \mathscr{S}(f_1, \dots, f_r)$ in $\mathbb{R}^n$ is compact. Then the preordering $PO(f_1, \dots, f_r)$ of $\mathbb{R}[x_1, \dots, x_n]$ contains every polynomial that is strictly positive on K.*

An equivalent formulation is:

THEOREM 2.1.2. *Let V be an affine $\mathbb{R}$-variety, and assume that the set $V(\mathbb{R})$ of $\mathbb{R}$-points on V is compact. Then every $f \in \mathbb{R}[V]$ that is strictly positive on $V(\mathbb{R})$ is a sum of squares in $\mathbb{R}[V]$.*

Schmüdgen's primary interest was in analysis rather than real algebra. In his original paper [Sm1] he deduced 2.1.1 from his solution to the multivariate moment problem in the compact case (see Theorem 5.4 below). The latter, in turn, was established by combining the Positivstellensatz 1.2.4 with operator-theoretic arguments from functional analysis.

For the algebraist, it seems more natural to proceed in a different order. Thanks to the Representation Theorem 1.5.10, an equivalent way of stating Theorem 2.1.1 is to say that the preordering in question is archimedean (c.f. 1.5.2):

THEOREM 2.1.3. *Let T be a finitely generated preordering of $\mathbb{R}[x_1, \ldots, x_n]$. Then T is archimedean if (and only if) the subset $\mathscr{S}(T)$ of $\mathbb{R}^n$ is compact.*

Around 1996, Wörmann was the first to use this observation for a purely algebraic proof of 2.1.1, resp. 2.1.3. His proof is simple and elegant, but not obvious, and can be found in [BW1], or in [PD] Thm. 5.1.17, [Ma] Thm. 4.1.1. In this way, Schmüdgen's theorem becomes a (non-obvious!) application of the Representation Theorem.

The proof can be applied in a more general setting, leading to the following more 'abstract' formulation:

THEOREM 2.1.4. *Let k be a ring containing $\frac{1}{2}$, let A be a finitely generated k-algebra, and let T be a preordering of A. Then*

T is weakly archimedean over k $\Leftrightarrow$ T is archimedean over k.

Moreover, if these equivalent conditions hold, and if the preordering $k \cap T$ of k is archimedean, then T is archimedean (over $\mathbb{Z}$). In particular, then, T contains every $f \in A$ with $f > 0$ on $\mathscr{X}_T$.

The second part is obvious (using the Representation Theorem), while the proof of '$\Rightarrow$' in the first part employs Wörmann's arguments from [BW1]. See [Ma2] Thm. 1.1 and [Sch4] Thm. 3.6. A sample application is the following very explicit characterization of (finitely generated) archimedean preorderings on affine varieties over totally archimedean fields (like number fields, for example):

COROLLARY 2.1.5. *Let k be a field with only archimedean orderings, and let V be an affine k-variety. Let T be a finitely generated preordering in $k[V]$. For every ordering P of k, let k_P be the real closure of (k, P), and let K_P be the basic closed semi-algebraic subset of $V(k_P)$ defined by T. Then the following are equivalent:*

(i) T is archimedean;

(ii) for every ordering P of k, K_P is semi-algebraically compact.[11]

[11] semi-algebraicAlly compact means closed and bounded (after some embedding into affine space).

2.1.6. Yet another approach to the Positivstellensatz 2.1.1, again purely algebraic, is due to Schweighofer [Sw1]. He reduces the proof to a celebrated classical result of Pólya (Theorem 2.2.1 below), thereby even making the proof largely algorithmic. To explain this in more detail, note first that, K being compact, there exists a real number $c > 0$ such that K is contained in the open ball of radius c around the origin. By the Positivstellensatz 1.2.4(a), there exist s, t in $T = PO(f_1, \dots, f_r)$ with

$$s\left(c^2 - \sum_i x_i^2\right) = 1 + t. \tag{2}$$

One may regard identity (2) as an explicit certificate for the compactness of K. Let now $f \in \mathbb{R}[x_1, \dots, x_n]$ be strictly positive on K. Starting from a compactness certificate (2), Schweighofer effectively constructs a representation

$$f = \sum_{e \in \{0,1\}^r} s_e \cdot f_1^{e_1} \cdots f_r^{e_r} \tag{3}$$

of f, where the s_e are sums of squares. Essentially, he does this by suitably pulling back a solution to Pólya's theorem (see 2.2.1 below).

2.1.7. Before we proceed to discuss Pólya's theorem, we sketch an application of Schmüdgen's Positivstellensatz to Hilbert's 17th problem. Consider a positive definite form[12] f in $\mathbb{R}[x_1, \dots, x_n]$. From Stengle's Positivstellensatz 1.2.4(a) it follows that f is a sum of squares of quotients of forms, where the denominators are positive definite (see Remark 1.2.5.4). This fact was given a refinement by Reznick in 1995 [Re1]. He showed that the denominators can be taken uniformly to be powers of $x_1^2 + \cdots + x_n^2$. In other words, the form $(x_1^2 + \cdots + x_n^2)^N \cdot f$ is a sum of squares of forms for sufficiently large N.

2.1.8. Reznick's uniform solution for positive definite forms can be considered as a particular case of 2.1.2. (The original proof in [Re1], however, is very different.) In fact, 2.1.7 can be generalized as follows. Given any two positive definite forms f and g in $\mathbb{R}[x_1, \dots, x_n]$ such that $\deg(g)$ divides $\deg(f)$, there exists $N \geq 1$ such that $g^N \cdot f$ is a sum of squares of forms. Indeed, setting $r = \frac{\deg(f)}{\deg(g)}$, it suffices to apply 2.1.2 to the rational function $\frac{f}{g^r}$. This is a strictly positive regular function on the complement V of the hypersurface $g = 0$ in $\mathbb{P}^{n-1}$. The $\mathbb{R}$-variety V is affine, and $V(\mathbb{R})$ is compact. By 2.1.2, therefore, $\frac{f}{g^r}$ is a sum of squares in $\mathbb{R}[V]$. For a generalization which goes still further see 2.5.7 below.

2.1.9. In general, the restriction to *strictly* positive forms in 2.1.7 and 2.1.8 is necessary: For any $n \geq 4$, there exists a positive semidefinite

[12] That is, f is homogeneous and strictly positive on $\mathbb{R}^n$ outside the origin.

form f in n variables together with a point $0 \neq p \in \mathbb{R}^n$ such that, whenever $h^2 f = \sum_i f_i^2$ with forms h and f_i, the form h vanishes in p. Such a point p has been called a *bad point* for f. The existence of bad points for suitable forms has long been known (Straus 1956[13]). Later, Choi-Lam and Delzell proved that bad points do not exist for forms in three variables ([CL], [De]). A much more recent result is that the uniform denominators theorem holds even for all *non-strictly* psd forms in three variables: See 3.3.8 below.

2.2. Pólya's theorem and preprimes. Pólya's theorem, proved in 1928, is as follows:

THEOREM 2.2.1. *Let $f \in \mathbb{R}[x_1, \dots, x_n]$ be a homogeneous polynomial that is strictly positive on the positive hyperoctant, i. e. that satisfies $f(x_1, \dots, x_n) > 0$ whenever $x_1, \dots, x_n \geq 0$ and $x_i \neq 0$ for at least one i. Then, for sufficiently large $N \geq 1$, all coefficients of the form*

$$(x_1 + \dots + x_n)^N \cdot f$$

are strictly positive.[14]

Note that, conversely, the conclusion of Theorem 2.2.1 implies the strict positivity condition for f. So Pólya's theorem is an example of a Positivstellensatz.

2.2.2. Pólya's theorem can be considered as a special case of the Representation Theorem 1.5.9, as shown in [BW1]: Write $h = x_1 + \dots + x_n$, and let V be the complement of the hyperplane $h = 0$ in projective space $\mathbb{P}^{n-1}$ over $\mathbb{R}$. Note that V is an affine $\mathbb{R}$-variety. The $\mathbb{R}$-preprime $P := PP_{\mathbb{R}}(\frac{x_1}{h}, \dots, \frac{x_n}{h})$ in $\mathbb{R}[V]$ is archimedean since $\sum_i \frac{x_i}{h} = 1$. Letting $d = \deg(f)$, $\frac{f}{h^d}$ is an element of $\mathbb{R}[V]$ which is strictly positive on $\mathscr{X}_P$ by the hypothesis on f. Therefore $\frac{f}{h^d} \in P$ by 1.5.9. This means that, for some $N \geq 1$, all coefficients of the form $h^N \cdot f$ are non-negative. This is a 'non-strict' version of 2.2.1, from which it is easy to derive the full statement above.

2.2.3. Quantitative versions of Pólya's theorem were studied by de Loera and Santos [LS]. Their results were improved by Powers and Reznick [PR2], who showed: Given $f = \sum_{|\alpha|=d} a_\alpha \mathrm{x}^\alpha$ as in 2.2.1, let $\lambda = \min_{x \in \Delta} f(x)$ and $L = \max_\alpha \frac{\alpha!}{d!} |a_\alpha|$. (Here $\Delta = \{x \colon \sum_i x_i = 1, x_i \geq 0\}$ is the standard $(n-1)$-simplex in $\mathbb{R}^n$, and $d = \deg(f)$.) Then for

$$N > \frac{d}{2}(d-1)\frac{L}{\lambda} - d,$$

the form $(x_1 + \dots + x_n)^N \cdot f$ has strictly positive coefficients. This bound is sharp for $d = 2$.

[13] According to a remark in [De].

[14] That is, every monomial of degree $N + \deg(f)$ appears in this product with a strictly positive coefficient.

2.2.4. It is an important and difficult problem to give complexity estimates for Schmüdgen's Positivstellensatz 2.1.1. Given a polynomial f which is strictly positive on K, one would like to have both upper and lower bounds for the degrees of the sums of squares s_e in a representation (3).

In a basic case (univariate polynomials $n = 1$, $T = PO((1-x^2)^3)$, a complete complexity analysis was carried out by Stengle [St2]. In the general case, upper complexity bounds have been given by Schweighofer [Sw3]. They are derived from his approach 2.1.6, using the bounds in 2.2.3 for Pólya's theorem.

Another related result is the following theorem by Handelman (1988):

THEOREM 2.2.5 ([Ha1]). *Let $f_1, \dots, f_r \in \mathbb{R}[1, \dots, x_n]$ be polynomials of degree one such that the polytope $K := \mathscr{S}(f_1, \dots, f_r)$ is compact and non-empty. Then the $\mathbb{R}$-preprime $PP_{\mathbb{R}}(f_1, \dots, f_r)$ contains every polynomial that is strictly positive on K.*

In other words, every polynomial f with $f|_K > 0$ admits a representation $f = \sum_i c_i f_1^{i_1} \cdots f_r^{i_r}$ (finite sum) with real numbers $c_i \geq 0$.

A classical argument, due to Minkowski, shows that the preprime in question is archimedean. Therefore Handelman's theorem can be seen as a particular case of the Representation Theorem 1.5.9 (c.f. also [BW1]). The argument mentioned in 2.2.2 shows also that Handelman's theorem directly implies Pólya's theorem. From the bound in 2.2.3, one can deduce a quantitative version of Handelman's theorem; see Powers and Reznick [PR2]. In the particular case where K is a simplex, this is immediate, whereas the general case is less explicit.

2.3. Quadratic modules. 2.3.1. Let $f_1, \dots, f_r \in \mathbb{R}[x_1, \dots, x_n]$ be such that the semi-algebraic set $K = \mathscr{S}(f_1, \dots, f_r)$ in $\mathbb{R}^n$ is compact. By 2.1.1, every polynomial f with $f > 0$ on K can be written

$$f = \sum_{e \in \{0,1\}^r} s_e f_1^{e_1} \cdots f_r^{e_r} \tag{4}$$

where the polynomials s_e are sums of squares.

Putinar raised the question whether all 2^r summands in (4) are needed, or whether one can get by with fewer of them. Specifically, he asked whether f can always be written

$$f = s_0 + s_1 f_1 + \cdots + s_r f_r \tag{5}$$

with sums of squares $s_0, \dots, s_r$, i.e., whether $f \in QM(f_1, \dots, f_r)$. He was drawn to such considerations by operator-theoretic ideas, and it was by such methods that he proved in 1993 that the answer is positive under suitable hypotheses on the f_i:

THEOREM 2.3.2 ([Pu] Lemma 4.1). *Let M be a finitely generated quadratic module in $\mathbb{R}[x_1, \dots, x_n]$, and assume that there is $g \in M$ for* *which $\{x \in \mathbb{R}^n : g(x) \geq 0\}$ is a compact set. Then M contains every polynomial that is strictly positive on $K = \mathscr{S}(M)$.*

Theorem 2.3.2 is known as Putinar's Positivstellensatz. In its light, Putinar's question 2.3.1 (in its strong form (5)) can be rephrased as follows: Given a finitely generated quadratic module M in $\mathbb{R}[x_1, \ldots, x_n]$ for which $K = \mathscr{S}(M)$ is compact, does M necessarily contain a polynomial g for which $\{g \geq 0\}$ is compact?

This question was subsequently answered in the negative by Jacobi and Prestel. Unlike Putinar, they worked purely algebraically, but they took up some of Putinar's ideas. A key step is the following extension of the Representation Theorem to archimedean quadratic modules, due to Jacobi ([Ja2] Thm. 4):

THEOREM 2.3.3. *Let A be a ring containing $\frac{1}{q}$ for some $q > 1$, and let M be a quadratic module in A which is archimedean. Then M contains every $f \in A$ with $f > 0$ on $\mathscr{X}_M$.*

2.3.4. In the geometric case ($A = \mathbb{R}[x_1, \ldots, x_n]$ and M finitely generated), Jacobi's theorem and Putinar's Positivstellensatz directly imply each other.[15] Thus, Jacobi's theorem can be considered to be a generalization of Putinar's theorem to an abstract setting.

Note that 2.3.3 is the exact analogue of Theorem 1.5.10, for quadratic modules instead of preprimes. The proof is more complicated than for preprimes (see also [PD] 5.3.6 or [Ma] 5.1.4) and requires the Positivstellensatz for archimedean quadratic modules 1.4.5(a). The above formulation of 2.3.3 is not the most general possible; we have assumed $\frac{1}{q} \in A$ for simplicity. Moreover, Jacobi generalizes his result to archimedean modules over generating preprimes ([Ja2] Thm. 3), which allows him to prove a 'higher level' analogue of 2.3.3 as well (see 2.5.2 below). A common generalization of Jacobi's result and the classical Representation Theorem 1.5.9 was found by Marshall [Ma3].

The next criterion, due to Jacobi and Prestel, gives a possible way to decide the answer to Putinar's question in concrete cases:

THEOREM 2.3.5 ([JP] Thm. 3.2). *Let $M = QM(f_1, \ldots, f_r)$ be a finitely generated quadratic module in $\mathbb{R}[x_1, \ldots, x_n]$ such that $K = \mathscr{S}(M)$ is compact. The following conditions are equivalent:*

(i) M is archimedean;

(ii) M contains every $f \in \mathbb{R}[x_1, \ldots, x_n]$ with $f > 0$ on K;

(iii) there is $g \in M$ with $\{g \geq 0\}$ compact;

(iv) there is $N \in \mathbb{Z}$ with $N - \sum_{i=1}^n x_i^2 \in M$;

[15] Putinar ⇒ Jacobi: Since M is archimedean, there is a real number $c > 0$ with $g := c - \sum_i x_i^2 \in M$. Since $\{g \geq 0\}$ is compact, 2.3.2 applies and gives the conclusion of 2.3.3. — Jacobi ⇒ Putinar: If there is $g \in M$ with $\{g \geq 0\}$ compact, then $QM(g) = PO(g)$ is archimedean by 2.1.3, and *a fortiori*, the larger quadratic module M is archimedean. So 2.3.3 applies to give the conclusion of 2.3.2.

(v) *for every prime ideal* $\mathfrak{p}$ *of* $\mathbb{R}[\mathrm{x}]$ *and every valuation* v *of* $\kappa(\mathfrak{p})$ *with real residue field and with* $v(x_i) < 0$ *for at least one index* $i \in \{1, \ldots, n\}$, *the quadratic form* $<1, f_1, \ldots, f_r>^*$ *over* $\kappa(\mathfrak{p})$ *has at least one residue form with respect to* v *which is weakly isotropic over* κ_v.

(See Section 1.4 for explanations of the terms occuring in (v).) The equivalence of (i)–(iv) is Putinar's theorem 2.3.2, but the algebraic condition (v) is new. The proof of (v) $\Rightarrow$ (i) uses 1.4.11 together with the fact that if every $S \in \mathscr{Y}_M$ is archimedean, then M is archimedean. (See also [PD] Thm. 5.1.18, Thm. 6.2.2.)

EXAMPLE 2.3.6. ([PD] 6.3.1) We illustrate the use of condition (v) from 2.3.5. Consider the quadratic module

$$M = QM(2x_1 - 1, \ldots, 2x_n - 1, 1 - x_1 \cdots x_n)$$

in $\mathbb{R}[x_1, \ldots, x_n]$. The associated semi-algebraic set $K = \mathscr{S}(M)$ is compact. But for $n \geq 2$, the module M is not archimedean (thus providing a negative answer to Putinar's question (5)). Using (v) from above, this can be verified as follows: The composition of $\mathbb{R}$-places

$$\mathbb{R}(x_1, \ldots, x_n) \xrightarrow{\lambda_1} \mathbb{R}(x_2, \ldots, x_n) \cup \infty \xrightarrow{\lambda_2} \cdots \mathbb{R}(x_n) \cup \infty \xrightarrow{\lambda_n} \mathbb{R} \cup \infty$$

(where λ_i sends x_i to ∞ and is the identity on $x_{i+1}, \ldots, x_n$) induces a valuation v on $\mathbb{R}(x_1, \ldots, x_n)$ whose value group is $\mathbb{Z}^n$, ordered lexicographically. With respect to v, every non-zero residue form of the quadratic form

$$<1, 2x_1 - 1, \ldots, 2x_n - 1, 1 - x_1 \cdots x_n>$$

has rank one, and is therefore not weakly isotropic. For example, the polynomial $c - \sum_i x_i^2$ is positive on K for $c > 2^{n-1}$, but is not contained in M.

At a first glance, one may expect it to be cumbersome to use condition (v) for showing that a given M is archimedean. However, the applications given in [JP] demonstrate the usefulness of the condition. Here are two samples:

THEOREM 2.3.7 ([JP] Thm. 4.1). *Let* $f_1, \ldots, f_r \in \mathbb{R}[x_1, \ldots, x_n]$, *let* $d_i = \deg(f_i)$, *and let* $\widetilde{f}_i$ *be the leading form of* f_i *(i. e. the homogeneous part of degree* d_i*). Assume that the following condition holds:*

$$\forall\, 0 \neq x \in \mathbb{R}^n \quad \exists\, i \in \{1, \ldots, r\} \quad \widetilde{f}_i(x) < 0. \tag{6}$$

Then the quadratic module $QM(f_i,\ f_i f_j \colon 1 \leq i < j \leq r)$ *is archimedean. Moreover, if* $d_i \equiv d_j \pmod 2$ *for all* i, j, *then even* $QM(f_1, \ldots, f_r)$ *is archimedean.*

(See also [PD] Thm. 6.3.4 for a slightly finer version.) Note that condition (6) implies compactness of $K := \mathscr{S}(f_1, \ldots, f_r)$, but in general, (6) is strictly stronger (as shown by 2.3.6, for example). However, if all f_i are linear, compactness of K implies (6), giving an easy example case to which the theorem applies.

While the strong version of Putinar's question has a negative answer, as we have seen in 2.3.6, the next result gives a positive answer in a weaker sense: Indeed not all 2^r summands in (4) are needed, rather a bit more than half of them is already enough:

THEOREM 2.3.8 ([JP] Thm. 4.4). *Let $f_1, \dots, f_r \in \mathbb{R}[x]$ such that $K = \mathscr{S}(f_1, \dots, f_r)$ is compact. Then there are 2^{r-1} (explicit) elements $h_1, \dots, h_{2^{r-1}}$ among the $2^r - 1$ products*

$$f_I := \prod_{i \in I} f_i, \qquad \varnothing \neq I \subset \{1, \dots, r\},$$

such that the quadratic module $QM(h_1, \dots, h_{2^{r-1}})$ is archimedean (and hence contains every polynomial f with $f|_K > 0$).

REMARKS 2.3.9.

1. If one enumerates the $2^r - 1$ products as

$$f_1,\ \dots,\ f_r,\ f_1 f_2,\ \dots,\ f_{r-1} f_r,\ f_1 f_2 f_3,\ \dots,\ f_1 \cdots f_r,$$

it suffices for 2.3.8 to take the first 2^{r-1} of them.

2. In particular, the answer to Putinar's question (5) is yes for $r \le 2$. On the other hand, the answer is no in general for $r \ge 3$, as is demonstrated by Example 2.3.6.

3. In [Ma3] one finds further sharpenings of some of the Jacobi-Prestel results. They use the general version of the Representation Theorem proved in this paper.

2.4. Rings of bounded elements. 2.4.1. Let A be a ring with $\frac{1}{2} \in A$ (for simplicity), and let T be a preordering of A. The subset

$$O_T(A) := \{a \in A\colon \exists\, n \in \mathbb{Z}\ \ n \pm a \in T\} = (T + \mathbb{Z}) \cap -(T + \mathbb{Z})$$

of *T-bounded elements* in A is a subring of A. Clearly, $O_T(A) = A$ holds if and only if A is archimedean. More generally, $O_T(A)$ is the largest subring B of A for which the preordering $T \cap B$ of B is archimedean.

2.4.2. We put

$$B_T(A) := O_{\mathrm{Sat}(T)}(A) = \{a \in A\colon \exists\, n \in \mathbb{Z}\ \ |a| \le n \text{ on } \mathscr{X}_T\},$$

and call $B_T(A)$ the ring of *weakly T-bounded elements* of A. Clearly, $B_T(A) = A$ holds if and only if the preordering T is weakly archimedean, i. e., if and only if the closed subset $\mathscr{X}_T$ of Sper A is bounded over $\mathbb{Z}$. In the case $T = \Sigma A^2$, the ring $B_T(A)$ is often called the *real holomorphy ring* of

A and denoted $H(A)$; thus $H(A)$ consists of the elements that are globally bounded by some integer in absolute value.

The study of these rings was initiated by Becker in the 1970s, in the case where $A = k$ is a field and $T = \Sigma k^2$. In this case, $B_T(k) = H(k)$ is the intersection of all valuation rings of k that have a real residue field. (It is this fact that motivated the name 'real holomorphy ring' for $H(k)$.) The rings $H(k)$ have important connections to quadratic form theory, to sums of higher powers and to real algebraic geometry. The article [BP] contains a good list of references for background reading.

2.4.3. Let T be a preordering of A. Starting from the preordered ring (A,T), we form the preordered ring

$$(A,T)' := \bigl(B_T(A),\, T \cap B_T(A)\bigr).$$

This step can be iterated. Thus we put $\bigl(B_T^n(A),\, T_n\bigr) := (A,T)^{(n)}$ for $n \geq 0$, where $(A,T)^{(0)} := (A,T)$ and $(A,T)^{(n)} := \bigl((A,T)^{(n-1)}\bigr)'$ for $n \geq 1$. These iterated rings of bounded elements were first studied by Becker and Powers [BP], in the case $T = \Sigma A^2$. (Note that $H(A) \cap \Sigma A^2 = \Sigma H(A)^2$, so in this case there is no need to keep track of the preordering.) The definition in the case of general preorderings is due to Schweighofer [Sw2].[16] A primary case of interest is when A is a finitely generated $\mathbb{R}$-algebra. However, there is a drawback: In general, the subalgebra $H(A)$ of A need not be finitely generated, not even noetherian. (See 2.4.7 below for more information.)

Clearly one has $A \supset B_T^1(A) \supset B_T^2(A) \supset \cdots$, and a priori it is not clear whether the iteration process stops. All rings $B_T^n(A)$ contain the ring $O_T(A)$.

We are now going to recall important work of Schweighofer, which extends former work by Becker–Powers and Monnier and, at the same time, generalizes Schmüdgen's Positivstellensatz. The following theorem generalizes the versions 2.1.3 or 2.1.4 of the latter. Now, the $\mathbb{R}$-algebra in question is no longer assumed to be finitely generated, rather only to have finite transcendence degree:[17]

THEOREM 2.4.4 ([Sw2] Thm. 4.13). *Let A be an $\mathbb{R}$-algebra of finite transcendence degree, and let T be a preordering of A. If T is weakly archimedean, then T is archimedean.*

In other words, $B_T(A) = A$ implies $O_T(A) = A$. Marshall [Ma1] has given an example of a preordering T of $A = \mathbb{R}[x_1, x_2, \dots]$ that is weakly archimedean (i.e., has $B_T(A) = A$) but satisfies $O_T(A) = \mathbb{R}$. This shows that the theorem breaks down completely if the transcendence degree is infinite.

[16] [Sw2] writes $H'(A)$ and $H(A)$ for our $O_T(A)$ and $B_T(A)$, respectively; we prefer to make the dependence on the preordering visible in the notation.

[17] By the transcendence degree of an $\mathbb{R}$-algebra A we mean the largest number of elements in A that are algebraically independent over $\mathbb{R}$.

On the other hand, the iteration process stops at a finite level:

THEOREM 2.4.5. *Let A be an $\mathbb{R}$-algebra of transcendence degree $d < \infty$, and let T be a preordering of A. Then $B_T^d(A) = B_T^{d+1}(A)$.*

Originally, this theorem is due to Becker and Powers, who proved it for the case $T = \Sigma A^2$ ([BP] Thm. 3.1). The extension to arbitrary preorderings was given by Schweighofer ([Sw2] Thm. 3.11). There are examples by Pingel showing that, in general, $B_T^{d-1}(A) \neq B_T^d(A)$, with $T = \Sigma A^2$ [Pi].

Combining Theorems 2.4.4 and 2.4.5, one gets the following corollary, conjectured by Monnier [Mo]:

COROLLARY 2.4.6 ([Sw2] Thm. 4.16). *Under the conditions of 2.4.5 one has $B_T^d(A) = O_T(A)$.*

Note that this corollary not only follows from 2.4.4 and 2.4.5, but also generalizes both of them at the same time. Interestingly, the results from [Sw2] have been used recently in an essential way in optimization, namely for an improved method to globally optimize polynomials [Sw5].

2.4.7. In geometric situations, rings of bounded elements can be approached by algebraic geometry methods, at least under suitable regularity conditions. Let V be an affine normal $\mathbb{R}$-variety and let K be a basic closed subset of $V(\mathbb{R})$. A Zariski open immersion $V \hookrightarrow X$ into a complete normal $\mathbb{R}$-variety X is called a *K-good compactification* of V if, for every irreducible component Z of $X - V$, the set $Z(\mathbb{R}) \cap \overline{K}$ is either empty or Zariski dense in Z.[18]

Let $B(K) \subset \mathbb{R}[V]$ be the ring of regular functions which are bounded on K. Given a K-good compactification X, let Y be the union of all irreducible components Z of $X - V$ with $Z(\mathbb{R}) \cap \overline{K} = \varnothing$, and let $U = X - Y$. Then $V \subset U \subset X$ are open immersions, and one has canonically $B(K) = \mathcal{O}_X(U)$.

If V is a curve, a K-good compactification always exists. The same is true when V is a surface and K is regular at infinity,[19] using resolution of singularities. By a theorem of Zariski, this implies that $B(K)$ is a finitely generated $\mathbb{R}$-algebra in this case. In higher dimensions, this is not true in general. (For $\dim(V) = 2$ it can also fail if K is not regular at infinity.) These and other related results can be found in [Pl].

2.5. Higher level. There exist analogues of the various stellensätze and representation theorems for sums of higher (even) powers, resp. for preorderings of higher level. I will briefly indicate a few results in this direction, but, unfortunately, cannot go much into details. The interested reader is referred to Chapter 7 of the book [PD], to [BW2] and to [Ja1].

2.5.1. Let A be a ring, and let $m \geq 1$ be a fixed integer. A *preordering of level $2m$* in A is a ($\mathbb{Z}$-) preprime (1.4.1) T of A which contains $\{a^{2m}:$

[18]Of course, $\overline{K}$ means the closure of K in $X(\mathbb{R})$.

[19]Meaning that K is the union of a compact set and the closure of an open set.

$a \in A\}$. The smallest preordering of level $2m$ in A is ΣA^{2m}. A *module of level* $2m$ in A is by definition a ΣA^{2m}-module in A (1.4.2).

It should be underlined that terminology is not uniform in the published literature. What we call 'level $2m$' here is sometimes referred to as 'level m' instead.

Jacobi's proof for the Representation Theorem 2.3.3 is given in a uniform way, which includes higher level analogues as well:

THEOREM 2.5.2. *Let M be a module of level $2m$ in A, and assume $\mathbb{Q} \subset A$ (for simplicity). If M is archimedean, then M contains every $f \in A$ with $f > 0$ on $\mathscr{X}_M$.*

See [Ja2] Thm. 4 or [PD] Thm. 7.3.2. This result is again complemented by a recognition theorem for the archimedean property which is similar to Theorem 2.3.5. The quadratic form $<1, f_1, \dots, f_r>$ (and its analogues over the residue fields of $\mathbb{R}[x]$) is now replaced by the diagonal form

$$X_0^{2m} + f_1 X_1^{2m} + \cdots + f_r X_r^{2m}$$

of degree $2m$. For more details see [PD] Thm. 7.3.9, or Jacobi's doctoral thesis [Ja1]. As an application, one can prove the higher level analogue of Theorem 2.3.7 above, see [PD] Thm. 7.3.11.

2.5.3. A thorough study of the archimedean property for preorderings of higher level was made by Berr and Wörmann. A key notion is the concept of *tame preorderings* introduced in [BW2]. Instead of giving the definition, we mention the two most important classes of examples:

- If m is odd, every preordering T of level $2m$ (with $\mathscr{X}_T \neq \varnothing$) is tame;
- for any m, the preordering ΣA^{2m} is tame (assuming $\operatorname{Sper} A \neq \varnothing$).

The main result is

THEOREM 2.5.4 ([BW2] Thm. 3.8). *Let A be a finitely generated algebra over a field k, and let T be a preordering of level $2m$ in A for which $k \cap T$ is archimedean. Assume that T is tame. Then, if the preordering $\operatorname{Sat}(T)$ is archimedean, so is T.*

By applying Wörmann's formulation of Schmüdgen's theorem (2.1.3) and the Representation Theorem 1.5.10, one gets:

COROLLARY 2.5.5 ([BW2] Cor. 4.2). *Let V be an affine $\mathbb{R}$-variety, and let $f_1, \dots, f_r \in \mathbb{R}[V]$ such that the subset $K = \mathscr{S}(f_1, \dots, f_r)$ of $V(\mathbb{R})$ is compact. Let $f \in \mathbb{R}[V]$ with $f > 0$ on K. Then for every odd $m \geq 1$, f lies in the preordering of level $2m$ generated by $f_1, \dots, f_r$ in $\mathbb{R}[V]$.*

If m is even, this is not necessarily true, due to the possible failure of the tameness condition. A basic example is the following ([BW2] Ex. 4.5): The preordering T of level 4 generated in $\mathbb{R}[x]$ by $1 - x^2$ does not contain $c - x$ for any real number c, as can be seen by inspecting the valuation at infinity.

COROLLARY 2.5.6 ([BW2] Cor. 4.6). *Let V be an affine $\mathbb{R}$-variety for which $V(\mathbb{R})$ is compact. Let $f \in \mathbb{R}[V]$ with $f > 0$ on $V(\mathbb{R})$. Then f is a sum of $2m$-th powers in $\mathbb{R}[V]$ for every $m \geq 1$.*

If $\mathbb{R}[V]$ is a domain, the converse holds as well: If $0 \neq f \in \mathbb{R}[V]$ is a sum of $2m$-th powers in $\mathbb{R}[V]$ for every $m \geq 1$, then f is *strictly* positive on $V(\mathbb{R})$ ([BW2] Cor. 4.10).

2.5.7. The last corollary allows to give uniform solutions to the 'higher level' analogue of Hilbert's 17th problem, for positive definite forms: If the forms f and g in $\mathbb{R}[x_1, \ldots, x_n]$ are positive definite, and if $\deg(g)$ divides $\deg(f)$, then for any $m \geq 1$ there exists an integer $N \geq 1$ such that $g^N \cdot f$ is a sum of $2m$-th powers of forms. The argument is similar to the one outlined in 2.1.8 (cf. [BW2] Thm. 4.13). The case $g = \sum x_i^2$ and $2m \mid \deg(f)$ is already in [Re1] (Thm. 3.15).

3. Nichtnegativstellensätze.

We now allow positive functions to have zeros, and we'll try to understand to what extent we can still find denominator-free representations (in terms of weighted sums of squares) Necessarily, the hypotheses will have to be more restrictive than in the previous section. For ease of exposition, we shall mainly stay in the 'geometric' setting (of finitely generated $\mathbb{R}$-algebras), although many of the results allow 'abstract' generalizations. For these, the reader is referred to the literature cited. Again, we stress the fact that the results depend on the archimedean property of the real closed base field.

3.1. General results.

Let V be an affine $\mathbb{R}$-variety (see 2.0 for general conventions). Let T be a finitely generated preordering of $\mathbb{R}[V]$ for which the set $K := \mathscr{S}(T)$ in $V(\mathbb{R})$ is compact, and let $f \in \mathbb{R}[V]$ with $f|_K \geq 0$. If $f|_K > 0$, then we know $f \in T$ by Schmüdgen's Positivstellensatz 2.1.1. We wish to find conditions which allow the same conclusion even when f has zeros in K. A very useful general criterion is the following:

THEOREM 3.1.1 ([Sch4] Thm. 3.13). *Assume that $K = \mathscr{S}(T)$ is compact, and let $f \in \mathbb{R}[V]$ with $f|_K \geq 0$. Let I be the ideal in $\mathbb{R}[V]$ consisting of all functions that vanish on $\mathscr{Z}(f) \cap K$. If $f \in T + I^n$ for every $n \geq 0$, then $f \in T$.*

3.1.2. We explain the condition in 3.1.1. First, recall that $\mathscr{Z}(f)$ denotes the zero set of f in $V(\mathbb{R})$. Write $A := \mathbb{R}[V]$. By 1.3.14 we have $I = \sqrt{\operatorname{supp}(T + fA)}$. Let $W_0 = \operatorname{Spec}(A/I)$, the reduced Zariski closure of $\mathscr{Z}(f) \cap K$ in V. Given an ideal J of A, let $W := \operatorname{Spec}(A/J)$ denote the closed subscheme of $V = \operatorname{Spec} A$ corresponding to J, and call the preordering $T|_W := (T + J)/J$ of $A/J = \mathbb{R}[W]$ the *restriction* of T to W. Then the condition in the theorem says: $f|_W \in T|_W$ for every closed subscheme W of V with $W_{\text{red}} = W_0$.

Loosely speaking, the theorem says therefore that the condition $f \in T$ can be localized to infinitesimal thickenings of the Zariski closure of the zeros of f in $K = \mathscr{S}(T)$. Schmüdgen's Positivstellensatz is contained in 3.1.1 as the case $\mathscr{Z}(f) \cap K = \varnothing$, i.e., $I = (1)$.

3.1.3. In [Sch4], an 'abstract' generalization of 3.1.1 is proved, in which A is an arbitrary ring containing $\frac{1}{2}$, T is an archimedean preordering of A and $I = \sqrt{\operatorname{supp}(T + fA)}$. Such generalizations are needed if one works over base fields (or rings) other than real closed fields. See [Sch4] Thm. 3.19 for an application to sums of squares on curves over number fields.

A version (of the abstract form) of 3.1.1 for quadratic modules instead of preorderings is given in [Sch5] 2.4. The compactness condition for K has to be replaced by the condition that the module is archimedean.

Theorem 3.1.1 becomes particularly useful when f has only isolated zeros in K:

Corollary 3.1.4 ([Sch4] Cor. 3.16). *Let V be an affine $\mathbb{R}$-variety and T a finitely generated preordering of $\mathbb{R}[V]$ for which $K = \mathscr{S}(T)$ is compact. Let $f \in \mathbb{R}[V]$ with $f|_K \geq 0$, and assume that f has only finitely many zeros $z_1, \ldots, z_r$ in K. If $f \in \widehat{T}_{z_i}$ for $i = 1, \ldots, r$, then $f \in T$.*

Here $\widehat{T}_z$ denotes the preordering generated by T in the completed local ring $\widehat{\mathbb{R}[V]_{\mathfrak{m}_z}}$, where $\mathfrak{m}_z$ is the maximal ideal of $\mathbb{R}[V]$ corresponding to z. Note that $\widehat{\mathbb{R}[V]_{\mathfrak{m}_z}}$ is a power series ring $\mathbb{R}[\![x_1, \ldots, x_d]\!]$ if z is a nonsingular point of V, and is always a quotient of such a ring by some ideal.

Corollary 3.1.4 is a perfect local-global principle for membership in T: There is a local condition corresponding to each zero of f in $K = \mathscr{S}(T)$.

Remarks 3.1.5.

1. For example, if $p \in \mathbb{R}[x_1, \ldots, x_n]$ is such that $K = \{p \geq 0\}$ is compact, if a polynomial f with $f|_K \geq 0$ is given which has only finitely many zeros $z_1, \ldots, z_r$ in K, and if $f|_{\partial K} > 0$ and $D^2 f(z_i) > 0$ for $i = 1, \ldots, r$, then f can be written $f = s + tp$ with s, t are sums of squares of polynomials ([Sch4] 3.18). For a stronger statement see 3.1.7 below.

2. Let z be a point in $V(\mathbb{R})$, let $\mathfrak{m}$ denote the maximal ideal of the completed local ring $\widehat{\mathbb{R}[V]_{\mathfrak{m}_z}}$. The local condition $f \in \widehat{T}_z$ can be checked algorithmically. Indeed, one can determine an integer N such that $f \in \widehat{T}_z + \mathfrak{m}^N$ implies $f \in \widehat{T}_z$. Testing whether $f \in \widehat{T}_z + \mathfrak{m}^N$ holds is a matter of linear algebra, which can be reformulated as a linear matrix inequality (LMI).

3.1.6. A local-global principle in the spirit of 3.1.4, but for a quadratic module M instead of a preordering T, was proved in [Sch5] (Thm. 2.8). Since M is not supposed to be multiplicatively closed, one needs to formulate the hypotheses in a stronger way. Instead of K compact one has to assume that M is archimedean, and instead of $|\mathscr{Z}(f) \cap K| < \infty$ one needs that the ideal $\operatorname{supp}(M + (f))$ has dimension ≤ 0. This last property is somewhat hard to control directly, but Prop. 3.4 of *loc. cit* provides sufficient geometric conditions which imply this property. Rather than including them here, we state a particularly useful and applicable case which was exhibited by Marshall:

THEOREM 3.1.7 ([Ma4] Thm. 2.3). *Let V be an affine $\mathbb{R}$-variety, and let M be a finitely generated quadratic module in $\mathbb{R}[V]$ which is archimedean. Let $f \in \mathbb{R}[V]$ such that $f \geq 0$ on $K = \mathscr{S}(M)$. For every $z \in \mathscr{Z}(f) \cap K$, assume that there are $t_1, \dots, t_m \in M$ satisfying*

(1) $t_1, \dots, t_m$ are part of a regular parameter system of V at z;[20]

(2) $(df)(z) = a_1(dt_1)(z) + \cdots + a_m(dt_m)(z)$ with real numbers $a_i > 0$;

(3) the restriction of f to the subvariety $t_1 = \cdots = t_m = 0$ of V has positive definite Hessian at z.

Then $f \in M$.

If $t_1, \dots, t_n$ (with $n \geq m$) is a regular parameter system of V at z, then (2) means that the linear term of the Taylor expansion of f by the t_i is $\sum_{i=1}^m a_i t_i$.

REMARKS 3.1.8.

1. Conditions (1)–(3) are called the *boundary Hessian conditions* in [Ma4]. Theorem 3.1.7 can be obtained as a direct application of [Sch5] Prop. 3.4(2). (The proof in [Ma4] proceeds in a different and more complicated way.)

2. For a sample application, let $f \in \mathbb{R}[x_1, \dots, x_n]$ be a polynomial for which the set $K = \mathscr{S}(f)$ in $\mathbb{R}^n$ is compact and convex. If $z = (z_1, \dots, z_n)$ is a non-degenerate boundary point of K (that is, $f(z) = 0$, $\frac{\partial f}{\partial x_i}(z) \neq 0$ for some i and the Hessian $D^2 f(z)$ negative definite), and $l = \sum_i \frac{\partial f}{\partial x_i}(z) \cdot (x_i - z_i)$ is the equation of the tangent hyperplane at z, then there is an identity $l = s + s'f$ with sums of squares s, s' in $\mathbb{R}[x_1, \dots, x_n]$. (The condition on the Hessian can be weakened, according to (3) in 3.1.7.)

3.1.9. The basic tool on which Theorem 3.1.1 and all its consequences are built is [Sch4] Prop. 2.5 (resp. [Sch5] Prop. 2.1 for quadratic modules). The original proof of this key result is based on Stone-Weierstraß approximation and some (easy) topological arguments. Meanwhile, other approaches have been found. One is due to Kuhlmann-Marshall-Schwartz and Marshall and relies on the 'Basic Lemma' 2.1 from [KMS] (see [KMS] Cor. 2.5, [Ma4] Thm. 1.3). Another one is due to Schweighofer and is based on a refined analysis of Pólya's theorem 2.2.1. Instead of mentioning [Sch4] Prop. 2.5 here, we prefer to state Schweighofer's version, which is significantly more general:

THEOREM 3.1.10 ([Sw4] Thm. 2). *Let A be a ring and P an archimedean preprime in A with $\frac{1}{q} \in P$ for some integer $q > 1$. Let*

[20] So z is, in particular, assumed to be a nonsingular point of V.

$X(P) := \{\alpha \in \mathrm{Hom}(A, \mathbb{R}) \colon \alpha|_P \geq 0\} = \mathscr{X}_P^{\max}$, *and let* $f \in A$ *with* $f \geq 0$ *on* $X(P)$. *Suppose there is an identity*

$$f = b_1 t_1 + \cdots + b_r t_r$$

with $b_i \in A$ *and* $t_i \in P$ *such that* $b_i > 0$ *on* $X(P) \cap \{f = 0\}$ $(i = 1, \ldots, r)$. *Then* $f \in P$.

Note that this theorem extends (the preprimes version of) the Representation Theorem: If $f > 0$ on $X(P)$, then $f = f \cdot 1$ is an identity as required in 3.1.10. As a (non-obvious) application of this criterion, Schweighofer gave a new proof to the following theorem, originally due to Handelman [Ha2]:

THEOREM 3.1.11. *Let* $f \in \mathbb{R}[x_1, \ldots, x_n]$ *be a polynomial such that* f^m *has non-negative coefficients for some* $m \geq 1$. *If* $f(1, \ldots, 1) > 0$, *then* f^r *has non-negative coefficients for all sufficiently large integers* r.

(Of course, the condition $f(1, \ldots, 1) > 0$ is only needed when m is even, and ensures that f is positive (instead of negative) on the open positive orthant.)

We now discuss a second powerful local-global principle. For simplicity, we give the formulation only in the geometric case and only for preorderings. (See [Sch7] for more general versions.)

THEOREM 3.1.12 ([Sch7] Cor. 2.10). *Let* V *be an affine* $\mathbb{R}$*-variety, let* T *be a finitely generated preordering of* $\mathbb{R}[V]$, *and assume that* $K = \mathscr{S}(T)$ *is compact. Let* $f \in \mathbb{R}[V]$ *with* $f \geq 0$ *on* K. *For every maximal ideal* $\mathfrak{m}$ *of* $\mathbb{R}[V]$ *with* $(f) + \mathrm{supp}(T) \subset \mathfrak{m}$, *assume that* $f \in T_{\mathfrak{m}}$. *Then* $f \in T$.

Here $T_{\mathfrak{m}}$ denotes the preordering generated by T in the local ring $\mathbb{R}[V]_{\mathfrak{m}}$. In particular, one deduces a local-global principle for saturatedness ([Sch7] Cor. 2.9):

COROLLARY 3.1.13. *Let* V *and* T *be as in 3.1.12. Then* T *is saturated if and only if* $T_{\mathfrak{m}}$ *is saturated (as a preordering of* $\mathbb{R}[V]_{\mathfrak{m}}$*) for every maximal ideal* $\mathfrak{m}$ *of* $\mathbb{R}[V]$.

It is interesting to compare 3.1.12 to the local-global principles mentioned earlier in this section. By the general result 3.1.1, the question whether f is in T gets localized to (nonreduced) 'thickenings' of the Zariski closure of $\mathscr{Z}(f) \cap K$. In general, these are still 'global' schemes. The most favorable case is when $\mathscr{Z}(f) \cap K$ is a finite set; then the question can be reduced to finitely many completed local rings (3.1.4). On the other hand, 3.1.12 always reduces the question to local rings, but not usually to complete local rings, and not just to local rings belonging to points in K, not even to local rings belonging to real points. (See [Sch7] 2.12 for why one has to take maximal ideals with residue field $\mathbb{C}$ into account.) Therefore, while both results give the global conclusion $f \in T$ from local versions thereof, they are in general not comparable.

The proof of Theorem 3.1.12 uses the 'Basic Lemma' 2.1 from [KMS] in a suitably generalized version ([Sch7] 2.4, 2.8). Applications of 3.1.12 will be given in Section 3.3.

We conclude this section with a negative result of general nature. It says that for a semi-algebraic set K of dimension at least three, there cannot be an unconditional Nichtnegativstellensatz 'without denominators'. This was realized quite early:

PROPOSITION 3.1.14 ([Sch2] Prop. 6.1). *Let V be an affine $\mathbb{R}$-variety and T a finitely generated preordering of $\mathbb{R}[V]$ for which the set $K = \mathscr{S}(T)$ has dimension ≥ 3. Then there exists $f \in \mathbb{R}[V]$ with $f \geq 0$ on $V(\mathbb{R})$ and $f \notin T$. In particular, T is not saturated.*

The proof is not hard. Basically, the reason is that there exist psd ternary forms that are not sums of squares of forms, like the Motzkin form. By 3.1.14, general denominator-free Nichtnegativstellensätze can only exist in dimensions at most two. Therefore we will now take a closer look at curves and surfaces.

3.2. Curves. 3.2.1. By an *affine curve* over a field k, we mean here an affine k-variety C (see 2.0) which is purely of Krull dimension one. Let us consider 3.1.4 in more detail in the one-dimensional case. If $V = C$ is an affine curve over $\mathbb{R}$ and $f \in \mathbb{R}[C]$, the assumption $|\mathscr{Z}(f) \cap K| < \infty$ of 3.1.4 is almost always satisfied. In particular, it holds whenever f is not a zero divisor in $\mathbb{R}[C]$. Therefore, 3.1.4 furnishes a characterization of the elements of T by local conditions, provided that K is compact.

3.2.2. These results can in fact be extended to cases where K is not necessarily compact. It suffices that K is virtually compact. We explain the notion of virtual compactness in the particular case where the curve C is integral (i.e., $\mathbb{R}[C]$ is an integral domain), and refer to [Sch4] for the general case:

Let C be an integral curve over $\mathbb{R}$. A closed semi-algebraic subset K of $C(\mathbb{R})$ is *virtually compact* if the curve C admits a Zariski open embedding into an affine curve C' such that the closure of K in $C'(\mathbb{R})$ is compact. It is equivalent that $B_C(K) \neq \mathbb{R}$ (see 2.4.7), i.e., there exists a non-constant function in $\mathbb{R}[C]$ which is bounded on K.

THEOREM 3.2.3 ([Sch4] Thm. 5.5). *Let C be an affine curve over $\mathbb{R}$, and let T be a finitely generated preordering of $\mathbb{R}[C]$ for which $K = \mathscr{S}(T)$ is virtually compact. Let $f \in \mathbb{R}[C]$ with $f|_K \geq 0$, and assume that f has only finitely many zeros $z_1, \ldots, z_r$ in K. If $f \in \widehat{T}_{z_i}$ for $i = 1, \ldots, r$, then $f \in T$.*

REMARK 3.2.4. Given f with $f|_K \geq 0$, the local condition $f \in \widehat{T}_z$ needs only to be checked in those points $z \in \mathscr{Z}(f) \cap K$ which are either singular points of the curve C or boundary points of the set K. Otherwise it holds automatically.

Another consequence is the characterization of saturated one-dimensional preorderings. We state here the nonsingular case, and refer to [Sch4] Thm. 5.15 for the general situation:

COROLLARY 3.2.5. *Let C be a nonsingular irreducible affine curve over $\mathbb{R}$, and let $T = PO(f_1, \dots, f_r)$ with $f_i \in \mathbb{R}[C]$. Assume that $K = \mathscr{S}(T)$ is virtually compact. Then the preordering T is saturated if and only if the following two conditions hold:*

(1) *For every boundary point z of K there is an index i with $\mathrm{ord}_z(f_i) = 1$;*
(2) *for every isolated point z of K there are indices i, j with $\mathrm{ord}_z(f_i) = \mathrm{ord}_z(f_j) = 1$ and $f_i f_j \le 0$ in a neighborhood of z on $C(\mathbb{R})$.* □

To summarize, the characteristic feature of virtually compact one-dimensional sets K is that (finitely generated) preorderings T with $\mathscr{S}(T) = K$ contain every function which they contain locally (in the completed local rings at points of K). Such preorderings are therefore as big as they are allowed by the local conditions.

Interestingly, the situation is *very* different in the remaining one-dimensional cases, namely when K is not virtually compact (with the exception of rational curves, see below):

THEOREM 3.2.6 ([Sch2] Thm. 3.5). *Let C be an irreducible nonsingular affine curve over $\mathbb{R}$ which is not rational. Let T be a finitely generated preordering of $\mathbb{R}[C]$, and assume that $K = \mathscr{S}(T)$ is not virtually compact. Then there exists $f \in \mathbb{R}[C]$ with $f \ge 0$ on $C(\mathbb{R})$ such that f is not contained in T. In particular, T is not saturated.*

The condition that C is not rational eliminates the case where C is $\mathbb{A}^1$ (the affine line) minus a finite set S of real points. See below for what happens in this case.

As a consequence of 3.2.6, we can characterize the one-dimensional sets K whose saturated preordering $\mathscr{P}(K)$ is finitely generated (under the restriction that K is contained in a nonsingular curve):

COROLLARY 3.2.7. *Let C be an irreducible nonsingular affine curve over $\mathbb{R}$ and K a closed semi-algebraic subset of $C(\mathbb{R})$.*

(a) The preordering $\mathscr{P}(K)$ is finitely generated if and only if K is virtually compact or C is an open subcurve of $\mathbb{A}^1$.
(b) If K is virtually compact, then $\mathscr{P}(K)$ can be generated by two elements, and even by a single element if K has no isolated points.

(b) follows from 3.2.5. On the other hand, when $C \subset \mathbb{A}^1$ and K is not virtually compact, the preordering $\mathscr{P}(K)$ is finitely generated but may need an arbitrarily large number of generators. There is an (elementary) analogue of 3.2.5 in this case, see [Sch4] 5.23 or [KM] Sect. 2.

The significance of the condition that $\mathscr{P}(K)$ is finitely generated is, of course, that it implies an unconditional Nichtnegativstellensatz without denominators for the set K.

3.2.8. In [Sch4] 5.21, a (partial) generalization of 3.2.7 is given for singular integral curves C and virtually compact sets K. It characterizes finite generation of $\mathscr{P}(K)$ (together with the number of generators) in terms of local conditions at the finitely many boundary points and singular points of K. A sample application is the following (*loc. cit.*, 5.26): Assume that C has no real singularities other than ordinary double points. If every double point in K is either an isolated or an interior point of K, then $\mathscr{P}(K)$ can be generated by at most four elements. If K contains a double point which is neither isolated nor interior, then $\mathscr{P}(K)$ cannot be finitely generated.

For the preordering of all sums of squares, stronger results have been proved:

THEOREM 3.2.9 ([Sch4] Thm. 4.17). *Let C be an integral affine curve over $\mathbb{R}$ which is not rational and for which $C(\mathbb{R})$ is not virtually compact. Then there exists $f \in \mathbb{R}[C]$ with $f > 0$ on $C(\mathbb{R})$ such that f is not a sum of squares.*

An essential ingredient for the proofs of Theorems 3.2.6 and 3.2.9 is an analysis of the Jacobians of (projective) curves over $\mathbb{R}$.

In particular, one can give a characterization of all irreducible affine curves C on which the preordering of sums of squares is saturated, i.e., on which every psd regular function is a sum of squares. We'll express this briefly by saying that 'psd = sos holds on C':

COROLLARY 3.2.10. *Let C be an irreducible affine curve over $\mathbb{R}$. Then* psd = sos *holds on C in each of the following cases:*

(1) $C(\mathbb{R}) = \varnothing$;
(2) C is an open subcurve of the affine line $\mathbb{A}^1$;
(3) C is reduced, $C(\mathbb{R})$ is virtually compact, and all real singular points are ordinary multiple points with independent tangents.

In all other cases we have psd $\neq$ sos *on C. That is, if at least one of the following conditions holds:*

(4) $C(\mathbb{R}) \neq \varnothing$ and C is not reduced;
(5) C is not an open subcurve of $\mathbb{A}^1$ and is not virtually compact;
(6) C has a real singular point which is not an ordinary multiple point with independent tangents.

The proof (see [Sch4] 4.18) uses the results discussed before, together with a study of sums of squares in one-dimensional local rings (Remark 4.6.3 below).

Note that if the curve C is integral and $C(\mathbb{R})$ has no isolated points, then the preordering of all psd regular functions on C is equal to $\mathbb{R}[C] \cap \Sigma\mathbb{R}(C)^2$, where $\mathbb{R}(C) = \operatorname{Quot}\mathbb{R}[C]$ is the function field of C. Therefore, the condition psd = sos on C means that every regular function on C which is a sum of squares of rational functions is in fact a sum of squares of regular functions.

Further results on finite generation of $\mathscr{P}(K)$ are contained in the thesis [Pl], in particular in the case when the curve C has several irreducible components. Summarizing, it seems fair to say that preorderings whose associated semi-algebraic set is one-dimensional are rather well understood.

3.2.11. We end the section on curves with the following very useful fact. Let C be an integral affine curve over $\mathbb{R}$ and M a finitely generated quadratic module in $\mathbb{R}[C]$ for which $K = \mathscr{S}(M)$ is compact. If K doesn't contain a singular point of C then M is automatically a preordering, i.e., closed under products. This is [Sch5] Cor. 4.4. Of course, the result doesn't extend to higher dimensions.

3.3. Surfaces. Given Hilbert's result on the existence of positive polynomials in $n \geq 2$ variables which are not sums of squares of polynomials, it came initially as a surprise that there exist two-dimensional semi-algebraic sets, both compact and non-compact ones, which allow an unconditional Nichtnegativstellensatz without denominators. Here we record the most important results in this direction. Although quite a bit is known by now, the understanding of the two-dimensional situation is still less complete than for one-dimensional sets.

In this section, by an *affine surface* over a field k we will always mean an *integral* affine k-variety V of Krull dimension two (integral meaning that $k[V]$ is a domain).

THEOREM 3.3.1 ([Sch7] Cor. 3.4). *Let V be a nonsingular affine surface over $\mathbb{R}$ for which $V(\mathbb{R})$ is compact. Then* psd = sos *holds on V: Every $f \in \mathbb{R}[V]$ with $f \geq 0$ on $V(\mathbb{R})$ is a sum of squares in $\mathbb{R}[V]$.*

This follows from the localization principle 3.1.12 and from the fact that psd = sos holds in every two-dimensional regular local ring (see 4.7 below). The result becomes false in general for singular V, even (surprisingly) when all real points of V are regular points ([Sch7] 3.8).

Theorem 3.3.1 has been generalized to compact basic closed sets K whose boundary is sufficiently regular. We give an example in the plane:

EXAMPLE 3.3.2. ([Sch7] Cor. 3.3) Let $p_1, \ldots, p_r$ be irreducible polynomials in $\mathbb{R}[x, y]$ such that $K = \mathscr{S}(p_1, \ldots, p_r)$ is compact. Assume that none of the curves $p_i = 0$ has a real singular point, that the real intersection points of any two of these curves are transversal, and that no three of them meet in a real point. Then the preordering $T = PO(p_1, \ldots, p_r)$ of $\mathbb{R}[x, y]$ is saturated.

See *loc. cit.* for other examples as well. Even cases like a polygon or a disk in the plane were initially quite unexpected. In order to apply 3.1.12 here, one needs to study the saturatedness of certain finitely generated preorderings in (regular) two-dimensional local rings. The case needed for 3.3.2 (generators are transversal and otherwise units) is an easy consequence of 4.7 below. But similar results have been proved as well (unpublished yet) in many cases where the boundary of K is much less regular. For this, a deeper study of saturation in local rings is necessary, as indicated in 4.16 and 4.17 below.

Closely related is the following result [Sch8]:

THEOREM 3.3.3. *Let V be a nonsingular affine surface, and let K be a basic closed compact subset of $V(\mathbb{R})$ which is regular (i. e., is the closure of its interior). Then the following are equivalent:*

(i) The saturated preordering $\mathscr{P}(K)$ in $\mathbb{R}[V]$ is finitely generated;

(ii) the saturated preordering of the trace of $\widetilde{K}$ in Sper $\mathbb{R}[V]_{\mathfrak{m}_z}$ *is finitely generated in the local ring $\mathbb{R}[V]_{\mathfrak{m}_z}$, for every point $z \in \partial K$ which is a singular point of the (reduced) Zariski closure of ∂K.*

Its essence is that the characterization of the two-dimensional sets K for which $\mathscr{P}(K)$ is finitely generated is a local matter (under the above side conditions on V and K), and is decided in the singular points of the boundary curve of K. Combined with local results like 4.17 below, this implies statements in the spirit of, but more general than, 3.3.2.

3.3.4. Results similar to 3.3.1 or 3.3.2 can be proved in situations of arithmetical nature as well. For example, if k is a number field and V is a nonsingular affine surface over k such that $V(\mathbb{R}, \sigma)$ is compact for every real place $\sigma : k \to \mathbb{R}$, then psd = sos holds in the coordinate ring $k[V]$ ([Sch7] 3.10).

It is even possible, to some extent, to relax the compactness hypothesis for K. In principle this is similar to the one-dimensional case, where compact could be replaced by virtually compact for most results (Section 3.2). Not surprisingly, matters become more difficult in dimension two. So far, a full and systematic understanding has not yet been reached. We content ourselves here with giving two examples:

THEOREM 3.3.5. *Let W be a nonsingular affine surface over $\mathbb{R}$ with $W(\mathbb{R})$ compact, and let C be a closed curve*[21] *on W. Then $V = W - C$ is an affine surface, and* psd = sos *holds in $\mathbb{R}[V]$ as well.*

In other words, psd = sos holds on every affine $\mathbb{R}$-surface V which admits an open immersion into a nonsingular affine surface with compact set of real points.

EXAMPLE 3.3.6. ([Sch7] 3.16) The preordering $T = PO(x, 1-x, y, 1-xy)$ in $\mathbb{R}[x, y]$ is saturated, although the associated set $K = \mathscr{S}(T)$ in $\mathbb{R}^2$ is unbounded (of dimension two). This is derived by elementary arguments from the saturatedness of $PO(u-u^2, v-v^2)$ in $\mathbb{R}[u, v]$ (corresponding to the unit square), via the substitutions $u = x$, $v = xy$.

Both in 3.3.5 and in 3.3.6, the ring $B(K)$ of K-bounded polynomials (with $K = V(\mathbb{R})$ in the first case) is large in the sense that it has transcendence degree two over $\mathbb{R}$. The other extreme would be the case where

[21] That is, a closed subvariety of W all of whose irreducible components have dimension one.

$B(K)$ consists only of the constant functions. Here one expects that $\mathscr{P}(K)$ cannot be finitely generated. The following theorem is a step in this direction. It has analogues relative to suitable basic closed sets K on surfaces (see also [Sch2] Rem. 6.7).

THEOREM 3.3.7 ([Sch2] Thm. 6.4). *Let V be an affine surface over $\mathbb{R}$ which admits a Zariski open embedding into a nonsingular complete surface X such that $(X - V)(\mathbb{R})$ is Zariski dense in $X - V$. Then the preordering of all psd elements in $\mathbb{R}[V]$ is not finitely generated. In particular, $\mathbb{R}[V]$ contains psd elements which are not sums of squares.*

Unfortunately, the general picture is not yet well understood. As an example we mention the following notorious open question: Is the preordering T generated by $1-x^2$ in $\mathbb{R}[x, y]$ saturated? It is not even known if T contains every polynomial which is strictly positive on the strip $K = [-1, 1] \times \mathbb{R}$. Note that $B(K) = \mathbb{R}[x]$ has transcendence degree one here.

We conclude the section with an application to Hilbert's 17th problem:

COROLLARY 3.3.8 ([Sch7] 3.12). *Let f, $h \in \mathbb{R}[x, y, z]$ be two positive semidefinite ternary forms, where h is positive definite. Then there exists an integer $N \geq 1$ such that $h^N \cdot f$ is a sum of squares of forms.*

In particular, $(x^2+y^2+z^2)^N \cdot f(x, y, z)$ is a sum of squares of forms for all $N \gg 0$. The proof is an application of Theorem 3.3.1 to the complement of the curve $g = 0$ in the projective plane. The remarkable point is that the assertion is true even when f has (non-trivial) zeros. As mentioned in 2.1.9, the corollary becomes false for forms in more than three variables.

3.3.9. Corollary 3.3.8 says that every definite ternary form h is a 'weak common denominator' for Hilbert's 17th problem, in the sense that, for every psd ternary form f, a suitable power of h can be used as a denominator in a rational sums of squares decomposition of f. This fact is nicely complemented by the following result of Reznick [Re3]. It says that there cannot be any common denominator in the 'strong' sense:

THEOREM 3.3.10. *Let finitely many non-zero forms $h_1, \ldots, h_N$ in $\mathbb{R}[x_1, \ldots, x_n]$ be given, where $n \geq 3$. Then there exists a psd form f in $\mathbb{R}[x_1, \ldots, x_n]$ such that none of the forms $fh_1, \ldots, fh_N$ is a sum of squares of forms.*

4. Local rings, Pythagoras numbers. Apart from being interesting by their own, results on local rings are often important as tools in the study of rings of global nature. This is evident from results like 3.1.4 or 3.1.12. Since the emphasis of this survey is on results from the last 15 years, I won't try to summarize earlier work. Too much would have to be mentioned otherwise. The reader is recommended to consult the original article [CDLR] by Choi, Dai, Lam and Reznick, a classic on the topic of sums of squares in rings, and Pfister's book [Pf] on quadratic forms. One may also consult the survey article [Sch1] from 1991.

4.1. Let A be a (commutative) ring. Given $a \in A$, we write $\ell(a)$ for the sum of squares length of a, i.e., the least integer n such that a is a sum of n squares in A. If $a \notin \Sigma A^2$ we put $\ell(a) = \infty$. The *level* of A is $s(A) := \ell(-1)$. Note that the level of A is finite if and only if the real spectrum of A is empty (e.g. by 1.3.8).

The *Pythagoras number* of A is defined as

$$p(A) := \sup\{\ell(a) : a \in \Sigma A^2\}.$$

This is a very delicate invariant which has received considerable attention in number theory, the theory of quadratic forms, and in real algebra and real geometry. We refer to [Pf] for further reading and for links to the literature.

Recall (1.3.12) that an element of A is called psd if it is non-negative with respect to every ordering of A.

4.2. Regarding the study of Pythagoras numbers in general (commutative) rings, a wealth of information and ideas is contained in the important paper [CDLR] by Choi, Dai, Lam and Reznick. One of the main results of this paper was the construction of rings with infinite Pythagoras number, like $k[x, y]$ (k a real field) or $\mathbb{Z}[x]$. From this, the authors deduced that $p(A)$ is infinite whenever A has a real prime ideal $\mathfrak{p}$ such that the local ring $A_{\mathfrak{p}}$ is regular of dimension ≥ 3. On the other hand, the finiteness of the Pythagoras number was proved for a variety of real rings of dimension one or two. In particular, it was shown that for any affine curve C over a real closed field R, the Pythagoras number of $R[C]$ is finite. This number can be arbitrarily large, at least for singular C.

4.3. A classical theorem of Pfister says that if R is a real closed field and K/R is a non-real field extension of transcendence degree d, then $s(K) \leq 2^d$. This result has been generalized to rings by Mahe: Given an R-algebra A of transcendence degree d with $s(A) < \infty$, he proved $s(A) \leq 2^{d+1} + d - 4$ if $d \geq 3$, and $s(A) \leq 3$ resp. $s(A) \leq 7$ if $d = 1$ resp. $d = 2$ [Mh].

On the other hand, Pfister has shown that any field extension K/R of transcendence degree d has Pythagoras number $p(K) \leq 2^d$. This was extended to semilocal rings by Mahé: If A is a semilocal R-algebra of transcendence degree d, then every psd unit in A is a sum of 2^d squares ([Mh] 6.1). Recently, this was generalized to totally (strictly) positive non-zero divisors (not necessarily units) in [Sch3] 5.10.

Coming now to more recent results, we start by reviewing [Sch3]. All results from this paper are based on the following lemma:

LEMMA 4.4 ([Sch3] Thm. 2.2). *Let A be a semilocal ring containing $\frac{1}{2}$, and let f be a psd element in A. If f is a sum of squares modulo the ideal (f^2), then f is a sum of squares.*

The following theorem is the (semi-) local analogue of Theorem 3.1.1 (or rather, of the case $T = \Sigma A^2$ of this theorem):

THEOREM 4.5 ([Sch3] Thm. 2.5). *Let A be semilocal, $\frac{1}{2} \in A$, and let $f \in A$ be a psd element which is not a sum of squares. Then there is an ideal J with $\sqrt{J} = \sqrt[\mathrm{re}]{(f)}$ such that f is not a sum of squares in A/J.*

REMARKS 4.6.

1. It is easy to generalize 4.5 from sums of squares to arbitrary preorderings, similar to 3.1.2.

2. A case which is particularly useful is when A is noetherian and $\sqrt[\mathrm{re}]{(f)}$ contains $\mathrm{Rad}(A)$, the intersection of the maximal ideals of A: From 4.5 it follows that if f is psd and is a sum of squares in $A/\mathfrak{m}^n$ for every maximal ideal $\mathfrak{m}$ and sufficiently large $n \geq 1$, then f is a sum of squares in A ([Sch3] 2.7). This result is the local analogue of Corollary 3.1.4 (c.f. also Remark 3.1.5.2).

3. The paper [Sch3] consists of applications of 4.5. It particular, local rings are studied in which psd = sos holds. Theorem 3.9 analyzes the one-dimensional case. If $\dim A = 1$ and the residue field $k = A/\mathfrak{m}$ is real closed, the answer is completely understood.[22] Namely, psd = sos holds if and only if

$$\widehat{A} \cong k[\![x_1, \ldots, x_n]\!]/(x_i x_j \colon 1 \leq i < j \leq n)$$

for some n. For two-dimensional local rings, the main application of 4.5 is

THEOREM 4.7 ([Sch3] Thm. 4.8). *Let A be a regular semilocal ring of dimension two. Then* psd = sos *holds in A.*

On the other hand, the proof of 4.4 is sufficiently explicit to provide more information, in particular on quantitative questions. The main result in this direction is

THEOREM 4.8 ([Sch3] Thm. 5.25). *Let A be a regular local ring of dimension two, with quotient field K. Then $p(K) \leq 2^r$ implies $p(A) \leq 2^{r+1}$.*

In particular, if $p(K)$ is finite, then so is $p(A)$, which answers a question from [CDLR].

4.9. A series of important results has been obtained by Ruiz and Fernando. They mostly study (real) local analytic rings, i.e., rings of the form

$$A = \mathbb{R}\{x_1, \ldots, x_n\}/I = \mathbb{R}\{\mathrm{x}\}/I,$$

[22] Under the (weak) technical assumption that A is a Nagata ring, for example an excellent ring.

where $\mathbb{R}\{x\}$ is the ring of convergent power series and I is an ideal. This is the ring of analytic functions germs on the real analytic space germ X defined by I in $(\mathbb{R}^n, 0)$.

We first consider analytic space germs with the property psd = sos. The one-dimensional germs with this property are all determined by Remark 4.6.3: For any $n \geq 1$, the germ $X_n = \{x_i x_j = 0,\ 1 \leq i < j \leq n\}$ is the unique curve germ of embedding dimension n with this property. For surface germs one has the following result, due to Fernando and Ruiz:

THEOREM 4.10. *The (unmixed) singular analytic surface germs of embedding dimension three with the property* psd = sos *are exactly the germs* $z^2 = f(x, y)$, *where* f *is one of the following:*

$$x^2,\ x^2y,\ x^2y + (-1)^m y^m\ (m \geq 3),\ x^2 + y^m\ (m \geq 2),$$
$$x^3 + xy^3,\ x^3 + y^4,\ x^3 + y^5.$$

This is Theorem 1.3 in [Fe1]. It subsumes and completes earlier work from [Rz] and [FR].

4.11. Fernando has also found several series of (irreducible) surface germs of arbitrarily large embedding dimensions for which psd = sos holds ([Fe3] Sect. 4). On the other hand, germs of dimension ≥ 3 never have the property psd = sos (see 4.14 below).

4.12. We now consider germs with finite Pythagoras number. For curve germs, the Pythagoras number is bounded by the multiplicity ([Or], [CR], [Qz]). For surface germs, it is bounded by a function of the multiplicity and the embedding dimension [Fe2]. Underlying this is the following result by Fernando ([Fe2] Thm. 3.10):

THEOREM 4.13. *Let* $K = \operatorname{Quot} \mathbb{R}\{x\}$ *(the field of convergent Laurent series in one variable), and let* A *be one of* $K[y]$, $\mathbb{R}\{x\}[y]$ *or* $\mathbb{R}\{x, y\}$. *If* B *is an* A*-algebra which is generated by* m *elements as an* A*-module, then* $p(B) \leq 2m$.

The analogous result was known before for the polynomial ring $A = \mathbb{R}[t]$, see [CDLR] Thm. 2.7. Fernando's theorem gives, for the above list of rings, a positive answer to the so-called 'Strong Question' from *loc. cit.*

A surprising fact, for which no direct proof has been given, is that the list of surface germs in $\mathbb{R}^3$ with psd = sos is also the list of surface germs in $\mathbb{R}^3$ with Pythagoras number two ([Rz], [Fe3]).

Every analytic surface germ of real dimension ≥ 3 has infinite Pythagoras number. This is a particular case of the following main result of [FRS1]:

THEOREM 4.14. *Let* A *be an excellent ring of real dimension* ≥ 3. *Then* psd $\neq$ sos *in* A, *and the Pythagoras number of* A *is* ∞.

Here the real dimension of A is defined as follows: Given a specialization $\beta \rightsquigarrow \alpha$ in $\operatorname{Sper} A$, one puts

$$\dim(\beta \rightsquigarrow \alpha) := \dim\bigl(A_{\operatorname{supp}(\alpha)} / \operatorname{supp}(\beta) A_{\operatorname{supp}(\alpha)}\bigr),$$

and defines the *real dimension* of A as

$$\dim_r(A) := \sup\{\dim(\beta \rightsquigarrow \alpha) \colon \beta \rightsquigarrow \alpha \text{ in } \operatorname{Sper} A\}.$$

The main result of [FRS2], which we won't state here, is a far generalization of Theorem 4.13. Instead we give the application to germs of dimension two:

THEOREM 4.15 ([FRS2] Thm. 2.9). *Let A be an excellent henselian local ring of dimension two, with residue field k. If the rational function field $k(t)$ has finite Pythagoras number, then so has A.*

All fields of geometric or arithmetic origin[23] have finite Pythagoras number. So 4.15 says that $p(A)$ is finite whenever k is such a field.

The results from [FRS2] give in fact more precise information. For A as in 4.15, the completion $\widehat{A}$ is a finite $k[\![x,y]\!]$-algebra, by Cohen's structure theorem. If $\widehat{A}$ is generated by n elements as a $k[\![x,y]\!]$-module, then $p(A) \le 2n \cdot \tau(k)$, where $\tau(k) = \sup\{s(E) \colon E/k \text{ finite}, s(E) < \infty\}$ is a power of two satisfying $\tau(k) < p(k(t))$.

As an aside, we remark that it is a well-known open problem whether $p(k) < \infty$ implies $p(k(t)) < \infty$.

With 4.15, the understanding of the Pythagoras numbers of (excellent) henselian local rings has become quite precise.

4.16. Above we have mentioned results that characterize local rings with the property psd = sos. A natural generalization (which is important for geometrical applications, e. g. by 3.1.13 or 3.3.3) is the study of saturatedness of finitely generated preorderings of local rings.

Let A be a local ring, e. g. the local ring of an algebraic surface over $\mathbb{R}$ in a real point, and let T be a finitely generated preordering of A. Let $\widehat{A}$ be the completion of A, and let $\widehat{T}$ be the preordering generated by T in $\widehat{A}$. The question whether T is saturated in A usually appears quite intractable at first sight. Its analogue in the completed case, that is, the question whether $\widehat{T}$ is saturated in $\widehat{A}$, is often more accessible. Thus it would be nice to reduce the saturatedness question from T to $\widehat{T}$. In [Sch8], a series of results is proved which achieve this under certain conditions of geometric nature. On the easier side, T saturated implies $\widehat{T}$ saturated. To get back is usually harder. Instead of explicitly stating such results here (and thus necessarily getting more technical), we prefer to show just one particular case as an illustration:

PROPOSITION 4.17. *Let $f \in \mathbb{R}[x,y]$ be a polynomial with $f = f_2 + f_3 + \cdots$, where f_d is homogeneous of degree d. Suppose that $f_2 \ne 0$. Let $A = \mathbb{R}[x,y]_{(x,y)}$, the local ring at the origin, and let T be the preordering of A generated by f. Then T is saturated if and only if the quadratic form f_2 has a strictly positive eigenvalue.*

[23] A more precise formulation is: all fields of finite virtual cohomological dimension (see [FRS2] Rem. 2.2).

A similarly complete result (but with a more complicated condition) exists when $f_2 = 0$ and $f_3 \neq 0$. Of course, when such results are combined with the localization principle 3.1.12, they give global results like 3.3.2 in which stronger singularities are allowed for the boundary.

5. Applications to moment problems. Results on sums of squares in polynomial rings over $\mathbb{R}$ have applications to a classical branch of analysis, the moment problem. In fact, some of the results reviewed in Section 2 were originally found (and proved) in this analytic setting, as has already been mentioned.

This section is *not* meant to be a survey of recent work on moment problems. Very important and substantial work related to moment problems will not be touched and not even be mentioned here. Rather, my guideline has been to referee such work (briefly) which has a direct and concrete relation to the title of this paper, and more specifically, to the subjects discussed in sections 2 and 3.

For more material on moment problems, in particular for work on truncated moment problems and relations to optimization, see Laurent's survey [Lt] in this volume.

5.1. We will often abbreviate $\mathbb{R}[\mathrm{x}] := \mathbb{R}[x_1, \ldots, x_n]$. Given a closed subset K of $\mathbb{R}^n$, the *K-moment problem* asks for a characterization of all (multi-) sequences $(m_\alpha)_{\alpha \in \mathbb{Z}_+^n}$ of real numbers which can be realized as the moment sequence of some positive Borel measure on K. In other words, the question is to characterize the *K-moment functionals* on $\mathbb{R}[\mathrm{x}]$, that is, the linear forms $L\colon \mathbb{R}[\mathrm{x}] \to \mathbb{R}$ for which there exists a positive Borel measure μ on K which satisfies

$$L(f) = \int_K f(x)\, d\mu =: L_\mu(f)$$

for every $f \in \mathbb{R}[\mathrm{x}]$. Let $\mathscr{M}(K)$ denote the set of all these K-moment functionals (considered as a convex cone in the dual linear space $\mathbb{R}[\mathrm{x}]^*$).

By a classical theorem of Haviland, L is a K-moment functional iff $L(f) \geq 0$ for every $f \in \mathscr{P}(K)$.[24] Given any subset M of $\mathbb{R}[\mathrm{x}]$, let $M^\vee = \{L \in \mathbb{R}[\mathrm{x}]^*\colon L|_M \geq 0\}$ denote the dual cone of M. Thus, Haviland's theorem says $\mathscr{M}(K) = \mathscr{P}(K)^\vee$. The bi-dual of M is

$$M^{\vee\vee} = \{f \in \mathbb{R}[\mathrm{x}]\colon \forall\, L \in M^\vee\ L(f) \geq 0\}.$$

If M is a convex cone[25] in $\mathbb{R}[\mathrm{x}]$, then by the Hahn-Banach separation theorem, $M^{\vee\vee} = \overline{M}$, the closure of M with respect to the *natural linear topology* on $\mathbb{R}[\mathrm{x}]$. By definition, a subset A of $\mathbb{R}[\mathrm{x}]$ is closed in this topology if and only if $A \cap U$ is closed in U for every finite-dimensional linear subspace

[24] Recall (1.3.7) that $\mathscr{P}(K)$ denotes the saturated preordering associated with K.

[25] Which, in the terminology of 1.4.2, is the same as an $\mathbb{R}_+$-module.

U of $\mathbb{R}[\mathbf{x}]$. The natural linear topology is the finest locally convex topology on $\mathbb{R}[\mathbf{x}]$.

5.2. A convex cone M in $\mathbb{R}[\mathbf{x}]$ is said to *solve the K-moment problem* if $M^\vee = \mathscr{M}(K)$, or equivalently, if $\overline{M} = \mathscr{P}(K)$. We shall adopt a convenient terminology introduced by Schmüdgen: A convex cone M has the *strong moment property (SMP)* if $M^\vee \subset \mathscr{M}(K)$ holds, where $K := \mathscr{S}(M) = \{x \in \mathbb{R}^n : \forall f \in M\ f(x) \ge 0\}$. As remarked before, this is equivalent to $\overline{M} = \mathscr{P}(K)$. Here we will only consider cones M that are quadratic modules, and usually only the case where M is finitely generated as a quadratic module. Under this condition we have $\mathscr{P}(K) = \mathrm{Sat}(M)$, and hence (SMP) is also equivalent to M being dense in its saturation.

5.3. Given K, it is a classically studied problem from analysis to exhibit 'finite' solutions to the K-moment problem, i. e., to find quadratic modules M which are finitely generated and solve the K-moment problem.[26] If M is given by explicit generators $f_1, \dots, f_r$, the condition $L \in M^\vee$ translates into positivity conditions for an explicit countable sequence of symmetric matrices.

The question whether M solves the K-moment problem depends usually on M, and not just on K. In other words, there may exist finitely generated quadratic modules (or even preorderings) M_1, M_2 with $\mathscr{S}(M_1) = \mathscr{S}(M_2) = K$ such that M_1 solves the K-moment problem, but M_2 does not.

If the saturated preordering $\mathscr{P}(K)$ happens to be finitely generated, then $M = \mathscr{P}(K)$ is a finite solution to the K-moment problem. But as a rule, the finite generation hypothesis is rarely fulfilled, and can anyway only hold if $\dim(K) \le 2$ (see 3.1.14). See 3.2.7 and Section 3.3 for detailed information on when $\mathscr{P}(K)$ is finitely generated.

Schmüdgen's Positivstellensatz 2.1.1 gives a general (finite) solution to the moment problem in the compact case:

THEOREM 5.4 ([Sm1]). *If T is a finitely generated preordering of $\mathbb{R}[\mathbf{x}]$ for which $\mathscr{S}(T)$ is compact, then T has the strong moment property (SMP).*

In other words, if $f_1, \dots, f_r \in \mathbb{R}[\mathbf{x}]$ are such that $K := \mathscr{S}(f_1, \dots, f_r)$ is compact, then the preordering $PO(f_1, \dots, f_r)$ solves the K-moment problem.

Originally, the order of argumentation was reversed: Schmüdgen [Sm1] first proved 5.4, using Stengle's Positivstellensatz 1.2.4(a) and combining it with operator-theoretic arguments on Hilbert spaces. From 5.4, he subsequently derived the Positivstellensatz 2.1.1.

5.5. Implicitly, [Sm1] contains a stronger statement than 5.4, which is valid regardless of the compactness hypothesis. This was observed by

[26] Clearly, such M can only exist when K is a basic closed semi-algebraic set.

Netzer. Indeed, let T be any finitely generated preordering of $\mathbb{R}[\mathrm{x}]$, and let $K = \mathscr{S}(T)$. Let $B(K)$ be the ring of polynomials which are bounded on K. Then $B(K) \cap \mathscr{P}(K) \subset \overline{T}$ holds ([Ne] Thm. 2.2). Of course, if K is compact, this reduces to $\mathscr{P}(K) \subset \overline{T}$, which means that T solves the K-moment problem.

5.6. In the situation of 5.4, it is not in general true that the quadratic module $M := QM(f_1, \dots, f_r)$ solves the K-moment problem. A counter-example is given by 2.3.6 (see [PD] p. 155). However, M will certainly solve the K-moment problem if M is archimedean, since then M contains every $f \in \mathbb{R}[\mathrm{x}]$ with $f|_K > 0$ (Putinar's Positivstellensatz 2.3.2). Therefore, the results by Putinar and Jacobi–Prestel, reviewed in Section 2.3, give sufficient conditions for M to solve the K-moment problem. In particular, $r \le 2$ is such a sufficient condition.

A refinement which is very useful in optimization applications has recently been considered by several mathematicians. The idea is that there is a version of Putinar's theorem which is well adapted to structured sparsity of the defining polynomials.

We write $\mathrm{x} = (x_1, \dots, x_n)$, and write $\mathrm{x}_I := (x_i)_{i \in I}$ for each subset I of $\{1, \dots, n\}$. Assume that we have sets $I_1, \dots, I_r$ whose union is $\{1, \dots, n\}$ and which satisfy the following *running intersection property*:

$$\forall\, i = 2, \dots, r \;\; \exists\, j < i \;\; I_i \cap \bigcup_{k<i} I_k \subset I_j.$$

For each index $j = 1, \dots, r$, suppose that M_j is a finitely generated archimedean quadratic module in the polynomial ring $\mathbb{R}[\mathrm{x}_{I_j}]$, with associated compact set $K_j := \mathscr{S}(M_j)$ in $\mathbb{R}^{I_j}$. Let $K := \{x \in \mathbb{R}^n \colon x_{I_j} \in K_j,\ j = 1, \dots, r\} \subset \mathbb{R}^n$.

THEOREM 5.7. *If $f \in \mathbb{R}[x_1, \dots, x_n]$ is strictly positive on K, and if f is sparse in the sense that $f \in \sum_j \mathbb{R}[\mathrm{x}_{I_j}]$, then f can be written $f = f_1 + \cdots + f_r$ with $f_j \in M_j$ $(j = 1, \dots, r)$.*

The theorem as stated is essentially due to Lasserre [La3]. It goes back to ideas of Waki, Kim, Kojima and Muramatsu, who demonstrated the usefulness of such a decomposition for polynomial optimization by numerical implementations. The above formulation is taken from [GNS], where an elementary proof is given which only uses a weak form of 2.3.2. The result is also proved in [KP] in considerably greater generality.

5.8. Putinar and Vasilescu [PV] have given a different twist to moment problems, which works regardless of whether K is compact or not. To describe it, we follow an algebraic approach due to Marshall and Kuhlmann ([Ma2], [KM] Sect. 4). Let $f_1, \dots, f_r \in \mathbb{R}[\mathrm{x}]$ be given and put $K = \mathscr{S}(f_1, \dots, f_r)$ and $T = PO(f_1, \dots, f_r)$, as before. Embed $\mathbb{R}^n \hookrightarrow \mathbb{R}^{n+1}$ by

$$(x_1, \dots, x_n) \mapsto \Big(x_1, \dots, x_n, \frac{1}{1 + x_1^2 + \cdots + x_n^2}\Big).$$

Algebraically, this means adjoining $1/p$ to $\mathbb{R}[\mathrm{x}]$, where we write $p = 1 + \sum_{i=1}^n x_i^2$. Let T' be the preordering generated by $f_1, \dots, f_r$ in $\mathbb{R}\bigl[\mathrm{x}, 1/p\bigr]$. A linear functional $L\colon \mathbb{R}[\mathrm{x}] \to \mathbb{R}$ is a K-moment functional if and only if it extends to a linear functional

$$L'\colon \mathbb{R}\bigl[\mathrm{x}, 1/p\bigr] \to \mathbb{R}$$

with $L'|_{T'} \ge 0$ ([KM] Cor. 4.6). This shows that a finite characterization of the *extended* moment sequences

$$\int_K \frac{x^\alpha}{(1+\|x\|^2)^m}\, d\mu \quad (\alpha \in \mathbb{Z}_+^n,\ m \ge 0)$$

(instead of the usual ones $\int_K x^\alpha\, d\mu$) is always possible. But note that this usually doesn't give a finite solution to the original K-moment problem.

5.9. To get a better understanding of the (original) moment problem in the non-compact case, Kuhlmann and Marshall [KM] introduced several variants of the (SMP) condition which are slightly stronger. Let M be a finitely generated quadratic module in $\mathbb{R}[\mathrm{x}]$. Particularly interesting is the following condition on M:[27]

(‡) $\forall f \in \mathrm{Sat}(M)\ \exists g \in \mathbb{R}[\mathrm{x}]\ \forall \varepsilon > 0\ \ f + \varepsilon g \in M.$

Clearly, (‡) implies $\mathrm{Sat}(M) = \overline{M}$, that is, (SMP) for M. The question whether conversely (SMP) implies (‡) was open for some time, until Netzer showed that (SMP) is strictly weaker than (‡) (unpublished so far): The preordering $T = PO(1-x^2,\, x+y,\, 1-xy,\, y^3)$ of $\mathbb{R}[x,y]$ satisfies (SMP), but not (‡).

5.10. We now describe several results, both positive and negative ones, which address the finite solvability of the moment problem in non-compact cases. Let a basic closed set $K \subset \mathbb{R}^n$ be given. For studying the K-moment problem, one can replace affine n-space by the Zariski closure V of K without essentially affecting the question. Since geometric properties of V are inherent to the discussion anyway, it is preferable to start with an affine $\mathbb{R}$-variety V and a basic closed subset K of $V(\mathbb{R})$. On $\mathbb{R}[V]$ we consider the natural linear topology as in 5.2 above. We say that a finitely generated quadratic module M in $\mathbb{R}[V]$ has the strong moment property (SMP) if $\overline{M} = \mathrm{Sat}(M)$, and that the K-moment problem is finitely solvable if there exists such M with $\mathscr{S}(M) = K$.

For many sets K, the K-moment problem can be shown to have no finite solution at all. A large supply of such cases comes from the following result:

THEOREM 5.11 ([PoSch] Thm. 2.14). *Assume that V admits a Zariski open embedding into a complete normal $\mathbb{R}$-variety X for which $\overline{K} \cap (X - V)(\mathbb{R})$ is Zariski dense in $X - V$. Then every finitely generated quadratic module M in $\mathbb{R}[V]$ with $\mathscr{S}(M) = K$ is closed.*[28] *In particular, the K-*

[27] In [Sm2], condition (‡) was called (SPS).

[28] With respect to the natural linear topology.

moment problem is not finitely solvable, unless the saturated preordering $\mathscr{P}(K)$ *is finitely generated.*

5.12. Cases in which $\mathscr{P}(K)$ is not finitely generated are when $\dim(K) \geq 3$ (3.1.14), or when $K \subset \mathbb{R}^n$ contains a two-dimensional cone ([Sch2] 6.7). In the latter case, the K-moment problem is not finitely solvable by 5.11 (see [KM] 3.10). More generally, the same is true whenever K contains a piece of a nonrational algebraic curve which is not virtually compact (cf. 3.2.6). See [PoSch] 3.10, and also [Sch4] 6.6.

In fact, the only known case where the condition of 5.11 holds and $\mathscr{P}(K)$ is finitely generated is when $V \subset \mathbb{A}^1$.

5.13. Let A be a finitely generated $\mathbb{R}$-algebra and M a finitely generated quadratic module in A. If (SMP) holds for M, and if I is any ideal of A, then (SMP) holds for $M+I$ as well, and hence for the quadratic module $(M+I)/I$ in A/I ([Sch6] 4.8). This shows that (SMP) is inherited by restriction of quadratic modules to closed subvarieties of affine $\mathbb{R}$-varieties (in the sense of 3.1.2). Thus, if V is an affine $\mathbb{R}$-variety, if K is a basic closed set in $V(\mathbb{R})$, and if there is a closed subvariety W of V such that the moment problem for $K \cap W(\mathbb{R})$ is not finitely solvable, then the moment problem for K is not finitely solvable either.

5.14. On the other hand, there is a series of positive results. If T is a (finitely generated) preordering of $\mathbb{R}[\mathrm{x}]$ such that $K = \mathscr{S}(T)$ has dimension one and is virtually compact (3.2.2), then T contains every f with $f|_K > 0$ (Theorem 3.2.3), and, in particular, solves the K-moment problem ([Sch4] 6.3). Together with the results mentioned before, this means that for one-dimensional sets K, the solutions of the K-moment problem (by preorderings) are largely understood. (Parts of the above discussion apply only when the Zariski closure V of K is irreducible. For what can happen when V is a curve with several irreducible components, see [Pl].)

From results in 3.3 (see Example 3.3.6) we get finite solutions of the K-moment problem for certain non-compact two-dimensional sets K. But in fact there exist non-compact sets of arbitrary dimension with finitely solvable moment problem. A case in point is cylinders with compact cross-section, which is due to Kuhlmann and Marshall:

Proposition 5.15 ([KM] Thm. 5.1). *Let* $f_1, \ldots, f_r \in \mathbb{R}[x_1, \ldots, x_n]$ *such that the set* $K_0 = \mathscr{S}(f_1, \ldots, f_r)$ *in* $\mathbb{R}^n$ *is compact. Let* T *be the preordering generated by* $f_1, \ldots, f_r$ *in* $\mathbb{R}[x_1, \ldots, x_n, y]$. *Then* T *solves the moment problem for* $K = K_0 \times \mathbb{R} \subset \mathbb{R}^{n+1}$.

In fact, what is shown in [KM] is that T has property (‡) (see 5.9), which is one of the reasons for the interest in this property. In general, one cannot do much better, since there may exist polynomials $p \notin T$ with $p|_K \geq \varepsilon > 0$ ([KM] 5.2).

In 2003, Schmüdgen proved a far-reaching and powerful generalization of 5.15:

THEOREM 5.16 ([Sm2]). *Let $T = PO(f_1, \dots, f_r)$ be a finitely generated preordering of $\mathbb{R}[x_1, \dots, x_n]$, let $K = \mathscr{S}(T)$, and let $h = (h_1, \dots, h_m)\colon \mathbb{R}^n \to \mathbb{R}^m$ be a polynomial map for which $h(K)$ is bounded. For each $y \in \mathbb{R}^m$ consider the preordering*

$$\begin{aligned} T_y &:= T + (h_1 - y_1, \dots, h_m - y_m) \\ &= PO\big(f_1, \dots, f_r, \pm(h_1 - y_1), \dots, \pm(h_m - y_m)\big) \end{aligned}$$

in $\mathbb{R}[x_1, \dots, x_n]$. If T_y has property (SMP) for every y, then T has property (SMP).

By 5.13, the fibre conditions in 5.16 are not only sufficient (for T to have (SMP)), but also necessary. Note that $\mathscr{S}(T_y) = K \cap h^{-1}(y) =: K_y$, the fibre of y in K. Thus the theorem reduces the question whether T solves the K-moment problem to the fibres of h. These will be of smaller dimension than K (except in degenerate cases), thus opening the door for inductive reasoning.

5.17. Schmüdgen's proof of 5.13 uses deep methods from operator theory. A simpler and more elementary proof was later given by Netzer [Ne], which uses only measure-theoretic arguments. (A similar approach was found independently by Marshall.) Without doubt, it would be instructive to have a purely algebraic approach to this important result, as was the case in the past with Schmüdgen's Positivstellensatz 2.1.1. It seems not clear, however, if one can reasonably expect such a proof to exist.

As Netzer points out, both his and the original proof actually give a somewhat sharper version of 5.16, namely

$$\overline{T} = \bigcap_{y \in \mathbb{R}^m} \overline{T_y}.$$

(Note that $T_y = \mathbb{R}[x_1, \dots, x_n]$ unless $y \in h(\mathbb{R}^n)$.)

5.18. Theorem 5.16 does not allow any stronger conclusions on T, even if the fibres T_y satisfy stronger hypotheses. For example, if T_y contains, for every y, all polynomials that are strictly positive on K_y, it need not be true that T contains every polynomial which is strictly positive on K ([KM] 5.2). Also, T may satisfy (‡) fibrewise without satisfying (‡) globally, as shown by Netzer's example 5.9 (take for h the projection $(x, y) \mapsto x$). This answers a question raised in [Sm2].

Proposition 5.15 is just one of the simplest applications of 5.16. Another application of 5.16 is to one-dimensional sets K which are virtually compact. One finds that any T with $\mathscr{S}(T) = K$ solves the K-moment problem, a result we have already mentioned (in stronger form) in 5.14. Beyond these examples, however, there is a plethora of examples covered by Theorem 5.16 which are completely new.

5.19. Moment problems with symmetries have been studied in [CKS]. Given a basic closed set $K \subset \mathbb{R}^n$ which is invariant under a subgroup G of the general linear group, one may ask for characterizations of $\mathscr{M}(K)^G$, the set of K-moment functionals which are G-invariant, among all G-invariant linear functionals. It is shown that such finite characterizations can exist in situations where the unrestricted K-moment problem is unsolvable. On the other hand, an equivariant version of the negative result 5.11 is proved. All these results require that the group G is compact.

5.20. Given a basic closed subset K of $V(\mathbb{R})$ as before, let $B(K)$ be the subring of $\mathbb{R}[V]$ consisting of all functions that are bounded on K. All known results support the feeling that the 'size' of the ring $B(K)$ directly influences the (finite) solvability question of the K-moment problem. The idea is roughly that the K-moment problem can only be solvable if $B(K)$ is sufficiently large. As a crude measure for the size one can take the transcendence degree of $B(K)$ (as an algebra over $\mathbb{R}$), for example.

For a brief informal discussion, assume that K is Zariski dense in V (see 5.10) and V is irreducible. On the one extreme, $B(K)$ is as large as possible (namely equal to $\mathbb{R}[V]$) if and only if K is compact, in which case the moment problem is always solvable (5.4). Netzer's remark 5.5 and Schmüdgen's fibre criterion 5.16 also supports the above idea. If V is a nonsingular irreducible curve, not rational, then the K-moment problem is solvable if and only if $B(K) \neq \mathbb{R}$ (5.12, 5.14). In the situation of 5.11 one has $B(K) = \mathbb{R}$. No example seems to be known where the K-moment problem is solvable and $B(K) = \mathbb{R}$, except when V is a rational curve or a point. If V is a nonsingular surface, the unsolvability of the K-moment problem is proved in [Pl] under a condition which is slightly stronger than $B(K) = \mathbb{R}$ (negative definiteness of the intersection matrix of the divisor $X - U$, where $V \subset U \subset X$ is as in 2.4.7).

5.21. We briefly discuss the concept of stability. Let $M = QM(f_1, \dots, f_r)$ be a finitely generated quadratic module in $\mathbb{R}[\mathrm{x}]$. Then M is called *stable* if there exists a function $\phi\colon \mathbb{N} \to \mathbb{N}$ such that the following holds: For every $d \in \mathbb{N}$ and every $f \in M$ with $\deg(f) \le d$, there exists an identity $f = s_0 + s_1 f_1 + \cdots + s_r f_r$ with sums of squares s_i such that $\deg(s_i) \le \phi(d)$. The existence of such ϕ depends only on M, and not on the choice of the generators of M.

Given M in terms of a generating system $f_1, \dots, f_r$, there are two natural computational problems. The recognition problem is to decide whether a given polynomial lies in M. The realization problem is to find an identity $f = s_0 + \sum_{i=1}^r s_i f_i$ explicitly, provided it is known to exist, together with explicit sums of squares decompositions of the s_i. If M is stable then both problems are a priori bounded, and thus (at least in principle) accessible computationally. On the other hand, if M is not stable, both problems are expected to be computationally hard.

The notion of stability was introduced in [PoSch]. There is a series of results showing that the question whether M is stable is linked to the geometry of the set $K = \mathscr{S}(M)$, and also to the solvability of the K-moment problem. The most important results are:

THEOREM 5.22. *Let M be a finitely generated quadratic module in $\mathbb{R}[x]$. If M is stable, then $\overline{M} = M + \sqrt{\mathrm{supp}(M)}$ holds, and this quadratic module is again stable.*

See [PoSch] Cor. 2.11 and [Sch6] Thm. 3.17. In particular, the closure of a stable quadratic module is always finitely generated. The proof of Theorem 5.11 above was in fact given by first showing that the M discussed there is necessarily stable, and by then applying 5.22.

THEOREM 5.23 ([Sch6] Thm. 5.4). *Let M be a finitely generated quadratic module in $\mathbb{R}[x]$, let $K = \mathscr{S}(M)$. If M has the strong moment property and $\dim(K) \geq 2$, then M cannot be stable.*

This means, unfortunately, that except when $\dim(K) \leq 1$, M can only be stable when M is quite far from being saturated. For example, M can never be stable when M is archimedean and K has dimension ≥ 2. Interestingly, there are examples where M is stable and K is compact of dimension 2 ([Sch6] 5.7).

The condition (SMP) for M in 5.23 can even be weakened to a condition called (MP) (otherwise not discussed here), which requires that $\overline{M}$ contains all polynomials which are psd on $\mathbb{R}^n$.

5.24. The moment problem from analysis has classically two aspects, the existence question and the uniqueness question. We have only discussed the former here, since until recently the latter had not been related to positivity and sums of squares questions. Given a closed set $K \subset \mathbb{R}^n$ and a K-moment functional $L \in \mathscr{M}(K)$ on $\mathbb{R}[x]$, let us say that L is determinate (for K) if there exists a unique positive Borel measure μ on K representing L. If there is more than one such μ, L is called indeterminate.

If K is compact, it is classically known that every $L \in \mathscr{M}(K)$ is determinate. In [PuSch] it is shown that the same is true more generally if the ring $B(K)$ of K-bounded polynomials separates the points of K. For example, this applies when K is one-dimensional and is virtually compact (see 3.2.2). On the other hand, when $\dim(K) = 1$ and K is not virtually compact, it is shown under certain geometric conditions that there exist indeterminate K-moment functionals. It is an open question whether this is always true for K of dimension one and not virtually compact. Even less is so far known in higher dimensions.

REFERENCES

[A] E. ARTIN, *Über die Zerlegung definiter Funktionen in Quadrate.* Abh. Math. Sem. Univ. Hamburg **5**:100–115 (1927).

[BP] E. BECKER AND V. POWERS, *Sums of powers in rings and the real holomorphy ring.* J. reine angew. Math. **480**:71–103 (1996).

[BS] E. BECKER AND N. SCHWARTZ, *Zum Darstellungssatz von Kadison-Dubois.* Arch. Math. **40**:421–428 (1983).

[BChR] CH. BERG, J.P.R. CHRISTENSEN, AND P. RESSEL, *Positive definite functions on abelian semigroups.* Math. Ann. **223**:253–272 (1976).

[BW1] R. BERR AND TH. WÖRMANN, *Positive polynomials on compact sets.* Manuscr. math. **104**:135–143 (2001).

[BW2] R. BERR AND TH. WÖRMANN, *Positive polynomials and tame preorders.* Math. Z. **236**:813–840 (2001).

[Bl] G. BLEKHERMAN, *There are significantly more nonnegative polynomials than sums of squares.* Israel J. Math. **153**:355–380 (2006).

[BCR] J. BOCHNAK, M. COSTE, AND M.-F. ROY, *Real Algebraic Geometry.* Erg. Math. Grenzgeb. **36**(3), Springer, Berlin, 1998.

[Br] L. BRÖCKER, *Zur Theorie der quadratischen Formen über formal reellen Körpern.* Math. Ann. **210**:233–256 (1974).

[CR] A. CAMPILLO AND J.M. RUIZ, *Some remarks on pythagorean real curve germs.* J. Algebra **128**:271–275 (1990).

[CDLR] M.D. CHOI, Z.D. DAI, T.Y. LAM, AND B. REZNICK, *The Pythagoras number of some affine algebras and local algebras.* J. reine angew. Math. **336**: 45–82 (1982).

[CL] M.D. CHOI AND T.Y. LAM, *An old question of Hilbert.* In: Conf. Quadratic Forms (Kingston, Ont., 1976), ed. G. Orzech, Queen's Papers Pure Appl. Math. **46**:385–405 (1977).

[CKS] J. CIMPRIČ, S. KUHLMANN, AND C. SCHEIDERER, *Sums of squares and moment problems in equivariant situations.* Trans. Am. Math. Soc., 2007 (to appear).

[De] CH.N. DELZELL, *A constructive, continuous solution to Hilbert's* 17th *problem, and other results in semi-algebraic geometry.* Doctoral Dissertation, Stanford University, 1980.

[Fe1] J.F. FERNANDO, *Positive semidefinite germs in real analytic surfaces.* Math. Ann. **322**:49–67 (2002).

[Fe2] J.F. FERNANDO, *On the Pythagoras numbers of real analytic rings.* J. Algebra **243**:321–338 (2001).

[Fe3] J.F. FERNANDO, *Analytic surface germs with minimal Pythagoras number.* Math. Z. **244**:725–752 (2003); Erratum, *ibid.*, **250**:967–969 (2005).

[FR] J.F. FERNANDO AND J.M. RUIZ, *Positive semidefinite germs on the cone.* Pacific J. Math. **205**:109–118 (2002).

[FRS1] J.F. FERNANDO, J.M. RUIZ, AND C. SCHEIDERER, *Sums of squares in real rings.* Trans. Am. Math. Soc. **356**:2663–2684 (2003).

[FRS2] J.F. FERNANDO, J.M. RUIZ, AND C. SCHEIDERER, *Sums of squares of linear forms.* Math. Res. Lett. **13**:947–956 (2006).

[GNS] D. GRIMM, T. NETZER, AND M. SCHWEIGHOFER, *A note on the representation of positive polynomials with structured sparsity.* Preprint 2006.

[Ha1] D. HANDELMAN, *Representing polynomials by positive linear functions on compact convex polyhedra.* Pacific J. Math. **132**:35–62 (1988).

[Ha2] D. HANDELMAN, *Polynomials with a positive power.* In: Symbolic dynamics and its applications (New Haven, CT, 1991), Contemp. Math. **135**, Am. Math. Soc., Providence, RI, 1992, pp. 229–230.

[HMPV] J.W. HELTON, S. MCCULLOUGH, M. PUTINAR, AND V. VINNIKOV, *Convex matrix inequalities versus linear matrix inequalities.* Preprint 2007.

[HP] J.W. Helton and M. Putinar, *Positive polynomials in scalar and matrix variables, the spectral theorem and optimization.* Preprint 2006.

[H1] D. Hilbert, *Über die Darstellung definiter Formen als Summe von Formenquadraten.* Math. Ann. **32**:342–350 (1888).

[H2] D. Hilbert, *Über ternäre definite Formen.* Acta math. **17**:169–197 (1893).

[Ja1] Th. Jacobi, *A representation theorem for certain partially ordered commutative rings.* Math. Z. **237**:259–273 (2001).

[Ja2] Th. Jacobi, *Über die Darstellung positiver Polynome auf semi-algebraischen Kompakta.* Doctoral Dissertation, Universität Konstanz, 1999.

[JP] Th. Jacobi and A. Prestel, *Distinguished representations of strictly positive polynomials.* J. reine angew. Math. **532**:223–235 (2001).

[KS] M. Knebusch and C. Scheiderer, *Einführung in die reelle Algebra.* Vieweg, Wiesbaden, 1989.

[Ko] J. Kollár, *Sharp effective Nullstellensatz.* J. Am. Math. Soc. **1**:963–975 (1988).

[Kr] J.L. Krivine, *Anneaux préordonnés.* J. Analyse Math. **12**:307–326 (1964).

[KM] S. Kuhlmann and M. Marshall, *Positivity, sums of squares and the multidimensional moment problem.* Trans. Am. Math. Soc. **354**:4285–4301 (2002).

[KMS] S. Kuhlmann, M. Marshall, and N. Schwartz, *Positivity, sums of squares and the multi-dimensional moment problem II.* Adv. Geom. **5**:583–606 (2005).

[KP] S. Kuhlmann and M. Putinar, *Positive polynomials on projective limits of real algebraic varieties.* Preprint, IMA Preprint Series 2162, 2007.

[La1] J.B. Lasserre, *A sum of squares approximation of nonnegative polynomials.* SIAM J. Optim. **16**:751–765 (2006).

[La2] J.B. Lasserre, *Sum of squares approximation of polynomials, nonnegative on a real algebraic set.* SIAM J. Optim. **16**:610–628 (2006).

[La3] J.B. Lasserre, *Convergent SDP-relaxations in polynomial optimization with sparsity.* SIAM J. Optim. **17**:822–843 (2006).

[LN] J.B. Lasserre and T. Netzer, *SOS approximations of nonnegative polynomials via simple high degree perturbations.* Math. Z. **256**:99–112 (2007).

[Lt] M. Laurent, *Sums of squares, moment matrices and optimization over polynomials.* This volume.

[LS] J.A. de Loera and F. Santos, *An effective version of Pólya's theorem on positive definite forms.* J. Pure Applied Algebra **108**:231–240 (1996); Erratum, *ibid.* **155**:309–310 (2001).

[Mh] L. Mahé, *Level and Pythagoras number of some geometric rings.* Math. Z. **204**:615–629 (1990); Erratum, *ibid.* **209**:481–483 (1992).

[Ma1] M. Marshall, *A real holomorphy ring without the Schmüdgen property.* Canad. Math. Bull. **42**:354–358 (1999).

[Ma] M. Marshall, *Positive Polynomials and Sums of Squares.* Istituti Editoriali e Poligrafici Internazionali, Pisa, 2000.

[Ma2] M. Marshall, *Extending the archimedean Positivstellensatz to the noncompact case.* Canad. Math. Bull. **44**:223–230 (2001).

[Ma3] M. Marshall, *A general representation theorem for partially ordered commutative rings.* Math. Z. **242**:217–225 (2002).

[Ma4] M. Marshall, *Representation of non-negative polynomials with finitely many zeros.* Ann. Fac. Sci. Toulouse (6) **15**:599–609 (2006).

[Mo] J.-Ph. Monnier, *Anneaux d'holomorphie et Positivstellensatz archimédien.* Manuscr. Math. **97**:269–302 (1998).

[Ne] T. Netzer, *An elementary proof of Schmüdgen's theorem on the moment problem of closed semi-algebraic sets.* Preprint, 2006.

[Or] J. Ortega, *On the Pythagoras number of a real irreducible algebroid curve.* Math. Ann. **289**:111–123 (1991).

[Pf] A. PFISTER, *Quadratic Forms with Applications to Algebraic Geometry and Topology.* London Math. Soc. Lect. Notes **217**, Cambridge, 1995.

[Pi] S. PINGEL, *Der reelle Holomorphiering von Algebren.* Doctoral Dissertation, Fernuniversität Hagen, 1998.

[Pl] D. PLAUMANN, *Bounded polynomials, sums of squares and the moment problem.* Doctoral dissertation, Univ. Konstanz, 2008.

[PR1] V. POWERS AND B. REZNICK, *Polynomials that are positive on an interval.* Trans. Am. Math. Soc. **352**:4677–4692 (2000).

[PR2] V. POWERS AND B. REZNICK, *A new bound for Pólya's theorem with applications to polynomials positive on polyhedra.* J. Pure Applied Algebra **164**:221–229 (2001).

[PoSch] V. POWERS AND C. SCHEIDERER, *The moment problem for non-compact semialgebraic sets.* Adv. Geom. **1**:71–88 (2001).

[Pr] A. PRESTEL, *Lectures on Formally Real Fields.* Lect. Notes Math. **1093**, Springer, Berlin, 1984.

[PD] A. PRESTEL AND CH.N. DELZELL, *Positive Polynomials.* Monographs in Mathematics, Springer, Berlin, 2001.

[Pu] M. PUTINAR, *Positive polynomials on compact semi-algebraic sets.* Indiana Univ. Math. J. **42**:969–984 (1993).

[PuSch] M. PUTINAR AND C. SCHEIDERER, *Multivariate moment problems: Geometry and indeterminateness.* Ann. Sc. Norm. Pisa (5) **5**:137–157 (2006).

[PV] M. PUTINAR AND F.-H. VASILESCU, *Solving moment problems by dimensional extension.* Ann. Math. **149**:1097–1107 (1999).

[Qz] R. QUAREZ, *Pythagoras numbers of real algebroid curves and Gram matrices.* J. Algebra **238**:139–158 (2001).

[Re1] B. REZNICK, *Uniform denominators in Hilbert's seventeenth problem.* Math. Z. **220**:75–97 (1995).

[Re2] B. REZNICK, *Some concrete aspects of Hilbert's 17th problem.* Proc. RAGOS Conf., Contemp. Math. **253**:251–272 (2000).

[Re3] B. REZNICK, *On the absence of uniform denominators in Hilbert's 17th problem.* Proc. Am. Math. Soc. **133**:2829–2834 (2005).

[Re4] B. REZNICK, *On Hilbert's construction of positive polynomials.* Preprint, 2007.

[Rz] J.M. RUIZ, *Sums of two squares in analytic rings.* Math. Z. **230**:317–328 (1999).

[Scha] W. SCHARLAU, *Quadratic and Hermitian Forms.* Grundl. Math. Wiss. **270**, Springer, Berlin, 1985.

[Sch1] C. SCHEIDERER, *Real algebra and its applications to geometry in the last ten years: Some major developments and results.* In: Real algebraic geometry (Rennes 1991), M. Coste, L. Mahé, M.-F. Roy (eds.), Lect. Notes Math. **1524**, Springer, Berlin, 1992, pp. 75–96.

[Sch2] C. SCHEIDERER, *Sums of squares of regular functions on real algebraic varieties.* Trans. Am. Math. Soc. **352**:1039–1069 (1999).

[Sch3] C. SCHEIDERER, *On sums of squares in local rings.* J. reine angew. Math. **540**:205–227 (2001).

[Sch4] C. SCHEIDERER, *Sums of squares on real algebraic curves.* Math. Z. **245**:725–760 (2003).

[Sch5] C. SCHEIDERER, *Distinguished representations of non-negative polynomials.* J. Algebra **289**:558–573 (2005).

[Sch6] C. SCHEIDERER, *Non-existence of degree bounds for weighted sums of squares representations.* J. Complexity **21**:823–844 (2005).

[Sch7] C. SCHEIDERER, *Sums of squares on real algebraic surfaces.* Manuscr. math. **119**:395–410 (2006).

[Sch8] C. SCHEIDERER, *Local study of preorderings* (in preparation).

[Sd] J. SCHMID, *On the degree complexity of Hilbert's 17th problem and the real Nullstellensatz.* Habilitationsschrift, Universität Dortmund, 1998.

[Sm1] K. SCHMÜDGEN, *The K-moment problem for compact semi-algebraic sets.* Math. Ann. **289**:203–206 (1991).

[Sm2] K. SCHMÜDGEN, *On the moment problem of closed semi-algebraic sets.* J. reine angew. Math. **558**:225–234 (2003).

[Sm3] K. SCHMÜDGEN, *Noncommutative real algebraic geometry — some basic concepts and first ideas.* This volume.

[Sw1] M. SCHWEIGHOFER, *An algorithmic approach to Schmüdgen's Positivstellensatz.* J. Pure Applied Algebra **166**:307–319 (2002).

[Sw2] M. SCHWEIGHOFER, *Iterated rings of bounded elements and generalizations of Schmüdgen's Positivstellensatz.* J. reine angew. Math. **554**:19–45 (2003).

[Sw3] M. SCHWEIGHOFER, *On the complexity of Schmüdgen's Positivstellensatz.* J. Complexity **20**:529–543 (2004).

[Sw4] M. SCHWEIGHOFER, *Certificates for nonnegativity of polynomials with zeros on compact semialgebraic sets.* Manuscr. math. **117**:407–428 (2005).

[Sw5] M. SCHWEIGHOFER, *Global optimization of polynomials using gradient tentacles and sums of squares.* SIAM J. Optim. **17**:920–942 (2006).

[St1] G. STENGLE, *A nullstellensatz and a positivstellensatz in semialgebraic geometry.* Math. Ann. **207**:87–97 (1974).

[St2] G. STENGLE, *Complexity estimates for the Schmüdgen Positivstellensatz.* J. Complexity **12**:167–174 (1996).

[S] B. STURMFELS, *Solving Systems of Polynomial Equations.* CBMS Reg. Conf. Ser. Math., Am. Math. Soc., Providence, RI, 2002.

NONCOMMUTATIVE REAL ALGEBRAIC GEOMETRY - SOME BASIC CONCEPTS AND FIRST IDEAS

KONRAD SCHMÜDGEN*

Abstract. We propose and discuss how basic notions (quadratic modules, positive elements, semialgebraic sets, Archimedean orderings) and results (Positivstellensätze) from real algebraic geometry can be generalized to noncommutative $*$-algebras. A version of Stengle's Positivstellensatz for $n \times n$ matrices of real polynomials is proved.

Key words. Noncommutative real algebraic geometry, quadratic module, sum of squares, $*$-representation, positivity, positive semidefinite matrices.

AMS(MOS) subject classifications. Primary: 13J30, 47L60, 15A48; Secondary: 11E25.

1. Introduction. In recent years various versions of noncommutative Positivstellensätze have been proved, about free polynomial algebras by J.W. Helton and his coworkers [9–11] and about the Weyl algebra [30] and enveloping algebras of Lie algebras [31] by the author. These results can be considered as very first steps towards a new mathematical field that might be called *noncommutative real algebraic geometry.*

The aim of this paper is to discuss how some basic concepts and results from real algebraic geometry should look like in the noncommutative setting. This article unifies a series of talks I gave during the past five years at various conferences (Pisa, Marseille, Palo Alto, Banff) and at other places. It should be emphasized that it represents the authors personal view and ideas on this topic. These concepts and ideas exist and will be presented at different levels of exactness and acceptance by the community. Some of them (quadratic modules, positivity by representations, Archimedean orderings) are more or less clear and accepted. The definition of a semialgebraic set seems to be natural as well. Possible formulations of Artin's theorem in the noncommutative case are at a preliminary stage and will be become clearer when more results are known, while others such as the definition of the preorder or a noncommutative formulation of Stengle's theorem require more work before satisfying formulations can be given.

In Section 2 we collect some general definitions and notations which are used throughout this paper. In Section 3 we set up basic axioms, concepts and examples for noncommutative real algebraic geometry. Roughly speaking, by passing to the noncommutative case the polynomial algebra and the points of $\mathbb{R}^d$ are replaced by a finitely generated $*$-algebra and a distinguished family of irreducible $*$-representations. In Section 4 we investigate and discuss possible formulations of Artin's theorem and Stengle's theorem in the noncommutative case. As the main new result of this

*Fakultät für Mathematik und Informatik, Universität Leipzig, Germany.

paper we obtain a version of Stengle's theorem for the algebra of $n \times n$ matrices with polynomial entries. Section 5 is devoted to $*$-algebras with Archimedean quadratic modules. We derive some properties and abstract Positivstellensätze for such $*$-algebras and develop a variety of examples. In Section 6 we show how pre-Hilbert $*$- bimodules can be used to transport quadratic modules from one algebra to another. This applies nicely to algebras of matrices and it might have further applications.

I thank Y. Savchuk for helpful discussions on the subject of this paper.

2. Definitions and notations. *Throughout this article $\mathcal{A}$ denotes a real or complex unital $*$-algebra.* By a *$*$-algebra* we mean an algebra $\mathcal{A}$ over the field $\mathbb{K} = \mathbb{R}$ or $\mathbb{K} = \mathbb{C}$ equipped with a mapping $a \rightarrow a^*$ of $\mathcal{A}$ into itself, called the *involution* of $\mathcal{A}$, such that $(\lambda a + \mu b)^* = \bar{\lambda} a^* + \bar{\mu} b^*, (ab)^* = b^* a^*$ and $(a^*)^* = a$ for $a, b \in \mathcal{A}$ and $\lambda, \mu \in \mathbb{K}$. The unit element of $\mathcal{A}$ is denoted by 1 and $\mathcal{A}_h := \{a \in \mathcal{A} : a = a^*\}$ is the set of *Hermitian elements* of $\mathcal{A}$.

As usual, $\mathbb{R}[t]=\mathbb{R}[t_1, \ldots, t_d]$ resp. $\mathbb{C}[t]=\mathbb{C}[t_1, \ldots, t_d]$ are the $*$-algebras of real resp. complex polynomials in d commuting Hermitian indeterminates $t_1, \ldots, t_d$. Set $\mathcal{N}(p) = \{s \in \mathbb{C}^d : p(s) = 0\}$ for $p \in \mathbb{C}[t]$. Let $\mathcal{M}_{k,n}(R)$ denote the $k \times n$-matrices over a ring R and set $\mathcal{M}_n(R) := \mathcal{M}_{n,n}(R)$.

If a is an operator on a Hilbert space, we denote by $\mathcal{D}(a)$ its *domain*, by $\mathcal{R}(a)$ its *range*, by $\mathcal{N}(a)$ its kernel, by $\bar{a}$ its *closure* and by a^* its *adjoint* (if they exist). A subset $\mathcal{E}$ of $\mathcal{D}(a)$ is called a *core* for a if for each $\varphi \in \mathcal{D}(a)$ there is a sequence of vectors $\varphi_n \in \mathcal{E}$ such that $\varphi_n \rightarrow \varphi$ and $a\varphi_n \rightarrow a\varphi$.

We now turn to some notions on $*$-representations, see [28] for a treatment of this subject. Let $\mathcal{D}$ be a pre-Hilbert space with scalar product $\langle \cdot, \cdot \rangle$. A *$*$-representation* of $\mathcal{A}$ on $\mathcal{D}(\pi):=\mathcal{D}$ is an algebra homomorphism π of $\mathcal{A}$ into the algebra of linear operators mapping $\mathcal{D}$ into itself such that $\pi(1)\varphi = \varphi$ and $\langle \pi(a)\varphi, \psi \rangle = \langle \varphi, \pi(a^*)\psi \rangle$ for all $\varphi, \psi \in \mathcal{D}$ and $a \in \mathcal{A}$. Two $*$-representation π_1 and π_2 are *(unitarily) equivalent* if there exists an isometric linear mapping U of $\mathcal{D}(\pi_1)$ onto $\mathcal{D}(\pi_2)$ such that $\pi_2(a) = U\pi_1(a)U^{-1}$ for $a \in \mathcal{A}$. A $*$-representation π is called *irreducible* if any decomposition of $\mathcal{D}(\pi)$ as an orthogonal sum of subspaces $\mathcal{D}_1$ and $\mathcal{D}_2$ such that $\pi(a)\mathcal{D}_1 \subseteq \mathcal{D}_1$ and $\pi(a)\mathcal{D}_2 \subseteq \mathcal{D}_2$ for all $a \in \mathcal{A}$ implies that $\mathcal{D}_1 = \{0\}$ or $\mathcal{D}_2 = \{0\}$.

A *state* of $\mathcal{A}$ is a linear functional f on $\mathcal{A}$ such that $f(1) = 1$ and $f(a^*a) \geq 0$ for all $a \in \mathcal{A}$. A state f of $\mathcal{A}$ is called *pure* if each state g satisfying $g(a^*a) \leq f(a^*a)$ for all $a \in \mathcal{A}$ is a multiple of f. If π is a $*$-representation of $\mathcal{A}$ and $\varphi \in \mathcal{D}(\pi)$ is a unit vector, then $f(\cdot) := \langle \pi(\cdot)\varphi, \varphi \rangle$ is a state of $\mathcal{A}$. These states are called *vector states* of π. Each state arises in this manner. That is, for each state f on $\mathcal{A}$ there exists a distinguished $*$-representation of $\mathcal{A}$, called the *GNS representation* of f and denoted by π_f (see [28], 8.6), and a vector $\varphi \in \mathcal{D}(\pi_f)$ such that $\mathcal{D}(\pi_f) = \pi_f(\mathcal{A})\varphi_f$ and

$$f(a) = \langle \pi_f(a)\varphi_f, \varphi_f \rangle \text{ for } a \in \mathcal{A}. \tag{2.1}$$

We assume some familiarity with "ordinary" real algebraic geometry (see e.g. the books [2, 16, 22] or [18]). To fix our notation, we recall two standard definitions. Suppose $\mathcal{B}$ is a commutative unital algebra over $\mathbb{R}$. Let $\hat{\mathcal{B}}$ denote the set of all algebra homomorphisms of $\mathcal{B}$ into $\mathbb{R}$. We write $f(s) := s(f)$ for $f \in \mathcal{B}$ and $s \in \hat{\mathcal{B}}$. If $\mathcal{B} = \mathbb{R}[t_1, \dots, t_d]$, then each element of $\hat{\mathcal{B}}$ is given by the evaluation at some point of $\mathbb{R}^d$, that is, $\hat{\mathcal{B}} \cong \mathbb{R}^d$.

If $f=(f_1, \cdots, f_k)$ is a k-tuple of elements $f_j \in \mathcal{B}$, the *basic closed semialgebraic set* $\mathcal{K}_f$ and the *preorder* $\mathcal{T}_f$ associated with f are defined by

$$\mathcal{K}_f = \{s \in \hat{\mathcal{B}} : f_1(s) \geq 0, \cdots, f_r(s) \geq 0\}, \tag{2.2}$$

$$\mathcal{T}_f = \{ \textstyle\sum_{\varepsilon_i \in \{0,1\}} \sum_{l=1}^{r} f_1^{\varepsilon_1} \cdots f_k^{\varepsilon_k} g_l^2 \, ; \; g_l \in \mathcal{B}, \; r \in \mathbb{N}\}. \tag{2.3}$$

3. Basic concepts of noncommutative real algebraic geometry.

3.1. Two main ingredients of noncommutative real algebraic geometry. The first main ingredient of real algebraic geometry is the algebra $\mathbb{R}[t_1, \dots, t_d]$ of polynomials or a finitely generated commutative unital $\mathbb{R}$-algebra $\mathcal{B}$. Its counter-part in the noncommutative case is a

- *finitely generated real or complex unital $*$-algebra $\mathcal{A}$.*

In real algebraic geometry elements of the algebras $\mathbb{R}[t_1, \dots, t_d]$ or $\mathcal{B}$ are evaluated at the points of $\mathbb{R}^d$ or at the real points of the associated affine algebraic variety V. As a noncommutative substitute for the set of point evaluations we assume that we have a given distinguished

- *family $\mathcal{R}$ of equivalence classes of irreducible $*$-representations of $\mathcal{A}$.*

Elements of $\mathcal{R}$ can be interpreted as "points" of a "noncommutative space".

One could also take a *family of pure states on $\mathcal{A}$* instead of representations. The GNS representation of a pure state is irreducible ([28], 8.6.8). The converse is valid for bounded representations on a Hilbert space.

If π is a $*$-representation of $\mathcal{A}$ and $a \in \mathcal{A}_h$, we write

$$\pi(a) \geq 0 \;\; if \; and \; only \; if \;\; \langle \pi(a)\varphi, \varphi \rangle \geq 0 \;\; for \; all \;\; \varphi \in \mathcal{D}(\pi).$$

This may be considered as a generalization of the positivity $f(t) \geq 0$ of the point evaluation of $f \in \mathcal{B}$ at $t \in V$.

Let us collect a number of important examples.

EXAMPLE 1. *Commutative polynomial algebras.*
The basic $*$-algebra for "ordinary" real algebraic geometry is the real $*$-algebra $\mathcal{A} = \mathbb{R}[t_1, \dots, t_d]$ with trivial involution $a^* = a$. One can also take the complex $*$-algebra $\mathcal{A} = \mathbb{C}[t_1, \dots, t_d]$ with involution defined by $p^*(t) := \sum_\alpha \overline{c}_\alpha \, t^\alpha$ for $p(t) = \sum_\alpha c_\alpha t^\alpha$.

Let $\mathcal{R} \cong \mathbb{R}^d$ be the set of point evaluations, that is, $\mathcal{R} = \{\pi_t : t \in \mathbb{R}^d\}$, where $\pi_t(p) = p(t)$ for $p \in \mathcal{A}$. ○

EXAMPLE 2. *Weyl algebras.*
Let $d \in \mathbb{N}$. The Weyl algebra $\mathcal{W}(d)$ is the unital $*$-algebra with generators $a_1, \dots, a_d, \; a_{-1}, \dots, a_{-d}$, defining relations

$$a_k a_{-k} - a_{-k} a_k = 1 \;\; \text{and} \;\; a_k a_l = a_l a_k \;\; \text{if} \;\; k \neq l,$$

and involution defined by $(a_k)^* = a_{-k}$ for $k=1,\ldots,d$.

For the Weyl algebra $\mathcal{W}(d)$ the set $\mathcal{R}$ consists of a single element π_0, the *Bargmann-Fock representation.* It is described by the actions of generators on an orthonormal basis $\{e_n; n \in \mathbb{N}_0^d\}$ of the Hilbert space given by

$$\pi_0(a_k)e_n = n_k^{1/2} e_{n-1_k}, \ \pi_0(a_{-k})e_n = (n_k+1)^{1/2} e_{n+1_k}$$

for $k=1,\ldots,d$ and $n=(n_1,\ldots,n_d) \in \mathbb{N}_0^d$. Here 1_k is the d-tuple with 1 at the k-th place and 0 otherwise and $e_{n-1_k} := 0$ when $n_k = 0$. The domain $\mathcal{D}(\pi_0)$ consist of all sums $\sum \varphi_n e_n$ such that $\sum n_1^r \ldots n_d^r |\varphi_n|^2 < \infty$ for all $r \in \mathbb{N}$. ∘

EXAMPLE 3. *Enveloping algebras.*
Let $\mathcal{E}(\mathcal{G})$ be the complex universal enveloping algebra of a finite dimensional real Lie algebra $\mathcal{G}$. Then $\mathcal{E}(\mathcal{G})$ is a $*$–algebra with involution given by $x^* = -x$ for $x \in \mathcal{G}$.

There is a simply connected Lie group G having $\mathcal{G}$ as its Lie algebra. Let $\hat{G}$ denote the unitary dual of G, that is, $\hat{G}$ is the set of equivalence classes of irreducible unitary representations of G. For each $\alpha \in \hat{G}$ we fix a representation U_α of the class α. Each representation U_α of G gives rise to an irreducible $*$-representation dU_α of the $*$-algebra $\mathcal{E}(\mathcal{G})$, see e.g. [28], 10.1, for details. The irreducibility of dU_α follows from [28], 10.2.18.

As $\mathcal{R}$ we take the family $\{dU_\alpha; \alpha \in \hat{G}\}$ of $*$-representations of $\mathcal{E}(\mathcal{G})$. ∘

EXAMPLE 4. *Free polynomial algebras.*
For $d \in \mathbb{N}$, let $\mathcal{A} = \mathbb{C}\langle t_1,\ldots,t_d\rangle$ be the free unital complex algebra with d generators $t_1,\ldots,t_d$. It is a $*$-algebra with involution determined by $t_j^* = t_j$, $j=1,\ldots,d$.

Let $\mathcal{R}$ be the equivalence classes of all irreducible representations by bounded operators on a Hilbert space. We may also take the families $\mathcal{R}_1$ of equivalence classes of π in $\mathcal{R}$ which act on a *fixed* (sufficiently large) Hilbert space or $\mathcal{R}_2$ of equivalence classes of finite dimensional representations in $\mathcal{R}$. Since the $*$-algebra is free, *each d*-tuple of bounded selfadjoint operators T_j on a Hilbert space $\mathcal{H}$ defines a $*$-representation π on $\mathcal{H}$ by $\pi(t_j) = T_j$.

Often it is convenient to use the $*$-algebra $\mathcal{A}_0 = \mathbb{C}\langle z_1,\ldots,z_d,w_1,\ldots,w_d\rangle$ with involution given by $z_j^* := w_j$. Clearly, $\mathcal{A}_0$ is $*$-isomorphic to $\mathcal{A}$ with a $*$-isomorphism determined by $z_j \to t_j + \mathrm{i}t_{d+j}$, $j=1,\ldots,d$. ∘

EXAMPLE 5. *Matrix algebras over commutative $*$-algebras.*
Suppose that $\mathcal{B}$ is a finitely generated commutative unital complex (or real) $*$-algebra and R is the set of $*$-homomorphisms of $\mathcal{B}$ into $\mathbb{C}$ (or $\mathbb{R}$). Let $\mathcal{A}$ be the complex (or real) matrix $*$-algebra $\mathcal{M}_n(\mathcal{B})$ with involution $(b_{ij})^* = (b_{ji}^*)$ and $\mathcal{R}$ the set $\{\rho_s; s \in R\}$ of $*$-representations $\rho_s : A \to A(s)$, where the matrix $A(s)$ acts as linear operator on the Hilbert space $\mathbb{K}^d$ in the usual way.

$*$-Subalgebras of matrix $*$-algebras $\mathcal{M}_n(\mathcal{B})$ provide a large class of interesting $*$-algebras for noncommutative real algebraic geometry.

Example 10 below is one of such examples. More can be found in the book [20]. ◦

How a possible new theory as noncommutative real algebraic geometry will further evolve depends essentially on what will be considered as typical examples and fundamental problems. Positivstellensätze should be one of the basic problems to be studied. PI–algebras and ∗-algebras which are finite dimensional over their centers (see Example 5) will lead to a theory that is close to real algebraic geometry. They might be studied first. Weyl algebras and enveloping algebras (and some algebras from Subsection 5.4) are interesting but challenging classes of examples. It is likely that a theory based on these examples will be very different from the classical theory.

3.2. Quadratic modules and orderings.

DEFINITION 1. *A* pre-quadratic module *of $\mathcal{A}$ is a subset $\mathcal{C}$ of $\mathcal{A}_h$ such that*

$$\mathcal{C}+\mathcal{C}\subseteq\mathcal{C},\quad \mathbb{R}_+\cdot\mathcal{C}\subseteq\mathcal{C}, \tag{3.1}$$

$$b^*\mathcal{C}b\in\mathcal{C} \text{ for all } b\in\mathcal{A}. \tag{3.2}$$

A quadratic module *of $\mathcal{A}$ is a pre-quadratic module $\mathcal{C}$ such that* $1\in\mathcal{C}$.

Quadratic modules are important in theory of ∗-algebras where they have been called *m-admissible wedges* ([28], p. 22). Following the terminology from real algebraic geometry we prefer to use the name "quadratic module".

Each quadratic module gives an ordering $\preceq$ on the real vector space $\mathcal{A}_h$ by defining $a\preceq b$ (and likewise $b\succeq a$) if and only if $a-b\in\mathcal{C}$.

If $\mathcal{X}$ is a subset of $\mathcal{A}_h$, then

$$\mathcal{C}_{\mathcal{X}} := \left\{\sum_{j=1}^{s}\sum_{l=1}^{k} a_{jl}^* x_l a_{jl};\ \ a_{jl}\in\mathcal{A},\ x_l\in\mathcal{X},\ s,k\in\mathbb{N}\right\}$$

is the pre-quadratic module of $\mathcal{A}$ generated by the set $\mathcal{X}$.

All elements a^*a, where $a\in\mathcal{A}$, are called *squares* of $\mathcal{A}$. The wedge

$$\sum\mathcal{A}^2 := \left\{\sum_{j=1}^{n} a_j^* a_j;\ \ a_1,\ldots,a_n\in\mathcal{A}, n\in\mathbb{N}\right\}$$

of finite sums of squares is obviously the smallest quadratic module of $\mathcal{A}$.

If $\mathcal{S}$ is a family of ∗-representations of $\mathcal{A}$, then

$$\mathcal{A}(\mathcal{S})_+ := \{a\in\mathcal{A}_h : \pi(a)\geq 0 \text{ for all } \pi\in\mathcal{S}\}$$

is a quadratic module of $\mathcal{A}$. The interplay between quadratic modules which are defined in algebraic terms (such as $\sum\mathcal{A}^2$) and those which are defined by means of ∗-representations (such as $\mathcal{A}(\mathcal{S})_+$) is one of the most interesting challenges for the theory.

The following polarization identities are useful. For $a, x, y \in \mathcal{A}$, we have

$$4x^*ay = (x+y)^*a(x+y) - (x-y)^*a(x-y) \\ - \mathrm{i}(x+\mathrm{i}y)^*a(x+\mathrm{i}y) + \mathrm{i}(x-\mathrm{i}y)^*a(x-\mathrm{i}y), \tag{3.3}$$

$$2(x^*ay + y^*ax) = (x+y)^*a(x+y) - (x-y)^*a(x-y). \tag{3.4}$$

From (3.4), applied with $a{=}y{=}1$, and (3.3) we easily conclude that

$$\mathcal{A}_h = \mathcal{C} - \mathcal{C}, \quad \mathcal{A} = (\mathcal{C} - \mathcal{C}) + \mathrm{i}(\mathcal{C} - \mathcal{C}). \tag{3.5}$$

for any quadratic module $\mathcal{C}$. Of course, for (3.3) and for the second equalitiy of (3.5) one has to assume that $\mathcal{A}$ is a *complex* $*$-algebra.

A quadratic module $\mathcal{C}$ is called *proper* if $\mathcal{C} \neq \mathcal{A}_h$. By (3.4), $\mathcal{C}$ is proper iff -1 is not in $\mathcal{C}$. A proper quadratic module $\mathcal{C}$ of $\mathcal{A}$ is called *maximal* if there is no proper quadratic module $\tilde{\mathcal{C}}$ of $\mathcal{A}$ such that $\mathcal{C} \subseteq \tilde{\mathcal{C}}$ and $\mathcal{C} \neq \tilde{\mathcal{C}}$.

If $\mathcal{C}$ is a maximal proper quadratic module of a commutative unital ring A, then $\mathcal{C} \cap (-\mathcal{C})$ is a prime ideal and $\mathcal{C} \cup (-\mathcal{C}) = A$. In the noncommutative case the second assertion is not true, for the first we have the following theorem due to J. Cimpric [3].

THEOREM 1. *Suppose $\mathcal{C}$ is a quadratic module of a complex $*$-algebra $\mathcal{A}$. Let $\mathcal{C}^0 := \mathcal{C} \cap (-\mathcal{C})$ and $\mathcal{I}_\mathcal{C} := \mathcal{C}^0 + \mathrm{i}\mathcal{C}^0$.*
(i) $\mathcal{I}_\mathcal{C}$ is a two-sided $*$-ideal of $\mathcal{A}$.
(ii) If $\mathcal{C}$ is a maximal proper quadratic module, $\mathcal{I}_\mathcal{C}$ is a prime ideal and

$$\mathcal{I}_\mathcal{C} = \{a \in \mathcal{A} : axx^*a^* \in \mathcal{C}^0 \text{ for all } x \in \mathcal{A}\}.$$

Proof. (i) Clearly, $\mathcal{I}_\mathcal{C}$ is $*$-invariant and $\mathcal{C}^0$ is a real vector subspace. If $a \in \mathcal{C}^0$ and $x \in \mathcal{A}$, then $(x^*{+}\mathrm{i}^k y)^*a(y^*{+}\mathrm{i}^k y) \in \mathcal{C}^0$ for $k{=}0,1,2,3$ by (3.2) and hence $4xay \in \mathcal{C}^0 + \mathrm{i}\mathcal{C}^0{=}\mathcal{I}_\mathcal{C}$ by (3.3). Thus $\mathcal{A} \cdot \mathcal{C}^0 \cdot \mathcal{A} \subseteq \mathcal{J}_\mathcal{C}$ and hence $\mathcal{A} \cdot \mathcal{I}_\mathcal{C} \cdot \mathcal{A} \subseteq \mathcal{I}_\mathcal{C}$.
(ii) [3], Theorem 1 and Remark on p. 5. □

3.3. Noncommutative semialgebraic sets. Let $\mathcal{R}$ be a family of (equivalence classes) of $*$-representations of $\mathcal{A}$.

DEFINITION 2. *A subset $\mathcal{K}$ of $\mathcal{R}$ is* semialgebraic *if it is a finite Boolean combination (that is, using unions, intersections and complements) of sets $\{\pi \in \mathcal{R} : \pi(f) \geq 0\}$ for $f \in \mathcal{A}_h$. It is* algebraic *if there is a finite subset $f{=}\{f_1, \ldots, f_k\}$ of $\mathcal{A}$ such that $\mathcal{K} = \mathcal{Z}(f){:=}\{\pi \in \mathcal{R} : \pi(f_1){=}0, \cdots, \pi(f_k){=}0\}$.*

Let $f = (f_1, \ldots, f_k)$ be a k-tuple of elements of $\mathcal{A}_h$, where $f_1 = 1$. We define the *basic closed semialgebraic set* associated with f by

$$\mathcal{K}(f) = \{\pi \in \mathcal{R} : \pi(f_1) \geq 0, \ldots, \pi(f_k) \geq 0\} \tag{3.6}$$

and the associated wedges by

$$\mathcal{P}(f) = \{a \in \mathcal{A}_h : \pi(a) \geq 0 \text{ for all } \pi \in \mathcal{K}(f)\}, \quad (3.7)$$

$$\mathcal{C}(f) = \{ \textstyle\sum_{j=1}^{s} \sum_{l=1}^{k} a_{jl}^* f_l a_{jl} : a_{jl} \in \mathcal{A},\ s \in \mathbb{N}\}. \quad (3.8)$$

Then $\mathcal{P}(f)$ and $\mathcal{C}(f)$ are quadratic modules such that $\mathcal{C}(f) \subseteq \mathcal{P}(f)$ and $\mathcal{C}(f)$ is the smallest quadratic module that contains all elements $f_1, \ldots, f_k$. In general we cannot add mixed products $f_j f_l$ to the wedge $\mathcal{C}(f)$, because $f_j f_l$ is not Hermitian if f_j and f_l do not commute.

EXAMPLE 6. *Commutative polynomial algebras.*
If $\mathcal{A}$ and $\mathcal{R}$ are as in Example 1, semialgebraic sets, algebraic sets, and basic closed semialgebraic sets according to the preceding definitions are just the ordinary ones in real algebraic geometry (see e.g. [2, 16, 22, 18]) and $\mathcal{P}(f)$ is the wedge of nonnegative real polynomials on $\mathcal{K}(f)$. Since $\mathcal{C}(f)$ is in general not closed under multiplication, it is not a preorder. If we replace f by the tuple $\tilde{f}$ of all products $f_{i_1} \cdots f_{i_r}$, where $1 \leq i_1 < i_2 < \cdots < i_r \leq k$, then $\mathcal{K}(f) = \mathcal{K}(\tilde{f})$ and $\mathcal{C}(\tilde{f})$ is the usual preorder $\mathcal{T}_f$. ∘

EXAMPLE 7. *Free polynomial algebras.*
Let $\mathcal{A} = \mathbb{C}\langle t_1, \ldots, t_d\rangle$ and $\mathcal{A}_0 = \mathbb{C}\langle z_1, \ldots, z_d, w_1, \ldots,, w_d\rangle$ be the $*$-algebras from Example 4 and let $\mathcal{R}$ be the family of all bounded $*$-representations of $\mathcal{A}$ resp. $\mathcal{A}_0$ on a separable Hilbert space. Then the basic semialgebraic set $\mathcal{K}(f)$ for $\mathcal{A}$ defined by (3.6) corresponds to the *positivity domain* $\mathcal{D}_f$ of f according to [10]. For the polynomial $f(z_1, \ldots, z_d) := z_1^* z_1 + \ldots + z_d^* z_d$ of $\mathcal{A}_0$ the algebraic set $\mathcal{Z}(f)$ corresponds to the spherical isometries in [11]. Many considerations on noncommutative real geometry based on *free* polynomial algebras by J.W. Helton and his coworkers fit nicely into the above concepts.
Let $f{=}(f_1, \ldots, f_{d+1})$, where $f_j(z_1, \ldots, z_d){=}(1{-}z_j^* z_j)^2$ for $j{=}1, \ldots, d$ and $f_{d+1}(z_1, \ldots, z_d){=}(1{-}\sum_{l=1}^{d} z_l z_l^*)^2$. Then the elements of the algebraic set $\mathcal{Z}(f)$ for $\mathcal{A}_0$ is in one-to-one correspondence with representations of the Cuntz algebra $\mathcal{O}_d$. ∘

3.4. The role of well-behaved unbounded representations. In this subsection we will show that in case of unbounded $*$-representations one has to select "good" $*$-representations rather than taking *all* (irreducible) $*$-representations as $\mathcal{R}$.

For this we restate a preliminary result (Proposition 3) proved in [27]. It can be derived from the following lemma which might be useful for other purposes as well. Lemma 2 is a slight generalization of Proposition 11.6.3 in [28] and is proved in the same manner (see e.g. Corollary 2.11 in [21]).

Let τ_{st} denote the finest locally convex topology on a vector space.

LEMMA 2. *Let $\mathcal{A}$ be a unital $*$-algebra which has a faithful $*$-representation π (that is, $\pi(a) = 0$ implies that $a = 0$). Assume that $\mathcal{A}$ is the union of a sequence of finite dimensional subspaces E_n, $n \in \mathbb{N}$,*

and that for each $n \in \mathbb{N}$ there exists a number $k_n \in \mathbb{N}$ such that the following is satisfied:
If $a \in \sum \mathcal{A}^2$ is in E_n, then we can write a as a finite sum $\sum_j a_j^* a_j$ such that all elements a_j are in E_{k_n}.
The the cone $\sum \mathcal{A}^2$ is closed in $\mathcal{A}$ with respect to the topology τ_{st}.

PROPOSITION 3. *If $\mathcal{A}$ is the commutative $*$-algebra $\mathbb{C}[t_1, \ldots, t_d]$, the Weyl algebra $\mathcal{W}(d)$, the enveloping algebra $\mathcal{E}(\mathcal{G})$ or the free $*$-algebra $\mathbb{C}\langle t_1, \ldots, t_d \rangle$ (see Examples 1-4), then the cone $\sum \mathcal{A}^2$ is τ_{st}-closed in $\mathcal{A}$.*

Proof. [27], Theorem 4.2, p. 95, see e.g. [28], Corollary 11.6.4. □

PROPOSITION 4. *Let $\mathcal{A}$ be a countably generated complex unital $*$-algebra such that $\sum \mathcal{A}^2$ is τ_{st}-closed in $\mathcal{A}$. For any $a \in \mathcal{A}_h$ the following are equivalent:*
(i): $a \in \sum \mathcal{A}^2$.
(ii): $\pi(a) \geq 0$ *for all $*$-representations π of $\mathcal{A}$.*
(iii): $\pi(a) \geq 0$ *for all irreducible $*$-representations π of $\mathcal{A}$.*
(iv): $f(a) \geq 0$ *for each state f of $\mathcal{A}$.*
(v): $f(a) \geq 0$ *for each pure state f of $\mathcal{A}$.*

Proof. (i)→(ii): $\langle \pi(\sum_j a_j^* a_j)\varphi, \varphi\rangle = \sum_j \langle \pi(a_j)\varphi, \pi(a_j)\varphi\rangle \geq 0$ for any $\varphi \in \mathcal{D}(\pi)$.
(ii)→(iv) and (iii)→(iv): We apply (ii) resp. (iii) to the GNS representation π_f of the state f and use formula (2.1). Note that the GNS representation π_f is irreducible if the state f is pure ([28], Corollary 8.6.8).
(iv)→(i): Assume to the contrary that a is not in $\sum \mathcal{A}^2$. By the separation theorem for convex sets (see e.g. [26], II.9.2), applied to the compact set $\{a\}$ and the closed (!) convex set $\sum \mathcal{A}^2$ of the locally convex space $\mathcal{A}_h[\tau_{st}]$, there exist a $\mathbb{R}$-linear functional g on $\mathcal{A}_h$ such that $g(a) < \inf \{g(c); c \in \sum \mathcal{A}^2\}$. Since $\sum \mathcal{A}^2$ is a wedge, the infimum is zero, so we have $g(a) < 0$ and $g(\sum \mathcal{A}^2) \geq 0$. The latter implies that the Cauchy-Schwarz inequality holds. Therefore, $0 \neq |g(a)|^2 \leq g(1)g(a^2)$ which yields $g(1) > 0$. Extending the $\mathbb{R}$-linear functional $g(1)^{-1}g$ on $\mathcal{A}_h$ to a $\mathbb{C}$-linear functional f on $\mathcal{A}$, we obtain a state f such that $f(a) < 0$.
(v)→(iv): Since $\mathcal{A}$ is countably generated, the assumptions of Theorem 12.4.7 in [28] are satisfied. By this theorem, each state of $\mathcal{A}$ is an integral over *pure* states. This in turn gives the implication (v)→(iv).
Since the implications (ii)→(iii) and (iv)→(v) are trivial, the equivalence of (i)–(v) is proved. □

Let $\mathcal{A}$ be one of $*$-algebras from Proposition 3. Since then $\sum \mathcal{A}^2$ is τ_{st}-closed, Proposition 4 applies and states that the sums of squares in $\mathcal{A}$ are precisely those elements which are nonnegative in *all* irreducible $*$-representations (or for *all* pure states) of $\mathcal{A}$. In particular, there is no difference between the commutative $*$-algebra $\mathbb{C}[t_1, \ldots, t_d]$ and the free $*$-algebra $\mathbb{C}\langle t_1, \ldots, t_d\rangle$ in this respect. In order to get an interesting theory in the spirit of classical real algebraic geometry one has to select a distinguished class $\mathcal{R}$ of *well-behaved* $*$-representations rather taking all irreducible $*$-representations. For the $*$-algebras $\mathbb{C}[t_1,\ldots,t_d]$, $\mathcal{W}(d)$ and $\mathcal{E}(\mathcal{G})$ families $\mathcal{R}$ of such representations have been chosen in Examples 1–3. It should be noted that there is no general procedure for finding well-behaved representations of arbitrary $*$-algebras.

Using essentially the τ_{st}-closedness of the cone $\sum \mathbb{C}\langle t_1,\ldots,t_d\rangle^2$ proved in [27] we give a short proof of the following result of Helton [9].

PROPOSITION 5. *Let* $\mathcal{A}{=}\mathbb{C}\langle t_1,\ldots,t_d\rangle$ *be the free complex* $*$*-algebra in d Hermitian indeterminates* $t_1,\ldots,t_d$ *and* $a \in \mathcal{A}_h$. *If* $\pi(a) \geq 0$ *for all finite dimensional* $*$*-representations* π *of* $\mathcal{A}$, *then* $a \in \sum \mathcal{A}^2$.

Proof. Let π be a $*$-representation of $\mathcal{A}$ and $\varphi \in \mathcal{D}(\pi)$. By Proposition 4,(ii)$\to$(i), it suffices to show that there is a *finite dimensional* $*$-representation ρ such that $\varphi \in \mathcal{D}(\rho)$ and $\langle \pi(a)\varphi,\varphi\rangle = \langle \rho(a)\varphi,\varphi\rangle$. This is easily done as follows.

Let $\mathcal{A}_k$ be the vector space of polynomials of degree less than k. We choose k such that $a \in \mathcal{A}_k$. Let P denote the projection of the Hilbert space $\mathcal{H}$ on the finite dimensional subspace $\pi(\mathcal{A}_k)\varphi$. Since $T_j{:=}P\pi(t_j)\lceil P\mathcal{H}$, $j{=}1,\ldots,d$, are selfadjoint operators on $P\mathcal{H}$, there is a $*$-representation ρ of of $\mathcal{A}$ on $P\mathcal{H}$ such that $\rho(t_j) = T_j$. By construction we have $\pi(b)\varphi = \rho(b)\varphi$ and hence $\langle \pi(b)\varphi,\varphi\rangle = \langle \rho(b)\varphi,\varphi\rangle$ for all $b \in \mathcal{A}_k$, in particular for $b = a$. □

4. Positivstellensätze for general $*$-algebras.

4.1. Artin's theorem for general $*$-algebras. Let us begin our discussion with the commutative case. By Artin's theorem for each nonnegative polynomial a on $\mathbb{R}^d$ there exists a nonzero polynomial $c \in \mathbb{R}[t]$ such that $c^2a \in \sum \mathbb{R}[t]^2$. For a noncommutative $*$-algebra $\mathcal{A}$ a natural guess is to generalize the latter to $c^*ac \in \sum \mathcal{A}^2$. (One might also think of $\sum_l c_l^* a c_l \in \sum \mathcal{A}^2$, but Proposition 17 below shows that such a condition corresponds to a Nichtnegativstellensatz rather than a Positivstellensatz.)

In the commutative case the relation $c^2a \in \sum \mathbb{R}[t]^2$ implies that the polynomial a is nonnegative on $\mathbb{R}^d$. However, in the noncommutative case such a converse is not true in general as the following examples show.

EXAMPLE 8. Let $\mathcal{A}$ be the Weyl algebra $\mathcal{W}(1)$ and $\mathcal{R}{=}\{\pi_0\}$, see Example 2. Set $N{=}a^*a$. Since $aa^*{-}a^*a{=}1$, we have $a(N{-}1)a^*{=}N^2{+}a^*a \in \sum \mathcal{A}^2$. But $\pi_0(N{-}1)$ is not nonnegative, since $\langle \pi_0(N{-}1)e_0, e_0\rangle = -1$ for the vacuum vector e_0. ∘

EXAMPLE 9. Let $\mathcal{A}$ be the $*$-algebra with a single generator a and defining relation $a^*a = 1$. Then $p_0 := 1{-}aa^*$ is a nonzero projection in $\mathcal{A}$ and we have $p_0axa^*p_0 = 0 \in \sum \mathcal{A}^2$ for arbitrary $x \in \mathcal{A}$. But elements of the form axa^* are in general not nonnegative in $*$-representations of $\mathcal{A}$. ∘

For a reasonable generalization of Artin's theorem one should add conditions which ensure that $\pi(a) \geq 0$ for $\pi \in \mathcal{R}$. In the commutative case c can be chosen such that the zero set $\mathcal{N}(c)$ is contained in the zero set $\mathcal{N}(a)$. (This follows, for instance, from Stengle's Positivstellensatz.) It seems to be natural to require a generalization of this condition in the noncommutative case as well. Thus, our *first version of a noncommutative generalization of Artin's theorem* for $\mathcal{A}$ and $\mathcal{R}$ is the following assertion:

For each $a \in \mathcal{A}_h$ *such that* $\pi(a) \geq 0$ *for all* $\pi \in \mathcal{R}$ *there exists an element* $c \in \mathcal{A}$ *such that*

$$c^*ac \in \sum \mathcal{A}^2, \tag{4.1}$$

$$\mathcal{N}(\pi(c)^*) \subseteq \mathcal{N}(\overline{\pi(a)}) \quad \text{for each } \pi \in \mathcal{R}. \tag{4.2}$$

Let $a \in \mathcal{A}_h$ and suppose conversely that there exists a $c \in \mathcal{A}$ such that (4.1) and (4.2) hold. For $\pi \in \mathcal{R}$ we put $\mathcal{E}_\pi := \pi(c)\mathcal{D}(\pi) + \mathcal{N}(\overline{\pi(a)})$.

LEMMA 6. *$\mathcal{E}_\pi$ is dense in $\mathcal{H}(\pi)$ and $\langle \overline{\pi(a)}\eta, \eta\rangle \geq 0$ for $\eta \in \mathcal{E}_\pi$.*

Proof. Since $\mathcal{H} = \overline{\mathcal{R}(\pi(c))} \oplus \mathcal{N}(\pi(c)^*)$ and $\mathcal{R}(\pi(c))$ is dense in $\overline{\mathcal{R}(\pi(c))}$,the linear subspace $\mathcal{E}_\pi$ is dense in $\mathcal{H}(\pi)$. For $\varphi \in \mathcal{N}(\pi(c)^*)$ and $\psi \in \mathcal{D}(\pi)$, using condition (4.2) we obtain

$$\langle \overline{\pi(a)}(\varphi + \pi(c)\psi), \varphi + \pi(c)\psi\rangle = \langle \pi(a)\pi(c)\psi, \pi(c)\psi\rangle = \langle \pi(c^*ac)\psi, \psi\rangle \geq 0,$$

where the last inequality follows at once from condition (4.1). □

Since $\mathcal{E}_\pi$ is dense in $\mathcal{H}(\pi)$, it is obvious that $\pi(a) \geq 0$ on $\mathcal{D}(\pi)$ when the operator $\pi(a)$ is *bounded.* If $\pi(a)$ is unbounded, it follows that $\pi(a) \geq 0$ on $\mathcal{D}(\pi)$ if we replace (4.2) by the following technical condition:

$$\mathcal{N}(\overline{\pi(a)}) + \pi(c)\mathcal{D}(\pi) \quad \text{is a core for } \overline{\pi(a)}. \tag{4.3}$$

In many cases it is difficult to decide whether or nor (4.2) can be satisfied. We now formulate another condition which is often easier to verify.

Let $\Psi(c)$ denote the set of all finite sums of linear functionals of the form $\langle \pi(\cdot)\pi(c)\varphi, \pi(c)\varphi\rangle$ on $\mathcal{A}$, where $\pi \in \mathcal{R}$ and $\varphi \in \mathcal{D}(\pi)$. By our *second version of a noncommutative generalization of Artin's theorem* we mean that assertion (4.1) and the following density condition (4.4) hold:

$$\begin{aligned}&\text{For each } \pi \in \mathcal{R} \text{ and } \psi \in \mathcal{D}(\pi) \text{ the functional } \langle \pi(\cdot)\psi, \psi\rangle \\ &\text{on } \mathcal{A} \text{ is the weak limit of a net of functionals from } \Psi(c).\end{aligned} \tag{4.4}$$

Since (4.1) obviously implies that $\pi(a) \geq 0$ on $\pi(c)\mathcal{D}(\pi)$, it follows from (4.1) and (4.4) that $\pi(a) \geq 0$ on $\mathcal{D}(\pi)$ for all $\pi \in \mathcal{R}$. For the latter statement it suffices to assume the condition in (4.4) evaluated at a and for vectors ψ from a core for $\overline{\pi(a)}$ rather than for all $\psi \in \mathcal{D}(\pi)$.

We give an example for this second version. This example is due to Y. Savchuk and details of proofs will appear in his forthcoming thesis.

EXAMPLE 10. Let $\mathcal{A}$ be the complex $*$-algebra with a generator a and defining relation $a^*a + aa^* = 1$. All $*$-representations of $\mathcal{A}$ act by bounded operators. Let $\mathcal{R}$ be the equivalence classes of all irreducible $*$-representations. They are formed by series $\rho_{\alpha,\varphi}$, where $\alpha \in [0,1/2), \varphi \in [0,2\pi)$, of 2-dimensional representations and ρ_φ, where $\varphi \in [0,2\pi)$, of 1-dimensional representations. These representations act on the generator a by

$$\rho_{\alpha,\varphi}(a) = \begin{pmatrix} 0 & e^{i\varphi}\sqrt{\alpha} \\ \sqrt{1-\alpha} & 0 \end{pmatrix} \quad \text{and} \quad \rho_\varphi(a) = \frac{e^{i\varphi}}{\sqrt{2}}.$$

For α=1/2 the matrix $\rho_{\alpha,\varphi}(a)$ defines *reducible* $*$-representation of $\mathcal{A}$. For the $*$-algebra $\mathcal{A}$ we have the following Positivstellensatz:
Suppose that $b \in \mathcal{A}_h$ and $\pi(b) \geq 0$ for all $\pi \in \mathcal{R}$. Then there exists an element $c = c^*$ of the center of $\mathcal{A}$ such that $c^2 a \in \sum \mathcal{A}^2$ and condition (4.4) is satisfied.

Let $\mathcal{B}$ denote the $*$-algebra of complex polynomials in three commuting indeterminates u=u^*, v, v^* satisfying the relation $u^2 + vv^* = 1$. The map

$$a \to \begin{pmatrix} 0 & v \\ u & 0 \end{pmatrix}$$

extends to a $*$-isomorphism of $\mathcal{A}$ onto a $*$-subalgebra of the matrix algebra $\mathcal{M}_2(\mathcal{B})$. If we consider $\mathcal{A}$ as a $*$-subalgebra of $\mathcal{M}_2(\mathcal{B})$, then the element c is a multiple $c_0 \cdot I$, where $c_0 \in \mathcal{B}_h$, of the unit matrix I. ○

4.2. Generalizations of Stengle's theorem to general $*$-algebras. As already noted in Subsection 3.3 the usual definition of the preorder does not make sense in the noncommutative case, because the product of noncommuting Hermitian elements is not Hermitian. For arbitrary $*$-algebras and semialgebraic sets I don't know how a proper generalization of the preorder might look like. In this subsection we propose one possible way to remedy this difficulty by reducing the problem to some appropriate *commutative* $*$-subalgebra. Our guiding examples for this method are $*$-subalgebras of matrix algebras $\mathcal{M}_n(\mathcal{B})$ (Example 5).

Let $\mathcal{Z}(\mathcal{A})$ be the center of $\mathcal{A}$ and let $\sum_{\mathcal{Z}}$ denote the set of nonzero elements of the wedge $\sum \mathcal{Z}(\mathcal{A})^2$. We shall assume the following:
If $az = 0$ for some $a \in \mathcal{A}$ and $z \in \mathcal{Z}(\mathcal{A})$, then $a = 0$ or $z = 0$.

Obviously, this is fulfilled if $\mathcal{A}$ has no zero divisors.

DEFINITION 3. *For $a_1, a_2 \in \mathcal{A}_h$, we write $a_1 \sim a_2$ if there exist elements $s_1, s_2 \in \sum_{\mathcal{Z}}$, $z \in \mathcal{Z}(\mathcal{A})$ and $x_\pm \in \mathcal{A}$ such that*

$$x_- x_+ = x_+ x_- = z \quad \text{and} \quad s_1\, a_1 = s_2\, x_+ a_2 x_+^*. \tag{4.5}$$

LEMMA 7. *"$\sim$" is an equivalence relation on $\mathcal{A}_h$.*

Proof. Suppose $a_1 \sim a_2$. Multiplying the second equation of (4.5) by x_- from the left and by x_-^* from the right and using the first equations we get

$$(s_2 z z^*)\, a_2 = s_1\, x_- a_1 x_-^*.$$

Since $s_2zz^* \in \sum_Z$, the latter means that $a_2 \sim a_1$.

Suppose $a_1 \sim a_2$ and $a_2 \sim a_3$. Then there are elements $s_1, s_2, s_3, s_4 \in \sum_Z$, $x_+, x_-, y_+, y_- \in \mathcal{A}$ and $z, w \in \mathcal{Z}(\mathcal{A})$ such that $x_-x_+ = x_+x_- = z$, $y_-y_+ = y_+y_- = w$, $s_1\ a_1 = s_2\ x_+a_2x_+^*$, $s_3\ a_2 = s_4\ y_+a_3y_+^*$. Setting $u_- := y_-x_-$, $u_+ := x_+y_+$, we have $u_-u_+ = u_+u_- = zw$ and $s_3s_1\ a_1 = s_2x_+(s_3a_2)x_+^* = s_2s_4x_+y_+a_3y_+^*x_+^* = s_2s_4u_+a_3u_+^*$, so that $a_1 \sim a_3$.

Since obviously $a \sim a$, "$\sim$" is an equivalence relation. □

Let $\mathcal{C}$ be a quadratic module of $\mathcal{A}$ and assume that $a_1 \sim a_2$. Then we have $s_1a_1 \in \mathcal{C}$ if and only if $s_2zz^*a_2 \in \mathcal{C}$. That is, up to multiples from the set $\sum_Z$, a_1 belongs to $\mathcal{C}$ if and only if a_2 is in $\mathcal{C}$.

The relation $\sim$ is extended to tuples $a=(a_1, \ldots, a_n)$ and $b=(b_1, \ldots, b_r)$ from $\mathcal{A}_h$ by defining $a \sim b$ if $a_j \sim b_l$ for all $j=1, \ldots, n$ and $l=1, \ldots, r$.

DEFINITION 4. *Suppose $a \sim b$. We shall write $a \sim^+ b$ if for any representation $\pi \in \mathcal{R}$, $\pi(a_j) \geq 0$ for all $j=1, \ldots, n$ implies that $\pi(b_l) \geq 0$ for all $l=1, \ldots, r$ and we write $a \overset{+}{\sim} b$ if $a \sim^+ b$ and $b \sim^+ a$.*

To begin with the setup for Stengle's theorem, let us fix a k-tuple $f = (f_1, \ldots, f_k)$ of elements $f_j \in \mathcal{A}_h$. Let a be an element of $\mathcal{A}_h$ which is nonnegative on the semialgebraic set $\mathcal{K}(f)$, that is, $\pi(a) \geq 0$ for $\pi \in \mathcal{K}(f)$.

Suppose there exist a finitely generated *commutative* real subalgebra $\mathcal{B}$ of $\mathcal{A}_h$ such that the following assumptions are fulfilled:

(I) *There exist finite tuples $c = (c_1, \ldots, c_m)$ and $b = (b_1, \ldots, b_r)$ of elements of $\mathcal{B}$ such that $a \overset{+}{\sim} c$ and $f \overset{+}{\sim} b$.*

(II) *For $j=1, \ldots, m$, we have $\pi(c_j) \geq 0$ for all $\pi \in \mathcal{K}(b)$ if and only if $c_j(s) \geq 0$ for all $s \in \mathcal{K}_b$.*

Recall that $\mathcal{K}(b)$ is the noncommutative semialgebraic set defined by (3.6) and $\mathcal{K}_b = \{s \in \hat{\mathcal{B}} : b_1(s) \geq 0, \ldots, b_r(s) \geq 0\}$ is the "ordinary" semialgebraic set for the commutative real algebra $\mathcal{B}$ defined by (2.2).

We now derive our noncommutative version of Stengle's theorem. By assumption (I), we have $f \overset{+}{\sim} b$ and $a \sim^+ c$. The relation $f \overset{+}{\sim} b$ implies that the two semialgebraic sets $\mathcal{K}(f)$ and $\mathcal{K}(b)$ of $\mathcal{A}$ coincide. Since $a \sim^+ c$, we have $\pi(c_j) \geq 0$ for all $\pi \in \mathcal{K}(f) = \mathcal{K}(b)$ and $j=1, \ldots, m$. Therefore, by assumption (II), $c_j \geq 0$ on $\mathcal{K}_b$. Let $\mathcal{T}_b$ denote the preorder (2.3) for the commutative algebra $\mathcal{B}$. By Stengle's theorem, applied to $\mathcal{K}_b$ and $\mathcal{T}_b$, there exist elements $g_j, h_j \in \mathcal{T}_b$ and numbers $n_j \in \mathbb{N}$ such that

$$g_jc_j = c_jg_j = c_j^{2n_j} + h_j. \tag{4.6}$$

Since $c_j \sim a$ and $b_l \sim f_l$, there exist elements $s_{1j}, s_{2j}, s_{3l}, s_{4l} \in \sum_Z$ and $x_{+j}, y_{+l} \in \mathcal{A}$ such that $s_{1j}c_j = s_{2j}x_{+j}ax_{+j}^*$ and $s_{3l}b_j = s_{4l}y_{+l}f_ly_{+l}^*$. Put $s_3 := s_{31} \cdots s_{3r}$. Multiplying (4.6) by the central element $s_3s_{1j}^{2n_j+1}$ we obtain

$$s_3g_j\ s_{1j}^{2n_j}s_{2j}\ x_{+j}ax_{+j}^* = s_3s_{1j}(s_{2j}x_{+j}ax_{+j}^*)^{2n_j} + s_{1j}^{2n_j+1}\ s_3h_j. \tag{4.7}$$

for $j=1,\ldots,m$. Set $p_j := s_3 g_j s_{1j}^{2n_j} s_{2j}$. Let $\mathcal{T}(f)$ denote the quadratic module of $\mathcal{A}$ generated by the preorder $\mathcal{T}_b$ of $\mathcal{B}$. Since $s_3 g_j$ and $s_3 h_j$ belong to $\mathcal{T}(f)$, $p_j \in \mathcal{T}(f)$. Hence the right-hand side of (4.7) is in $\mathcal{T}(f)$, so we have

$$p_j x_{+j} a x_{+j}^* = x_{+j} a x_{+j}^* p_j \in \mathcal{T}(f) \text{ for } j = 1, \ldots, m. \quad (4.8)$$

That is, there exist elements $p_j \in \mathcal{T}(f)$ and $x_{+j} \in \mathcal{A}$ such that (4.8) holds. We consider this statement (and likewise the more precise equalities (4.7)) as a *noncommutative version of Stengle's theorem.*

We now turn to the converse direction, that is, we show that our version of Stengle's theorem implies that $\pi(a) \geq 0$ for all $\pi \in \mathcal{K}(f)$. For suppose that (4.7) is satisfied for $j = 1, \ldots, m$ with $s_{1j}, s_{2j}, s_3, g_j, h_j, x_{+j}$ as above. Then, since (4.7) is nothing but (4.6) multiplied by $s_{1j}^{2n_j+1} s_3$, equation (4.6) holds, so each element c_j is nonnegative on the set $\mathcal{K}_b$. Therefore, by assumption (II) we have $\pi(c_j) \geq 0$ for all $\pi \in \mathcal{K}(b) = \mathcal{K}(f)$. Since $c \sim^+ a$ by assumption (I), it follows that $\pi(a) \geq 0$ for all $\pi \in \mathcal{K}(f)$.
We close this subsection by discussing assumption (II). The following simple example shows that it is not always satisfied.

EXAMPLE 11. Let $\mathcal{B}$ be the $*$- algebra $\mathbb{C}[t]$ of complex polynomials in one Hermitian indeterminate t. From the moment problem theory it is known that there exists a state f on $\mathcal{B}$ such that $f(t^3 p\bar{p}) \geq 0$ for all $p \in \mathcal{B}$ and $f(tp_0^2) < 0$ for some $p_0 \in \mathcal{B}_h$. For the GNS representation π_f of f we then have $\pi_f(t^3) \geq 0$ and $\pi_f(t) \not\geq 0$. Therefore, if the family $\mathcal{R}$ contains π_f, then $t \notin \mathcal{K}(t^3)$, but $t \in \mathcal{K}_{t^3} = [0, \infty)$.

The converse direction fails if $\mathcal{R}$ is to "small". For instance, if we take $\mathcal{R}=\{\pi_s; s \in [0,1]\}$, where $\pi_s(p) = p(s)$, then $t-1 \in \mathcal{K}(t)$, but $t-1 \notin \mathcal{K}_t$. ∘

To give a sufficient condition for assumption (II), we fix Hermitian generators $y_1, \ldots, y_d$ of $\mathcal{B}$. Then $\hat{\mathcal{B}}$ becomes a subset of $\mathbb{R}^d$ by identifying a character with its values at the generators. If π is a bounded $*$-representation of $\mathcal{B}$ on a Hilbert space , then the d-tuple of commuting bounded selfadjoint operators $\pi(y_j)$ has a unique spectral measure E_π.

PROPOSITION 8. *Suppose $\mathcal{R}$ is family of bounded $*$-representation of $\mathcal{B}$ on Hilbert spaces. Let $b=(b_1, \ldots, b_r)$ be an r–tuple of elements of $\mathcal{B}$ such that $\mathcal{K}_b$ is the union of all its subsets of the form* supp E_π, *where $\pi \in \mathcal{R}$. Then for any $a \in \mathcal{B}$, we have $a \in \mathcal{K}(b)$ if and only if $a \in \mathcal{K}_b$.*

Proof. Since $\pi \in \mathcal{R}$ is bounded, it is a direct sum of cyclic representations. Hence we can assume without loss of generality that each $\pi \in \mathcal{R}$ has a cyclic vector φ_π. Then $\mu_\pi(\cdot) := \langle E_\pi(\cdot)\varphi_\pi, \varphi_\pi \rangle$ defines a positive Borel measure on $\mathbb{R}^d$ such that supp E_π = supp μ_π. From the spectral theorem we obtain

$$\langle \pi(p(y))\varphi_\pi, \varphi_\pi \rangle = \int_{\mathbb{R}^d} p(s)\, \mathrm{d}\mu_\pi(s) \text{ for } p \in \mathbb{C}[y_1, ..., y_d]. \quad (4.9)$$

Suppose that $c \in \mathcal{B}_h \cong \mathbb{R}[y_1, \ldots, y_d]$. For $p \in \mathbb{C}[y_1, \ldots, y_d]$, we have

$$\langle \pi(c)\pi(p)\varphi_\pi, \pi(p)\varphi_\pi \rangle = \int_{\mathbb{R}^d} c(s)|p(s)|^2\, \mathrm{d}\mu_\pi(s).$$

Since the polynomials are uniformly dense in the continuous functions on the compact set $\operatorname{supp}\mu_\pi$, it follows that $\pi(c) \geq 0$ if and only if $\operatorname{supp}\mu_\pi \subseteq \mathcal{K}_c$. This implies that $\mathcal{K}(b) = \{\pi \in \mathcal{R} : \operatorname{supp}\mu_\pi \subseteq \mathcal{K}_b\}$. Therefore, if $a \in \mathcal{K}_b$, then $a \in \mathcal{K}(b)$ by (4.9). Conversely, if $a \in \mathcal{K}(b)$, then $\operatorname{supp}\mu_\pi \subseteq \mathcal{K}_a$ for all $\pi \in \mathcal{R}$ such that $\operatorname{supp}\mu_\pi \subseteq \mathcal{K}_b$. By assumption $\mathcal{K}_b$ is the union of all sets $\operatorname{supp}\mu_\pi = \operatorname{supp}E_\pi$ which are contained in $\mathcal{K}_b$. Hence $\mathcal{K}_b \subseteq \mathcal{K}_a$ which in turn yields $a \in \mathcal{K}_b$. □

4.3. Diagonalization of matrices with polynomial entries. In the rest of this section, $\mathcal{A}$ is the real $*$-algebra $\mathcal{M}_n(\mathbb{R}[t])$ of $n \times n$-matrices over $\mathbb{R}[t] = \mathbb{R}[t_1, \ldots, t_d]$ with involution given by the transposed matrix A^t of A and $\mathcal{R}$ is the set $\{\rho_s; s \in \mathbb{R}^d\}$ of irreducible $*$-representations $\rho_s: A \to A(s)$, see Example 5. Then $\mathcal{A}_h$ is the set $\mathcal{S}_n(\mathbb{R}[t])$ of symmetric matrices and the unit of $\mathcal{A}$ is the unit matrix I. Clearly, $\rho_s(A) = A(s) \geq 0$ if and only if the matrix $A(s)$ is positive semidefinite.

We begin with some notation. If $i=(i_1, \ldots, i_p)$ and $j=(j_1, \ldots, j_p)$ are p-tuples of integers such that $1 \leq i_1 < \cdots < i_p \leq n$ and $1 \leq j_1 < \cdots < j_p \leq n$ and $A \in \mathcal{A}$, then $M_j^i = M_j^i(A)$ denotes the principal minor of A with columns i_k and rows j_k. If $i_1=j_1=1, \ldots, i_p=j_p=p$, we write M_p instead of M_j^i.

For $\lambda = (\lambda_1, \ldots, \lambda_p)$, where $p \leq n$, let $D(\lambda) = D(\lambda_1, \ldots, \lambda_p)$ denote the $n \times n$ diagonal matrix with diagonal entries $\lambda_1, \ldots, \lambda_p, 0, \ldots, 0$.

Now let $A \in \mathcal{S}_n(\mathbb{R}[t])$, $A \neq 0$, $n \geq 2$, and assume that A has rank r and that $M_1(A) \neq 0, \ldots, M_r(A) \neq 0$. If the latter is true, we say that A has *standard form*. For such a matrix A we define two lower triangular $n \times n$-matrices $Y_\pm = (y_{ij}^\pm)$ with entries given by the rational functions

$$
\begin{aligned}
&y_{ij}^\pm = \pm M_{(1,\ldots,j-1,j)}^{(1,\ldots,j-1,i)} M_j^{-1} \text{ for } j=1, \ldots, r,\ i=j+1, \ldots, n,\\
&y_{ii}^\pm = 1 \text{ for } i=1, \ldots, n,\\
&y_j^\pm = 0 \text{ otherwise } (j=r+1, \ldots, n, i=j+1, \ldots, n \text{ and } i \geq j, i, j=1, \ldots, n).
\end{aligned}
$$

By Satz 6.2 in [7], p. 64, we have $A = Y_+ D(M_1, M_2 M_1^{-1}, \ldots, M_r M_{r-1}^{-1}) Y_+^t$. Since obviously $Y_+^{-1} = Y_-$, the latter yields

$$
D(M_1, M_2 M_1^{-1}, \ldots, M_r M_{r-1}^{-1}) = Y_- A Y_-^t. \tag{4.10}
$$

Set $D = M_1 \cdots M_{r-1} D(M_1, M_2 M_1^{-1}, \ldots, M_r M_{r-1}^{-1})$ and $X_\pm = M_1 \cdots M_{r-1} Y_\pm$. Clearly, D and $X_\pm$ are in $\mathcal{M}_n(\mathbb{R}[t])$. From the relations $Y_-^{-1} = Y_+$ and (4.10) we obtain

$$
X_+ X_- = X_- X_+ = (M_1 \cdots M_{r-1})^2 I, \tag{4.11}
$$

$$
(M_1 \cdots M_{r-1})^4 A = X_+ D X_+^t, \quad D = X_- A X_-^t. \tag{4.12}
$$

That is, we have shown that *for any matrix $A \in \mathcal{S}_n(\mathbb{R}[t])$ in standard form (that is, rank $A = r$ and $M_1 \neq 0, \ldots, M_r \neq 0$) there exists a diagonal* *matrix $D \in \mathcal{M}_n(\mathbb{R}[t])$ such that $A \sim D$ and the corresponding matrices $X_\pm$ can be chosen to be lower triangular.*

We now turn to arbitrary matrices in $\mathcal{S}_n(\mathbb{R}[t])$. Our aim is to prove Proposition 9 below. The main technical ingredient for this proof is the following procedure for block matrices over a ring R. We write a matrix $A \in \mathcal{S}_n(R)$, $n \geq 2$, as

$$A = \begin{pmatrix} \alpha & \beta \\ \beta^t & C \end{pmatrix}, \text{ where } C \in \mathcal{S}_{n-1}(R),\ \beta \in \mathcal{M}_{1,n-1}(R),$$

and put

$$X_\pm = \begin{pmatrix} \alpha & 0 \\ \pm\beta^t & \alpha I \end{pmatrix}, \widetilde{A} = \begin{pmatrix} \alpha^3 & 0 \\ 0 & \alpha(\alpha C - \beta^t\beta) \end{pmatrix}.$$

Then we have

$$X_+X_- = X_-X_+ = \alpha^2 \cdot I, \tag{4.13}$$

$$\alpha^4 A = X_+\widetilde{A}X_+^t,\ \widetilde{A} = X_-AX_-^t. \tag{4.14}$$

PROPOSITION 9. *Let $A \in \mathcal{S}_n(\mathbb{R}[t]), A \neq 0$. Then there exist diagonal matrices $D_l \in \mathcal{M}_n(\mathbb{R}[t])$, matrices $X_{\pm,l} \in \mathcal{M}_n(\mathbb{R}[t])$ and polynomials $z_l \in \sum \mathbb{R}[t]^2$, $l = 1, \ldots, m$, such that:*

(i) $X_{+l}X_{-l} = X_{-l}X_{+l} = z_jI$, $D_l = X_{-l}AX_{-l}^t$, $z_lA = X_{+l}D_lX_{+l}^t$,

(ii) *For $s \in \mathbb{R}^d$, $A(s) \geq 0$ if and only if $D_l(s) \geq 0$ for all $l = 1, \ldots, m$.*

Proof. Let $i, j \in \{1, \ldots, n\}$, $i \leq j$. Put $\widetilde{a}_{ii} = a_{ii}$ and $\widetilde{a}_{ij} = a_{ij} + \frac{1}{2}(a_{ii} + a_{jj})$ if $i < j$. We first show that there is an orthogonal matrix $T_{ij} \in \mathcal{M}_n(\mathbb{R})$ such that

$$T_{ij}AT_{ij}^t = \begin{pmatrix} \widetilde{a}_{ij} & * \\ * & * \end{pmatrix}. \tag{4.15}$$

For $l \in \{2, \ldots, n\}$, let P_l denote the permutation matrix which permutes the first row and the l-th row. Setting $T_{11} = I$ and $T_{ii} = P_i$ for $i = 2, \ldots, n$, (4.15) holds for $i = j$. Now suppose $i < j$. Let $S = (s_{kl}) \in \mathcal{M}_n(\mathbb{R})$ be the matrix with $s_{ii} = s_{ij} = s_{ji} = 2^{-1/2}$, $s_{jj} = -2^{1/2}$, $s_{ll} = 1$ if $l \neq i, j$ and $s_{kl} = 0$ otherwise. Set $T_{ij} = T_{ii}S$. One easily checks that T_{ij} is orthogonal and (4.15) is satisfied for $i < j$.

Now we apply the above procedure to the block matrix $A_{ij} := T_{ij}AT_{ij}^t$, $i \leq j$, given by (4.15). Let $\widetilde{A}_{ij}$, $X_{\pm,ij}$ denote the corresponding matrices. Then there is a matrix $B_{ij} \in \mathcal{S}_{n-1}(\mathbb{R}[t])$ such that

$$\widetilde{A}_{ij} = \begin{pmatrix} \widetilde{a}_{ij}^3 & 0 \\ 0 & B_{ij} \end{pmatrix}. \tag{4.16}$$

We claim that for any $s \in \mathbb{R}^d$, $A(s) \geq 0$ if and only if $\widetilde{a}_{ij}(s) \geq 0$ and $B_{ij}(s) \geq 0$ for all $i, j \in \{1, \ldots, n\}$, $i \leq j$.

Indeed, if $A(s) \geq 0$, then $A_{ij}(s) \geq 0$ by (4.15) and hence $\widetilde{A}_{ij} = X_{-,ij}A_{ij}(s)X^t_{-,ij} \geq 0$ by (4.14), so $\widetilde{a}_{ij}(s) \geq 0$ and $B_{ij}(s) \geq 0$ by (4.16). Conversely, assume that $\widetilde{a}_{ij}(s) \geq 0$ and $B_{ij}(s) \geq 0$ for all i, j, $i \leq j$. Then $\widetilde{A}_{ij}(s) \geq 0$ for all i, j. If $\widetilde{a}_{ij}(s) = 0$ for all i, j, then $a_{ij}(s) = 0$ for all i, j and hence $A(s) = 0$. If $\widetilde{a}_{ij}(s) > 0$ for some i, j, we conclude that $A_{ij}(s) = \widetilde{a}_{ij}(s)^{-4}X_{+,ij}\widetilde{A}_{ij}(s)X^t_{+,ij} \geq 0$ by (4.14) and so $A(s) = T^t_{ij}A_{ij}(s)T_{ij} \geq 0$. This completes the proof of the claim.

Applying the same reasoning to the matrices B_{ij} instead of A and proceeding by induction we obtain after at most $n-1$ steps a finite sequence of diagonal matrices having the desired properties. □

COROLLARY 10. *For each matrix $A \in \mathcal{S}_n(\mathbb{R}[t]), A \neq 0$, there exist nonzero polynomials $b, d_j \in \mathbb{R}[t]$, $j=1, \ldots, r, r \leq n$, and matrices X_+, $X_- \in \mathcal{M}_n(\mathbb{R}[t])$ such that*

$$X_+X_- = X_-X_+ = bI,\ b^2A = X_+DX^t_+,\ D = X_-AX^t_-,$$

where D is the diagonal matrix $D(d_1, \ldots, d_r)$. In particular, $A \sim D$.

Proof. Since $A \neq 0$, $a_{ij} \neq 0$ for some i, j. We apply the above procedure to the matrix B_{ij} from (4.16) and proceed by induction until the corresponding matrix B_{ij} is identically zero. □

REMARK. Suppose that $A \in \mathcal{S}_n(\mathbb{R}[t])$, $A \neq 0$. Because the rank r of A is the column rank and the row rank it follows that A has a non-zero principal minor of order r. Hence, there exists a permutation matrix P such that $M_r(PAP^t) \neq 0$. But it may happen that all principal minors of order $r-1$ of PAP^t vanish, so PAP^t is *not* in standard form. A simple example is

$$\begin{pmatrix} 1 & -1 & 1 \\ -1 & 1 & 1 \\ 1 & 1 & 1 \end{pmatrix}.$$

4.4. Artin's theorem and Stengle's theorem for matrices of polynomials. From Corollary 10 and Proposition 9 we easily derive versions of Artin's theorem and Stengle's theorem for matrices of polynomials.

The next proposition is Artin's theorem for matrices of polynomials. It was first proved in [8] and somewhat later also in [23].

PROPOSITION 11. *Let $A \in \mathcal{S}_n(\mathbb{R}[t])$. If $A(t) \geq 0$ for all $t \in \mathbb{R}^d$, then there exist a polynomial $c \in \mathbb{R}[t]$, $c \neq 0$, such that $c^2A \in \sum \mathbb{R}[t]^2$.*

Proof. Let $D = D(d_1, \ldots, d_r)$ be the diagonal matrix from Corollary 10. Since $D = X_-AX^t_-$ and $A(t) \geq 0$ on $\mathbb{R}^d$, we have $D(t) \geq 0$ and hence $d_j(t) \geq 0$ on $\mathbb{R}^d$. Since $b^2A = X_+DX^t_+$, the assertion follows at once by applying Artin's theorem for polynomials to the diagonal entries $d_1, \ldots, d_r$ and multiplying by the product of denumerators. □

Let $c \in \mathbb{R}[t]$, $c \neq 0$. Since the set $\{s \in \mathbb{R}^d : c(s) \neq 0\}$ is dense in $\mathbb{R}^d$, each $s \in \mathbb{R}^d$ is limit of a sequence of points s_n such that $c(s_n) \neq 0$. Then each vector state of ρ_s is weak limit of vector states of ρ_{s_n} with vectors from $c(s_n)\mathbb{R}^d$. Since these functionals belong to the set $\Psi(c)$, condition (4.4) is fulfilled and the second version of Artin's theorem holds.

We now turn to Stengle's theorem and apply the setup of Subsection 4.2 to the $*$-algebra $\mathcal{A} = \mathcal{M}_n(\mathbb{R}[t])$ and its commutative $*$-subalgebra $\mathcal{B}$ of diagonal matrices. Let $F = (F_1, \ldots, F_k)$ be a k-tuple of elements from $\mathcal{S}_n(\mathbb{R}[t])$ and let $\mathcal{K}(F)=\{\rho_s : s \in \mathbb{R}^d, F_1(s) \geq 0, \ldots, F_k(s) \geq 0\}$ be the corresponding noncommutative semialgebraic set. Suppose that $A \in \mathcal{S}_n(\mathbb{R}[t])$ and $\rho_s(A) = A(s) \geq 0$ for all $s \in \mathcal{K}(F)$. We will show that assumption (I) and (II) from Subsection 4.2 are satsified.

By Proposition 9 there exists an m-tuple $C = (C_1, \ldots, C_m)$ of diagonal matrices such that $A \overset{+}{\sim} C$. Applying Proposition 9 to each matrix F_j we obtain a finite sequence of diagonal matrices. Let $B = (B_1, \cdots, B_r)$ denote r-tuple formed by all these diagonal matrices and all diagonal matrices obtained by permutations of their diagonal entries for j=1, ..., k. By Proposition 9, we then have $F \overset{+}{\sim} B$, so assumption (I) is satisfied. The set $\hat{\mathcal{B}}$ of characters of $\mathcal{B}$ consists of all functionals $h_{i,s}$, where $s \in \mathbb{R}^d$ and $j=1, \ldots, n$, given by $h_{j,s}(D(d_1, \ldots, d_n)) = d_j(s)$. Let $b_{j1}, \ldots, b_{jn}$ be the diagonal entries of B_j. Since B contains all permuted diagonal matrices, the set $\mathcal{K}:=\{s \in \mathbb{R}^d : b_{1l}(s) \geq 0, \ldots, b_{rl}(s) \geq 0\}$ does not depend on $l=1, \ldots, n$. Hence we have $\mathcal{K}(B) = \{\rho_s : s \in \mathcal{K}\}$ and $\mathcal{K}_B = \{h_{i,s} : s \in \mathcal{K}\}$ which implies that assumption (II) is fulfilled. Therefore the version of Stengle's theorem stated in Subsection 4.2 is valid. Recall that $\mathcal{T}(F)$ is the quadratic module of $\mathcal{A}$ generated by all products $B_{i_1} \cdots B_{i_l}$, where $1 \leq i_1 < i_2 < \ldots i_l \leq r$. We may consider $\mathcal{T}(F)$ as a noncommutative substitute of the preorder associated with F. Since $\mathcal{T}(F)$ depends on the particular diagonalizations of F_j, it is neither uniquely nor canonically associated with F.

5. Archimedean quadratic modules. In this section $\mathcal{A}$ is *complex* unital $*$-algebra. Then we have $\mathcal{A} = \mathcal{A}_h + \mathrm{i}\mathcal{A}_h$ be writing $a \in \mathcal{A}$ as

$$a = a_1 + \mathrm{i}a_2, \text{ where } a_1 = \mathrm{Re}\, a := (a^* + a)/2, \quad a_2 = \mathrm{Im}\, a := \mathrm{i}(a^* - a)/2.$$

5.1. Definition and simple properties. Let $\mathcal{C}$ be a quadratic module of $\mathcal{A}$. We denote by $\mathcal{A}_b(\mathcal{C})$ the set of all elements $a \in \mathcal{A}$ for which exists a a number $\lambda_a > 0$ such that

$$\lambda_a \cdot 1 \pm \mathrm{Re}\, a \in \mathcal{C} \text{ and } \lambda_a \cdot 1 \pm \mathrm{Im}\, a \in \mathcal{C}. \tag{5.1}$$

The set $\mathcal{A}_b(\mathcal{C})$ was introduced in [30] and in [4] where the following proposition was proved.

PROPOSITION 12. (i) *$\mathcal{A}_b(\mathcal{C})$ is a unital $*$-subalgebra of $\mathcal{A}$.*
(ii) *An element $a \in \mathcal{A}$ is in $\mathcal{A}_b(\mathcal{C})$ if and only if a^*a is in $\mathcal{A}_b(\mathcal{C})$.*

We call $\mathcal{A}_b(\mathcal{C})$ the *$*$-subalgebra of $\mathcal{C}$-bounded elements* of $\mathcal{A}$. In the case $\mathcal{C} = \sum \mathcal{A}^2$ we denote $\mathcal{A}_b(\mathcal{C})$ by $\mathcal{A}_b$. Note that $\mathcal{A}_b(\mathcal{C})$ is the counter-part of the ring of bounded elements (see [32, 19]) in real algebraic geometry.

The main notion in this section is the following.

DEFINITION 5. *A quadratic module $\mathcal{C}$ of $\mathcal{A}$ is called* Archimedean *if for each element $a \in \mathcal{A}_h$ there exists a $\lambda > 0$ such that $\lambda{\cdot}1 - a \in \mathcal{C}$ and $\lambda{\cdot}1 + a \in \mathcal{C}$.*

Let $\mathcal{C}$ be a quadratic module of $\mathcal{A}$. By the definition of $\mathcal{A}_b(\mathcal{C})$ the quadratic module $\mathcal{C}_b{:=}\mathcal{C} \cap \mathcal{A}_b(\mathcal{C})$ of the $*$-algebra $\mathcal{A}_b(\mathcal{C})$ is Archimedean. Obviously, $\mathcal{C}$ is Archimedean if and only if $\mathcal{A}_b(\mathcal{C}) = \mathcal{A}$. In order to prove that a quadratic module $\mathcal{C}$ is Archimedean, by Proposition 12(i) it suffices to show that a set of generators of $\mathcal{A}$ is in $\mathcal{A}_b(\mathcal{C})$. This fact is essentially used in proving Archimedeaness for all corresponding examples in this section.

Clearly, $\mathcal{C}$ is Archimedean if and only if 1 is an order unit (see [14]) of the corresponding ordered vector space $(\mathcal{A}_h, \succeq)$.

Recall that a point x of a subset M of a real vector space E is called an *internal point* of M if for any $y \in E$ there exists a number $\varepsilon_y > 0$ such that $x + \lambda y \in M$ for all $\lambda \in \mathbb{R}$, $|\lambda| \leq \varepsilon_y$. Let M° denote the set of internal points of M.

Since order units and internal points coincide [14], $\mathcal{C}$ is Archimedean if and only if 1 is an internal point of $\mathcal{C}$. The existence of an internal point is the crucial assumption for Eidelheit's separation theorem for convex sets. Let us say that a $*$-representation π is *$\mathcal{C}$-positive* if $\pi(c) \geq 0$ for all $c \in \mathcal{C}$.

LEMMA 13. *Let $\mathcal{C}$ be an Archimedean quadratic module of $\mathcal{A}$. Suppose that $\mathcal{B}$ is a convex subset of $\mathcal{A}_h$ such that $\mathcal{C}^\circ \cap \mathcal{B} = \emptyset$. Then there exists a state F of the $*$-algebra $\mathcal{A}$ such that the GNS representation π_F is $\mathcal{C}$-positive and $F(b) \leq 0$ for all $b \in \mathcal{B}$. In particular, F is $\mathcal{C}$-positive.*

Proof. By Eidelheit's theorem (see e.g. [14], 0.2.4) there exists a $\mathbb{R}$-linear functional $f \neq 0$ on $\mathcal{A}_h$ such that $\inf\{f(c); c \in \mathcal{C}\} \geq \sup\{f(b); b \in \mathcal{B}\}$. Because $\mathcal{C}$ is a wedge, $f(c) \geq 0$ for all $c \in \mathcal{C}$ and $f(b) \leq 0$ for $b \in \mathcal{B}$. Since $1 \in \mathcal{C}^\circ$ and $f \neq 0$, $f(1) > 0$. We extend $F := f(1)^{-1} f$ to a $\mathbb{C}$-linear functional on $\mathcal{A}$ which is denoted again by F. Since $\sum \mathcal{A}^2 \subseteq \mathcal{C}$, F is a state of the $*$-algebra $\mathcal{A}$. Let π_F denote the GNS representation of F. For $c \in \mathcal{C}$ and $a \in \mathcal{A}$, we have $a^*ca \in \mathcal{C}$ and hence $F(a^*ca) = f(1)^{-1}f(a^*ca) \geq 0$. Therefore, using formula (2.1) we obtain

$$\langle \pi_F(c)\pi_F(a)\varphi_F, \pi_F(a)\varphi_F \rangle = \langle \pi_F(a^*ca)\varphi_F, \varphi_F \rangle = F(a^*ca) \geq 0,$$

that is, $\pi_F(c) \geq 0$ and π_F is $\mathcal{C}$-positive. □

LEMMA 14. *If $\mathcal{C}$ is an Archimedean quadratic module and π is a $\mathcal{C}$-positive $*$-representation of $\mathcal{A}$, then all operators $\pi(a)$, $a \in \mathcal{A}$, are bounded.*

Proof. Let $a \in \mathcal{A}$. Since $\mathcal{C}$ is Archimedean, by Proposition 12(ii) there exists a positive number λ such that $\lambda{\cdot}1 - a^*a \in \mathcal{C}$. Therefore,

$$\langle (\pi(\lambda \cdot 1 - a^*a)\varphi, \varphi \rangle = \lambda \|\varphi\|^2 - \|\pi(a)\varphi\|^2 \geq 0$$

and hence $\|\pi(a)\varphi\| \leq \lambda^{1/2}\|\varphi\|$ for all $\varphi \in \mathcal{D}(\pi)$. □

DEFINITION 6. *A $*$-algebra $\mathcal{A}$ is called* algebraically bounded *if the quadratic module $\sum \mathcal{A}^2$ is Archimedean.*
Since $*$-representations are always $\sum \mathcal{A}^2$-positive, each $*$-representation of an algebraically bounded $*$-algebra acts by bounded operators.

5.2. Abstract positivstellensätze for Archimedean quadratic modules. For the following three propositions we assume that $\mathcal{C}$ is an Archimedean quadratic module of $\mathcal{A}$.

PROPOSITION 15. *For any element $a \in \mathcal{A}_h$ the following are equivalent:*
(i) $a+\varepsilon\cdot 1 \in \mathcal{C}$ *for each* $\varepsilon > 0$.
(ii) $\pi(a) \geq 0$ *for each $\mathcal{C}$-positive $*$-representation π of $\mathcal{A}$.*
(iii) $f(a) \geq 0$ *for each $\mathcal{C}$-positive state f on $\mathcal{A}$.*

Proof. The implications (i)→(ii)→(iii) are clear. To prove that (iii) implies (i) let us assume to the contrary that $a+\varepsilon\cdot 1$ is not in $\mathcal{C}$ for some $\varepsilon > 0$. Applying Lemma 13 with $\mathcal{B} := \{a+\varepsilon\cdot 1\}$ yields a $\mathcal{C}$-positive state f such that $f(a+\varepsilon\cdot 1) \leq 0$. Then we have $f(a) < 0$ which contradicts (iii). □

PROPOSITION 16. *For $a \in \mathcal{A}_h$ the following conditions are equivalent:*
(i) *There exists $\varepsilon > 0$ such that $a-\varepsilon\cdot 1 \in \mathcal{C}$.*
(ii) *For each $\mathcal{C}$-positive $*$-representation π of $\mathcal{A}$ there exists a number $\delta_\pi > 0$ such that $\pi(a-\delta_\pi\cdot 1) \geq 0$.*
(iii) *For each $\mathcal{C}$-positive state f of $\mathcal{A}$ there exists a number $\delta_f > 0$ such that $f(a-\delta_f\cdot 1) \geq 0$.*

Proof. As above, (i)→(ii)→(iii) is obvious. We prove (iii)→(i). Assume that (i) does not hold. We apply Lemma 13 to the Archimedean quadratic module $\tilde{\mathcal{C}} = \mathbb{R}_+\cdot 1 + \mathcal{C}$ and $\mathcal{B} = \{a\}$ and obtain a $\tilde{\mathcal{C}}$-positive state f on $\mathcal{A}$ such that $f(a) \leq 0$. Since f is also $\mathcal{C}$-positive, this contradicts (iii). □

The assertion of next proposition is due to J. Cimpric [3]. It was contained in his talk at the Marseille conference, March 2005.

PROPOSITION 17. *For $a \in \mathcal{A}_h$ the following are equivalent:*
(i) *There exist nonzero elements $x_1, \ldots, x_r$ of $\mathcal{A}$ such that $\sum_{k=1}^r x_k^* a x_k$ belongs to $1+\mathcal{C}$.*
(ii) *For any $\mathcal{C}$-positive $*$-representation π of $\mathcal{A}$ there exists a vector η such that $\langle \pi(a)\eta, \eta\rangle > 0$.*

Proof. (i)→(ii): Suppose that $\sum_k x_k^* a x_k = 1 + c$ with $c \in \mathcal{C}$. If π is a $\mathcal{C}$-positive $*$-representation and $\varphi \in \mathcal{D}(\pi)$, $\varphi \neq 0$, then

$$\begin{aligned}&\textstyle\sum_k \langle \pi(a)\pi(x_k)\varphi, \pi(x_k)\varphi\rangle = \sum_k \langle \pi(x_k^* a x_k)\varphi, \varphi\rangle \\ &= \langle \pi(1+c)\varphi, \varphi\rangle \geq \langle \pi(1)\varphi, \varphi\rangle = \|\varphi\|^2 > 0.\end{aligned}$$

Hence at least one summand $\langle \pi(a)\pi(x_k)\varphi, \pi(x_k)\varphi\rangle$ is positive.

(ii)$\to$(i): Let $\mathcal{B}$ be the set of finite sums of elements x^*ax, where $x \in \mathcal{A}$, and let $\tilde{\mathcal{C}}:=1+\mathcal{C}$. If (i) does not hold, then $\mathcal{B}\cap\tilde{\mathcal{C}} = \emptyset$. By Lemma 13 there exists a state f of $\mathcal{A}$ such that the GNS representation π_f is $\tilde{\mathcal{C}}$-positive and $f(\mathcal{B}) \le 0$. The latter means that $f(x^*ax) = \langle \pi_f(a)\pi_f(x)\varphi_f, \pi(x)\varphi_f\rangle \le 0$ for all $x \in \mathcal{A}$. Since $\mathcal{D}(\pi_f) = \pi_f(\mathcal{A})\varphi_f$ (see e.g. [28], 8.6), the condition in (ii) is not satisfied for the GNS representation π_f. □

5.3. The Archimedean positivstellensatz for compact semialgebraic sets. Suppose that $f{=}(f_1,\cdots,f_k)$ is a k-tuple of polynomials $f_j \in \mathbb{R}[t_1,\cdots,t_d]$. Recall that $\mathcal{K}_f$ denotes the basic closed semialgebraic set (2.2) and $\mathcal{T}_f$ the preorder (2.3) associated with f (and $\mathcal{B} = \mathbb{R}[t_1,\cdots,t_d]$). Then $\mathcal{T}_f$ is a quadratic module of the complex $*$-algebra $\mathcal{A}{=}\mathbb{C}[t_1,\ldots,t_d]$. Recall that "$\preceq$" denotes the order relation defined by $\mathcal{T}_f$.

PROPOSITION 18. *If the set $\mathcal{K}_f$ is compact, then $\mathcal{T}_f$ is Archimedean.*

Proof. Let $p \in \mathbb{R}[t]$ and fix a positive number λ such that $\lambda^2{-}p^2 > 0$ on the compact set $\mathcal{K}_f$. By Stengle's Positivstellensatz, applied to the positive polynomial $\lambda^2{-}p^2$ on $\mathcal{K}_f$, there exist $g, h \in \mathcal{T}_f$ such that

$$g(\lambda^2 - p^2) = 1 + h. \tag{5.2}$$

For $n \in \mathbb{N}_0$, we have $p^{2n}(1+h) \in \mathcal{T}_f$. Therefore, using (5.2) it follows that $p^{2n+2}g = p^{2n}\lambda^2 g{-}p^{2n}(1+h) \preceq p^{2n}\lambda^2 g$. By induction we get

$$p^{2n}g \preceq \lambda^{2n}g. \tag{5.3}$$

Since $p^{2n}(h+gp^2) \in \mathcal{T}_f$, using first (5.2) and then (5.3) we obtain

$$p^{2n} \preceq p^{2n} + p^{2n}(h+gp^2) = p^{2n}\lambda^2 g \preceq \lambda^{2n+2}g. \tag{5.4}$$

Now we out $p := (1+t_1^2)\cdots(1+t_d^2)$. If $|\alpha| \le k$, $k \in \mathbb{N}$, we have

$$\pm 2t^\alpha \preceq t^{2\alpha} + 1 \preceq \sum_{|\beta|\le k} t^{2\beta} = p^k. \tag{5.5}$$

Hence there exist numbers $c > 0$ and $k \in \mathbb{N}$ such that $g \preceq 2cp^k$. Combining the latter with (5.4), we get $p^{2k} \preceq 2c\lambda^{2k+2}p^k$ and so $(p^k{-}\lambda^{2k+2}c)^2 \preceq (\lambda^{2k+2}c)^2{\cdot}1$. Therefore, by Proposition 12(ii), $p^k{-}\lambda^{2k+2}c \in \mathcal{A}_b(\mathcal{T}_f)$ and so $p^k \in \mathcal{A}_b(\mathcal{T}_f)$. Since $\pm t_j \preceq p^k$ by (5.5), we have $t_j \in \mathcal{A}_b(\mathcal{T}_f)$ for $j{=}1,\cdots,d$. From Proposition 12(i) it follows that $\mathcal{A}_b(\mathcal{T}_f) = \mathcal{A}$ which means that $\mathcal{T}_f$ is Archimedean. □

Using the preceding result we now give a new and an (almost) *elementary* proof of the author's Positivstellensatz [29].

THEOREM 19. *Let $q \in \mathbb{R}[t_1,\cdots,t_d]$. If $q(s) > 0$ for all $s \in \mathcal{K}_f$ and $\mathcal{K}_f$ is compact, then $q \in \mathcal{T}_f$.*

Proof. Assume to the contrary that q is not in $\mathcal{T}_f$. By Proposition 18, $\mathcal{T}_f$ is Archimedean. Therefore, by Lemma 13 there exists a $\mathcal{T}_f$-positive state F on $\mathcal{A}$ such that $F(q) \leq 0$. Let $\| p \|$ denote the supremum of $p \in \mathbb{R}[t]$ on the compact set $\mathcal{K}_f$. Our first aim is to show that F is $\| \cdot \|$-continuous.

For let $p \in \mathbb{R}[t]$. Fix $\varepsilon>0$ and put $\lambda := \| p \| + \varepsilon$. We define a state F_1 on the polynomials in one Hermitian indeterminate x by $F_1(x^n) := F(p^n)$, $n \in \mathbb{N}_0$. By the solution of the Hamburger moment problem there exists a positive Borel measure ν on $\mathbb{R}$ such that $F_1(x^n) = \int s^n d\nu(s)$, $n \in \mathbb{N}_0$. For $\gamma > \lambda$ let χ_γ denote the characteristic function of $(-\infty, -\gamma] \cup [\gamma, +\infty)$. Since $\lambda^2 - p^2 > 0$ on $\mathcal{K}_f$, we have $p^{2n} \preceq \lambda^{2n+2} g$ by equation (5.3) of the preceding proof. Using the $\mathcal{T}_f$-positivity of F we derive

$$\gamma^{2n} \int \chi_\gamma \, d\nu \leq \int s^{2n} d\nu(s) = F_1(x^{2n}) = F(p^{2n}) \leq \lambda^{2n+2} F(g)$$

for all $n \in \mathbb{N}$. Since $\gamma > \lambda$, the preceding implies that $\int \chi_\gamma \, d\nu = 0$. Therefore, supp $\nu \subseteq [-\lambda, \lambda]$. Using the Cauchy-Schwarz inequality for F we obtain

$$|F(p)|^2 \leq F(p^2) = F_1(x^2) = \int_{[-\lambda,\lambda]} s^2 \, d\nu(s) \leq \lambda^2 = (\| p \| + \varepsilon)^2.$$

Letting $\varepsilon \to 0$, we get $|F(p)| \leq \| p \|$. That is, F is $\| \cdot \|$-continuous on $\mathbb{R}[t]$.

Since $q > 0$ on the compact set $\mathcal{K}_f$, there is a positive number δ such that $q - \delta \geq 0$ on $\mathcal{K}_f$. By the classical Weierstrass theorem the continuous function $\sqrt{q(s) - \delta}$ on $\mathcal{K}_f$ is uniform limit of a sequence of polynomials $p_n \in \mathbb{R}[t]$. Then $\lim_n \| p_n^2 - q + \delta \| = 0$ and hence $\lim_n F(p_n^2 - q + \delta) = 0$ by the continuity of the functional F. But since $F(p_n^2) \geq 0$ and $F(q) \leq 0$, we have $F(p_n^2 - q + \delta) \geq \delta > 0$ which is the desired contradiction. □

REMARK. 1. The fact that for compact sets $\mathcal{K}_f$ the preorder $\mathcal{T}_f$ is Archimedean was first shown by T. Wörmann [34]. An algorithmic proof of Theorem 19 was given by M. Schweighofer [33].
2. Shortly after the Positivstellensatz [29] appeared, A. Prestel observed that there is a small gap in the proof. (It has to be shown that the functional G_{n+1} occuring therein is nontrivial.) This was immediately repaired by the author and it was the reasoning used in the above proof of Proposition 18 that filled this gap.
3. Having Proposition 18 there are various ways to prove Theorem 19. One can use the spectral theorem as in [29], the Kadison-Dubois theorem as in [34], Jabobi's theorem [12] or Proposition 20 below.

Let $g=(g_1, \cdots, g_n)$ be a tuple of polynomials $g_j \in \mathbb{R}[t_1, \cdots, t_d]$ and let

$$\mathcal{M}_g := \{s_0 + s_1 g_1 + \cdots + s_n g_n;\ s_0, \ldots, s_n \in \sum \mathbb{C}[t_1, \ldots, t_d]^2\}$$

be the quadratic module of $\mathcal{A}=\mathbb{C}[t_1, \ldots, t_d]$ generated by g. The following Proposition is due to M. Putinar [24].

PROPOSITION 20. *Suppose that the set $\mathcal{K}_g$ is compact and the quadratic module $\mathcal{M}_g$ is Archimedean. If $q \in \mathbb{R}[t_1, \cdots, t_d]$ and $q(s) > 0$ for all $s \in \mathcal{K}_g$, then $q \in \mathcal{M}_g$.*

Proof. Assume that $q \notin \mathcal{M}_g$. By Lemma 13, there exists a state F on $\mathcal{A}$ such that $F(q) \leq 0$ and the GNS representation π_F is $\mathcal{M}_g$-positive. By Lemma 14, the symmetric operators $\pi_F(t_1), \ldots, \pi_F(t_d)$ are bounded. The closures of these operators are pairwise commuting bounded selfadjoint operators on the Hilbert space. By the multidimensional spectral theorem (see e.g. [1], 6.5) there is a spectral measure E on a compact subset $\mathcal{Q}$ of $\mathbb{R}^d$ such that $F(p) = \int_{\mathcal{Q}} p(s) d\mu(s)$ for all $p \in \mathbb{C}[t]$, where $\mu(\cdot) := \langle E(\cdot)1, 1\rangle$.

Now we use a standard argument based on the Weierstrass approximation theorem to show that supp $\mu \subseteq \mathcal{K}_g$. Indeed, suppose that $s_0 \notin \mathcal{K}_g$. Then there are a k and a ball $\mathcal{U}$ with radius $\rho>0$ around s_0 such that $g_k < 0$ on $\mathcal{U}$. Define a continuous function f on $\mathbb{R}^d$ by $f(s) = \sqrt{2\rho - ||s||}$ for $||s|| \leq 2\rho$ and $f(s) = 0$ otherwise. From the Weierstrass theorem, there is a sequence of polynomials $p_n \in \mathbb{R}[t]$ which converges to f uniformly on the compact set $\mathcal{Q}$. Then we obtain $\lim_n F(g_k p_n^2) = \int_{\mathcal{U}} g_k(s)(2\rho - ||s||) d\mu(s)$. Since F is $\mathcal{M}_g$-positive, we have $F(g_k p_n^2) \geq 0$, but $g_k(s)(2\rho - ||s||) < 0$ on $\mathcal{U}$. Therefore, $\mu(\mathcal{U}) = 0$. This proves that supp $\mu \subseteq \mathcal{K}_g$.

Since $F(q) = \int_{\mathcal{K}_g} q(s)\ d\mu(s) \leq 0$ and $q > 0$ on $\mathcal{K}_g$, we arrive at a contradiction. □

REMARK. 1. The main part of the preceding proof of Proposition 20 is just the functional-analytic part of the proof of Theorem 1 in [29]. Another more algebraic proof is due to Jacobi [12], see e.g. [18], p. 46.
2. Setting $g_1 = f_1, \ldots, g_k = f_k, g_{k+1} = f_1 f_2, \ldots, g_{2^k - 1} = f_1 \cdots f_k$, we have $\mathcal{K}_g = \mathcal{K}_f$ and $\mathcal{M}_g = \mathcal{T}_f$. Therefore, Theorem 19 follows by combining Propositions 18 and 20.
3. Clearly, the quadratic module $\mathcal{M}_g$ is Archimedean if for some $N \in \mathbb{N}$ the polynomial $N - \sum_{k=1}^d t_k^2$ or the polynomials $N \pm t_k$, $k=1, \ldots, d$, are $\mathcal{M}_g$. In particular, $\mathcal{M}_g$ is Archimedean if one polynomial g_j is $N - \sum_{k=1}^d t_k^2$. On the other hand, if $g_1 = 2t_1 - 1$, $g_2 = 2t_2 - 1$, $g_3 = 1 - t_1 t_2$, then the set $\mathcal{K}_g$ is compact, but $\mathcal{M}_g$ is not Archimedean (see [22], p. 146).
4. The question when the quadratic module $\mathcal{M}_g$ is Archimedean or how many mixed products $f_{i_1} \cdots f_{i_r}$ are needed in order to get the conclusion of Theorem 19 was answered by T. Jacobi and A. Prestel [13], see e.g. [22], Theorems 6.2.2 and 6.3.4.

5.4. Examples of Archimedean quadratic modules.

EXAMPLE 12. *Veronese Map.*
Let $\mathcal{A}$ be the complex $*$–algebra of rational functions generated by

$$x_{kl} := x_k x_l (1 + x_1^2 + \cdots + x_d^2)^{-1}\ ,\ k, l = 0, 1, \ldots, d,$$

where $x_0 := 1$. Since $1 = \sum_{r,s} x_{rs}^2 \succeq x_{kl}^2 \succeq 0$ for $k, l = 0, \ldots, d$, it follows from Proposition 12 that $x_{kl} \in \mathcal{A}_b$ and hence $\mathcal{A}_b = \mathcal{A}$. That is, the quadratic module $\sum \mathcal{A}^2$ is Archimedean and $\mathcal{A}$ is algebraically bounded. This algebra has been used by M.Putinar and F. Vasilescu in [25]. ∘

A large class of algebraically bounded $*$–algebras is provided by coordinate $*$–algebras of compact quantum groups and quantum spaces.

EXAMPLE 13. *Compact quantum group algebras.*
Any compact quantum group algebra $\mathcal{A}$ (see e.g. [15], p. 415) is linear span of matrix elements of finite dimensional unitary corepresentations. These matrix elements v_{kl} with respect an orthonormal basis satisfy the relation $\sum_{l=1}^{d} v_{kl}^* v_{kl} = 1$ for all k ([15], p. 401). Hence each v_{kl} is in $\mathcal{A}_b$ and so $\mathcal{A}_b = \mathcal{A}$. That is, each compact quantum group algebra $\mathcal{A}$ is algebraically bounded and $\sum \mathcal{A}^2$ is Archimedean.

The simplest example is the quantum group $SU_q(2)$, $q \in \mathbb{R}$. The corresponding $*$–algebra has two generators a and c and defining relations

$$ac = qca, \quad c^*c = cc^*, \quad aa^* + q^2cc^* = 1 \ , \quad a^*a + c^*c = 1.$$

From the last relation we see that a and c are in $\mathcal{A}_b$. Hence $\mathcal{A}_b = \mathcal{A}$. ∘

EXAMPLE 14. *Compact quantum spaces.*
Many compact quantum spaces have algebraically bounded coordinate $*$–algebras $\mathcal{A}$. Famous examples are the so-called quantum spheres, see e.g. [15], p. 449. One of the defining relations of the $*$-algebra $\mathcal{A}$ is $\sum_{k=1}^{n} z_k z_k^* = 1$ for the generators $z_1, \dots, z_n$. Hence we have $\mathcal{A}_b = \mathcal{A}$.
The simplest example is the $*$-algebra $\mathcal{A}$ with generators a and defining relation $aa^* + qa^*a = 1$, where $q > 0$. ∘

Weyl algebras and enveloping algebras are not algebraically bounded, but they do have algebraically bounded fraction $*$-algebras. The fraction algebras of the next two examples have been the main technical tools in the proofs of a strict Positivstellensatz in [30] and in [31].

EXAMPLE 15. *A fraction algebra for the Weyl algebra.*
Let $\mathcal{W}(d)$ be the Weyl algebra (Example 2) and set $N = a_1^*a_1 + \cdots + a_d^*a_d$. Let us fix real number α which is not an integer. Let $\mathcal{A}$ be the $*$-subalgebra of the fraction algebra of $\mathcal{W}(d)$ generated by the elements

$$x_{kl} := a_k a_l (N + \alpha 1)^{-1}, \ k, l = 0, \dots, d, \ \text{ and } \ y_n := (N + (\alpha + n)1)^{-1}, \ n \in Z,$$

where $a_0 := 1$. Then $\mathcal{A}$ is algebraically bounded ([30], Lemma 3.1). ∘

EXAMPLE 16. *A fraction algebra for enveloping algebras.*
Let $\mathcal{E}(\mathcal{G})$ be the complex universal enveloping algebra of a Lie algebra $\mathcal{G}$ (Example 3). We fix a basis $\{x_1, \dots, x_d\}$ of the real vector space $\mathcal{G}$ and put $a := 1 + x_1^*x_1 + \cdots + x_d^*x_d$. Let $\mathcal{A}$ be the unital $*$–subalgebra of the fraction algebra of $\mathcal{E}(\mathcal{G})$ generated by the elements $x_{kl} := x_k x_l a^{-1}$, $k, l = 0, \dots, d$, where $x_0 := 1$. As shown in [31], $\mathcal{A}$ is algebraically bounded. ∘

6. Transport of quadratic modules by pre-Hilbert $*$-bimodules. Let $\mathcal{A}$ and $\mathcal{B}$ be complex unital $*$-algebras. We shall show how $\mathcal{A}$–$\mathcal{B}$–bimodules equipped with $\mathcal{A}$– and $\mathcal{B}$–valued sesquilinear forms can be used to move quadratic modules from one algebra to the other. Our assumptions (i)–(vii) are close to the axioms of equivalence bimodules in the theory of Hilbert C^*-modules (see [17], 1.5.3).

Let $\mathcal{X}$ be a left $\mathcal{A}$-module and a right $\mathcal{B}$-module such that $(a{\cdot}x){\cdot}b = a{\cdot}(x{\cdot}b)$ for $a \in \mathcal{A}$, $b \in \mathcal{B}$ and $x \in \mathcal{X}$. Suppose that there is a sesquilinear map $\langle\cdot,\cdot\rangle_{\mathcal{B}} : \mathcal{X} \times \mathcal{X} \to \mathcal{B}$ which is conjugate linear in the first variable and satisfies the following conditions for $x, y \in \mathcal{X}$, $b \in \mathcal{B}$, and $a \in \mathcal{A}$:

(i) $\langle x, y\rangle_{\mathcal{B}}^* = \langle y, x\rangle_{\mathcal{B}}$,
(ii) $\langle x, y{\cdot}b\rangle_{\mathcal{B}} = \langle x, y\rangle_{\mathcal{B}}\, b$,
(iii) $\langle a{\cdot}x, x\rangle_{\mathcal{B}} = \langle x, a^*{\cdot}x\rangle_{\mathcal{B}}$.

For a pre-quadratic module $\mathcal{C}$ of $\mathcal{A}$, let $\mathcal{C}_{\mathcal{X}}$ be the set of all finite sums of elements $\langle a{\cdot}x, x\rangle_{\mathcal{B}}$, where $a \in \mathcal{C}$ and $x \in \mathcal{X}$.

LEMMA 21. *$\mathcal{C}_{\mathcal{X}}$ is a pre-quadratic module of the $*$-algebra $\mathcal{B}$.*

Proof. From (i) and (iii) it follows that $\mathcal{C}_{\mathcal{X}}$ is contained in $\mathcal{B}_h$. Obviously, $\mathcal{C}_{\mathcal{X}}$ satisfies (3.1). Let $b \in \mathcal{B}$, $c \in \mathcal{C}$ and $x \in \mathcal{X}$. Using conditions (ii) and (i) and the bimodule axiom $(a{\cdot}x){\cdot}b = a{\cdot}(x{\cdot}b)$ we obtain

$$b^*\langle a{\cdot}x, x\rangle_{\mathcal{B}} b = b^*\langle a{\cdot}x, x{\cdot}b\rangle_{\mathcal{B}} = (\langle x{\cdot}b, a{\cdot}x\rangle_{\mathcal{B}} b)^* =$$
$$(\langle x{\cdot}b, (a{\cdot}x){\cdot}b\rangle_{\mathcal{B}})^* = \langle x{\cdot}b, a{\cdot}(x{\cdot}b)\rangle_{\mathcal{B}})^* = \langle a{\cdot}(x{\cdot}b), (x{\cdot}b)\rangle_{\mathcal{B}}.$$

Hence $\mathcal{C}_{\mathcal{X}}$ satisfies (3.2) and $\mathcal{C}_{\mathcal{X}}$ is a pre-quadratic module. □

Suppose that $\langle\cdot,\cdot\rangle_{\mathcal{A}} : \mathcal{X} \times \mathcal{X} \to \mathcal{A}$ is a sesquilinear map which is conjugate linear in the second variable such that for $x, y \in \mathcal{X}$, $b \in \mathcal{B}$ and $a \in \mathcal{A}$:

(iv) $\langle x, y\rangle_{\mathcal{A}}^* = \langle y, x\rangle_{\mathcal{A}}$,
(v) $\langle a{\cdot}x, y\rangle_{\mathcal{A}} = a\, \langle x, y\rangle_{\mathcal{A}}$,
(vi) $\langle x{\cdot}b, x\rangle_{\mathcal{A}} = \langle x, x{\cdot}b^*\rangle_{\mathcal{A}}$.

For a pre-quadratic module $\mathcal{P}$ of $\mathcal{B}$, let ${}_{\mathcal{X}}\mathcal{P}$ denote the finite sums of elements $\langle y, y{\cdot}b\rangle_{\mathcal{A}}$, where $b \in \mathcal{P}$ and $y \in \mathcal{X}$. A similar reasoning as in the proof of Lemma 21 shows *${}_{\mathcal{X}}\mathcal{P}$ is a pre-quadratic module for $\mathcal{A}$.*

Finally, we assume the following compatibility condition:

(vii) $\langle x, y\rangle_{\mathcal{A}} \cdot z = x \cdot \langle y, z\rangle_{\mathcal{B}}$ for all $x, y, z \in \mathcal{X}$.

PROPOSITION 22. *If $\mathcal{C}$ is a pre-quadratic module of $\mathcal{A}$ and $\mathcal{P}$ is a pre-quadratic module of $\mathcal{B}$, then we have ${}_{\mathcal{X}}(\mathcal{C}_{\mathcal{X}}) \subseteq \mathcal{C}$ and $({}_{\mathcal{X}}\mathcal{P})_{\mathcal{X}} \subseteq \mathcal{P}$.*

Proof. We prove only the first inclusion. Since ${}_{\mathcal{X}}(\mathcal{C}_{\mathcal{X}})$ consists of sums of elements of the form $\langle y, y{\cdot}\langle a{\cdot}x, x\rangle_{\mathcal{B}}\rangle_{\mathcal{A}}$, where $a \in \mathcal{C}$ and $x, y \in \mathcal{X}$, it suffices to show that these elements are in $\mathcal{C}$. We compute

$$\langle y, y{\cdot}\langle a{\cdot}x, x\rangle_{\mathcal{B}}\rangle_{\mathcal{A}} = \langle y{\cdot}\langle x, a{\cdot}x\rangle_{\mathcal{B}}, y\rangle_{\mathcal{A}} = \langle \langle y, x\rangle_{\mathcal{A}}{\cdot}(a{\cdot}x), y\rangle_{\mathcal{A}}$$
$$= \langle y, x\rangle_{\mathcal{A}}\langle a{\cdot}x, y\rangle_{\mathcal{A}} = (\langle x, y\rangle_{\mathcal{A}})^*\, a\, \langle x, y\rangle_{\mathcal{A}},$$

where the first equality follows from assumptions (vi) and (i), the second from (vii), the third from (v), and the fourth from (iv) and (v). By (3.2) the right hand side of the preceding equations is in $\mathcal{C}$. □

We illustrate these general constructions by an important example.

EXAMPLE 17. *Quadratic modules of k-positive $n \times n$ matrices.* Let R be a unital $*$-algebra. Set $\mathcal{A}=\mathcal{M}_k(R)$, $\mathcal{B}=\mathcal{M}_n(R)$, and $\mathcal{X}=\mathcal{M}_{kn}(R)$. Then $\mathcal{X}$ is an $\mathcal{A}$–$\mathcal{B}$–bimodule with module operations defined by the left resp. right multiplications of matrices and equipped with $\mathcal{B}$–resp. $\mathcal{A}$-valued "scalar products" $\langle x,y\rangle_{\mathcal{B}} := x^*y$ and $\langle x,y\rangle_{\mathcal{A}} := xy^*$ for $x,y \in \mathcal{X}$. With these definitions all assumptions (i)–(vii) are satisfied.

If $\mathcal{C}$ and $\mathcal{P}$ are pre-quadratic modules for $\mathcal{A}=\mathcal{M}_k(R)$ and $\mathcal{B}=\mathcal{M}_n(R)$, respectively, then the pre-quadratic modules $\mathcal{C}_{\mathcal{X}}$ and ${}_{\mathcal{X}}\mathcal{P}$ are given by

$$\begin{aligned} \mathcal{C}_{n,k} &:= \mathcal{C}_{\mathcal{X}} = \{\textstyle\sum_{l=1}^{s} x_l^* a_l x_l;\ a_l \in \mathcal{C},\ x_l \in \mathcal{M}_{kn}(R),\ s \in \mathbb{N}\},. \\ \mathcal{P}_{k,n} &:= {}_{\mathcal{X}}\mathcal{P} = \{\textstyle\sum_{l=1}^{s} y_l b_l y_l^*;\ b_l \in \mathcal{P},\ y_l \in \mathcal{M}_{kn}(R),\ s \in \mathbb{N}\}. \end{aligned} \tag{6.1}$$

Now we specialize the preceding by setting $R=\mathbb{C}[t_1,\ldots,t_d]$. Let $\mathcal{C}$ be the quadratic module $\mathcal{M}_k(\mathbb{C}[t])_+$ of Hermitian $k \times k$ matrices over $\mathbb{C}[t]$ which are positive semidefinite for all $s \in \mathbb{R}^d$. Put $\mathcal{C}_{n,0} := \sum \mathcal{B}^2$. From (6.1) we obtain an increasing chain of quadratic modules

$$\mathcal{C}_{n,0} \subseteq \mathcal{C}_{n,1} \subseteq \mathcal{C}_{n,2} \subseteq \cdots \subseteq \mathcal{C}_{n,n} \tag{6.2}$$

of $\mathcal{B}=\mathcal{M}_n(\mathbb{C}[t])$. Matrices belonging to $\mathcal{C}_{n,k}$ will be called *k-positive.*

If $d=1$ and $a \in \mathcal{C}_{n,n}$, then the matrix a is positive semidefinite on $\mathbb{R}$ and hence of the form $a = b^*b$ for some $b \in \mathcal{M}_n(\mathbb{C}[t])$ [5]. Therefore all quadratic modules in (6.2) coincide with $\sum \mathcal{M}_n(\mathbb{C}[t])^2$.

Suppose now that $d \geq 2$. As shown in [6], the matrix

$$\begin{pmatrix} 1+t_1^4t_2^2 & t_1t_2 \\ t_1t_2 & 1+t_1^2t_2^4 \end{pmatrix}.$$

is in $\mathcal{C}_{2,2}$, but not in $\mathcal{C}_{2,1}$. For $d \geq 2$ we have a sequence $\mathcal{C}_{n,k}$ of intermediate quadratic modules between the two extremes $\mathcal{C}_{n,0} = \sum \mathcal{M}_n(\mathbb{C}[t])^2$ and $\mathcal{C}_{n,n} = \mathcal{M}_n(\mathbb{C}[t])_+$. These quadratic modules are used in Hilbert space representation theory to characterize k-positive representations of the polynomial algebra $\mathbb{C}[t_1,\ldots,t_d]$ (see [6] and [28], Proposition 11.2.5).

REFERENCES

[1] M.S. BIRMAN AND M.Z. SOLOMJAK, *Spectral Theory of Self-Adjoint Operators in Hilbert Space*, D. Reidel Publishing Co., Dordrecht, 1987.

[2] J. BOCHNAK, M. COSTE, AND M.-F. ROY, *Real Algebraic Geometry*, Springer-Verlag, Berlin, 1998.

[3] J. CIMPRIC, *Maximal quadratic modules on $*$-rings*, Algebr. Represent. Theory (2007), DOI 10.1007/S10468-007-9076-z.

[4] J. CIMPRIC, *A representation theorem for quadratic modules on $*$-rings*, 2005, to appear in Canadian Mathematical Bulletin.

[5] D.Z. DJOKOVIC, *Hermitean matrices over polynomial rings*, J. Algebra **43** (1976), pp. 359–374.

[6] J. FRIEDRICH AND K. SCHMÜDGEN, *n-Positivity of unbounded $*$-representations*, Math. Nachr. **141** (1989), pp. 233–250.

[7] F.R. Gantmacher, *Matrizentheorie*, DVW, Berlin, 1986.
[8] D. Gondard and P. Ribenboim, *Le 17e probleme de Hilbert pour les matrices*, Bull. Sci. Math. **98** (1974), pp. 49–56.
[9] J.W. Helton, *Positive noncommutative polynomials are sums of squares*, Ann. Math. **156** (2002), pp. 675–694.
[10] J.W. Helton and S. McCullough, *A Positivstellensatz for non-commutative polynomials*, Trans. Amer. Math. Soc. **356** (2004), pp. 3721–3737.
[11] J.W. Helton, S. McCullough, and M. Putinar, *A non-commutative Positivstellensatz on isometries*, J. Reine Angew. Math. **568** (2004), pp. 71–80.
[12] T. Jacobi, *A representation theorem for certain partially ordered commutative rings*, Math. Z. **237** (2001), pp. 259–273.
[13] T. Jacobi and A. Prestel, *Distingiushed representations of strictly positive polynomials*, J. reine angew. Math. **532** (2001), 223–235.
[14] G. Jameson, *Ordered Linear Spaces*, Lecture Notes in Math. No. **141**, Spinger-Verlag, Berlin, 1970.
[15] A. Klimyk and K. Schmüdgen, *Quantum Groups and Their Representations*, Springer-Verlag, Berlin, 1997.
[16] M. Knebusch and C. Scheiderer, *Einführung in die reelle Algebra*, Vieweg-Verlag, Braunschweig, 1989.
[17] V.M. Manuilov and E.V. Troitsky, *Hilbert C^*-Modules*, Amer. Math. Soc., Transl. Math. Monographs **226**, 2005.
[18] M. Marshall, *Positive Polynomials and sums of Squares*, Univ. Pisa, Dipart. Mat. Istituti Editoriali e Poligrafici Internaz., 2000.
[19] M. Marshall, *Extending the Archimedean Positivstellensatz to the non-compact case*, Can. Math. Bull. **14**(2001), 223–230.
[20] V. Ostrovskyj and Yu. Samoilenko, *Introduction to the Theory of Finitely Presented $*$-Algebras*, Harwood Acad. Publ., 1999.
[21] V. Powers and C. Scheiderer, *The moment problem for non-compact semialgebraic sets*, Adv. Geom.**1** (2001), pp. 71–88.
[22] A. Prestel and Ch.N. Delzell, *Positive Polynomials*, Spinger-Verlag, Berlin, 2001.
[23] C. Procesi and M. Schacher, *A Non-Commutative Real Nullstellensatz and Hilbert's 17th Problem*, Ann Math. **104** (1976), pp. 395–406.
[24] M. Putinar, *Positive polynomials on compact semi-algebraic sets*, Indiana Univ. Math. J. **42**, pp. 969–984.
[25] M. Putinar and F. Vasilescu, *Solving moment problems by dimension extension*, Ann. Math. **149**, pp. 1087–1107.
[26] H. Schäfer, *Topological Vector Spaces*, Springer-Verlag, Berlin, 1972.
[27] K. Schmüdgen, *Graded and filtrated topological $*$-algebras II. The closure of the positive cone*, Rev. Roum. Math. Pures et Appl. **29** (1984), pp. 89–96.
[28] K. Schmüdgen, *Unbounded Operator Algebras and Representation Theory*, Birkhäuser-Verlag, Basel, 1990.
[29] K. Schmüdgen, *The K-moment problem for compact semi-algebraic sets*, Math. Ann. **289** (1991), pp. 203–206.
[30] K. Schmüdgen, *A strict Positivstellensatz for the Weyl algebra*, Math. Ann. **331** (2005), pp. 779–794.
[31] K. Schmüdgen, *A strict Positivstellensatz for enveloping algebras*, Math. Z. **254** (2006), pp. 641–653. Erratum: DOI 10.1007/s00209-007-0295-0.
[32] M. Schweighofer, *Iterated rings of bounded elements and generalizations of Schmüdgen's theorem*, J. Reine Angew. Math. **554** (2003), pp. 19–45.
[33] M. Schweighofer, *An algorithmic approach to Schmüdgen's Positivstellensatz*, J. Pure Appl. Algebra **166** (2002), pp. 307–309.
[34] T. Wörmann, *Strict positive Polynome in der semialgebraischen Geometrie*, Dissertation, Universität Dortmund, 1998.

OPEN PROBLEMS IN ALGEBRAIC STATISTICS

BERND STURMFELS*

Abstract. Algebraic statistics is concerned with the study of probabilistic models and techniques for statistical inference using methods from algebra and geometry. This article presents a list of open mathematical problems in this emerging field, with main emphasis on graphical models with hidden variables, maximum likelihood estimation, and multivariate Gaussian distributions. These are notes from a lecture presented at the IMA in Minneapolis during the 2006/07 program on Applications of Algebraic Geometry.

Key words. Algebraic statistics, contingency tables, hidden variables, Schur modules, maximum likelihood, conditional independence, multivariate Gaussian, gaussoid.

AMS(MOS) subject classifications. 13P10, 14Q15, 62H17, 65C60.

1. Introduction. This article is based on a lecture given in March 2007 at the workshop on *Statistics, Biology and Dynamics* held at the Institute for Mathematics and its Applications (IMA) in Minneapolis as part of the 2006/07 program on *Applications of Algebraic Geometry*. In four sections we present mathematical problems whose solutions would likely become important contributions to the emerging interactions between algebraic geometry and computational statistics. Each of the four sections starts out with a "specific problem" which plays the role of representing the broader research agenda. The latter is summarized in a "general problem".

Algebraic statistics is concerned with the study of probabilistic models and techniques for statistical inference using methods from algebra and geometry. The term was coined in the book of Pistone, Riccomagno and Wynn [25] and subsequently developed for biological applications in [24]. Readers from statistics will enjoy the introduction and review recently given by Drton and Sullivant [8], while readers from algebra will find various points of entry cited in our discussion and listed among our references.

2. Graphical models with hidden variables. Our first question concerns three-dimensional contingency tables (p_{ijk}) whose indices i, j, k range over a set of four elements, such as the set {A, C, G, T} of DNA bases.

Specific problem: *Consider the variety of* $4\times4\times4$*-tables of tensor rank at most* 4. *There are certain known polynomials of degree at most nine which vanish on this variety. Do they suffice to cut out the variety?* This particular open problem appears in [24, Conjecture 3.24], and it here serves as a placeholder for the following broader direction of inquiry.

General problem: *Study the geometry and commutative algebra of graphical models with hidden random variables. Construct these varieties by gluing familiar secant varieties, and by applying representation theory.*

*University of California, Berkeley, CA 94720, USA (bernd@math.berkeley.edu).

We are interested in statistical models for discrete data which can be represented by polynomial constraints. As is customary in algebraic geometry, we consider varieties over the field of complex numbers, with the tacit understanding that statisticians mostly care about points whose coordinates are real and non-negative. The model referred to in the Specific Problem lives in the 64-dimensional space $\mathbb{C}^4 \otimes \mathbb{C}^4 \otimes \mathbb{C}^4$ of 4×4×4-tables (p_{ijk}), where $i, j, k \in \{\mathtt{A}, \mathtt{C}, \mathtt{G}, \mathtt{T}\}$. It has the parametric representation

$$p_{ijk} \quad = \quad \begin{array}{l} \rho_{\mathtt{A}i} \cdot \sigma_{\mathtt{A}j} \cdot \theta_{\mathtt{A}k} + \rho_{\mathtt{C}i} \cdot \sigma_{\mathtt{C}j} \cdot \theta_{\mathtt{C}k} + \\ \rho_{\mathtt{G}i} \cdot \sigma_{\mathtt{G}j} \cdot \theta_{\mathtt{G}k} + \rho_{\mathtt{T}i} \cdot \sigma_{\mathtt{T}j} \cdot \theta_{\mathtt{T}k}. \end{array} \qquad (2.1)$$

Our problem is to compute the homogeneous prime ideal I of all polynomials which vanish on this model. The desired ideal I lives in the polynomial ring $\mathbb{Q}[p_{\mathtt{AAA}}, p_{\mathtt{AAC}}, p_{\mathtt{AAT}}, \ldots, p_{\mathtt{TTG}}, p_{\mathtt{TTT}}]$ with 64 unknowns. In principle, one can compute generators of I by applying Gröbner bases methods to the parametrization (2.1). However, our problem has 64 probabilities and 48 parameters, and it is simply too big for the kind of computations which were performed in [24, §3.2] using the software package `Singular` [13].

Given that Gröbner basis methods appear to be too slow for any problem size which is actually relevant for real data, skeptics may wonder why a statistician should bother learning the language of ideals and varieties. One possible response to the practitioner's legitimate question *"Why (pure) mathematics?"* is offered by the following quote due to Henri Poincaré:

"Mathematics is the Art of Giving the Same Name to Different Things".

Indeed, our prime ideal I gives the same name to the following things:

- the set of 4×4×4-tables of tensor rank ≤ 4,
- the mixture of four models for three independent random variables,
- the naive Bayes model with four classes,
- the conditional independence model $[X_1 \perp\!\!\!\perp X_2 \perp\!\!\!\perp X_3 \,|\, Y\,]$,
- the fourth secant variety of the Segre variety $\mathbb{P}^3 \times \mathbb{P}^3 \times \mathbb{P}^3$,
- the general Markov model for the phylogenetic tree $K_{1,3}$,
- superposition of four pure states in a quantum system [4, 14].

These different terms have been used in the literature for the geometric object represented by (2.1). The concise language of commutative algebra and algebraic geometry can be an effective channel of communication for the different communities of statisticians, computer scientists, physicists, engineers and biologists, all of whom have encountered formulas like (2.1).

The generators of lowest degree in our ideal I have degree five, and the known generators of highest degree have degree nine. The analysis of Landsberg and Manivel in [20, Proposition 6.3] on 3×3×4-tables of tensor rank four implies the existence of additional ideal generators of degree six in I. This analysis had been overlooked by the authors of [24] when they formulated their Conjecture 3.24. Readers of [24, Chapter 3] are herewith kindly asked to replace *"of degree* 5 *and* 9*"* by *"of degree at most* 9*"*.

In what follows we present the known minimal generators of degree five and nine in our prime ideal I, and we postpone a more detailed discussion of the Landsberg-Manivel sextics in [20, Proposition 6.3] to a future study.

Consider any $3\times4\times4$-subtable (p_{ijk}) and let A, B, C be the 4×4-slices gotten by fixing i. To be precise, the entry of the 4×4-matrix A in row j and column k equals $p_{\mathbf{A}jk}$, the entry of B in row j and column k equals $p_{\mathbf{C}jk}$, and the entry of C in row j and column k equals $p_{\mathbf{G}jk}$. We can check that the following identity of 4×4-matrices holds for all tables in our model, provided the matrix B is invertible:

$$A\cdot B^{-1}\cdot C \;=\; C\cdot B^{-1}\cdot A.$$

After clearing the denominator $\det(B)$, we can write this identity as

$$A\cdot \mathrm{adj}(B)\cdot C \,-\, C\cdot \mathrm{adj}(B)\cdot A \;=\; 0, \tag{2.2}$$

where $\mathrm{adj}(B)=\det(B)\cdot B^{-1}$ is the adjoint matrix of B. The matrix entries on the left hand side give 16 quintic polynomials which lie in our prime ideal I. Each matrix entry is a polynomial with 180 terms which involve only 30 of the 64 unknowns. For example, the upper left entry looks like this:

$$\begin{gathered}
p_{\mathbf{AAC}}p_{\mathbf{CCA}}p_{\mathbf{CGG}}p_{\mathbf{CTT}}p_{\mathbf{GAA}} - p_{\mathbf{AAC}}p_{\mathbf{CCA}}p_{\mathbf{CGT}}p_{\mathbf{CTG}}p_{\mathbf{GAA}} \\
- p_{\mathbf{AAC}}p_{\mathbf{CCG}}p_{\mathbf{CGA}}p_{\mathbf{CTT}}p_{\mathbf{GAA}} + p_{\mathbf{AAC}}p_{\mathbf{CCT}}p_{\mathbf{CGA}}p_{\mathbf{CTG}}p_{\mathbf{GAA}} \\
+ \quad \cdots\cdots (175 \text{ terms})\cdots\cdots \quad - p_{\mathbf{ATA}}p_{\mathbf{CAG}}p_{\mathbf{CCC}}p_{\mathbf{CGA}}p_{\mathbf{GAT}}.
\end{gathered}$$

We note that there are no non-zero polynomials of degree ≤ 4 in the ideal I. This follows from general results on secant varieties [5, 17].

An explicit linear algebra computation reveals that all polynomials of degree five in I are gotten from the above construction by relabeling and considering all subtables of format $3\times4\times4$, format $4\times3\times4$ and format $4\times4\times3$, and by applying the natural action of the group $GL(\mathbb{C}^4)\times GL(\mathbb{C}^4)\times GL(\mathbb{C}^4)$ on $4\times4\times4$-tables. This action leaves the ideal I fixed. We identify the representation of this group on the space of quintics in I.

PROPOSITION 2.1. *The space of quintic polynomials in the prime ideal I of (2.1) has dimension* 1728. *As a $GL(\mathbb{C}^4)^3$-module, it is isomorphic to*

$$\begin{gathered}
S_{311}(\mathbb{C}^4)\otimes S_{2111}(\mathbb{C}^4)\otimes S_{2111}(\mathbb{C}^4)\\
\oplus\; S_{2111}(\mathbb{C}^4)\otimes S_{311}(\mathbb{C}^4)\otimes S_{2111}(\mathbb{C}^4)\\
\oplus\; S_{2111}(\mathbb{C}^4)\otimes S_{2111}(\mathbb{C}^4)\otimes S_{311}(\mathbb{C}^4).
\end{gathered}$$

Here $S_\lambda(\mathbb{C}^4)$ denotes the *Schur modules* which are the irreducible representations of $GL(\mathbb{C}^4)$. We refer to [10] for the relevant basics on representation theory of the general linear group, and to [17, 18, 19] for more detailed information about the specific modules under consideration here.

The known invariants of degree nine are also obtained by a similar construction. Consider any $3 \times 3 \times 3$-subtable (p_{ijk}) and denote the three slices of that table by A, B and C. We now consider the 3×3-determinant

$$\det(A \cdot B^{-1} \cdot C - C \cdot B^{-1} \cdot A). \tag{2.3}$$

The denominator of the rational function (2.3) is $\det(B)^2$ and not $\det(B)^3$ as one might think on first glance. The numerator of (2.3) is a homogeneous polynomial of degree nine with 9216 terms which remains invariant under permuting A, B and C. This homogeneous polynomial of degree nine lies in the ideal I and is known as the *Strassen invariant.*

PROPOSITION 2.2. *The $GL(\mathbb{C}^4)^3$-submodule of the degree 9 component I_9 generated by the Strassen invariant is not contained in the ideal $\langle I_5 \rangle$ generated by the quintics in Proposition 2.1. This module has vector space dimension 8000 and it is isomorphic to the representation*

$$S_{333}(\mathbb{C}^4) \otimes S_{333}(\mathbb{C}^4) \otimes S_{333}(\mathbb{C}^4).$$

The first appearance of the Strassen invariant in algebraic statistics was [11, Proposition 22]. A conceptual study of the matrix construction $AB^{-1}C - CB^{-1}A$ was undertaken by Landsberg and Manivel in [18].

The Specific Problem at the beginning of this section plays a pivotal role also in algebraic phylogenetics [1, 2, 3]. Our model (2.1) is known there as the general Markov model on a tree with three leaves branching off directly from the root. Allman and Rhodes [2, §6] showed that phylogenetic invariants which cut out the general Markov model on any larger binary rooted tree can be constructed from the generators of our ideal I by a gluing process. The invariants of degree five and nine arising from (2.2) and (2.3) are therefore basic building blocks for phylogenetic invariants on arbitrary trees whose nodes are labeled with the four letters **A**, **C**, **G** and **T**.

In her lecture at the same IMA conference in March 2007, Elizabeth Allman [1] offered an extremely attractive prize for the resolution of the Specific Problem. She offered to personally catch and smoke wild salmon from the Copper River, located in her "backyard" in Alaska, and ship it to anyone who will determine a minimal generating set of the prime ideal I.

In Propositions 2.1 and 2.2, we emphasized the language of representation theory in characterizing the defining equations of graphical statistical models. This methodology is a main focus in the forthcoming book by J.M. Landsberg and Jason Morton, which advocates the idea of using Schur modules $S_\lambda(\mathbb{C}^n)$ in the description of such models. Morton's key insight is that this naturally generalizes conditional independence, the current language of choice for characterizing graphical models. Conditional independence statements can be interpreted as a convenient shorthand for large systems of quadratic equations; see [12, §4.1] or [27, Proposition 8.1].

In the absence of hidden random variables, the quadratic equations expressed implicitly by conditional independence are sufficient to characterize graphical models. This is the content of the *Hammersley-Clifford Theorem* (see e.g. [12, Theorem 4.1] or [24, Theorems 1.30 and 1.33]). However, when some of the random variables in a graphical model are hidden then the situation becomes much more complicated. We believe that representation theory of the general linear group can greatly enhance the conditional independence calculus which is so widely used by graphical models experts. The representation-theoretic notation was here illustrated for a tiny graphical model, having three observed random variables and one hidden random variable, all four having the same state space $\{\mathbf{A}, \mathbf{C}, \mathbf{G}, \mathbf{T}\}$.

3. Maximum likelihood estimation. In this section we discuss topics concerning the algebraic approach to maximum likelihood estimation [24, §3.3]. The following open problem was published in [16, Problem 13].

Specific problem: *Find a geometric characterization of those projective varieties whose maximum likelihood degree (ML degree) is equal to one.*

This question and others raised in [6, 16] are just the tip of an iceberg:

General problem: *Study the geometry of maximum likelihood estimation for algebraic statistical models.*

Here algebraic statistical models are regarded as projective varieties. A model has ML degree one if and only if its maximum likelihood estimator is a rational function of the data. Models which have this property tend to be very nice. For instance, in the special context of undirected graphical models (Markov random fields), the property of having ML degree one is equivalent to the statement that the graph is decomposable [12, Theorem 4.4]. For toric varieties, our question was featured in [27, Problem 8.23].

It is hoped that the ML degree is related to convergence properties of numerical algorithms used by statisticians, such as iterative proportional scaling or the EM algorithm, but no systematic study in this direction has yet been undertaken. In general, we wish to learn how statistical features of a model relate to geometric properties of the corresponding variety.

Here are the relevant definitions for our problems. We fix the complex projective space $\mathbb{P}^n$ with coordinates $(p_0 : p_1 : \cdots : p_n)$. The coordinate p_i represents the probability of the ith event. The n-dimensional probability simplex is identified with the set $\mathbb{P}^n_{\geq 0}$ of points in $\mathbb{P}^n$ which have non-negative real coordinates. The data comes in the form of a non-negative integer vector $(u_0, u_1, \ldots, u_n) \in \mathbb{N}^{n+1}$. Here u_i is the number of times the ith event was observed. The corresponding *likelihood function* is defined as

$$L(p_0, p_1, \ldots, p_n) \quad = \quad \frac{{p_0}^{u_0} \cdot {p_1}^{u_1} \cdot {p_2}^{u_2} \cdot \cdots \cdot {p_n}^{u_n}}{(p_0 + p_1 + \cdots + p_n)^{u_0 + u_1 + \cdots + u_n}}. \tag{3.1}$$

Statistical computations are typically done in affine n-space specified by $p_0 + p_1 + \cdots + p_n = 1$, where the denominator of L can be ignored. However, the denominator is needed in order for L to be a well-defined rational function on $\mathbb{P}^n$. The unique critical point of the likelihood function L is at $(u_0 : u_1 : \cdots : u_n)$, and this point is the global maximum of L over $\mathbb{P}^n_{\geq 0}$. By a *critical point* we mean any point at which the gradient of L vanishes.

An *algebraic statistical model* is represented by a subvariety $\mathcal{M}$ of the projective space $\mathbb{P}^n$. The model itself is the intersection of $\mathcal{M}$ with the probability simplex $\mathbb{P}^n_{\geq 0}$. The *ML degree* of the variety $\mathcal{M}$ is the number of complex critical points of the restriction of the likelihood function L to $\mathcal{M}$. Here we disregard singular points of $\mathcal{M}$, we only count critical points that are not poles or zeros of L, and $u_0, u_1, \ldots, u_n$ are assumed to be generic. If $\mathcal{M}$ is smooth and the divisor on $\mathcal{M}$ defined by L has normal crossings then there is a geometric characterization of the ML degree, derived in the paper [6] with Catanese, Hoşten and Khetan. The assumptions of smoothness and normal crossing are very restrictive and almost never satisfied for models of statistical interest. In general, to understand the ML degree will require invoking some resolution of singularities and its algebraic underpinnings.

We illustrate the computation of the ML degree for the case when $\mathcal{M}$ is a plane curve. Here $n = 2$ and $\mathcal{M}$ is the zero set of a homogeneous polynomial $F(p_0, p_1, p_2)$. Using Lagrange multipliers or [16, Proposition 2], we derive that the condition for $(p_0 : p_1 : p_2)$ to be a critical point of the restriction of L to $\mathcal{M}$ is equivalent to the system of two equations

$$F(p_0, p_1, p_2) \;=\; \det \begin{pmatrix} u_0 & p_0 & p_0 \cdot \partial F/\partial p_0 \\ u_1 & p_1 & p_1 \cdot \partial F/\partial p_1 \\ u_2 & p_2 & p_2 \cdot \partial F/\partial p_2 \end{pmatrix} \;=\; 0.$$

For a general polynomial F of degree d, these equations will have $d(d+1)$ solutions, by Bézout's Theorem. Moreover, all of these solutions satisfy

$$p_0 \cdot p_1 \cdots p_n \cdot (p_0 + p_1 + \cdots + p_n) \;\neq\; 0, \tag{3.2}$$

and we conclude that the ML degree of a general plane curve of degree d is equal to $d(d+1)$. However, that number can drop considerably for special curves. For instance, while the ML degree of a general plane quadric equals six, the special quadric $\{p_1^2 = \lambda p_0 p_2\}$ has ML degree two for $\lambda \neq 4$, and it has ML degree one for $\lambda = 4$. Thus, returning to the Special problem, our first example of a variety of ML degree one is the plane curve defined by

$$F \;=\; \det \begin{pmatrix} 2p_0 & p_1 \\ p_1 & 2p_2 \end{pmatrix}. \tag{3.3}$$

Biologists know this as the *Hardy-Weinberg curve*, with the parametrization

$$p_0 \,=\, \theta^2, \quad p_1 \,=\, 2\theta(1-\theta), \quad p_2 \,=\, (1-\theta)^2. \tag{3.4}$$

The unique critical point of the likelihood function L on this curve equals

$$\big((2u_0 + u_1)^2 \,:\, 2(2u_0{+}u_1)(u_1{+}2u_2) \,:\, (u_1 + 2u_2)^2 \big).$$

Determinantal varieties arise naturally in statistics. They are the models $\mathcal{M}$ that are specified by imposing rank conditions on a matrix of unknowns. A first example is the model (3.4) for two i.i.d. binary random variables. For a second example we consider the general 3×3-matrix

$$P \quad = \quad \begin{pmatrix} p_{00} & p_{01} & p_{02} \\ p_{10} & p_{11} & p_{12} \\ p_{20} & p_{21} & p_{22} \end{pmatrix} \tag{3.5}$$

which represents two ternary random variables. The independence model for these two random variables is the variety of rank one matrices. This model also has ML degree one, i.e., the maximum likelihood estimator is a rational function in the data. It is given by the 3×3-matrix whose entry in row i and column j equals $(u_{i0} + u_{i1} + u_{i2}) \cdot (u_{0j} + u_{1j} + u_{2j})$.

By contrast, consider the *mixture model* based on two ternary random variables. It consists of all matrices P of rank at most two. Thus this model is the hypersurface defined by the cubic polynomial $F = \det(P)$. Explicit computation shows that the ML degree of this hypersurface is ten. In general, it remains an open problem to find a formula, in terms of m, n and r, for the ML degree of the variety of $m \times n$-matrices of rank $\leq r$.

The first interesting case arises when $m = n = 4$ and $r = 2$. At present we are unable to solve the likelihood equations for this case symbolically. The following concrete biology example was proposed in [24, Example 1.16]:

"Our data are two aligned DNA sequences ...

ATCACCAAACATTGGGATGCCTGTGCATTTGCAAGCGGCT
ATGAGTCTTAAACGCTGGCCATGTGCCATCTTAGACAGCG

... test the hypothesis that these two sequences were generated by DiaNA using one biased coin and four tetrahedral dice ... "

Here the model $\mathcal{M}$ consists of all (positive) 4×4-matrices (p_{ij}) of rank at most two. In the given alignment, each match occurs four times and each mismatch occurs two times. Hence the likelihood function (3.1) equals

$$L \;=\; \Big(\prod_i p_{ii}\Big)^4 \cdot \Big(\prod_{i \neq j} p_{ij}\Big)^2 \cdot \Big(\sum_{i,j} p_{ij}\Big)^{-40}.$$

Based on experiments with the EM algorithm, we conjectured that the matrix $(\hat{p}_{ij}) = \frac{1}{40}\begin{pmatrix} 3 & 3 & 2 & 2 \\ 3 & 3 & 2 & 2 \\ 2 & 2 & 3 & 3 \\ 2 & 2 & 3 & 3 \end{pmatrix}$ is a global maximum of the likelihood function L. In the *Nachdiplomsverlesung* (postgraduate course) which I held at ETH Zürich in the summer of 2005, I offered a cash prize of 100 Swiss Francs for the resolution of this very specific conjecture, and this prize remains unclaimed and is still available at this time (August 2007).

The state of the art on this *100 Swiss Francs Conjecture* is the work of Hersh which originated in March 2007 at the IMA. She proved a range of constraints on the maximum likelihood estimates of determinantal models, especially when the data u_{ij} have symmetry. A discussion of these ideas appears in Hersh's paper with Fienberg, Rinaldo and Zhou [9]. That paper gives an exposition of MLE for determinantal models aimed at statisticians.

4. Gaussian conditional independence models. The early literature on algebraic statistics, including the book [24], dealt primarily with discrete random variables (binary, ternary,...). The set-up was as described in the previous two sections. We now shift gears and consider multivariate Gaussian distributions. For continuous random variables, we must work in the space of model parameters in order to apply algebraic geometry. The following concrete problem concerns Gaussian distributions on $\mathbb{R}^5$.

Specific problem: *Which sets of almost-principal minors can be zero for a positive definite symmetric* 5×5*-matrix?*

The general question behind this asks for characterization of all conditional independence models which can be realized by Gaussians on $\mathbb{R}^n$.

General problem: *Study the geometry of conditional independence models for multivariate Gaussian random variables.*

The state of the art on these problems appears in the work of František Matúš and his collaborators. In particular, Matúš' recent paper with Lněnička [20] on *representation of gaussoids* solves our Specific Problem for symmetric 4×4-matrices. Sullivant's construction in [28] complements that work. For more information see also the article by Šimeček [26].

Let us begin, however, with some basic definitions. Our aim is to discuss these problems in a self-contained manner. A *multivariate Gaussian* distribution on $\mathbb{R}^n$ with mean zero is specified by its covariance matrix $\Sigma = (\sigma_{ij})$. The $n\times n$-matrix Σ is symmetric and it is *positive definite*, which means that all its 2^n principal minors are positive real numbers.

An *almost-principal minor* of Σ is a subdeterminant which has row indices $\{i\}\cup K$ and column indices $\{j\}\cup K$ for some $K \subset \{1,\ldots,n\}$ and $i,j \in \{1,\ldots,n\}\backslash K$. We denote this subdeterminant by $[\, i \perp\!\!\!\perp j \,|K\,]$. For example, if $n=5$, $i=2, j=4$ and $K=\{1,5\}$ then the corresponding almost-principal minor of the symmetric 5×5-matrix Σ equals

$$[\,2 \perp\!\!\!\perp 4\,|\{1,5\}\,] \quad = \quad \det\begin{pmatrix} \sigma_{24} & \sigma_{12} & \sigma_{25} \\ \sigma_{14} & \sigma_{11} & \sigma_{15} \\ \sigma_{45} & \sigma_{15} & \sigma_{55} \end{pmatrix}.$$

Our notation for almost-principal minors is justified by their intimate connection to conditional independence, expressed in the following lemma. We note that the almost-principal minors are referred to as *partial covariance* (or, if renormalized, *partial correlations*) in the statistics literature.

LEMMA 4.1. *The subdeterminant* $[i \perp\!\!\!\perp j \,|\, K]$ *is zero for a positive definite symmetric* $n \times n$*-matrix* Σ *if and only if, for the Gaussian random variable* X *on* $\mathbb{R}^n$ *with covariance matrix* Σ*, the random variable* X_i *is independent of the random variable* X_j *given the joint variable* X_K*.*

Proof. See [7, Equation(5)], [22, Section 1], or [28, Proposition 2.1]. □

Let PD_n denote the $\binom{n+1}{2}$-dimensional cone of positive definite symmetric $n \times n$-matrices. Note that this cone is open. A *Gaussian conditional independence model*, or *GCI model* for short, is any semi-algebraic subset of the cone PD_n which can be defined by polynomial equations of the form

$$[i \perp\!\!\!\perp j \,|\, K] \quad = \quad 0. \tag{4.1}$$

In algebraic geometry, we simplify matters by studying the complex algebraic varieties defined by equations of the form (4.1). Of course, what we are particularly interested in is the real locus of such a complexified GCI model, and how it intersects the positive definite cone PD_n and its closure.

As an illustration of algebraic reasoning for Gaussian conditional independence models, we examine an example taken from [28]. Let $n = 5$ and consider the GCI model given by the five quadratic polynomials

$$\begin{aligned}
[1 \perp\!\!\!\perp 2 \,|\, \{3\}] &= \sigma_{12}\sigma_{33} - \sigma_{13}\sigma_{23} \\
[2 \perp\!\!\!\perp 3 \,|\, \{4\}] &= \sigma_{23}\sigma_{44} - \sigma_{24}\sigma_{34} \\
[3 \perp\!\!\!\perp 4 \,|\, \{5\}] &= \sigma_{34}\sigma_{55} - \sigma_{35}\sigma_{45} \\
[4 \perp\!\!\!\perp 5 \,|\, \{1\}] &= \sigma_{45}\sigma_{11} - \sigma_{14}\sigma_{15} \\
[5 \perp\!\!\!\perp 1 \,|\, \{2\}] &= \sigma_{15}\sigma_{22} - \sigma_{25}\sigma_{12}.
\end{aligned}$$

This variety is a complete intersection (it has dimension ten) in the 15-dimensional space of symmetric 5×5-matrices. Primary decomposition reveals that it is the union of precisely two irreducible components, namely,

- the linear space $\{\sigma_{12} = \sigma_{23} = \sigma_{34} = \sigma_{45} = \sigma_{15} = 0\}$, and
- the toric variety defined by the five quadrics plus the extra equation

$$\sigma_{11}\sigma_{22}\sigma_{33}\sigma_{44}\sigma_{55} = \sigma_{13}\sigma_{14}\sigma_{24}\sigma_{25}\sigma_{35}. \tag{4.2}$$

All matrices in the open cone PD_5 satisfy the inequalities $\sigma_{ii} > 0$ and

$$\sigma_{11}\sigma_{33} > \sigma_{13}^2\,,\ \sigma_{22}\sigma_{44} > \sigma_{24}^2\,,\ \sigma_{33}\sigma_{55} > \sigma_{35}^2\,,\ \sigma_{44}\sigma_{11} > \sigma_{14}^2\,,\ \sigma_{55}\sigma_{22} > \sigma_{25}^2.$$

Multiplying the left hand sides and right hand sides respectively, we find

$$\sigma_{11}^2\sigma_{22}^2\sigma_{33}^2\sigma_{44}^2\sigma_{55}^2 > \sigma_{13}^2\sigma_{14}^2\sigma_{24}^2\sigma_{25}^2\sigma_{35}^2.$$

This is a contradiction to the equation (4.2), and we conclude that the intersection of our GCI model with PD_5 is contained in the linear space $\{\sigma_{12} = \sigma_{23} = \sigma_{34} = \sigma_{45} = \sigma_{15} = 0\}$. The vanishing of the off-diagonal entry σ_{ij} means that X_i is independent of X_j, or, in symbols, $[i \perp\!\!\!\perp j]$. Our algebraic computation thus implies the following axiom for GCI models.

COROLLARY 4.1. *Suppose the conditional independence statements* $[1 \perp\!\!\!\perp 2 \mid \{3\}]$, $[2 \perp\!\!\!\perp 3 \mid \{4\}]$, $[3 \perp\!\!\!\perp 4 \mid \{5\}]$, $[4 \perp\!\!\!\perp 5 \mid \{1\}]$, $[5 \perp\!\!\!\perp 1 \mid \{2\}]$ *hold for some multivariate Gaussian distribution. Then also the following five statements must hold:* $[1 \perp\!\!\!\perp 2]$, $[2 \perp\!\!\!\perp 3]$, $[3 \perp\!\!\!\perp 4]$, $[4 \perp\!\!\!\perp 5]$ *and* $[5 \perp\!\!\!\perp 1]$.

Let us now return to the question *"which almost-principal minors can simultaneously vanish for a positive definite symmetric* $n \times n$*-matrix?"* Corollary 4.1 gives a necessary condition for $n = 5$. We next discuss the answer to our question for $n \leq 4$. For $n = 3$, the necessary and sufficient conditions are given (up to relabeling) by the following four axioms:

(a) $[1 \perp\!\!\!\perp 2]$ and $[1 \perp\!\!\!\perp 3 \mid \{2\}]$ implies $[1 \perp\!\!\!\perp 3]$ and $[1 \perp\!\!\!\perp 2 \mid \{3\}]$,
(b) $[1 \perp\!\!\!\perp 2 \mid \{3\}]$ and $[1 \perp\!\!\!\perp 3 \mid \{2\}]$ implies $[1 \perp\!\!\!\perp 2]$ and $[1 \perp\!\!\!\perp 3]$,
(c) $[1 \perp\!\!\!\perp 2]$ and $[1 \perp\!\!\!\perp 3]$ implies $[1 \perp\!\!\!\perp 2 \mid \{3\}]$ and $[1 \perp\!\!\!\perp 3 \mid \{2\}]$,
(d) $[1 \perp\!\!\!\perp 2]$ and $[1 \perp\!\!\!\perp 2 \mid \{3\}]$ implies $[1 \perp\!\!\!\perp 3]$ **or** $[2 \perp\!\!\!\perp 3]$.

The necessity of these axioms can be checked by simple calculations involving almost-principal minors of positive definite symmetric 3×3-matrices:

(a) $\sigma_{12} = \sigma_{13}\sigma_{22} - \sigma_{12}\sigma_{23} = 0$ implies $\sigma_{13} = \sigma_{12}\sigma_{33} - \sigma_{13}\sigma_{23} = 0$,
(b) $\sigma_{12}\sigma_{33} - \sigma_{13}\sigma_{23} = \sigma_{13}\sigma_{22} - \sigma_{12}\sigma_{23} = 0$ implies $\sigma_{12} = \sigma_{13} = 0$,
(c) $\sigma_{12} = \sigma_{13} = 0$ implies $\sigma_{12}\sigma_{33} - \sigma_{13}\sigma_{23} = \sigma_{13}\sigma_{22} - \sigma_{12}\sigma_{23} = 0$,
(d) $\sigma_{12} = \sigma_{12}\sigma_{33} - \sigma_{13}\sigma_{23} = 0$ implies $\sigma_{13} = 0$ **or** $\sigma_{23} = 0$.

The sufficiency of these axioms was noted in [22, Example 1].

For arbitrary $n \geq 3$, a collection of almost-principal minors is called a *gaussoid* if it satisfies the axioms (a)-(d), after relabeling and applying Schur complements. For instance, axiom (a) is then written as follows: $[i \perp\!\!\!\perp j \mid L]$ and $[i \perp\!\!\!\perp k \mid \{j\} \cup L]$ implies $[i \perp\!\!\!\perp k \mid L]$ and $[i \perp\!\!\!\perp j \mid \{k\} \cup L]$. This axiom is known as the *semigraphoid axiom*. See [23] for a discussion.

A gaussoid is *representable* if it is the set of vanishing almost-principal minors of some matrix in $\mathbf{PD}_n$. For $n = 3$ every gaussoid is representable by [22, Example 1]. For $n = 4$, a complete classification of the representable gaussoids was given in [20]. We are here asking for the extension to $n = 5$.

We now introduce a conceptual framework for our General problem. For each subset S of $\{1, 2, \ldots, n\}$ we introduce one unknown H_S, and we define the *submodular cone* to be the solution set in $\mathbb{R}^{2^n}$ of the system of linear inequalities

$$H_{\{i\} \cup K} + H_{\{j\} \cup K} \leq H_{\{i,j\} \cup K} + H_K, \tag{4.3}$$

where K is any subset of $\{1, \ldots, n\}$ and $i, j \in \{1, \ldots, n\} \backslash K$. We denote this cone by $\texttt{SubMod}_n \subset \mathbb{R}^{2^n}$. Note that $\texttt{SubMod}_n$ is a polyhedral cone living in a high-dimensional space while $\texttt{PD}_n$ is a non-polyhedral cone in a low-dimensional space. Between these two cones we have the *entropy map*

$$H : \texttt{PD}_n \to \texttt{SubMod}_n,$$

which is given by the logarithms of all 2^n principal minors of a positive definite matrix $\Sigma = (\sigma_{ij})$. Namely, the coordinates of the entropy map are

$$H(\Sigma)_I = -\log\det(\Sigma_I),$$

where I is any subset of $\{1,\ldots,n\}$ and Σ_I the corresponding principal minor. Note that the entropy map is well-defined because of the inequality

$$\det(\Sigma_{\{i\}\cup K})\cdot\det(\Sigma_{\{j\}\cup K}) \;\geq\; \det(\Sigma_{\{i,j\}\cup K})\cdot\det(\Sigma_K). \tag{4.4}$$

A matrix $\Sigma \in \mathrm{PD}_n$ satisfies (4.1) if and only if equality holds in (4.4) if and only if equality holds in (4.3). This implies the following result.

PROPOSITION 4.1. *The Gaussian conditional independence models are those subsets of the positive definite cone* PD_n *that arise as inverse images of the faces of the submodular cone* SubMod_n *under the entropy map* H.

The importance of the submodular cone for probabilistic inference with discrete random variables was highlighted in [23]. Here we are concerned with Gaussian random variables, and it is the geometry of the entropy map which we must study. We can thus paraphrase our problem as follows.

General problem: *Characterize the image of the entropy map* H *and how it intersects the various faces of* SubMod_n. *Study the fibers of this map.*

One approach to this problem is to work with the algebraic equations satisfied by the principal minors of a symmetric matrix. A characterization of these relations in terms of *hyperdeterminants* was proposed in [15]. What we are interested in here is the logarithmic image (or *amoeba*) of the positive part of the hyperdeterminantal variety of [15]. A reasonable first approximation to this amoeba is the tropicalization of that variety. More precisely, we seek to compute the *positive tropical variety* [24, §3.4] parametrically represented by the principal minors of a symmetric $n\times n$-matrix.

5. Bonus problem on rational points. Section 4 dealt with conditional independence (CI) models for Gaussians. Our bonus problem concerns CI models for discrete random variables, thus returning to the setting of Section 2. Consider n discrete random variables $X_1, X_2, \ldots, X_n$ with $d_1, d_2, \ldots, d_n$ states. Any collection of CI statements $X_i \perp\!\!\!\perp X_j \,|X_K$ specifies a determinantal variety in the space of tables

$$\mathbb{C}^{d_1}\otimes\mathbb{C}^{d_2}\otimes\cdots\otimes\mathbb{C}^{d_n}. \tag{5.1}$$

We call such a variety a *CI variety*. It is the zero set of a large collection of 2×2-determinants. These constraints are well-known and listed explicitly in [12, §4.1] or [27, Proposition 8.1]. The corresponding *strict CI variety* is the set of tables for which the given CI statements hold but all other CI statements do not hold. Thus a strict CI variety is a constructible subset of (5.1) which is Zariski open in a CI variety. The corresponding *strict CI model* is the intersection of the strict CI variety with the positive orthant. It consists of all positive $d_1\times d_2\times\cdots\times d_n$-tables that lie in a common equivalence class, where two tables are equivalent if precisely the same CI statements $X_i \perp\!\!\!\perp X_j \,|X_K$ are valid (resp. not valid) for both tables.

Bonus problem: *Does every strict CI model have a* $\mathbb{Q}$*-rational point?*

This charming problem was proposed by F. Matúš in [21, p. 275]. It suggests that algebraic statistics has something to offer also for arithmetic geometers. One conceivable solution to the Bonus Problem might say that CI models with no rational points exist but that rational points always appear when the number of states grows large, that is, for $d_1, d_2, \ldots, d_n \gg 0$. But that is pure speculation. At present we know next to nothing.

6. Brief conclusion. This article offered a whirlwind introduction to the emerging field of algebraic statistics, by discussing a few of its numerous open problems. Aside from the Bonus Problem above, we had listed three Specific Problems whose solution might be particularly rewarding:

- Consider the variety of $4\times4\times4$-tables of tensor rank at most 4. Do the known polynomial invariants of degree at most nine suffice to define this variety? Set-theoretically? Ideal-theoretically?
- Characterize all projective varieties whose maximum likelihood degree is equal to one.
- Which sets of almost-principal minors can be simultaneously zero for a positive definite symmetric 5×5-matrix?

REFERENCES

[1] E. ALLMAN, Determine the ideal defining $\mathrm{Sec}^4(\mathbb{P}^3 \times \mathbb{P}^3 \times \mathbb{P}^3)$. Phylogenetic explanation of an *Open Problem* at www.dms.uaf.edu/~eallman/salmonPrize.pdf.

[2] E. ALLMAN AND J. RHODES, Phylogenetic ideals and varieties for the general Markov model, *Advances in Applied Mathematics*, to appear.

[3] E. ALLMAN AND J. RHODES, Phylogenetics, in R. Laubenbacher (ed.): *Modeling and Simulation of Biological Networks*, Proceedings of Symposia in Applied Mathematics, American Mathematical Society, 2007, pp. 1–31.

[4] D. BRODY AND J. HUGHSTON, Geometric quantum mechanics, *J. Geom. Phys.* **38** (2001), 1953.

[5] L. CATALANO-JOHNSON, The homogeneous ideals of higher secant varieties, *Journal of Pure and Applied Algebra* **158** (2001), 123–129.

[6] F. CATANESE, S. HOŞTEN, A. KHETAN, AND B. STURMFELS, The maximum likelihood degree, *American Journal of Mathematics* **128** (2006), 671–697.

[7] M. DRTON, B. STURMFELS, AND S. SULLIVANT, Algebraic factor analysis: tetrads, pentads and beyond, *Probability Theory and Related Fields* **138** (2007), 463–493

[8] M. DRTON AND S. SULLIVANT, Algebraic statistical models, *Statistica Sinica* **17** (2007), 1273–1297.

[9] S. FIENBERG, P. HERSH, A. RINALDO, AND Y. ZHOU, Maximum likelihood estimation in latent class models for contingency table data, preprint, arXiv:0709.3535.

[10] W. FULTON AND J. HARRIS, *Representation Theory. A First Course*, Graduate Texts in Mathematics, **129**, Springer-Verlag, 1991.

[11] L. GARCIA, M. STILLMAN, AND B. STURMFELS, Algebraic geometry of Bayesian networks, *Journal of Symbolic Computation* **39** (2005), 331–355.

[12] D. GEIGER, C. MEEK, AND B. STURMFELS, On the toric algebra of graphical models, *Annals of Statistics* **34** (2006), 1463–1492

[13] G.-M. GREUEL, G. PFISTER, AND H. SCHÖNEMANN, *Singular 3.0. A Computer Algebra System for Polynomial Computations*, Centre for Computer Algebra, University of Kaiserslautern, http://www.singular.uni-kl.de, 2005.

[14] H. Heydari, General pure multipartite entangled states and the Segre variety, *J. Phys. A: Math. Gen.* **39** (2006), 9839–9844

[15] O. Holtz and B. Sturmfels, Hyperdeterminantal relations among symmetric principal minors, *Journal of Algebra* **316** (2007), 634–648.

[16] S. Hoşten, A. Khetan, and B. Sturmfels, Solving the likelihood equations, *Foundations of Computational Mathematics* **5** (2005), 389–407.

[17] J.M. Landsberg and L. Manivel, On the ideals of secant varieties of Segre varieties, *Foundations of Computational Mathematics* **4** (2004), 397–422.

[18] J.M. Landsberg and L. Manivel, Generalizations of Strassen's equations for secant varieties of Segre varieties. *Communications in Algebra*, to appear.

[19] J.M. Landsberg and J. Weyman, On the ideals and singularities of secant varieties of Segre varieties, Bulletin of the London Math. Soc. **39** (2007), 685–697.

[20] R. Lněnička and F. Matúš, On Gaussian conditional independence structures, *Kybernetika* **43** (2007), 327–342.

[21] F. Matúš, Conditional independences among four random variables III: Final conclusion *Combinatorics, Probability and Computing* **8** (1999), 269–276.

[22] F. Matúš, Conditional independences in Gaussian vectors and rings of polynomials. *Proceedings of WCII 2002* (eds. G. Kern-Isberner, W. Rdder, and F. Kulmann) LNAI 3301, Springer-Verlag, Berlin, 152–161, 2005.

[23] J. Morton, L. Pachter, A. Shiu, B. Sturmfels, and O. Wienand, Convex rank tests and semigraphoids, preprint, `ArXiv:math.CO/0702564`.

[24] L. Pachter and B. Sturmfels, *Algebraic Statistics for Computational Biology*, Cambridge University Press, 2005.

[25] G. Pistone, E. Riccomagno, and H. Wynn, *Algebraic Statistics: Computational Commutative Algebra in Statistics*, Chapman & Hall/CRC, 2000.

[26] P. Šimeček, *Classes of Gaussians, discrete and binary representable independence models that have no finite characterization*, Proceedings of Prague Stochastics 2006, pp. 622–632.

[27] B. Sturmfels, *Solving Systems of Polynomial Equations, CBMS Regional Conference Series in Mathematics*, Vol. **97**, Amer. Math. Society, Providence, 2002.

[28] S. Sullivant, Gaussian conditional independence relations have no finite complete characterization, preprint, `ArXiv:0704.2847`, 2007.

LIST OF WORKSHOP PARTICIPANTS

Workshop on Optimization and Control
January 16-20, 2007

- Cheonghee Ahn, Department of Mathematics, Yonsei University
- Suliman Al-Homidan, Department of Mathmatical Sciences, King Fahd University of Petroleum and Minerals
- Elizabeth S. Allman, Department of Mathematics and Statistics, University of Alaska
- Miguel F. Anjos, Department of Management Sciences, University of Waterloo
- D. Gregory Arnold, AFRL/SNAT, US Air Force Research Laboratory
- Douglas N. Arnold, Institute for Mathematics and its Applications, University of Minnesota Twin Cities
- Donald G. Aronson, Institute for Mathematics and its Applications, University of Minnesota Twin Cities
- Michel Baes, ESAT/SISTA, Katholieke Universiteit Leuven Joseph A. Ball, Department of Mathematics, Virginia Polytechnic Institute and State University
- Chunsheng Ban, Department of Mathematics, Ohio State University
- Alexander Barvinok, Department of Mathematics, University of Michigan
- Saugata Basu, School of Mathematics, Georgia Institute of Technology
- Daniel J. Bates, Institute for Mathematics and its Applications, University of Minnesota Twin Cities
- Carolyn Beck, Department of General Engineering, University of Illinois at Urbana-Champaign
- Dimitris Bertsimas, Sloan School of Management, Massachusetts Institute of Technology
- Yermal Sujeet Bhat, Institute for Mathematics and its Applications, University of Minnesota Twin Cities
- Vctor Blanco Izquierdo, Department of Statistics and Operational Research, University of Sevilla
- Cristiano Bocci, Department of Mathematics, Università di Milano
- Tristram Bogart, Department of Mathematics, University of Washington
- Hartwig Bosse, PNA1 (Algorithms, Combinatorics and Optimization), Center for Mathematics and Computer Science (CWI)
- Christopher J. Budd, Department of Mathematical Sciences, University of Bath

- Constantine M. Caramanis, Department of Electrical and Computer Engineering, University of Texas
- Enrico Carlini, Dipartimento di Matematica, Politecnico di Torino
- Dong Eui Chang, Department of Applied Mathematics, University of Waterloo
- Graziano Chesi, Department of Electrical and Electronic Engineering, University of Hong Kong
- Hi Jun Choe, Department of Mathematics, Yonsei University
- Ionut Ciocan-Fontanine, Institute for Mathematics and its Applications, University of Minnesota Twin Cities
- Raul Curto, Department of Mathematics, University of Iowa
- Etienne de Klerk, Department of Econometrics and Operations Research, Katholieke Universiteit Brabant (Tilburg University)
- Jesus Antonio De Loera, Department of Mathematics University of California
- Mauricio de Oliveira, Department of Mechanics and Aerospace Engineering, University of California, San Diego
- Xuan Vinh Doan, Operations Research Center, Massachusetts Institute of Technology
- John C. Doyle, Control and Dynamical Systems, California Institute of Technology
- Kenneth R. Driessel, Department of Mathematics, Iowa State University
- Michael Dritschel, Department of Mathematics, University of Newcastle upon Tyne
- Mathias Drton, Department of Statistics, University of Chicago
- Laurent El Ghaoui, Department of Electrical Engineering and Computer Science, University of California
- Lingling Fan, Midwest ISO
- Makan Fardad, Department of Electrical and Computer Engineering, University of Minnesota Twin Cities
- Maryam Fazel, Control and Dynamical Systems, California Institute of Technology
- Eric Feron, School of Aerospace Engineering, Georgia Institute of Technology
- Lawrence A. Fialkow, Department of Computer Science, SUNY at New Paltz
- Stephen E. Fienberg, Department of Statistics, Carnegie-Mellon University
- Pedro Forero, Department of Electrical Engineering, University of Minnesota Twin Cities
- Ioannis Fotiou, Automatic Control Laboratory, Eidgenssische TH Zürich-Hönggerberg
- Dennice Gayme, Department of Control and Dynamical Systems, California Institute of Technology
- Tryphon T. Georgiou, Department of Electrical Engineering, University of Minnesota Twin Cities

- Sonja Glavaski, Honeywell Advanced Technology, Honeywell Systems and Research Center
- Jason E. Gower, Institute for Mathematics and its Applications, University of Minnesota Twin Cities
- Carlos R. Handy, Department of Physics, Texas Southern University
- Bernard Hanzon, National University of Ireland, University College Cork
- Gloria Haro Ortega, Institute for Mathematics and its Applications, University of Minnesota Twin Cities
- Christoph Helmberg, Fakultät für Mathematik, Technische Universität Chemnitz-Zwickau
- J. William Helton, Department of Mathematics, University of California, San Diego
- Didier Henrion, LAAS, Centre National de la Recherche Scientifique (CNRS)
- Milena Hering, Institute for Mathematics and its Applications, University of Minnesota Twin Cities
- Christopher Hillar, Department of Mathematics, Texas A M University
- Jean-Baptiste Hiriart-Urruty, Institute of Mathematics, MIP laboratory, Université de Toulouse III (Paul Sabatier)
- Sung-Pil Hong, Department of Industrial Engineering, Seoul National University
- Serkan Hosten, Department of Mathematics, San Francisco State University
- Benjamin J. Howard, Institute for Mathematics and its Applications, University of Minnesota Twin Cities
- Evelyne Hubert, Project CAFE, Institut National de Recherche en Informatique Automatique (INRIA)
- Farhad Jafari, Department of Mathematics, University of Wyoming
- Amin Jafarian, Department of Electrical and Computer Engineering, University of Texas
- Anders Nedergaard Jensen, Institut for Matematiske Fag, Aarhus University
- Steve Kaliszewski, Department of Mathematics and Statistics, Arizona State University
- Tapan Kumar Kar, Faculty of Environment and Information Sciences, Yokohama National University
- Mordechai Katzman, Department of Pure Mathematics, University of Sheffield
- Edward D. Kim, Department of Mathematics, University of California
- Si-Jo Kim, Department of Chemicial Engineering, Andong National University

- Sunyoung Kim, Department of Mathematics, Ewha Womans University
- Henry C. King, Department of Mathematics, University of Maryland
- Masakazu Kojima, Department of Mathematical and Computing Sciences, Tokyo Institute of Technology
- Salma Kuhlmann, Research Center for Algebra, Logic and Computation, University of Saskatchewan
- Nuri Kundak, Department of Aerospace Engineering and Mechanics, University of Minnesota Twin Cities
- Song-Hwa Kwon, Institute for Mathematics and its Applications, University of Minnesota Twin Cities
- Sanjay Lall, Department of Aeronautical and Astronautical Engineering, Stanford University
- Andrew Lamperski, Department of Control and Dynamical Systems, California Institute of Technology
- Jean Bernard Lasserre, LAAS, Centre National de la Recherche Scientifique (CNRS) and Institute of Mathematics
- Niels Lauritzen, Institut for Matematiske Fag, Aarhus University
- Anton Leykin, Institute for Mathematics and its Applications, University of Minnesota Twin Cities
- Hstau Y Liao, Institute for Mathematics and its Applications, University of Minnesota Twin Cities
- James Lu, Department of Inverse Problems, Johann Radon Institute for Computational and Applied Mathematics
- Tom Luo, Department Electrical and Computer Engineering, University of Minnesota Twin Cities
- Gennady Lyubeznik, School of Mathematics, University of Minnesota Twin Cities
- Hannah Markwig, Institute for Mathematics and its Applications, University of Minnesota Twin Cities
- Thomas Markwig, Department of Mathematics, Universität Kaiserslautern
- Scott McCullough, Department of Mathematics, University of Florida
- Alexandre Megretski, Department of Electrical Engineering and Computer Science, Massachusetts Institute of Technology
- Lisa A. Miller, Department of Mechanical Engineering, University of Minnesota Twin Cities
- Richard B. Moeckel, School of Mathematics, University of Minnesota Twin Cities
- Uwe Nagel, Department of Mathematics, University of Kentucky
- Jiawang Nie, Institute of Mathematics and its Application, University of Minnesota Twin Cities

- Antonis Papachristodoulou, Department of Engineering Science, University of Oxford
- Pablo A. Parrilo, Laboratory for Information and Decision Systems, Massachusetts Institute of Technology
- Gabor Pataki, Department of Operations Research, University of North Carolina
- Helfried Peyrl, Automatic Control Laboratory Victoria Powers, Department of Mathematics and Computer Science, Emory University
- Mihai Putinar, Department of Mathematics, University of California, Santa Barbara
- Jacob Quant, University of Minnesota Twin Cities
- Bharath Rangarajan, Department of Mechanical Engineering, University of Minnesota Twin Cities
- Seid Alireza Razavi Majomard, Department of Electrical Engineering and Computer Science, University of Minnesota Twin Cities
- Ben Recht, Center for the Mathematics of Information, California Institute of Technology
- Victor Reiner School of Mathematics University of Minnesota Twin Cities
- Franz Rendl, Institut fr̈ Mathematik, Universitẗ Klagenfurt
- James Renegar, School of Operations Research and Industrial Engineering, Cornell University
- John A. Rhodes, Department of Mathematics and Statistics, University of Alaska
- Joel Roberts, School of Mathematics, University of Minnesota Twin Cities
- Marie Rognes, University of Oslo
- Philipp Rostalski, Automatic Control Laboratory, Eidgenössische TH Zürich-Hönggerberg
- Bjarke Hammersholt Roune, Department of Mathematics, Aarhus University
- Marie-Francoise Roy, IRMAR, Université de Rennes I
- Christopher Ryan, Sauder School of Business, University of British Columbia
- Arnd Scheel, Institute for Mathematics and its Applications, University of Minnesota Twin Cities
- Markus Schweighofer, Fachbereich Mathematik und Statistik, Universität Konstanz
- Parikshit Shah, Department of Electrical Engineering and Computer Science, Massachusetts Institute of Technology
- Chehrzad Shakiban, Institute of Mathematics and its Application, University of Minnesota Twin Cities
- Kartik K. Sivaramakrishnan, Department of Mathematics, North Carolina State University

- Steven Sperber, School of Mathematics, University of Minnesota Twin Cities
- Dumitru Stamate, School of Mathematics, University of Minnesota Twin Cities
- Bernd Sturmfels, Department of Mathematics, University of California
- Jie Sun, NUS Business School, National University of Singapore
- Bridget Eileen Tenner, Massachusetts Institute of Technology
- Tamas Terlaky, Department of Mathematics and Statistics, McMaster University
- Rekha R. Thomas, Department of Mathematics, University of Washington
- Carl Toews, Institute for Mathematics and its Applications, University of Minnesota Twin Cities
- Ufuk Topcu, Department of Mechanical Engineering, University of California
- Levent Tuncel, Department of Cominatorics and Optimization, University of Waterloo
- Victor Vinnikov, Department of Mathematics, Ben Gurion University of the Negev
- John Voight, Institute for Mathematics and its Applications, University of Minnesota Twin Cities
- Martin J. Wainwright, Department of Electrical Engineering and Computer Science, University of California
- Angelika Wiegele, Department of Mathematics, Universität Klagenfurt
- Henry Wolkowicz, Department of Cominatorics and Optimization, University of Waterloo
- Gregory Emmanuel Yawson, Department of Engineering Technology, Lawrence Technological University
- Josephine Yu, Department of Mathematics, University of California
- Hongchao Zhang, Institute for Mathematics and its Applications, University of Minnesota Twin Cities
- Lihong Zhi, Mathematics Mechanization Research Center, Chinese Academy of Sciences
- Yuriy Zinchenko, Advanced Optimization Lab, CAS, McMaster University

Workshop on Applications in Biology, Dynamics, and Statistics
March 5-9, 2007

- Elizabeth S. Allman, Department of Mathematics and Statistics, University of Alaska
- Douglas N. Arnold, Institute for Mathematics and its Applications, University of Minnesota
- Donald G. Aronson, Institute for Mathematics and its Applications, University of Minnesota
- Hlne Barcelo, Department of Mathematics and Statistics, Arizona State University
- Brandon Barker, University of Kentucky Daniel J. Bates, Institute for Mathematics and its Applications, University of Minnesota
- Niko Beerenwinkel, Program for Evolutionary Dynamics, Harvard University
- Jeremy Bellay, School of Mathematics, University of Minnesota
- Wicher Bergsma, London School of Economics and Political Science
- Vctor Blanco Izquierdo, Department of Statistics and Operational Research, University of Sevilla
- Cristiano Bocci, Department of Mathematics, Università di Milano
- Tristram Bogart, Department of Mathematics, University of Washington
- Joseph P. Brennan, Department of Mathematics, University of Central Florida
- Marta Casanellas, Departament de Matematica Aplicada I, Polytechnical University of Catalua (Barcelona)
- Hegang Chen, School of Medicine, Division of Biostatistics and Bioinformatics, University of Maryland at Baltimore
- Vera Cherepinsky, Department of Mathematics and Computer Science, Fairfield University
- Ionut Ciocan-Fontanine, Institute for Mathematics and its Applications, University of Minnesota
- Barry Cipra,
- Eduardo Corel, Laboratoire Statistique et Génome, Centre National de la Recherche Scientifique (CNRS)
- Gheorghe Craciun, Department of Mathematics and Department of Biomolecular Chemistry, University of Wisconsin
- Alicia Dickenstein, Departamento de Matematica - FCEyN, University of Buenos Aires
- Kequan Ding, Chinese Academy of Sciences
- Jianping Dong, Department of Mathematical Sciences, Michigan Technological University
- Kenneth R. Driessel, Department of Mathematics, Iowa State Uni-

versity
- Mathias Drton, Department of Statistics, University of Chicago
- Vanja Dukic, Department of Health Studies (Biostatistics), University of Chicago
- Ryan S. Elliott, University of Michigan
- Nicholas Eriksson, Department of Statistics, University of Chicago
- Makan Fardad, Department of Electrical and Computer Engineering, University of Minnesota
- Martin Feinberg, Department of Chemical and Biomolecular Engineering, Ohio State University
- David Fernández-Baca, Department of Computer Science, Iowa State University
- Stephen E. Fienberg, Department of Statistics, Carnegie Mellon University
- Robert M. Fossum, Department of Mathematics, University of Illinois at Urbana-Champaign
- Surya Ganguli, Sloan-Swartz Center for Theoretical Neurobiology, University of San Francisco
- Martin Golubitsky, Department of Mathematics, University of Houston
- Jason E. Gower, Institute for Mathematics and its Applications, University of Minnesota
- Mansoor Haider, Department of Mathematics, North Carolina State University
- Michael Hardy, School of Mathematics, University of Minnesota
- Gloria Haro Ortega, University of Minnesota
- Christine E. Heitsch, School of Mathematics, Georgia Institute of Technology
- J. William Helton, Department of Mathematics, University of California, San Diego
- Milena Hering, Institute for Mathematics and its Applications, University of Minnesota
- Patricia Hersh, Department of Mathematics, Indiana University
- Benjamin J. Howard, University of Minnesota
- Evelyne Hubert, Project CAFE, Institut National de Recherche en Informatique Automatique (INRIA)
- Peter Huggins, Department of Mathematics, University of California
- Satish Iyengar, Department of Statistics, University of Pittsburgh
- Farhad Jafari, Department of Mathematics, University of Wyoming
- Abdul Salam Jarrah, Virginia Bioinformatics Institute, Virginia Polytechnic Institute and State University
- Anders Nedergaard Jensen, Institut for Matematiske Fag, Aarhus University

- Renfang Jiang, Department of Mathematical Sciences, Michigan Technological University
- Ben Jordan, Department of Mathematics, University of Minnesota
- Steve Kaliszewski, Department of Mathematics and Statistics, Arizona State University
- Mordechai Katzman, Department of Pure Mathematics, University of Sheffield
- Michael Kerber, Universität Kaiserslautern
- Markus Kirkilionis, Department of Mathematics, University of Warwick
- Elizabeth Kleiman, Department of Mathematics, Iowa State University
- Song-Hwa Kwon, University of Minnesota
- Batool Labibi, Department of Electrical and Computer Engineering, University of Alberta
- Reinhard Laubenbacher, Virginia Bioinformatics Institute, Virginia Polytechnic Institute and State University
- Niels Lauritzen, Institut for Matematiske Fag, Aarhus University
- Reiner Lauterbach, Fachbereich Mathematik, Universität Hamburg
- Juyoun Lee, Department of Statistics, Pennsylvania State University
- Namyong Lee, Department of Mathematics, Minnesota State University
- Anton Leykin, Institute for Mathematics and its Applications, University of Minnesota
- Hstau Y Liao, University of Minnesota
- Gennady Lyubeznik, School of Mathematics, University of Minnesota
- James Madden, Department of Mathematics, Louisiana State University
- Thomas Markwig, Department of Mathematics, Universität Kaiserslautern
- Frantiek Matú, Institute of Information Theory and Automation, Czech Academy of Sciences (AVR)
- Richard B. Moeckel, School of Mathematics, University of Minnesota
- Jason Morton, Department of Mathematics, University of California
- Anca Mustata, Department of Mathematics, University of Illinois at Urbana-Champaign
- Uwe Nagel, Department of Mathematics, University of Kentucky
- Yuval Nardi, Department of Statistics, Carnegie Mellon University
- Jiawang Nie, University of Minnesota
- Ignacio Ojeda Martinez de Castilla, Departamento de Matemáti-

cas, University of Extremadura
- Sarah Olson, Department of Biomathematical Sciences, North Carolina State University
- Michael E. O'Sullivan, Department of Mathematics and Statistics, San Diego State University
- Lior Pachter, Department of Mathematics, University of California
- Casian Pantea, Department of Mathematics, University of Wisconsin
- Sonja Petrovic, Department of Mathematics, University of Kentucky
- Giovanni Pistone, Dipartimento di Matematica, Politecnico di Torino
- Mihai Putinar, Department of Mathematics, University of California
- Kristian Ranestad, Department of Mathematics, University of Oslo
- Victor Reiner, School of Mathematics, University of Minnesota
- John A. Rhodes, Department of Mathematics and Statistics, University of Alaska
- Eva Riccomagno, Dipartimento di Matematica, Universit di Genova
- Donald Richards, Department of Statistics, Pennsylvania State University
- Thomas S. Richardson, Department of Statistics, University of Washington
- Alessandro Rinaldo, Department of Statistics, Carnegie Mellon University
- Joel Roberts, School of Mathematics, University of Minnesota
- Daniel Robertz, Lehrstuhl B fur Mathematik, RWTH Aachen
- Marie Rognes, University of Oslo
- Bjarke Hammersholt Roune, Department of Mathematics, Aarhus University
- Arnd Scheel, Institute for Mathematics and its Applications, University of Minnesota
- Chehrzad Shakiban, Institute of Mathematics and its Application, University of Minnesota
- Anne Shiu, Department of Mathematics, University of California
- Aleksandra B. Slavković, Department of Statistics, Pennsylvania State University
- Jim Smith, Department of Statistics, University of Warwick
- Frank Sottile, Department of Mathematics, Texas A and M University
- Steven Sperber, School of Mathematics, University of Minnesota
- David Speyer, Department of Mathematics, University of Michigan
- Sundararajan Srinivasan Sr., Department of Electrical Engineering, Mississippi State University

- Dumitru Stamate, School of Mathematics, University of Minnesota
- Russell Steele, Department of Mathematics and Statistics, McGill University
- Brandilyn Stigler, Mathematical Biosciences Institute, Ohio State University
- Michael E. Stillman, Department of Mathematics, Cornell University

- Bernd Sturmfels, Department of Mathematics, University of California, Berkeley
- Seth Sullivant, Department of Mathematics, Harvard University
- Rebecca Swanson, Department of Mathematics, Indiana University
- Berhanu Tameru, Center for Computational Epidemiology, Bioinformatics and Risk Analysis (CCEBRA), Tuskegee University
- Amelia Taylor, Colorado College
- Glenn Tesler, Department of Mathematics, University of California, San Diego
- Thorsten Theobald, Department of Computer Science and Mathematics, Johann Wolfgang Goethe-Universität Frankfurt
- Rekha R. Thomas, Department of Mathematics, University of Washington
- Carl Toews, University of Minnesota
- Ya-lun Tsai, Department of Mathematics, University of Minnesota
- Caroline Uhler, Department of Mathematics, Universität Zürich
- Jose Vargas, CIIDIR-Oaxaca, Instituto Politecnico Nacional, Mexico
- Martha Paola Vera-Licona, Virginia Bioinformatics Institute, Virginia Polytechnic Institute and State University
- Alberto Vigneron-Tenorio, Facultad de Ciencias Sociales y de la Comunicacion, University of Cádiz
- John Voight, University of Minnesota
- Jaroslaw Wisniewski, Department of Mathematics, University of Warsaw
- Matthias Wolfrum, Weierstra-Institut für Angewandte Analysis und Stochastik (WIAS)
- Seongho Wu, Department of Statistics, University of Minnesota
- Alexander Yong, Department of Mathematics, University of Minnesota
- Ruriko Yoshida, Department of Statistics, University of Kentucky
- Josephine Yu, Department of Mathematics, University of California
- Cornelia Yuen, Department of Mathematics, University of Kentucky
- Debbie Yuster, Department of Mathematics, Columbia University
- Hongchao Zhang, Institute for Mathematics and its Applications, University of Minnesota
- Mengyuan Zhao, Department of Statistics, University of Pittsburgh
- Yi Zhou, Department of Machine Learning, Carnegie Mellon University

The IMA Volumes in Mathematics and its Applications

1 **Volume 142: Compatible Spatial Discretizations**
Editors: Douglas N. Arnold, Pavel B. Bochev, Richard B. Lehoucq, Roy A. Nicolaides, and Mikhail Shashkov

2 **Volume 143: Wireless Communications**
Editors: Prathima Agrawal, Daniel Matthew Andrews, Philip J. Fleming, George Yin, and Lisa Zhang

3 **Volume 144: Symmetries and Overdetermined Systems of Partial Differential Equations**
Editors: Michael Eastwood and Willard Miller, Jr.

4 **Volume 145: Topics in Stochastic Analysis and Nonparametric Estimation**
Editors: Pao-Liu Chow, Boris Mordukhovich, and George Yin

5 **Volume 146: Algorithms in Algebraic Geometry**
Editors: Alicia Dickenstein, Frank-Olaf Schreyer, and Andrew Sommese

6 **Volume 147: Symmetric Functionals on Random Matrices and Random Matchings Problems**
by Grzegorz A. Rempała and Jacek Wesołowski

7 **Volume 148: Software for Algebraic Geometry**
Editors: Michael E. Stillman, Nobuki Takayama, and Jan Verschelde

8 **Volume 149: Emerging Applications of Algebraic Geometry**
Editors: Mihai Putinar and Seth Sullivant

The full list of IMA books can be found at the Web site of Institute for Mathematics and its Applications:
http://www.ima.umn.edu/springer/volumes.html